Decrypted Secrets

Springer
Berlin
Heidelberg
New York
Barcelona
Hong Kong
London
Milan
Paris
Tokyo

Friedrich L. Bauer

Decrypted Secrets

Methods and Maxims of Cryptology

Third, Revised and Updated Edition

With 167 Figures, 26 Tables,
and 16 Color Plates

Springer

Dr. rer. nat. Dr. ès sc. h.c. Dr. rer. nat. h.c. mult. Friedrich L. Bauer
Professor Emeritus of Mathematics and Computer Science
Munich Institute of Technology
Department of Computer Science
Arcisstrasse 21
80333 Munich, Germany

ACM Computing Classification (1998): E.3, D.4.6, K.6.5, E.4
Mathematics Subject Classification (1991): 94A60, 68P25

ISBN 3-540-42674-4 Springer-Verlag Berlin Heidelberg New York

ISBN 3-540-66871-3 2nd ed. Springer-Verlag Berlin Heidelberg New York

Library of Congress Cataloging-in-Publication Data applied for

Die Deutsche Bibliothek – CIP-Einheitsaufnahme

Bauer, Friedrich L.: Decrypted Secrets : Methods and Maxims of Cryptology / Friedrich L. Bauer. – Third, Revised and Updated Edition – Berlin; Heidelberg; New York; Barcelona; Hong Kong; London; Milan; Paris; Tokyo: Springer, 2002
ISBN 3-540-42674-4

Springer-Verlag Berlin Heidelberg NewYork,
a member of BertelsmannSpringer Science+Business Media GmbH
http://www.springer.de

Cover Design: Design & Concept E. Smejkal, Heidelberg
Color Photos: Reinhard Krause, Deutsches Museum München
Typesetting: By the author in TEX

Printed on acid-free paper SPIN 10853358 45/3142 ud – 5 4 3 2 1 0

Preface

Towards the end of the 1960s, under the influence of the rapid development of microelectronics, electromechanical cryptological machines began to be replaced by electronic data encryption devices using large scale integrated circuits. This promised more secure encryption at lower prices. Then, in 1976, Diffie and Hellman opened up the new cryptological field of public key systems. Cryptography, hitherto cloaked in obscurity, was emerging into the public domain. Additionally, ENIGMA revelations awoke the public interest.

Computer science was a flourishing new field, too, and computer scientists became interested in several aspects of cryptology. But many of them were not well enough informed about the centuries-long history of cryptology and the high level it had attained. I saw some people starting to reinvent the wheel, and others who had an incredibly naive belief in safe encryption, and I became worried about the commercial and scientific development of professional cryptology among computer scientists and about the unstable situation with respect to official security services.

This prompted me to offer lectures on this subject at the Munich Institute of Technology. The first series of lectures in the winter term 1977/78, backed by the comprehensive and reliable book *The Codebreakers* (1967) by David Kahn, was held under the code name 'Special Problems of Information Theory' and therefore attracted neither too many students nor too many suspicious people from outside the university.

Next time, in the summer term 1981, my lectures on the subject were announced under the open title 'Cryptology'. This was seemingly the first publicly announced lecture series under this title at a German, if not indeed a Continental European, university.

The series of lectures was repeated a few times, and in 1986/87 lecture notes were printed which developed into Part I of this book. Active interest on the side of the students led to a seminar on cryptanalytic methods in the summer term 1988, from which Part II of the present book originated.

The 1993 first edition of my book *Kryptologie*, although written mainly for computer science students, found lively interest also outside the field. It was reviewed favorably by some leading science journalists, and the publisher followed the study book edition with a 1995 hardcover edition under the title *Entzifferte Geheimnisse* [Decrypted Secrets], which gave me the opportunity to round out some subjects. Reviews in American journals recommended also an English version, which led to the present book.

It has become customary among cryptologists to explain how they became acquainted with the field. In my case, this was independent of the Second World War. In fact, I was never a member of any official service—and I consider this my greatest advantage, since I am not bound by any pledge of secrecy. On the other hand, keeping eyes and ears open and reading between the lines, I learned a lot from conversations (where my scientific metier was a good starting point), although I never know exactly whether I am allowed to know what I happen to know.

It all started in 1951, when I told my former professor of formal logic at Munich University, Wilhelm Britzelmayr, of my invention of an error-correcting code for teletype lines[1]. This caused him to make a wrong association, and he gave me a copy of Sacco's book, which had just appeared[2]. I was lucky, for it was the best book I could have received at that time—although I didn't know that then. However, I devoured the book. Noticing this, my dear friend and colleague Paul August Mann, who was aware of my acquaintance with Shannon's redundancy-decreasing encoding, gave me a copy of the now-famous paper by Claude Shannon called *Communication Theory of Secrecy Systems*[3] (which in those days was almost unavailable in Germany as a Bell System Technical Report). I was fascinated by this background to Shannon's information theory, which I was already familiar with. This imprinted my interest in cryptology as a subfield of coding theory and formal languages theory, fields that held my academic interest for many years to come.

Strange accidents—or maybe sharper observation—then brought me into contact with more and more people once close to cryptology, starting with Willi Jensen (Flensburg) in 1955, Karl Stein (Munich) in 1955, Hans Rohrbach, my colleague at Mainz University in 1959, as well as Helmut Grunsky, Gisbert Hasenjäger, and Ernst Witt. In 1957, I became acquainted with Erich Hüttenhain (Bad Godesberg), but our discussions on the suitability of certain computers for cryptological work were in the circumstances limited by certain restrictions. Among the American and British colleagues in numerical analysis and computer science I had closer contact with, some had been involved with cryptology in the Second World War; but no one spoke about that, particularly not before 1974, the year when Winterbotham's book *The Ultra Secret* appeared. In 1976, I heard B. Randall and I. J. Good revealing some details about the Colossi in a symposium in Los Alamos. As a science-oriented civilian member of cryptological academia, my interest in cryptology was then and still is centered on computerized cryptanalysis. Other aspects

---

[1] DBP No. 892767, application date January 21, 1951.

[2] Général Luigi Sacco, *Manuel de Cryptographie*. Payot, Paris 1951.

[3] Bell Systems Technical Journal **28**, Oct. 1949, pp. 656–715.

of signals intelligence ('SIGINT'), for example traffic analysis and direction finding, are beyond the scope of the book; the same holds for physical devices screening electromechanical radiation emitted by cipher machines.

The first part of this book presents cryptographic methods. The second part brings on cryptanalysis, above all, the facts that are important for judging cryptographic methods and are intended to save the user from unexpected pitfalls. This follows from Kerckhoffs' maxim: Only a cryptanalyst can judge the security of a crypto system. A theoretical course on cryptographic methods alone seems to me to be bloodless. But a course on cryptanalysis is problematic: Either it is not conclusive enough, in which case it is useless, or it is conclusive, but touches a sensitive area. There is little clearance in between. I have tried to cover at least all the essential facts that are in the open literature or can be deduced from it. No censorship took place.

Cryptology is a discipline with an international touch and a particular terminology. It may therefore be helpful sometimes to give in the book references to terms in foreign language.

My intellectual delight in cryptology found an application in the collection 'Informatik und Automatik' of the Deutsches Museum in Munich which I built up in 1984–1988, where there is a section on cryptological devices and machines. My thanks go to the Deutsches Museum for providing color plates of some of the pieces on exhibit there.

And thanks go to my former students and co-workers in Munich, Manfred Broy, Herbert Ehler, and Anton Gerold for continuing support over the years, moreover to Hugh Casement for linguistic titbits, and to my late brother-in-law Alston S. Householder for enlightenment on my English. Karl Stein and Otto Leiberich gave me details on the ENIGMA story, and I had fruitful discussions and exchange of letters with Ralph Erskine, Heinz Ulbricht, Tony Sale, Frode Weierud, Kjell-Ove Widman, Otto J. Horak, Gilbert Bloch, Arne Fransén, and Fritz-Rudolf Güntsch. Great help was given to me by Kirk H. Kirchhofer from the Crypto AG, Zug (Switzerland). Hildegard Bauer-Vogg supplied translations of difficult Latin texts, Martin Bauer, Ulrich Bauer and Bernhard Bauer made calculations and drawings. Thanks go to all of them.

The English version was greatly improved by J. Andrew Ross, with whom working was a pleasure. In particular, my sincere thanks go to David Kahn who encouraged me ("The book is an excellent one and deserves the widest circulation") and made quite a number of proposals for improvements of the text. Finally, I have to thank once more Hans Wössner from Springer-Verlag for a well functioning cooperation of long standing. The publisher is to be thanked for the fine presentation of the book. And I shall be grateful to readers who are kind enough to let me know of errors and omissions.

Grafrath, Autumn 2001 F. L. Bauer

List of Color Plates

Plate A The disk of Phaistos

Plate B Brass cipher disks

Plate C The ‘Cryptograph’ of Wheatstone

Plate D The U.S. Army cylinder device M-94

Plate E The U.S. strip device M-138-T4

Plate F The cipher machine of Kryha

Plate G The Hagelin ‘Cryptographer’ C-36

Plate H The U.S. Army M-209, Hagelin licensed

Plate I The cipher machine ENIGMA with four rotors

Plate K Rotors of the ENIGMA

Plate L The British rotor machine TYPEX

Plate M *Uhr* box of the German *Wehrmacht*

Plate N Cipher teletype machine Lorenz SZ 42

Plate O Russian one-time pad

Plate P Modern crypto board

Plate Q CRAY Supercomputers

Contents

Part I: Cryptography 1

1 Introductory Synopsis 8
1.1 Cryptography and Steganography 8
1.2 Semagrams 9
1.3 Open Code: Masking 12
1.4 Cues 16
1.5 Open Code: Veiling by Nulls 18
1.6 Open Code: Veiling by Grilles 22
1.7 Classification of Cryptographic Methods 24

2 Aims and Methods of Cryptography 25
2.1 The Nature of Cryptography 25
2.2 Encryption 31
2.3 Cryptosystems 33
2.4 Polyphony 35
2.5 Character Sets 37
2.6 Keys 40

3 Encryption Steps: Simple Substitution 42
3.1 Case $V^{(1)} \dashrightarrow W$ (Unipartite Simple Substitutions) 42
3.2 Special Case $V \longleftrightarrow V$ (Permutations) 44
3.3 Case $V^{(1)} \dashrightarrow W^m$ (Multipartite Simple Substitutions) 51
3.4 The General Case $V^{(1)} \dashrightarrow W^{(m)}$, Straddling 53

4 Encryption Steps: Polygraphic Substitution and Coding . 56
4.1 Case $V^2 \dashrightarrow W^{(m)}$ (Digraphic Substitutions) 56
4.2 Special Cases of Playfair and Delastelle: Tomographic Methods 62
4.3 Case $V^3 \dashrightarrow W^{(m)}$ (Trigraphic Substitutions) 66
4.4 The General Case $V^{(n)} \dashrightarrow W^{(m)}$: Codes 66

5 Encryption Steps: Linear Substitution 78
5.1 Self-reciprocal Linear Substitutions 80
5.2 Homogeneous Linear Substitutions 80
5.3 Binary Linear Substitutions 84
5.4 General Linear Substitutions 84

5.5 Decomposed Linear Substitutions 85
5.6 Decimated Alphabets 88
5.7 Linear Substitutions with Decimal and Binary Numbers 89

6 Encryption Steps: Transposition 91
6.1 Simplest Methods 91
6.2 Columnar Transpositions 95
6.3 Anagrams 98

7 Polyalphabetic Encryption: Families of Alphabets 101
7.1 Iterated Substitutions 101
7.2 Shifted and Rotated Alphabets 102
7.3 Rotor Crypto Machines 105
7.4 Shifted Standard Alphabets: Vigenère and Beaufort 114
7.5 Unrelated Alphabets 117

8 Polyalphabetic Encryption: Keys 126
8.1 Early Methods with Periodic Keys 126
8.2 'Double Key' 128
8.3 Vernam Encryption 129
8.4 Quasi-nonperiodic Keys 131
8.5 Machines that Generate Their Own Key Sequences 132
8.6 Off-Line Forming of Key Sequences 143
8.7 Nonperiodic Keys 144
8.8 Individual, One-Time Keys 148
8.9 Key Negotiation and Key Management 151

9 Composition of Classes of Methods 155
9.1 Group Property 155
9.2 Superencryption 157
9.3 Similarity of Encryption Methods 159
9.4 Shannon's 'Pastry Dough Mixing' 160
9.5 Confusion and Diffusion by Arithmetical Operations 166
9.6 DES and IDEA® 170

10 Open Encryption Key Systems 179
10.1 Symmetric and Asymmetric Encryption Methods 180
10.2 One-Way Functions 182
10.3 RSA Method 189
10.4 Cryptanalytic Attack upon RSA 191
10.5 Secrecy Versus Authentication 194
10.6 Security of Public Key Systems 196

11 Encryption Security 197
11.1 Cryptographic Faults 197
11.2 Maxims of Cryptology 205
11.3 Shannon's Yardsticks 210
11.4 Cryptology and Human Rights 211

Part II: Cryptanalysis 217

12 Exhausting Combinatorial Complexity 220
12.1 Monoalphabetic Simple Encryptions 221
12.2 Monoalphabetic Polygraphic Encryptions 222
12.3 Polyalphabetic Encryptions 225
12.4 General Remarks on Combinatorial Complexity 227
12.5 Cryptanalysis by Exhaustion 227
12.6 Unicity Distance 229
12.7 Practical Execution of Exhaustion 231
12.8 Mechanizing the Exhaustion 234

13 Anatomy of Language: Patterns 235
13.1 Invariance of Repetition Patterns 235
13.2 Exclusion of Encryption Methods 237
13.3 Pattern Finding 238
13.4 Finding of Polygraphic Patterns 242
13.5 The Method of the Probable Word 242
13.6 Automatic Exhaustion of the Instantiations of a Pattern 247
13.7 Pangrams 249

14 Polyalphabetic Case: Probable Words 251
14.1 Non-Coincidence Exhaustion of Probable Word Position 251
14.2 Binary Non-Coincidence Exhaustion of Probable Word Position 254
14.3 The De Viaris Attack 255
14.4 Zig-Zag Exhaustion of Probable Word Position 263
14.5 The Method of Isomorphs 264
14.6 Covert Plaintext-Cryptotext Compromise 270

15 Anatomy of Language: Frequencies 271
15.1 Exclusion of Encryption Methods 271
15.2 Invariance of Partitions 272
15.3 Intuitive Method: Frequency Profile 274
15.4 Frequency Ordering 275
15.5 Cliques and Matching of Partitions 278
15.6 Optimal Matching 284
15.7 Frequency of Multigrams 286
15.8 The Combined Method of Frequency Matching 291
15.9 Frequency Matching for Polygraphic Substitutions 297
15.10 Free-Style Methods 298
15.11 Unicity Distance Revisited 299

16 Kappa and Chi 301
16.1 Definition and Invariance of Kappa 301
16.2 Definition and Invariance of Chi 304
16.3 The Kappa-Chi Theorem 306
16.4 The Kappa-Phi Theorem 307
16.5 Symmetric Functions of Character Frequencies 309

17 Periodicity Examination ... 311
17.1 The Kappa Test of Friedman ... 312
17.2 Kappa Test for Multigrams ... 313
17.3 Cryptanalysis by Machines ... 314
17.4 Kasiski Examination ... 320
17.5 Building a Depth and Phi Test of Kullback ... 326
17.6 Estimating the Period Length ... 329

18 Alignment of Accompanying Alphabets ... 331
18.1 Matching the Profile ... 331
18.2 Aligning Against Known Alphabet ... 335
18.3 Chi Test: Mutual Alignment of Accompanying Alphabets ... 339
18.4 Reconstruction of the Primary Alphabet ... 344
18.5 Kerckhoffs' Symmetry of Position ... 346
18.6 Stripping off Superencryption: Difference Method ... 351
18.7 Decryption of Code ... 354
18.8 Reconstruction of the Password ... 354

19 Compromises ... 356
19.1 Kerckhoffs' Superimposition ... 356
19.2 Superimposition for Encryptions with a Key Group ... 358
19.3 In-Phase Superimposition of Superencrypted Code ... 373
19.4 Cryptotext-Cryptotext Compromises ... 376
19.5 A Method of Sinkov ... 381
19.6 Cryptotext-Cryptotext Compromise: Indicator Doubling ... 388
19.7 Plaintext-Cryptotext Compromise: Feedback Cycle ... 403

20 Linear Basis Analysis ... 413
20.1 Reduction of Linear Polygraphic Substitutions ... 413
20.2 Reconstruction of the Key ... 414
20.3 Reconstruction of a Linear Shift Register ... 415

21 Anagramming ... 418
21.1 Transposition ... 418
21.2 Double Columnar Transposition ... 421
21.3 Multiple Anagramming ... 421

22 Concluding Remarks ... 424
22.1 Success in Breaking ... 425
22.2 Mode of Operation of the Unauthorized Decryptor ... 430
22.3 Illusory Security ... 435
22.4 Importance of Cryptology ... 436

Appendix: Axiomatic Information Theory ... 439

Bibliography ... 449

Index ... 453

Photo Credits ... 474

Part I: Cryptography

ars ipsi secreta magistro
[An art secret even for the master]
Jean Robert du Carlet, 1644

Protection of sensitive information is a desire reaching back to the beginnings of human culture.
Otto Horak, 1994

The People

W. F. Friedman

M. Rejewski

A. M. Turing

Only a few years ago one could say that cryptology, the study of secret writing and its unauthorized decryption, was a field that flourished in concealment—flourished, for it always nurtured its professional representatives well. Cryptology is a true science: it has to do with knowledge (Latin *scientia*), learning and lore. By its very nature it not only concerns secretiveness, but remains shrouded in secrecy itself—occasionally even in obscurity. It is almost a secret science. The available classic literature is scant and hard to track down: under all-powerful state authorities, the professional cryptologists in diplomatic and military services were obliged to adopt a mantle of anonymity or at least accept censorship of their publications. As a result, the freely available literature never fully reflected the state of the art—we can assume that things have not changed in that respect. Nations vary in their reticence: whereas the United States of America released quite generous information on the situation in the Second World War, the Soviet Union cloaked itself in silence. That was not surprising; but Great Britain has also pursued a policy of secretiveness which sometimes appears excessive—as in the COLOSSUS story. At least one can say that the state of cryptology in Germany was openly reported after the collapse of the Reich in 1945.[1]

Cryptology as a science is several thousand years old. Its development has gone hand in hand with that of mathematics, at least as far as the persons are concerned—names such as François Viète (1540–1603) and John Wallis (1616–1703) occur. From the viewpoint of modern mathematics, it shows traits of statistics (William F. Friedman, 1920), combinatory algebra (Lester S. Hill, 1929), and stochastics (Claude E. Shannon, 1941). The Second World War finally brought mathematicians to the fore: for example, Hans Rohrbach (1903–1993) in Germany and Alan Mathison Turing (1912–1954) in England; A. Adrian Albert (1905–1972) and Marshall Hall (1910–1990) were engaged in the field in the United States, also J. Barkley Rosser, Willard Van Orman Quine, Andrew M. Gleason, and the applied mathematicians Vannevar Bush (1890–1974) and Warren Weaver (1894–1978). And there was Arne Beurling (1905–1986) in Sweden, Marian Rejewski (1905–1980) in Poland, Maurits deVries in the Netherlands, Ernst S. Selmer (b. 1920) in Norway.

The mathematical disciplines that play an important part in the current state of cryptology include number theory, group theory, combinatory logic, com-

---

[1] Hans Rohrbach (1948), *Mathematische und maschinelle Methoden beim Chiffrieren und Dechiffrieren.* In: FIAT Review of German Science 1939–1941: Applied Mathematics, Part I, Wiesbaden 1948.

plexity theory, ergodic theory, and information theory. The field of cryptology can already be practically seen as a subdivision of applied mathematics and computer science. Conversely, for the computer scientist cryptology is gaining increasing practical importance in connection with access to operating systems, data bases and computer networks, including data transmission.

One could also mention a few present-day mathematicians who have been engaged in official cryptology for a time. Some would prefer to remain incognito.

Quite generally, it is understandable if intelligence services do not reveal even the names of their leading cryptologists. Admiral Sir Hugh P. F. Sinclair, who became in 1923 chief of the British Secret Intelligence Service (M.I.6), had the nickname 'Quex'. Semi-officially, he and his successor General Sir Stewart Graham Menzies (1890–1968), were traditionally known only as "C". Under them were a number of 'Passport Control Officers' at the embassies as well as the cryptanalytic unit at Bletchley Park. And the name of Ernst C. Fetterlein (dec. 1944), who was till the October Revolution head of a Russian cryptanalytic bureau (covername Popov) and served the Government Code and Cypher School of the British Foreign Office since June 1918, was mentioned in the open cryptological literature only incidentally in 1985 by Christopher Andrew and in 1986 by Nigel West.

Professional cryptology is far too much at risk from the efforts of foreign secret services. It is important to leave a potential opponent just as much in the dark about one's own choice of methods ('encryption philosophy') as about one's ability ('cryptanalytic philosophy') to solve a message that one is not meant to understand. If one does succeed in such unauthorized decryption—as the British did with ENIGMA-enciphered messages from 1940 till 1945—then it is important to keep the fact a secret from one's opponents and not reveal it by one's reactions. As a result of British shrewdness, the relevant German authorities, although from time to time suspicious, remained convinced until the approaching end of the war (and some very stubborn persons until 1974) that the ciphers produced by their ENIGMA machines were unbreakable.

The caution the Allies applied went so far that they even risked disinformation of their own people: Capt. Laurance F. Safford, U.S. Navy, Office of Naval Communications, Cryptography Section, wrote in an internal report of March 18, 1942, a year after the return of Capt. Abraham Sinkov and Lt. Leo Rosen from an informative visit in February 1941 to Bletchley Park: "Our prospects of ever [!] breaking the German 'Enigma' cipher machine are rather poor." This did not reflect his knowledge. But he was addressing U.S. Navy leadership.

In times of war, matériel and even human life must often be sacrificed in order to avoid greater losses elsewhere. In 1974, Group Captain Winterbotham said Churchill let Coventry be bombed because he feared defending it would reveal that the British were reading German ENIGMA-enciphered messages. This story was totally false: As the targets were indicated by changing code-words, this would not in fact have been possible. However, the British were initially very upset when, in mid-1943, the Americans began systematically to destroy all the tanker U-boats, whose positions they had learnt as a result of cracking the 4-rotor ENIGMA used by the German submarine command. The British were justifiably concerned that the Germans would suspect what had happened and would greatly modify their ENIGMA system again. In fact they did not, instead ascribing the losses (incorrectly) to treachery. How legitimate the worries had been became clear when the Allies found out that for May 1, 1945, a change in the ENIGMA keying procedures was planned that would have made all existing cryptanalytic approaches useless. This change "could probably have been implemented much earlier if it had deemed worthwhile" (Ralph Erskine).

This masterpiece of security work officially comprised "intelligence resulting from the solution of high-grade codes and ciphers". It was named by the British for short "special intelligence" and codenamed ULTRA, which also meant its security classification. The Americans similarly named MAGIC the information obtained from breaking the Japanese cipher machines they dubbed PURPLE. Both ULTRA and MAGIC remained hidden from Axis spies.

Cryptology also has points of contact with criminology. References to cryptographic methods can be found in several textbooks on criminology, usually accompanied by reports of successfully cryptanalyzed secret messages from criminals still at large—smugglers, drug dealers, gun-runners, blackmailers, or swindlers—and some already behind bars, usually concerning attempts to free them or to suborn crucial witnesses. In the law courts, an expert assessment by a cryptologist can be decisive in securing their conviction. During the days of Prohibition in the U.S.A., Elizebeth S. Friedman née Smith (1892–1980), wife of the famous William Frederick Friedman (1891–1969)[2] and herself a professional cryptologist, performed considerable service in this line. She did not always have an easy time in court: counsel for the defence expounded the theory that anything could be read into a secret message, and that her cryptanalysis was nothing more than "an opinion". The Swedish cryptologist Yves Gyldén (1895–1963), a grandson of the astronomer Hugo Gyldén, assisted the police in catching smugglers in 1934. Only a few criminal cryptologists are known, for example the New Yorker Abraham P. Chess in the early 1950s.

Side by side with state cryptology in diplomatic and military services have stood the amateurs, especially since the 19th century. Starting with the revelation of historic events by retired professionals such as Étienne Bazeries[3], to the after-dinner amusements practised by Wheatstone[4] and Babbage[5],

[2] Friedman, probably the most important U.S. American cryptologist of modern times, introduced in 1920 the *Index of Coincidence*, the sharpest tool of modern cryptanalysis.

[3] Étienne Bazeries (1846–1931), probably the most versatile French cryptologist of modern times, author of the book *Les chiffres secrets dévoilés* (1901).

[4] Sir Charles Wheatstone (1802–1875), English physicist, professor at King's College, London, best known for Wheatstone's bridge (not invented by him).

[5] Charles Babbage (1791–1871), Lucasian Professor of Mathematics at the University of Cambridge, best known for his Difference Engine and Analytical Engine.

with a journalistic cryptanalytic background ranging from Edgar Allan Poe to the present-day *Cryptoquip* in the *Los Angeles Times*, accompanied by excursions into the occult, visiting Martians, and terrorism, cryptology shows a rich tapestry interwoven with tales from one of the oldest of all branches of cryptology, the exchange of messages between lovers.

The letter-writer's guides that appear around 1750 soon offered cryptographic help, like *De geheime brieven-schryver, angetoond met verscheydene voorbeelden* by a certain G. v. K., Amsterdam 1780, and *Dem Magiske skrivekunstner*, Copenhagen 1796. A century later, we find *Sicherster Schutz des Briefgeheimnisses*, by Emil Katz, 1901, and *Amor als geheimer Bote. Geheimsprache für Liebende zu Ansichts-Postkarten*, presumably by Karl Peters, 1904.

Mixed with sensational details from the First and Second World Wars, an exciting picture of cryptology in a compact, consolidated form first reached a broad public in 1967 in David Kahn's masterpiece of journalism and historical science *The Codebreakers*. In the late 1970s there followed several substantial additions from the point of view of the British, whose wartime files were at last (more or less) off the secret list, among the earliest *The Secret War* by Brian Johnson, later *The Hut Six Story* by Gordon Welchman. Cryptology's many personalities make its history a particularly pleasurable field.

Commercial interest in cryptology after the invention of the telegraph concentrated on the production of code books, and around the turn of the century on the design and construction of mechanical and electromechanical ciphering machines. Electronic computers were later used to break cryptograms, following initial (successful) attempts during the Second World War. A programmable calculator is perfectly adequate as a ciphering machine. But it was not until the mid-1970s that widespread commercial interest in encrypting private communications became evident ("Cryptology goes public," Kahn 1979); the options opened up by integrated circuits coincided with the requirements of computer transmission and storage. Further contributing to the growth of cryptology were privacy laws and fears of wiretapping, hacking and industrial espionage. The increased need for information security has given cryptology a hitherto unneeded importance. Private commercial applications of cryptology suddenly came to the fore, and led to some unorthodox keying arrangements, in particular asymmetric public keys first proposed publicly in 1976 by Whitfield Diffie and Martin Hellman. More generally, lack of adequate copyright protection for computer programs has encouraged the use of encryption methods for software intended for commercial use.

However, the demand for "cryptology for everyman" raises contradictions and leads to a conflict of interests between the state and scientists. When cryptology use becomes widespread and numerous scientists are occupied in public with the subject, problems of national security arise. Typically, authorities in the United States began to consider whether private research into cryptology should be prohibited—as private research into nuclear weapons was. On May 11, 1978, two years after the revolutionary article by Diffie and Hellman, a high ranking judicial officer, John M. Harmon, assistant attorney general, Office of Legal Counsel, Department of Justice, wrote to Dr. Frank Press, science advisor to the President: "The cryptographic research and development of scientists and mathematicians in the private sector is known as 'public cryptography'. As you know, the serious concern expressed by the academic community over government controls of public cryptography led the Senate Select Committee on Intelligence to conduct a recently concluded study of certain aspects of the field." These aspects centered around the question of whether restraints based on the International Traffic in Arms Regulation (ITAR) "on dissemination of cryptographic information developed independent of government supervision or support by scientists and mathematicians in the private sector" are unconstitutional under the First Amendment, which guarantees freedom of speech and of the press. It was noted: "Cryptography is a highly specialized field with an audience limited to a fairly select group of scientists and mathematicians ... a temporary delay in communicating the results of or ideas about cryptographic research therefore would probably not deprive the subsequent publication of its full impact."

Cryptological information is both vital and vulnerable to an almost unique degree. Once cryptological information is disclosed, the government's interest in protecting national security is damaged and may not be repaired. Thus, as Harmon wrote in 1978, "a licensing scheme requiring prepublication submission of cryptographic information" might overcome a presumption of unconstitutionality. Such a scheme would impose "a prepublication review requirement for cryptographic information, if it provided necessary procedural safeguards and precisely drawn guidelines," whereas "a prior restraint on disclosure of cryptographic ideas and information developed by scientists and mathematicians in the private sector is unconstitutional."

Furthermore, in the 1980s, the Department of Justice warned that export controls on cryptography presented "sensitive constitutional issues".

Let us face the facts: cryptosystems are not only considered weapons by the U.S. Government—and not only by the U.S. Government—they *are* weapons, weapons for defense and weapons for attack. The Second World War has taught us this lesson.

Harmon wrote moreover: "Atomic energy research is similar in a number of ways to cryptographic research. Development in both fields has been dominated by government. The results of government created or sponsored research in both fields have been automatically classified because of the imminent danger to security flowing from disclosure. Yet meaningful research in the field may be done without access to government information. The results of both atomic energy and cryptographic research have significant nongovernmental uses in addition to military use. The principal difference between the fields is that many atomic energy researchers must depend upon the government to obtain radioactive source material necessary in their research. Cryptographers, however, need only obtain access to an adequate computer." In other words, cryptology invites dangerous machinations even more than atomic energy. At least the crypto weapon does not kill directly—but it may cover up crimes.

The responsibility of the government and the scientists in view of the nimbleness of cryptological activities is reflected in the Computer Security Act of the U.S. Congress of 1987 (Public Law 100-235). It established a Computer System Security and Privacy Advisory Board (CSSPAB), composed of members of the federal government and the computer industry. While a latent conflict did exist, its outbreak seemed to have been avoided in the U.S.A. till 1993 by voluntary restraint on the part of cryptologists (exercised by the Public Cryptography Study Group).

In 1993, however, a *crypto war* broke out between the government and civil rights groups, who felt provoked by the announcement in April 1993—which came also as a surprise to the CSSPAB—and the publication in February 1994 of an Escrowed Encryption Standard (EES), a Federal Information Processing Standards Publication (FIPS 185). The standard makes mandatory an escrow system for privately used keys. While this persistent conflict is not scientific, but rather political, it still could endanger the freedom of science.

Things look better in liberal, democratic Europe; prospects are lower that authorities would be successful everywhere in restraining scientific cryptology. In the European Union, discussions started in 1994 under the keyword "Euro-Encryption", and these may also lead in the end to a regulation of the inescapable conflict of interests of state authorities with scientists. France dropped in 1999 its escrow system. In the former Soviet Union, the problem was of course easily settled within the framework of the system, but in today's Russia, in China and in Israel strong national supervision continues.

* * *

Cryptography and cryptanalysis are the two faces of cryptology; each depends on the other and each influences the other in an interplay of improvements to strengthen cryptanalytic security on the one side and efforts to mount more efficient attacks on the other side. Success is rather rare, failures are more common. The silence preserved by intelligence services helps, of course, to cover up the embarrassments. All the major powers in the Second World War succeeded—at least occasionally—in solving enemy cryptosystems, but all in turn sometimes suffered defeats, at least partial. Things will not be so very different in the 21st century – thanks to human stupidity and carelessness.

1 Introductory Synopsis

En cryptographie, aucune règle n'est absolue.
[In cryptography, no rule is absolute.]
Étienne Bazeries (1901)

1.1 Cryptography and Steganography

We must distinguish between cryptography (Greek *kryptos* hidden) and steganography (Greek *steganos*, covered). The term *cryptographia*, to mean *secrecy in writing*, was used in 1641 by John Wilkins, a founder with John Wallis of the Royal Society; the word 'cryptography' was coined in 1658 by Thomas Browne, famous English physician and writer. It is the aim of cryptography to render a message incomprehensible to an unauthorized reader: *ars occulte scribendi*. One speaks of *overt secret writing*: overt in the sense of being obviously recognizable as secret writing.

The term *steganographia* was also used in this sense by Caspar Schott, a pupil of Athanasius Kircher, in the title of his book *Schola steganographia*, published in Nuremberg in 1665; however, it had already been used by Trithemius in his first (and amply obscure) work *Steganographia*, which he began writing in 1499, to mean 'hidden writing'. Its methods have the goal of concealing the very *existence* of a message (however that may be composed)—communicating without incurring suspicion (Francis Bacon, 1623: *ars sine secreti latentis suspicione scribendi*). By analogy, we can call this *covert secret writing* or indeed 'steganography'.

Cryptographic methods are suitable for keeping a private diary or notebook—from Samuel Pepys (1633–1703) to Alfred C. Kinsey (1894–1956)—or preventing a messenger understanding the dispatch he bears; steganographic methods are more suitable for smuggling a message out of a prison—from Sir John Trevanion (Fig. 13), imprisoned in the British Civil War, to the French bank robber Pastoure, whose conviction is described by André Langie, and Klaus Croissant, the lawyer and Stasi collaborator who defended the Baader-Meinhof terrorist gang. The imprisoned Christian Klar used a book cipher.

Steganography falls into two branches, linguistic steganography and technical steganography. Only the first is closely related to cryptography. The technical aspect can be covered very quickly: invisible inks have been in use since Pliny's time. Onion juice and milk have proved popular and effective through the ages (turning brown under heat or ultraviolet light). Other classical props are hollow heels and boxes with false bottoms.

Among the modern methods it is worth mentioning high-speed telegraphy, the spurt transmission of stored Morse code sequences at 20 characters per second, and frequency subband permutation ('scrambling') in the case of telephony, today widely used commercially. In the Second World War, the *Forschungsstelle* (research post) of the *Deutsche Reichspost* (headed by *Postrat* Kurt E. Vetterlein) listened in from March 1942 to supposedly secure radio telephone conversations between Franklin D. Roosevelt and Winston Churchill, including one on July 29, 1943, immediately before the cease-fire with Italy, and reported them via Schellenbergs *R.S.H.A. Amt VI* to Himmler.

Written secret messages were revolutionized by microphotography; a *microdot* the size of speck of dirt can hold an entire quarto page—an extraordinary development from the macrodot of Histiæus[1], who shaved his slave's head, wrote a message on his scalp, and then waited for the hair to grow again. Microdots were invented in the 1920s by Emanuel Goldberg. The Russian spy Rudolf Abel produced his microdots from spectroscopic film which he was able to buy without attracting attention. Another Soviet spy, Gordon Arnold Lonsdale, hid his microdots in the gutters of bound copies of magazines. The microdots used by the Germans in the Second World War were of just the right size to be used as a full stop (period) in a typewritten document.

1.2 Semagrams

Linguistic steganography recognizes two methods: a secret message is either made to appear innocent in an *open code*, or it is expressed in the form of visible (though often minute) graphical details in a script or drawing, in a *semagram*. This latter category is especially popular with amateurs, but leaves much to be desired, since the details are too obvious to a trained and wary eye. The young Francis Bacon (1561–1626) invented the use of two typefaces to convey a secret message (Fig. 1), described in the Latin translation *De dignitate et augmentis scientiarum* (1623) of his 1605 book *Proficience and Advancement*. It has never acquired any great practical importance (but see Sect. 3.3.3 for the binary code he introduced on this occasion).

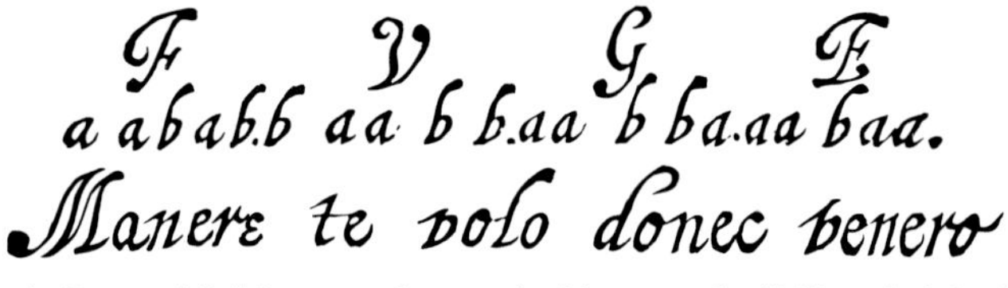
Manere te volo donec venero

Fig. 1. Francis Bacon: Visible concealment of a binary code ('biliteral cipher') by means of different types of script. Note the different forms of /e/ in the word *Manere*.

The same steganographic principle appears to have been known in Paris at the same time, and is mentioned by Vigenère in 1586. Despite its clumsiness

---

[1] Kahn spells the name Histiaeus on p. 81, Histaeius on p. 780, and Histaieus in the index of his book. Verily an example of *ars occulte scribendi* in an otherwise very reliable book!

In Königsberg i. Pr. gabelt sich der Pregel und umfließt eine Insel, die *Kneiphof* heißt. In den dreißiger Jahren des achtzehnten Jahrhunderts wurde das Problem gestellt, ob es wohl möglich wäre, in einem Spaziergang jede der sieben Königsberger Brücken genau einmal zu überschreiten.

Daß ein solcher Spaziergang unmöglich ist, war für L. EULER der Anlaß, mit seiner anno 1735 der Akademie der Wissenschaften in St. Petersburg vorgelegten Abhandlung *Solutio problematis ad geometriam situs pertinentis* (Commentarii Academiae Petropolitanae 8 (1741) 128-140) einen der ersten Beiträge zur Topologie zu liefern.

Das Problem besteht darin, im nachfolgend gezeichneten Graphen einen einfachen Kantenzug zu finden, der alle Kanten enthält. Dabei repräsentiert die Ecke vom Grad 5 den Kneiphof und die beiden Ecken vom Grad 2 die Krämerbrücke sowie die Grüne Brücke.

Fig. 2. Semagram in a 1976 textbook on combinatory logic (the passage deals with the famous Königsberg bridges problem). The lowered letters give the message "*nieder mit dem sowjetimperialismus*" [down with Soviet imperialism].

it has lasted well: the most recent uses known to me are A. van Wijngaarden's alleged usage of roman and italic full stops in the ALGOL 68 report.

A second steganographic principle consists of marking selected characters in a book or newspaper, for example by dots or by dashes. It is much more conspicuous than the above-mentioned method—unless an invisible ink is used—but simpler to implement. A variant (in a book on combinatory logic) uses an almost imperceptible lowering of the letters concerned (Fig. 2).

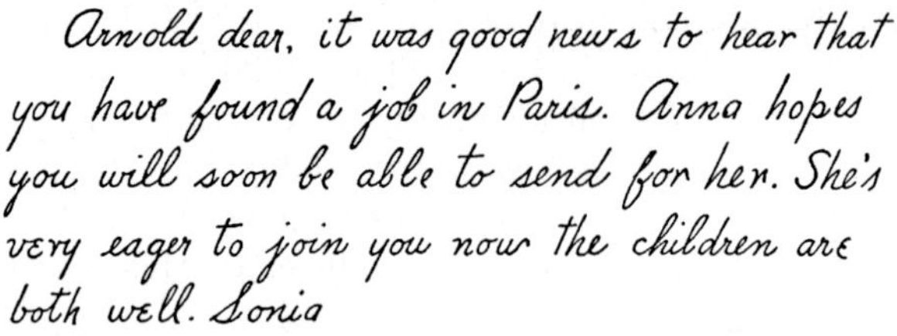
Arnold dear, it was good news to hear that you have found a job in Paris. Anna hopes you will soon be able to send for her. She's very eager to join you now the children are both well. Sonia

Fig. 3. Visible concealment of a numeric code by spacing the letters (Smith)

A third principle uses spaces between letters within a word (Fig. 3). In this example, it is not the letter before or after the space that is important, but the number of letters between successive letters ending with an upward stroke, 3 3 5 1 5 1 4 1 2 3 4 3 3 3 5 1 4 5 ... . In 1895, A. Boetzel and Charles O'Keenan demonstrated this steganographic principle, also using a numeric code, to the French authorities (who remained unconvinced of its usefulness, not without reason). It appears to have been known before then in Russian anarchist circles, combined with the "Nihilist cipher" (Sect. 3.3.1). It was also used by German U-boat officers in captivity to report home on the Allies' anti-submarine tactics.

Fig. 4. Secret message solved by Sherlock Holmes (AM HERE ABE SLANEY), from *The Adventure of the Dancing Men* by Arthur Conan Doyle

All these are examples of semagrams (visibly concealed secret writing). And there are many more. In antiquity Æneas used the astragal, in which a cord threaded through holes symbolized letters. A box of dominoes can conceal a message (by the positions of the spots), as can a consignment of pocket watches (by the positions of the hands). Sherlock Holmes' dancing men (Fig. 4) bear a message just as much as hidden Morse code (Fig. 5): "compliments of CPSA MA to our chief Col. Harold R. Shaw on his visit to San Antonio May 11th 1945" (Shaw had been head of the Technical Operations Division of the U.S. government's censorship division since 1943).

Fig. 5. Semagram. The message is in Morse code, formed by the short and long stalks of grass to the left of the bridge, along the river bank and on the garden wall

A maze is a good example of a clear picture hidden in a wealth of incidental detail: the tortuous paths of Fig. 6 reduce to a graph which can be taken in at a glance. Autostereograms which require the viewer to stare or to squint in order to see a three-dimensional picture (Fig. 7) are also eminently suitable for concealing images, at least for a while.

Of greater interest are those methods of linguistic steganography that turn a secret message into one that is apparently harmless and easily understood, although wrongly (open code). The principle is closer to that of cryptography. Again, there are two subcategories: masking and veiling.

Where will the balloon land, A or B?

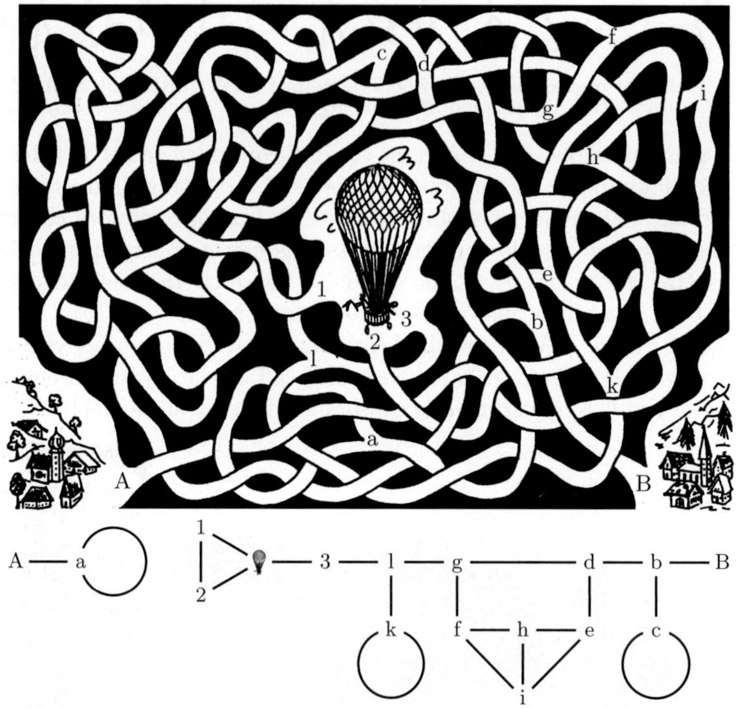

Fig. 6. Maze and its associated graph

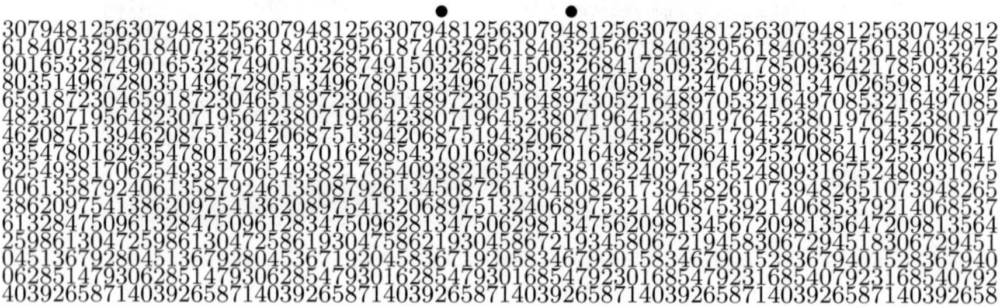

Bernhard Bauer

Fig. 7. Autostereogramm

1.3 Open Code: Masking

A secret writing or message masked as an open communication requires a prior agreement as to the true meaning of seemingly harmless phrases. This is probably the oldest form of secrecy technique—it is to be found in all cultures. Oriental and Far Eastern dealers and gamesters (and some western ones) are reputed to be masters in the use of gestures and expressions. The following system is said to be common among American card cheats. The manner of holding a cigarette or scratching one's head indicates the suit or value of the cards held. A hand on the chest with the thumb extended means "I'm going to take this game. Anybody want to partner me?" The right hand, palm down, on the table means "Yes", a clenched fist, "No, I'm working single, and I discovered this guy first, so scram!" The French conjurer Robert Houdin (1805–1871) is said to have used a similar system around 1845, with I, M, S, V standing for *cœur, carreau, trèfle, pique* : *il fait chaud* or *il y a du monde* means "I have hearts", as it starts with /I/. Things were no more subtle in English whist clubs in Victorian days; "Have you seen old Jones in the past fortnight?" would mean hearts, as it starts with /H/. The British team was suspected of exchanging signals at the world bridge championships in Buenos Aires in 1965—nothing could be proved, of course.

Sometimes, a covert message can be transmitted masked in an innocent way by using circumstances known only to the sender and the recipient. This may happen in daily life. A famous example was reported by Katia Mann: In March 1933, she phoned from Arosa in Switzerland her daughter Erika in Munich and said: "*Ich weiß nicht, es muß doch jetzt bei uns gestöbert werden, es ist doch jetzt die Zeit*". But Erika replied "*Nein, nein, außerdem ist das Wetter so abscheulich. Bleibt ruhig noch ein bissel dort, ihr versäumt ja nichts*" [No, no, anyway, the wheather is so atrocious. Stay a little while, you are not missing anything here]. After this conversation, it became clear to Katia and Thomas Mann, that they could not return to Germany without risk.

Fig. 8. Tramps' secret marks ('tines'), warning of a policeman's house and an aggressive householder (Central Europe, around 1930)

Secret marks have been in use for centuries, from the itinerant scholars of the Middle Ages to the present-day vagrants, tramps, hoboes and loafers. Figure 8 shows a couple of secret marks, such as could still be seen in a provincial town of Central Europe in the 1930s, Figure 9 shows a few used in the midwestern United States in the first half of the 20th century. Tiny secret marks are also used in engravings for stamps or currency notes as a distinguishing mark for a particular engraver or printer.

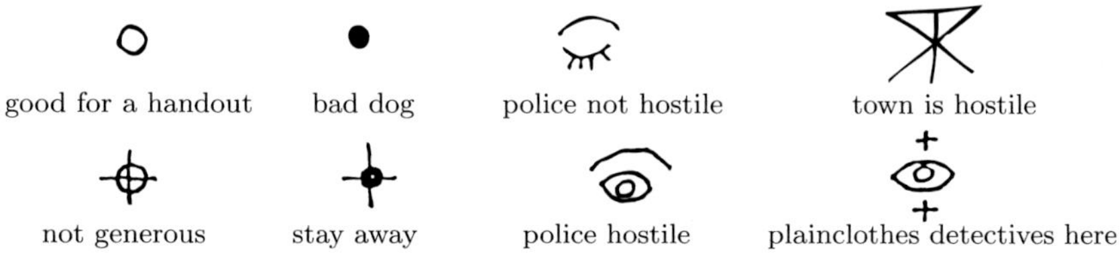

Fig. 9. Hoboes' secret marks for 'police not hostile' and other messages (Midwestern United States, first half of 20th century)

Languages specific to an occupation or social class, collectively known as jargon, above all the kinds used by beggars, vagabonds, and other rascals, variously called *argot* (France, U.S.A.), cant (England), thieves' Latin (England), *rotwelsch* (Germany), *fourbesque* (Italy), *alemania* (Spain), or *calão* (Portugal), and which serve to shield (and keep intact) a social group, often make use of masking. Masked secret writing is therefore called *jargon code*.

The oldest papal code in the 14th century used *Egyptians* for the Ghibellines, *Sons of Israel* for the Guelphs. One French code in the 17th century used jargon exclusively: *Jardin* for Rome, *La Roze* for the Pope, *Le Prunier* for the Cardinal de Retz, *La Fenestre* for the King's brother, *L'Écurie* (meaning either stable or gentry) for Germany, *Le Roussin* for the Duke of Bavaria and so on. A simple masking of names was used in a Bonapartist plot in 1831.

The languages of the criminal underworld are of particular steganographic interest. French argot offers many examples, some of which have become normal colloquial usage: *rossignol* (nightingale) for skeleton key, known since 1406; *mouche* (fly) for informer ('nark' in British slang), since 1389. Alliterative repetition is common: *rebecca* for rebellion, *limace* (slug) for *lime* (file)—which in turn is fourbesque for shirt; *marquise* for *marque* (mole or scar)—which in turn is alemania for a girl; *frisé* (curly) for Fritz (a popular name for a German). Not quite so harmless are metaphors: *château* for hospital, *mitraille* (bullet) for small change, or the picturesque but pejorative *marmite* (cooking pot) for a pimp's girlfriend, *sac à charbon* (coal sack) for a priest. Sarcastic metaphors such as *mouthpiece* for a lawyer are not confined to the underworld.

Some jargon is truly international: 'hole' – *trou* – *Loch* for prison; 'snow' – *neige* – *Schnee* or 'sugar' – *sucre* for cocaine; 'hot' – *heiß* for recently stolen goods; 'clean out' – *nettoyer* – *abstauben* for rob; 'rock' – *galette* (*galet* = gravel) for money. All kinds of puns and plays on words find their place here. The British 'Twenty Committee' in the Second World War, which specialized in double agents, took its name from the Roman number XX for double cross.

Well masked secret codes for more or less *universal* use are hard to devise and even harder to use properly—the practised censor quickly spots the stilted language. The abbot Johannes Trithemius (1462–1516), in his *Polygraphiæ*, six books printed in 1508–1518 (Fig. 10), presents a collection of Latin words as codes for individual letters (Fig. 11), the *Ave Maria* code. "Head", for example, could be masked as "ARBITER MAGNUS DEUS PIISSIMUS" . In fact, there were 384 such alphabets in the first book, to be used successively—a remarkable case of an early polyalphabetic encryption (Sect. 2.3.3).

It could be that present-day censors are not sufficiently well versed in Latin to cope with that. A favourite trick in censorship is to reformulate a message, preserving the semantics. In the First World War a censor altered a despatch from "Father is dead" to "Father is deceased." Back came the message "Is father dead or deceased?"

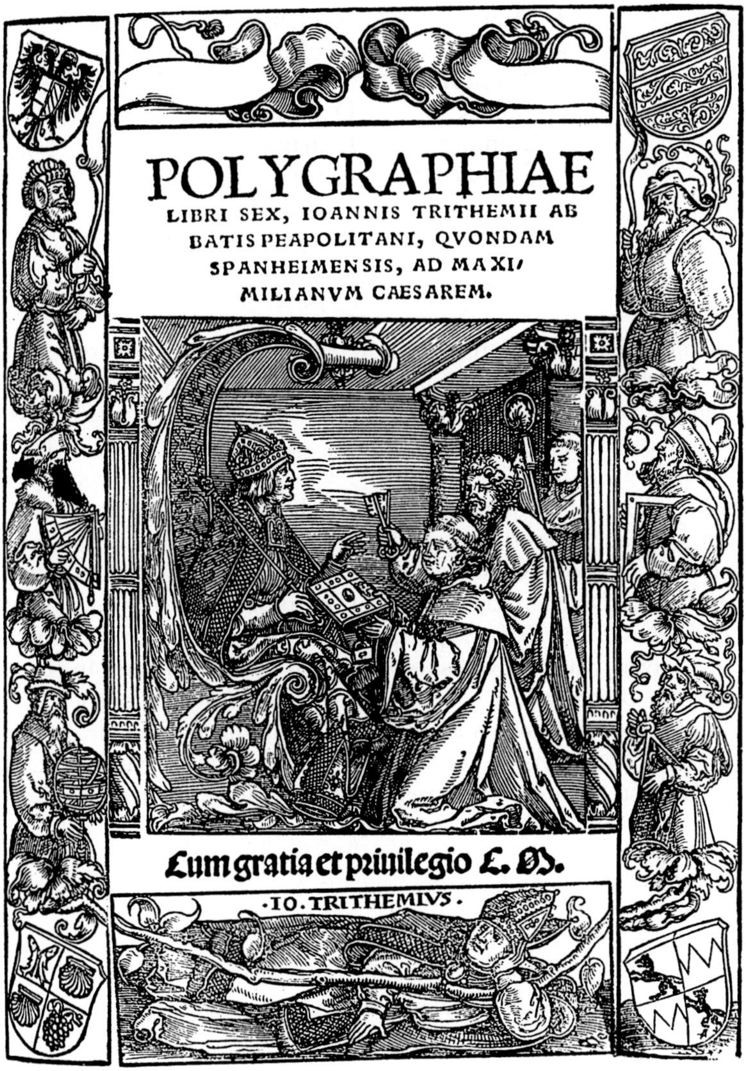

Fig. 10. Title page (woodcut) of the first printed book on cryptography (1508)

Allegorical language is of little help here. In Louis XV's diplomatic service, Chevalier Douglas was sent on a secret mission to Russia in 1755 with an allegorical arsenal from the fur trade, with *le renard noir était cher* for "the influence of the English party is increasing", *le loup-cervier avait son prix* for "the Austrian party (under Bestuchev) retains its dominant influence." Bestuchev himself, who was friendly to Prussia, was *le loup-cervier*, while *une peau de petit-gris* meant 3000 mercenaries in the pay of the British.

It is to be hoped that the chevalier was more subtle in the use of his allegorical code than the German spies, in the guise of Dutch merchants, who—as told by

A Deus	A clemens
B Creator	B clementiſsimus
C Conditor	C pius
D Opifex	D pijſsimus
E Dominus	E magnus
F Dominator	F excelſus
G Conſolator	G maximus
H Arbiter	H optimus

Fig. 11. The first entries of Trithemius' *Ave Maria* cipher

Major-General Kirke—ordered cigars in batches of thousands from Plymouth one day, Portsmouth the next, then Gravesend and so on—1000 coronas stood for one battleship. Their inadequate system brought their lives to a premature end on July 30, 1915. Luck was on the side of Velvalee Dickinson, a Japanophile woman in New York City, who kept up a lively correspondence on broken dolls in 1944. Things came to light when a letter to an address in Portland, Oregon, was returned, and the sender's name turned out to be false. The lady really did sell exquisite dolls from a shop in Madison Avenue. Technical Operations Division, the agency for detecting especially hard to find hidden messages, and the FBI managed to produce evidence for the prosecution, but she got away with ten years in prison and a $ 10 000 fine. In the Audrey Hepburn film *Breakfast at Tiffany's*, Miss Holly Golightly spent a night behind bars because she helped a gangster conduct his cocaine dealership from his prison cell by means of "weather reports"—it did occur to her, she admitted, that "snow in New Orleans" sounded somewhat improbable.

1.4 Cues

The most important special case of masking, i.e., of a jargon-code message, concerns the use of a *cue* (French *mot convenu*), a pre-arranged phrase or verse to mean a particular message. The importance of the message is linked to the time of transmission; the message serves as an alarm or acknowledgement. Large numbers of messages were broadcast by the BBC to the French *résistance* during the Second World War. It therefore attracted little attention when some masked messages with an importance several orders of magnitude greater than the others were broadcast, for example on June 1, 1944, when the 9 o'clock news was followed by a string of "personal messages", including the first half of the first verse of the poem *Chanson d'Automne* by Paul Verlaine (translated: "The long sobs of the autumn violins"). The second half (translated: "wound my heart with a monotonous languor") followed June 5th. The German command structure had already in January 1944 been informed by Admiral Canaris' *Abwehr* of the jargon code and its significance. When the 15th Army picked up the expected cue (Fig. 12), German command posts were warned, but for reasons that have not been fully explained to this

Tag Uhrzeit Ort und Art der Unterkunft	Darstellung der Ereignisse (Dabei wichtig: Beurteilung der Lage [Feind- und eigene], Eingangs- und Abgangszeiten von Meldungen und Befehlen)
5.6.44	Am 1., 2. und 3.6.44 ist durch die Nast innerhalb der "Messages personelles" der französischen Sendungen des britischen Rundfunks folgende Meldung abgehört worden : "Les sanglots longs des violons de l'automme ". Nach vorhandenen Unterlagen soll dieser Spruch am 1. oder 15. eines Monats durchgegeben werden, nur die erste Hälfte eines ganzen Spruches darstellen und ankündigen, dass binnen 48 Stunden nach Durchgabe der zweiten Hälfte des Spruches, gerechnet von 00.00 Uhr des auf die Durchsage folgenden Tages ab, die anglo-amerikanische Invasion beginnt.
21.15 Uhr	Zweite Hälfte des Spruches "Blessent mon coeur d'une longeur monotone" wird durch Nast abgehört.
21.20 Uhr	Spruch an Ic-AO durchgegeben. Danach mit Invasionsbeginn ab 6.6. 00.00 Uhr innerhalb 48 Stunden zu rechnen. Überprüfung der Meldung durch Rückfrage beim Militärbefehlshaber Belgien/Nordfrankreich in Brüssel (Major von Wangenheim).
22.00 Uhr	Meldung an O.B. und Chef des Generalstabes.
22.15 Uhr	Weitergabe gemäss Fernschreiben (Anlage 1) an Generalkommandos. Mündliche Weitergabe an 16. Flak-Division.

Fig. 12. Extract from a log kept by the 15th Army's radio reconnaissance section (Lt. Col. Helmuth Meyer, Sgt. Walter Reichling). *automme* is to be read *automne*, *longeur* is to be read *langueur*.

day the alarm did not reach the 7th Army, on whose part of the coast the invasion took place within 48 hours, on June 6, 1944.

The Japanese used a similar system in 1941. For example HIGASHI NO KAZE AME (east wind, rain), inserted in the weather report in the overseas news and repeated twice, was to announce "war with the U.S.A." The U.S. Navy intercepted a diplomatic radio message to that effect on November 19, 1941 and succeeded in solving it by the 28th. As tension mounted, numerous reconnaissance stations in the U.S.A. were monitoring Japanese radio traffic for the cue. It came on December 7th—hours after the attack on Pearl Harbor—in the form NISHI NO KAZE HARE (west wind, clear), indicating the commencement of hostilities with Great Britain, which came as very little surprise by then. Perhaps the whole thing was a Japanese double cross.

Technically, masked secret writing shows a certain kinship with enciphered secret writing (Sect. 2.2), particularly with the use of substitutions (Sect. 3) and codes (Sect. 4.4).

In a different category are secret writings or messages veiled as open ones (invisibly concealed secret writing). Here, the message to be transmitted is somehow embedded in the open, harmless-looking message by adding nulls.

In order to be able to reconstruct the real message, the place where it is concealed must be arranged beforehand (*concealment cipher*). There are two obvious possibilities for using *garbage-in-between*: by specifying rules (*null cipher*, *open-letter cipher*) or by using a *grille* (French for 'grating').

1.5 Open Code: Veiling by Nulls

Rules for veiled messages are very often of the type "the nth character after a particular character", e.g., the next letter after a space ("family code", popular among soldiers in the Second World War, to the great displeasure of the censors); better would be the third letter after a space, or the third letter after a punctuation mark. Such secret messages are called acrostics. A practised censor usually recognizes immediately from the stilted language that something is amiss, and his sharp eye will certainly detect what

P̌RESIDENT'S ĚMBARGO ŘULING ŠHOULD ȞAVE ǏMMEDIATE
ŇOTICE. ǦRAVE ŠITUATION ǍFFECTING ǏNTERNATIONAL ĽAW.
ŠTATEMENT F̌ORESHADOWS ŘUIN ǑF M̌ANY ŇEUTRALS. Y̌ELLOW
J̌OURNALS ǓNIFYING ŇATIONAL ĚXCITEMENT ǏMMENSELY

means—a message intercepted in the First World War.

If necessary, it can help to write out the words one below the other:

```
    ↓
I N S P E C T
D E T A I L S
F O R
T R I G L E T H
A C K N O W L E D G E
T H E
B O N D S
F R O M
F E W E L L
```

The disguise falls away; the plain text "jumps out of the page".

Sir John Trevanion, who lived during the time of Oliver Cromwell (1599–1658), saved himself from execution by using his imagination. In a letter from his friend R. T. he discovered the message "panel at east end of chapel slides" – and found his way out of captivity (Fig. 13).

There is a story of a soldier in the U.S. Army who arranged with his parents that he would tell them the name of the place he had been posted to by means of the initial letter of the first word (after the greeting) in consecutive letters home—from a cryptographic and steganographic point of view not such a bad idea. However, his cover was blown when his parents wrote back "Where is Nutsi? We can't find it in our atlas." The poor fellow had forgotten to date his letters.

Worthie Sir John: — Hop̌e, thăt is ye beste comfort of ye afflicted, caňnot much, I fĕar me , hel̆p you now. Thăt I would saye to you, is t̆his only: if ĕver I may be able to requite that I do owe you, stănd not upon asking me. 'Tiš not much that I can do: but̆ what I can do, beĕ ye verie sure I wille. I kňowe that, if d̆ethe comes, if ŏrdinary men fear it, it f̆rights not you, acčounting it for a high honour, to h̆ave such a rewarde of your loyalty. Prăy yet that you may be spared this soe bitter, cup̌. I fĕar not that you will grudge any sufferings; onl̆y if bie submission you can turn them away, 'tiš the part of a wise man. Tel̆l me, an ĭf you can, to d̆o for you anythinge that you wolde have done. Thĕ general goes back on Wednesday. Reštinge your servant to command. — R. T.

Fig. 13. Message to Sir John Trevanion: panel at east end of chapel slides (third letter after punctuation mark)

The technique of acrostics even found its way into belletristic literature. In the classical acrostic, it was the initial letters, syllables, or words of successive lines, verses, sections, or chapters which counted. Words, author's names, or sentences (Fig. 14) were enciphered in this way. Acrostics also served as an insurance against omissions and insertions: an early example of the present-day parity checks or error-detecting codes.

In a similar way, the chronogram conceals a (Roman) numeral in an inscription; usually it is a date, for example the year when the plaque was erected:

In the baroque church of the former Cistercian monastery Fürstenfeld near Munich, in 1766 a statue of the Wittelsbachian founder Ludwig der Strenge (1229–1294) was placed, below there is a tablet with the chronogram

LVDoVICVs seVerVs DVX baVarVs aC paLatInVs,
hIC In sanCta paCe qVIesCIt .

(Ludwig the Severe, Duke of Bavaria and Count Palatine, rests here in holy peace.)

If the chronogram consists of a verse, then the technical term is a chronostichon—or chronodistichon for a couplet.

Composers have concealed messages in their compositions, either in the notes of a musical theme (a famous example[2] is B A C H), or indirectly by means of a numerical alphabet: if the i-th note of the scale occurs k times, then the k-th letter of the alphabet is to be entered in the i-th position. Johann Sebastian Bach was fond of this cipher; in the organ chorale "*Vor deinen Thron*", written in 1750 in the key of G major, g occurs twice (B), a once (A), b three times (C), and c eight times (H) in the four-bar flourish.

Nulls are also used in many jargons: simply appending a syllable (parasitic suffixing) is the simplest and oldest system. In French, for example,

floutiere for *flou*, argot for 'go away!'; *girolle* for *gis*, argot for 'yes'; *mezis* for *me*; *icicaille* for *ici*

[2] In German, b is used for b flat, h for b. In G major, g is first, a second, h third, etc.

and there are hundreds of similar forms. Cartouche (18th century) has

vousierge trouvaille bonorgue ce gigotmouche

where the nulls are underlined.

Fast writing method

He must have had a special trick, said Robert K. Merton, for he wrote such an amazing quantity of material that his friends were simply astonished at his prodigious output of long manuscripts, the contents of which were remarkable and fascinating, from the first simple lines, over fluently written pages where word after word flowed relentlessly onward, where ideas tumbled in a riot of colorful and creative imagery, to ends that stopped abruptly, each script more curiously charming than its predecessors, each line more whimsically apposite, yet unexpected, than the lines on which it built, ever onward, striving toward a resolution in a wonderland of playful verbosity. Fuller could write page after page so fluently as to excite the envy of any writers less gifted and creative than he. At last, one day, he revealed his secret, then died a few days later. He collected a group of acolytes and filled their glasses, then wrote some words on a sheet of paper, in flowing script. He invited his friends to puzzle a while over the words and departed. One companion took a pen and told the rest to watch. Fuller returned to find the page filled with words of no less charm than those that graced his own writings. Thus the secret was revealed, and Fuller got drunk. He died, yet still a space remains in the library for his collected works.

Ludger Fischer / J. Andrew Ross

Fig. 14. Self-describing acrostic

Tut Latin, a language of schoolchildren, inserts TUT between all the syllables. Such school jargons seem to be very old; as early as 1670 there are reports from Metz (Lorraine) of a 'stuttering' system, where for example *undreque foudreque* stood for *un fou.*

The Javanais language is also in this class:

jave for *je*; *laveblavanc* for *le blanc*; *navon* for *non* ;
chavaussavurave for *chaussure* .

Other systems use dummy syllables with duplicated vowels, such as B talk in German:

GABARTEBENLAUBAUBEBE for *gartenlaube* (bower)

or Cadogan in French:

CADGADODGOGADGAN for *cadogan* .

Joachim Ringelnatz (1883–1934) wrote a poem in *Bi* language (Fig. 15) .

Simple reversing of the letters, called back slang, occurs in cant: OCCABOT for 'tobacco', KOOL for 'look', YOB for 'boy', SLOP for 'police'. Permutation

Gedicht in Bi-Sprache

Ibich habibebi dibich,
Lobittebi, sobi liebib.
Habist aubich dubi mibich
Liebib? Neibin, vebirgibib.

Nabih obidebir febirn,
Gobitt seibi dibir gubit.
Meibin Hebirz habit gebirn
Abin dibir gebirubiht.

Fig. 15. Poem in the *Bi* language by Joachim Ringelnatz

of the syllables is found in the French *Verlan* (from *l'envers*): NIBERQUE for *bernique* ("nothing doing", said to be related to *bernicles*, tiny shells); LONTOU for *Toulon*, LIBRECA for *calibre* (in the sense of a firearm); DREAUPER for *perdreau* (partridge, to mean a policeman); RIPOU for *pourri* (rotten); BEUR for *rebeu* (Arab). More recent are FÉCA for *café*, TÉCI for *cité*.

More complicated systems involve shuffling the letters, i.e., a transposition (Sect. 6.1). Criminal circles were the origin of the Largonji language:

leudé for *deux* [francs]; *linvé* for *vingt* [sous]; *laranqué* for *quarante* [sous];

with the phonetic variants

linspré for *prince* (Vidocq, 1837); *lorcefée* for *La Force*, a Paris prison;

and of the Largonjem language:

lombem for *bon* (1821); *loucherbem* for *boucher*; *olrapem* for *opéra* (1883) .

The name Largonji is itself formed in this way from 'jargon'.

A variant with suppression of the initial consonant is the Largondu language:

lavedu for *cave*; *loquedu* for *toque*; *ligodu* for *gigo(t)* .

Similar formation rules lie behind the following:

locromuche for *maquerau* (pimp) ; *leaubiche* for *beau*;
nebdutac for *tabac* (1866); *licelargu* for *cigare* (1915) .

These systems also have parallels in East Asia (Hanoï, Haïphong).

Pig Latin, another school language, puts AY at the end of a cyclically permuted word: third becomes IRDTHAY. Cockneys have a rhyming slang with nulls: TWIST AND TWIRL for *girl*, JAR OF JAM for *tram*, STORM AND STRIFE for *wife*, BOWL OF CHALK for *talk*, FLEAS AND ANTS for *pants*, APPLES AND PEARS for *stairs*, BULL AND COW for *row*, CAIN AND ABEL for *table*, FRANCE AND SPAIN for *rain*, PLATES OF MEAT for *feet*, LOAF OF BREAD for *head*. The actual rhyming word is usually omitted—the initiated can supply it from memory. Some of these expressions have entered the language (lexicalization): few people are aware of the origin of "use your loaf" or "mind your plates".

Jonathan Swift (1667–1745) was not overcautious in his *Journal to Stella*, who in fact was Esther Johnson (1681–1728): in a letter on Feb. 24, 1711 he merely inserted a null as every second character.

And there was mounting in hot haste the steed,
The mustering squadron and the clattering car,
And swiftly forming in the ranks of war;
And deep the thunder peal on peal afar;
And near, the beat of the alarming drum
Roused up the soldier ere the morning star
While thronged the citizens with terror dumb
Or whispering, with white lips, — 'the
foe! they come, they come!'

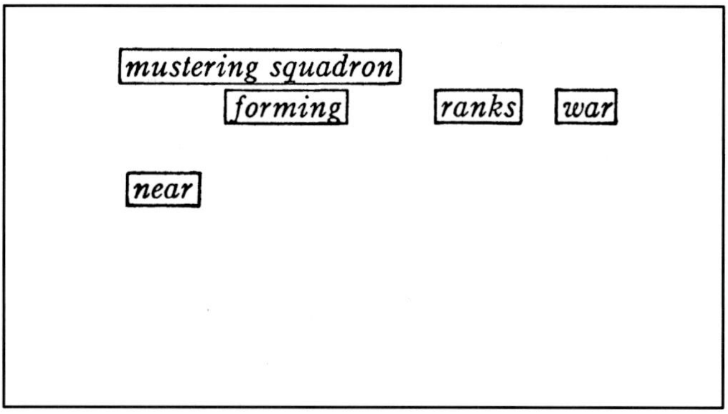

Fig. 16. Lord Byron's hypothetical message

1.6 Open Code: Veiling by Grilles

The method of the grille, which goes back to Geronimo Cardano (in *De Subtilitate, 1550*), is simple to understand, but suffers from the disadvantage that both sides must possess and retain the grille—in the case of a soldier in the field or a prisoner not something that can be taken for granted. It is also awfully hard to compose a letter using it. If Lord Byron (1788–1824)—admittedly no ordinary soldier—had used the method, his talents would have come in extremely handy for composing a poem such as that in Fig. 16. He would presumably also have been able to lay it out so attractively that the plain text fitted the windows of the grille without calling attention.

Cardano, incidentally, insisted on copying out the message three times, to remove any irregularities in the size or spacing of the letters. The method was occasionally used in diplomatic correspondence in the 16th and 17th centuries. Cardinal Richelieu is said to have made use of it. The modern literature also mentions some more cunning rules, for example to convey binary numbers (in turn presumably used to encipher a message), in which a word with an even number of vowels represents the digit 0, or an odd number the digit 1.

Veiled secret writing is a concealment cipher. In professional use, it is usually considered as enciphered secret writing (Sect. 2.2), it shows a certain kinship particularly in the use of nulls (Sect. 2.3.1) and of transposition (Sect. 6.1).

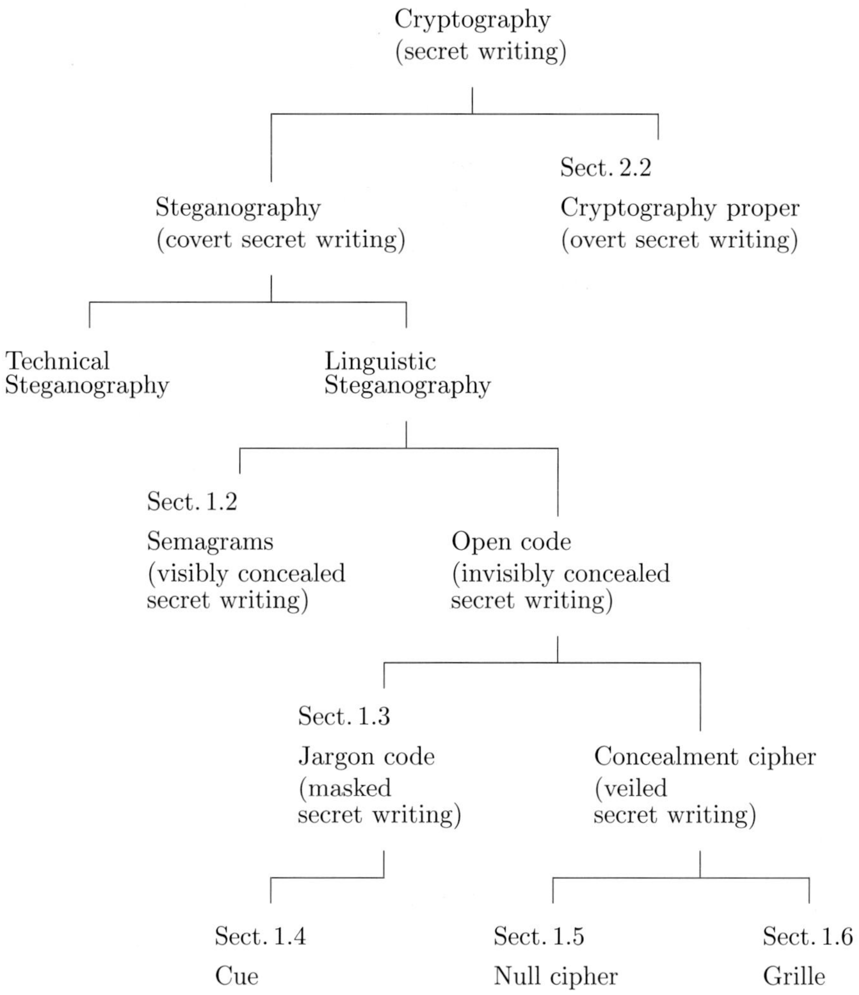

Fig. 17. Classification of steganographic and cryptographic methods

1.7 Classification of Cryptographic Methods

Figure 17 shows a diagrammatic summary of the classification of methods of steganography and cryptography proper as given in this and the next chapter.

Masking and veiling have been treated in detail here because they provide a methodical guide: masking leads to substitution, veiling leads to transposition. These are the two basic elements of cryptography proper. We shall introduce them in the next chapter.

Steganography also reveals an important maxim: natural language – spoken, written, or in gestures—has its own particular rules, and it is even harder to imitate them (as in steganography) than to suppress them (as in cryptography).

Linguistic steganography is therefore treated with caution by pure cryptographers; it is a censor's job to combat it. By its very nature, an amateur steganogram can be rendered harmless by suppressing or revealing it. For the censorship, the actual solution is often of little importance (except, perhaps, to provide evidence for a subsequent court case).

The professional use of linguistic steganography can be justified only in special cases–unless it represents a concealment of a cryptographic method.

Steganography and cryptography proper fall under the concept of cryptology. The term *cryptologia* was used, like *cryptographia*, by John Wilkins in 1641, to mean *secrecy in speech.* In 1645, 'cryptology' was coined by James Howell, who wrote "cryptology, or epistolizing in a clandestine way, is very ancient." The use of the words 'cryptography', *cryptographie, crittografia, Kryptographie* has until recently dominated the field, even when cryptanalysis was included.

Claude Shannon (1916-2001)

Claude Shannon, in 1945, still called his confidential report on safety against unauthorized decryption *A Mathematical Theory of Cryptography.* Within book titles, *cryptologue* was used by Yves Gyldén (1895–1963) in 1932 and in more modern time 'cryptologist' by William F. Friedman (1891–1969) in 1961; 'cryptology' shows up in the title of an article by David Kahn in 1963; it was used internally by Friedman and Lambros D. Callimahos (1911–1977) in the 1950s. With Kahn's *The Codebreakers* of 1967, the word 'cryptology' was firmly established to involve both cryptography and cryptanalysis and is widely accepted now.

With the widespread availability of sufficiently fast computer-aided image manipulation, steganography nowadays sees a revival. By subtle algorithms, messages are hidden within pictures.

2 Aims and Methods of Cryptography

> Nearly every inventor of a cipher system has been convinced of the unsolvability of his brainchild.
>
> *David Kahn*

A survey of the known cryptographic methods is given in this chapter from the point of view of securing[1] established channels of communication against (passive) eavesdropping and (active) falsification (ISO 7498). Security against breaking the secrecy in the sense of confidentiality and privacy is the classic goal, whereas security against forgery and spurious messages, that is to say authentication of the sender, has only recently acquired much importance.

Besides mathematical questions, philological ones play an important part in cryptology. A kindred topic is the decryption of ancient scripts in extinct languages[2], an appealing field bordering on both archæology and linguistics. Plate A shows an example, the disc of Phaistos.

2.1 The Nature of Cryptography

The objective of cryptography is to make a message or record incomprehensible to unauthorized persons. This can easily be overdone, thereby making the message indecipherable to the intended recipient—who has not experienced being unable to read a hastily written note a few weeks (or even days) later?

Seriously speaking, it is fatal if an encryption error is made or if radio communications have been garbled or corrupted by atmospheric disturbances. Any attempt to re-encipher and retransmit the same message—correctly, this time—represents a serious security risk for reasons to be discussed in Chapter 10 and Part II. Therefore, encryption discipline forbids this strictly; the text has to be edited, without altering the content, of course. This is easier said than done—the road to doom is usually paved with good intentions.

---

[1] Since the discoveries of Shannon and Hamming in about 1950, mere garbling and corruption of communication channels by physical or technical means has been countered by error-detecting and error-correcting codes, which need not be considered here.

[2] Johannes Friedrich, *Extinct Languages*, New York 1957.

2.1.1 Therefore, the encryption and decryption method must not be too complicated: it must be appropriate to the intelligence and circumstances of the people who have to use it. The lowest standards apply on the battlefield. Immediately after that comes the field of diplomacy, at any rate if the ambassador is expected to carry out the encryption and decryption himself. When Wheatstone demonstrated the method now known as PLAYFAIR (Sect. 4.2) at the British Foreign Office in 1854, saying that three out of four boys from the nearest school could master it, the Under-Secretary of State remarked drily "that is very possible, but you could never teach it to attachés!"

It should also be borne in mind that many messages need be kept secret only for as long as the events they refer to have not in any case become public knowledge. Admittedly, it may be wiser to keep diplomatic messages secret for decades afterwards. The British—not to mention the Russians—are unsurpassable in this respect, as is the veil of secrecy which they cast over their entire cryptographic system. At any rate, we need only know how long, at least, a cryptanalyst must work on a message to read it—to break the cipher—and then it becomes pointless to maintain that a particular method is absolutely secure. A code used by the Germans on the Western Front in 1917 (known by the French as KRUSA, because all the code groups began with one of these five letters) was based on 'planned obsolescence'. The code sequence changed every month, but the French had usually worked it out after two weeks—often after only two days.

However, a quantitative assessment of encryption methods was only made possible by the pioneering ideas of Claude E. Shannon (see Appendix). The suitability of the various methods was still very imperfectly understood in the First World War, as was shown when *Le Matin* revealed in 1914 that the French were reading German messages. The German General Staff made a sudden and radical change to its encryption system on November 18th. The change from a double columnar transposition (Sect. 6.2.4) to a VIGENÈRE with key ABC (Sect. 8.1.2) and subsequent simple columnar transposition was a *complication illusoire* (illusory complication) for the French decryptors.

Evidently the lovers, who used to declare their feelings for each other in coded messages in the 'personal' columns of British newspapers about 1850, had every confidence in their cryptosystem. Eavesdropping on these messages provided pleasure for a section of London society; this included Charles Babbage and also Charles Wheatstone and Lyon Playfair, who broke into one such correspondence with a suitable message of their own, thus prompting the reaction "Dear Charlie: write no more. Our cipher is discovered." from the young lady. Incidentally, in spite—or possibly as a result—of Shannon's commendable elucidation, coded messages are said to be a regular feature of the 'agony columns' to this day.

Another lady was greatly impressed by a man to whom she had given an enciphered recipe for making gold; she alone knew the key. The man not only informed her that he had deciphered it, but also told her the key word.

He must be a magician, she thought. As he could obviously read her mind, it was best to give him the key to her heart. The year was 1757, she was the affluent Madame d'Urfé , and the cavalier (who abandoned her soon afterwards) was Giacomo Girolamo Casanova, Chevalier de Seingalt, whose cryptanalytical zeal is evidently not sufficiently well known.

Casanova

Marie Antoinette also knew how to combine love with cryptography, as did King Edward VIII (the later Duke of Windsor). Besides its diplomatic and military uses, cryptography thus has its private and civil applications, not to mention the commercial ones, such as bookseller's price cipher, or the packaging date for butter (Sect. 3.1.1), or the markings on car tires (Fig. 18).

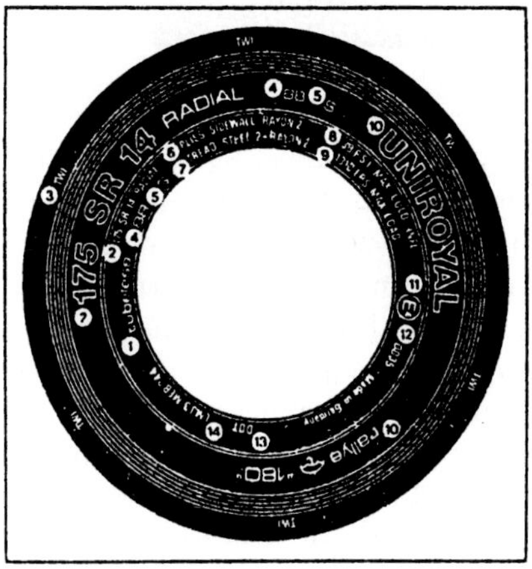

Tire marks

(1) tubeless

(2) 175 is the width of the tire in mm
S stands for speed
(up to 180 km/h in the case of summer tires)
R means radial plies (omitted for crossply tyres)
14 is the diameter of the wheel rim in inches

(3) TWI = tread wear indicator
(six ribs which appear in the tread pattern when it has worn down to 1/16 inch)

(4, 5) additional markings for Europe:
88 is the (coded) maximum load per wheel
S again means 180 km/h

(6) sidewall consists of two layers of rayon fibres

(7) tread has two layers of steel and two of rayon

(8) maximum cold inflation pressure
(applies to U.S.A. only)

(9) maximum load per wheel
(applies to U.S.A. only)

(10) brand name

(11, 12) tested to European standards
4 is the country where the test took place
(in this case Holland).

(13) DOT = Department of Transportation
(the US transport ministry)

(14) manufacturer's codes:
LM = factory; J3 = size; MEB = type;
344 = date (34th production week of 1974)

Fig. 18.
Coding system for car tires

There are some amusing stories about the supposed unbreakability of ciphers. Over-assessment of one's own cleverness is a regular source of advantage to the opposing side. Sometimes the exaggeration is the work of others. It was said of Paul Schilling von Cannstatt, one of the inventors of the electromagnetic telegraph (1832), that "for the Russian ministry he compiled such a secret alphabet, the so-called chiffre, that even so ingenious a secret cabinet as the Austrians possessed could not have penetrated it in fifty years" (F.P. Fonton, after A.V. Jarozkij). And as late as 1917, the respected periodical *Scientific American* declared Vigenère's method (Sect. 7.4) to be unbreakable.

It was an ironic twist of fate that Étienne Bazeries, the great French cryptologist (1846–1931), who shattered a whole series of supposedly unbreakable cryptosystems that had been presented to the French security agency, was himself presumptuous enough to believe he had found an absolutely secure method (*je suis indéchiffrable*, see Fig. 19). His antagonist the Marquis de Viaris—the first modern cryptologist, incidentally, to make use of mathematics—derived no small pleasure in taking his revenge by breaking several ciphers which Bazeries sent him (Sects. 7.5.3, 14.3.1).

Fig. 19. Bazeries' cylinder with the message *je suis indéchiffrable*

2.1.2 The invention and financial exploitation of enciphering and deciphering machines is a lucrative branch of cryptography. Until the 19th century they were mechanical; from the beginning of the 20th century automation made its appearance, around the middle of the century came electronics and more recently microelectronic miniaturization. Towards the end of 1939, Konrad Zuse, the 30-year old German computer pioneer, stationed as an infantryman on the *Siegfried Line*, also invented an enciphering machine, an attachment to a teleprinter. He was not able to persuade the German War Office of the advantages of his invention, which used the Vernam principle (Sect. 8.3); he was given to understand that the authorities already possessed good equipment of a similar nature. They were referring to Lorenz SZ 40 and Siemens T 52, not ENIGMA, as Zuse incorrectly recently assumed.

Today's microcomputers—roughly the size, weight, and price of a pocket calculator—have a performance as good as the best enciphering machines from the Second World War. That restores the earlier significance of good methods, which had been greatly reduced by the presence of 'giant' computers in cryptanalysis centres. More than that, a normal commercial microcomputer costing about $100—not to mention a PC—can carry out a much more complex encryption than the classic machines were capable of.

In assessing a method based on any kind of documentation and encryption apparatus, it must be borne in mind that such objects could fall into the opponent's hands (Shannon's maxim, Sect. 11.2.3). A microcomputer fed with a program or data on magnetic card possesses no telltale cryptographic structure of its own—except, possibly, an alphabetic keyboard and display.

In the case of the public keys propagated for commercial communication links, even the encryption and decryption methods are published. It is only the key for decryption which remains secret. Shannon's maxim "The enemy knows the system" is thus carried to extremes. At the same time, the increasing use of public communications channels has led to authentication becoming as much a declared objective of cryptography as secrecy is (Sects. 10.5, 10.6).

2.1.3 Cryptological techniques are occasionally used in literature. Intricately woven literary works such as *Zettels Traum* by Arno Schmidt (1970) just ask to be 'decrypted'. Ostensibly secret messages represent a particular problem. In *La physiologie du mariage* (1829), Honoré de Balzac has this passage: *"La Bruyère a dit très spirituellement: 'C'est trop contre un mari que la dévotion et la galanterie: une femme devrait opter.' L'auteur pense que La Bruyère s'est trompé. En effet: ~~~ ."* What then follows is a higgledy-piggledy jumble of letters, as if a type-case had been spilt on the page. Four editions of the book, three of them printed in Balzac's lifetime, in fact contain four different versions. The author must have been playing a practical joke on the reader. Nevertheless, Bazeries investigated such a cryptogram in 1901 and found that it did not fit any known scheme; it was *une facétie de l'auteur*.

There was much ado in 1878 when Ignatius Donelly, an American provincial politician and imaginative pseudo-scientist who had already speculated on Atlantis and a collision between the Earth and a meteor, set about finding steganographic proof in the works of Shakespeare that the author was in fact Sir Francis Bacon (Georg Cantor, the founder of modern set theory, also hunted this chimæra for many years). Now if you take a long enough text, and declare enough characters as irrelevant (perhaps also permuting the ones that remain), then you can read anything into it—Lord Byron's hypothetical message in Sect. 1.6 could serve as an example. So Donelly was apparently successful. A flood of amateurs joined in the search. None of this would have been very remarkable, had not a certain William Frederick Friedman[3], who had studied genetics, been hired by a rich textile merchant in Geneva near Chicago, Colonel George Fabyan. Besides funding laboratories for biology, chemistry, and acoustics, Fabyan employed cryptologists who were supposed to prove that Bacon was Shakespeare. Friedman was attracted by cryptology and also by Elizebeth Smith, a young cryptologist working there. He espoused himself to both, and became the most successful American cryptologist.

2.1.4 The official cryptological services in the 20th century go by mysterious-sounding names, in keeping with the spirit of the times. They are usually embedded in the secret services concerned with counter-espionage and intelligence-gathering beyond their own borders. The most famous are M.I.6, the British Secret Intelligence Service (S.I.S.) which is directly answerable to the Foreign Office; and in the U.S.A. since 1947 the Central Intelligence Agency (C.I.A.), which is, together with the Defense Intelligence Agency (D.I.A.), subordinate to the U.S. Intelligence Board (U.S.I.B.) and therefore

---

[3] Born Wolfe Friedmann in Kishinev (Moldavia) on Sep. 24, 1891. The family emigrated to the U.S.A. the following year. He died on Nov. 2, 1969 and was buried in Arlington.

controlled by legislatives and executives. Post-war Germany has its *Bundesnachrichtendienst* (BND), directly answerable to the Chancellor's Office.

The actual cryptological services, especially the departments for cryptanalysis, are frequently divided into diplomatic and military parts. That may have good organizational reasons, but it often hinders the exchange of experience. In wartime Britain, the Admiralty (O.I.C., Operational Intelligence Centre) and the Foreign Office (Department of Communications) were forced into close cooperation by the desperate situation in 1940; Winston Churchill chipped in and set up a powerful superordinate authority in the form of the London Controlling Section (L.C.S.), headed by Colonel John Henry Bevan and directly answerable to the Prime Minister. Responsibility for cryptanalysis was held within M.I.8 by its G.C.H.Q. (Government Communications Headquarters), with various nicknames, some of them with historical significance: G.C. & C.S. (Government Code and Cypher[4] School), War Station, Station X, Room 47 Foreign Office; it was also often called B.P. (Bletchley Park), after the place where it was housed from 1939. Even within B.P. a certain distinction was maintained between AI (Air Intelligence) and MI (Military Intelligence) on the one hand, and the Navy, steeped in tradition, on the other. Both looked back on their successes in the First World War, gained by M.I.1(b) (Military Intelligence Division) of the War Office and Room 40 at the Admiralty. Postwar G.C.H.Q. is located in Cheltenham.

After the United States entered the First World War in 1917, rapid expansion of military cryptology became necessary. As part of the A.E.F. (American Expeditionary Forces), G.2 A.6 (General Staff, Intelligence Section, Military Information Division, Radio Intelligence Section) and the Code Compilation Section of the Signal Corps found themselves under the supervision of MI-8 (Cryptological Section of Military Intelligence Division), headed by Herbert Osborne Yardley (1889–1958). Rivalry between the Army and the Navy continued throughout the Second World War: OP-20-G was the naval cryptological organization with its cryptanalysis department OP-20-GY, while SIS (Signal Intelligence Service) was the army counterpart which Yardley had built up and which had been headed by William Friedman since 1929. The experience gained in the Second World War led to a concentration of resources: within G.2: the Army Signal Security Agency merged with the cryptanalysis department of the Signal Corps in 1945 to produce the A.S.A. (Army Security Agency), then in 1949 the A.F.S.A. (Armed Forces Security Agency), and in 1952 the N.S.A. (National Security Agency) under the Secretary of Defense, led 1977–1981 by the legendary Bobby Ray Inman. Important subdivisions are the Defense Intelligence Agency and I.D.A., the Institute for Defense Analyses, which is freer and loosely connected to some universities. N.S.A. is located on Fort George G. Meade in Maryland.

In the German *Reich*, too, the services were split: the *Auswärtiges Amt* (foreign office) on the one hand, and the army and navy on the other, re-

[4] *Cypher* is an older form of *cipher*, still current in Great Britain.

ceived further competition in the Second World War from the *Reichsluftfahrtministerium* and the *Sicherheitsdienst* (SD, "security service"). The future General Erich Fellgiebel (1886–1944) had introduced in 1934 a common enciphering machine, the ENIGMA; but the coordination of all operations, which OKW/Chi, the Cipher Branch of the Armed Forces Supreme Command constantly demanded, was still blocked in the autumn of 1943 by Ribbentrop, Göring, and Himmler. When coordination by WNV/Chi (*Wehrmachtnachrichtenverbindungen Chiffrierwesen*) was finally achieved in November 1944 by order of the Führer, intelligence was firmly in the hands of Walter Schellenberg (1910–1952), who was ambitious and always feigned devotion to his leaders Himmler and Hitler, and advanced to major-general in the SS. He died of a liver complaint after serving a sentence of only six years passed by the Nuremberg tribunal—the fashion designer Coco Chanel paid his funeral costs.

In post-war Germany, in 1953, an authority was established in Bad Godesberg near Bonn whose cover name—*Bundesstelle für Fernmeldestatistik* (Federal Office for Telecommunications Statistics)—was something of an understatement. In fact, it was a cryptanalytical subdivision of BND; its true name was *Zentralstelle für das Chiffrierwesen* (Central Office for Cryptology). A reorganization ('*Amt für Militärkunde*') took place in 1990 by splitting off the BSI, which deals with questions of public cryptography.

In France, 2bis (a street number in the Avenue de Trouville) was the *nom de guerre* for the S.R. (*Service de Renseignement*) with its cryptanalytic bureau (*section de transmission et décryptement*). The Swedish cryptanalytical agency was known by the abbreviation FRA (*Försvarets Radioanstalt*), while in Italy it was SIM (*Servizio degli Informazione Militare*). In Japan, *Tokumu Han* (espionage department) is the name for the cryptanalytical department of the admiralty staff intelligence group, set up in 1925, and *Angō Kenkyū Han* (cipher research department) for that of the foreign ministry.

Spets otdel ('Special Department') was the name of the cryptographical and cryptanalytical service of the Union of Soviet Socialist Republics, established in 1921 by orders of Vladimir Ilyich Lenin and for some time under the command of Lev Davidovich Trotzki.

2.2 Encryption

To summarize: cryptology is the science of (overt) secret writing (cryptography), of its unauthorized decryption (cryptanalysis), and of the rules which are in turn intended to make that unauthorized decryption more difficult (encryption security).

2.2.1 Vocabulary, character set. The set of characters, V, used to formulate the plaintext[5] is the plaintext vocabulary or plaintext character set. The set of characters, W, used to formulate the ciphertext or codetext is the cryptotext vocabulary or cryptotext character set. The individual

[5] In contrast to *cleartext*, which means a text transmitted without encryption.

characters in W can also be logograms, special symbols representing a word or phrase, such as &, %, $, £, ©; also shorthand symbols. V and W can be different, or overlapping, or identical sets.

2.2.1.1 Let V^*, W^* be the set of words constructed from V, W (plaintext words, cryptotext words). ε indicates the empty word. $Z^n \subseteq Z^*$ is the set of all words of length n; $Z^{(n)}$ denotes $\{\varepsilon\} \cup Z \cup Z^2 \cup Z^3 \ldots \cup Z^n$. V^* is called the plaintext space, W^* the cryptotext space.

2.2.1.2 In all practical cases, V and W are nonempty finite sets. Theoretically, however, we could allow denumerable sets V and W; then V^n and W^n are also denumerable.

2.2.2 Encryption and decryption. An encryption is defined as a relation $\mathbf{X} : V^* \dashrightarrow W^*$. The converse relation $\mathbf{X}^{-1} : V^* \dashleftarrow W^*$, defined by $x \leftarrow\!\dashv y$ if and only if $x \mapsto y$, is then called a decryption.

2.2.2.1 The intended recipient of an encrypted message should be able to reconstitute the original message without ambiguities. An encryption therefore as a rule is *injective*, i.e., unambiguous from right to left (left-univalent):

$$(x \mapsto z) \wedge (y \mapsto z) \Rightarrow (x = y) \quad .$$

We define $\mathcal{H}_x = \{y \in W^* : x \overset{\mathbf{X}}{\mapsto} y\}$ as the fiber of $x \in V^*$.

As a rule it is also a requirement that the encryption $\mathbf{X}$ be *total* (definal); that is to say, $\mathcal{H}_x$ is nonempty for all $x \in V^*$.

2.2.2.2 The encryption $\mathbf{X} : V^* \dashrightarrow W^*$ is implemented by means of Hilbert's non-deterministic 'choice operator' η, where $\mathbf{X}(x) = \eta\,\mathcal{H}_x$. The elements of $\mathcal{H}_x$ (assuming there are more than one) are called variants, also homophones of x. Thus, variants are different cryptotext words assigned to the same plaintext word in the encryption relation $\mathbf{X} : V^* \dashrightarrow W^*$.

If the relation $V^* \dashrightarrow W^*$ is also unambiguous from left to right (right-univalent); i.e., $\mathcal{H}_x$ contains at most one element for all $x \in V^*$, then the encryption is *functional*, $V^* \dashrightarrow W^*$ is a function $V^* \longrightarrow W^*$ and, if it is in addition *surjective*, even becomes a one-to-one function $V^* \longleftrightarrow W^*$.

In the functional case there are no variants; the encryption is deterministic and thus a one-to-one mapping of plaintext space into cryptotext space.

2.2.2.3 As a rule, $\varepsilon \overset{\mathbf{X}}{\mapsto} \varepsilon$. If $\mathcal{H}_\varepsilon$ also contains elements other than ε which are homophones for $\varepsilon \in V^*$, these are called null texts or dummy texts.

Note that the set of all encryptions $V^* \dashrightarrow W^*$ (in the case of fixed nonempty V, W) is non-denumerable.

2.2.3 Inductive definitions. An encryption $\mathbf{X} : V^* \dashrightarrow W^*$ is said to be finite if the set of all pairs in the relation is finite. Then $\mathbf{X} : V^{(n)} \dashrightarrow W^{(m)}$ for suitable natural numbers n, m .

But how can a relation $V^* \dashrightarrow W^*$ be defined and specified? Even if it is finite, it may very well not be practicable to list all the pairs. For that reason, inductive rules are frequently used. This is studied in the next paragraph.

2.3 Cryptosystems

Let M, the encryption system, be a nonempty, as a rule *finite* set $\{\chi_0, \chi_1, \chi_2, \ldots, \chi_{\theta-1}\}$ of (injective) relations $\chi_i : V^{(n_i)} \dashrightarrow W^{(m_i)}$. Each χ_i is called an encryption step. An encryption system together with a corresponding decryption system is a cryptosystem.

An encryption $\mathbf{X} = [\,\chi_{i_1}, \chi_{i_2}, \chi_{i_3}, \ldots\,]$ is called finitely generated (by means of the encryption system M), if it is *induced* by a (terminating or infinite) sequence $(\chi_{i_1}, \chi_{i_2}, \chi_{i_3}, \ldots)$ of encryption steps $\chi_i \in M$ under the concatenation $.\star.$, i.e.,

$x \overset{\mathbf{X}}{\mapsto} y$ holds for $x \in V^*$, $y \in W^*$ if and only if there exist decompositions

$x = x_1 \star x_2 \star x_3 \star \ldots \star x_k$ and $y = y_1 \star y_2 \star y_3 \star \ldots \star y_k$ with[6]

$$x_j \overset{\chi_{i_j}}{\mapsto} y_j \quad \text{for } j = 1, 2, \ldots k .$$

Example:

$\chi_i : V^{(n_i)} \dashrightarrow V^{(n_i)}$: cyclic transposition of n_i elements $(\theta = 4)$;

$n_1 = 3,\ n_2 = 5,\ n_0 = 2,\ n_3 = 6$

n e a	r l y e v	e r	y i n v e n	$(\chi_1, \chi_2, \chi_0, \chi_3)$
e a n	l y e v r	r e	i n v e n y	

2.3.1 Basic concepts. $\theta = |M|$ denotes the cardinal number of the encryption system. An encryption step $\chi_i : V^{(n_i)} \dashrightarrow W^{(m_i)}$ is a generating relation; the number n_i is called the (maximal) plaintext encryption width, the number m_i the (maximal) crypt width of χ_i. The relation χ_i may be nondeterministic. The encryption step is said to be endomorphic if $V = W$.

Speaking of homophones and variants (also optional substitutes, multiple substitutes) and of nulls (also dummies, French *nonvaleurs*, German *Blender, Blindsignale*), in most cases those of the encryption step are meant. If the cryptotext character set of the encryption step contains words of different length, the encryption step is called "straddling" (German *gespreizt*).

A generated encryption is not necessarily injective, even if the generating encryption steps are:

Assume

a $\mapsto$ ·–

i $\mapsto$ ··

l $\mapsto$ ·–··

belong to an injective $V^1 \longrightarrow W^{(4)}$, then in $V^* \dashrightarrow W^*$

ai $\mapsto$ ·–·· and l $\mapsto$ ·–·· ;

this means that injectivity is violated (by sloppy radio operators).

---

6 Every $x \in V^*$ is taken to be suitably filled up by meaningless symbols.

2.3.2 Ciphering and coding. An encryption step $\chi_i : V^{(n_i)} \dashrightarrow W^{(m_i)}$ is by its very nature finite, provided V and W are finite; it can be specified in principle by enumeration (encryption table). An actual enumeration is often called a code or cipher (French *chiffre*); the encryption step is then called the encoding step or enciphering step. The terminological boundary between 'cipher', 'encipher', 'decipher' and 'code', 'encode', 'decode' is fuzzy and essentially determined by historical usage (see also Sect. 4.4). The terms 'cipher' and 'code', and more generally 'crypt', are also used for the elements of $W^{(m_i)}$.

2.3.2.1 An encryption $\mathbf{X} = [\chi_{i_1}, \chi_{i_2}, \chi_{i_3}, \ldots]$, finitely generated by M, is *monoalphabetic* if it comprises or uses a single encryption step ('alphabet'). Otherwise it is called polyalphabetic. (If M is a singleton ($\theta = 1$), then every encryption generated by means of M is monoalphabetic.)

2.3.2.2 A finitely generated encryption is said to be *monographic* if all the n_i of the encryption steps used equal 1, polygraphic otherwise. In a special case of particular interest for encryption by machines, all encryption steps of M show equal maximal encryption width n and equal maximal crypt width m. Then M is necessarily finite. If even

$$\chi_i : V^n \dashrightarrow W^m$$

holds for all $\chi_i \in M$, which means that no encryption step is straddling, $[\chi_{i_1}, \chi_{i_2}, \chi_{i_3}, \ldots]$ is a block encryption; a word from V^n is an encryption block. In a suitable vocabulary of character n-tuples, a block encryption can be interpreted theoretically as a monographic encryption.

Encryption systems with $\chi_i : V^n \dashrightarrow W^m$ for $n = 2,\ 3,\ 4$ establish bigram, trigram, tetragram encryptions, which for $m = 1,\ 2,\ 3$ are called unipartite, bipartite, tripartite (French *bifide, trifide*). Frequently $V = W$ and $m = n$ are chosen, to give us a block encryption in the narrow sense.

2.3.3 Text streams. A stream $(z_1, z_2, z_3, \ldots)$ is an infinite sequence of blocks of characters. There is a one-to-one correspondence between the stream $(z_1, z_2, z_3, \ldots)$ and an infinite sequence $((z_1), (z_1 \star z_2), (z_1 \star z_2 \star z_3), \ldots)$ of words, the segments $(z_1 \star z_2 \star \ldots \star z_i)$ of the stream.

A plaintext stream is an infinite sequence of blocks $(p_1, p_2, p_3, \ldots)$, where $p_j \in V^n$; correspondingly, an infinite sequence of blocks $(c_1, c_2, c_3, \ldots)$, where $c_j \in W^m$, is a cryptotext stream. A stream encryption is a block encryption of segments of a fictitious plaintext stream to segments of a likewise fictitious cryptotext stream.

An encryption $\mathbf{X} = [\chi_{i_1}, \chi_{i_2}, \chi_{i_3}, \ldots]$, finitely generated by M, is called *periodic* (repeated key) or *nonperiodic* ('aperiodic', running key), depending on whether the infinite sequence $(\chi_{i_1}, \chi_{i_2}, \chi_{i_3}, \ldots)$ finally is periodic or not.

A monoalphabetic encryption is trivially periodic. A nonperiodic (running key) encryption therefore is necessarily polyalphabetic. This will be given more attention in Sect. 8.7.

Every periodic block encryption of period r can be interpreted theoretically as a *monoalphabetic* encryption, with

$$\chi_0 \ : \ V^{n \cdot r} \dashrightarrow W^{m \cdot r}$$

as the sole encryption step. For running key encryptions this is not the case. They belong fundamentally to a more powerful category of methods. There is a one-to-one correspondence of the sequence $(\chi_{i_1}, \chi_{i_2}, \chi_{i_3}, \ldots)$, $\chi_i \in M$, and a real number represented in a number system to the basis θ by the fraction $0.i_1\ i_2\ i_3\ \ldots$. For fixed M, a subset of the denumerable set of rational numbers corresponds to the set of periodic block encryptions; the set of nonperiodic (running key) encryptions thus corresponds to the non-denumerable set of irrational numbers between 0 and 1.

An up-to-date example of a monoalphabetic, polygraphic block encryption is the DES cryptosystem, a block encryption (and decryption) method propagated by the National Bureau of Standards of the U.S.A. since 1977; the encryption step (in the ECB mode) is a one-to-one endomorphic encryption, chosen among 2^{56} possibilities (key length 56, Sect. 9.6.1.1), a $V^8 \longleftrightarrow V^8$ permutation with a vocabulary $V = Z_2^8$ of 256 different 8-bit words. An encryption step of this size cannot be documented by enumeration but is defined algorithmically. Algorithmic definitions, however, are unsympathetic to the use of homophones and encourage the restriction to block encryption.

There is an example of a polyalphabetic, polygraphic encryption which is not a block encryption: plaintext encrypted word by word using a number of code books in some periodically or nonperiodically changing order. This is not very practical in computerized cryptography. Polyalphabetic polygraphic block encryption is the domain of present-day computers.

2.4 Polyphony

Use of homophones and nulls has been standard in cryptography since 1400. Around 1500, encipherings with cipher elements of different length began to be used, and the importance of the left-uniqueness condition for straddling encryption steps (Sect. 3.4) was recognized at the latest around 1580 by the papal secretary of ciphers Giovanni Battista Argenti and his nephew Matteo Argenti. The modern Fano condition ("no cipher element is head of another cipher element") is a sufficient condition the Argentis were apparently familiar with. For unstraddling encryption steps, the hiatus and thus the right decomposition can be found by counting.

2.4.1 Polyphones. They are unblushingly used in English when, e.g., both the phonemes \ā\ as in \brāk\ and \ē\ as in \frēk\ are printed 'ea'. Cryptographically, polyphonic enciphering steps, in which several plaintext words are assigned to one and the same cryptotext word, violate injectivity and are rare.

The 'SA Cipher', a code used by the British Admiralty in 1918 (Sect. 4.4.3, Fig. 37), and the Duchesse de Berry's cipher which used as a substitution alphabet LEGOUVERNEMENTPROVISOIRE (Sect. 3.2.5), are among the very few examples of genuine polyphony. In practice, there is sufficient semantic information to avoid ambiguity if, say, 'Diesel oil', 'Corporal', and 'Paris', or 'runway', 'General', and 'ground fog' are polyphones. The idea seems to occur to amateurs more than anyone. A loving couple in England provided Babbage with a tough nut to crack in 1853, with a polyphonic cipher using the digits $0 \dots 9$, in which, e.g.,

1 stood for t and u, 8 for h and i, 2 for m and o, 4 for e and r.

The message began with

1821 82734 29 30 84541

which (allowing for two enciphering errors) meant "thou image of my heart". It seems that the lovers derived special pleasure from the unnecessary complication.

However, polyphonic ciphers were used in the ancient civilizations between the Nile and the Euphrates. As the letters of the alphabet also served as number symbols, it was a popular pastime to add up the values of the symbols representing a secret word (*gematria*). In this way, the *isopsephon* 666 mentioned in the Apocalypse (Rev. 13.18) has been taken to represent the Emperor Nero (Fig. 20). There are said to be people who refuse to accept a car registration involving the "number of the beast" 666.

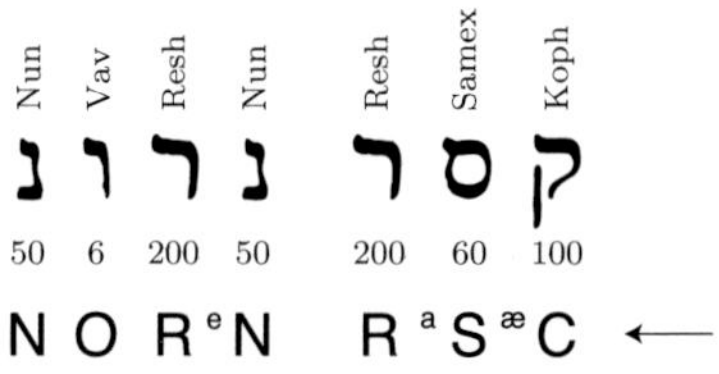

Fig. 20.
Value 666 associated with the Hebrew letters for *Cæsar Nero* (courtesy Ralf Steinbrüggen)

Seen from the point of view of common European languages, Arabic script (without vowels) is also polyphonic. The puzzle of what "Pthwndxrclzp" in James Joyce's *Finnegan's Wake* means is something that will keep historians of literature (and undertakers) busy for many years.

From a technical standpoint, Bazeries' cylinder (Fig. 19), dealt with in Sect. 7.5.3, operates with both homophones and polyphones. However, injectivity is effectively maintained because the 'illegal' polyphonic texts are almost certainly meaningless (Fig. 21). Polyphony may also cause difficulties in certain ways of cryptanalysis. Polyphone plaintexts belonging to the same cryptotext are called variants there.

2.4.2 Word spacing. The suppression of word spacing and of punctuation, one of the basic rules of classical professional cryptography (i.e., of "formal ciphers"), is strictly speaking polyphony. In some cases genuine ambiguity

```
G X Y Y S X D B R Z Z B G B B G S I C U
H Z Q X R V P I Y D L D L C C N O U H S
I A R V O T R E B I S G O D D F N A V T
J E S U I S I N D E C H I F F R A B L E
K I T T Q R J H E U O J R G G T B C B L
L O V S P Q U U T P U K E J H H C F D A
M U X R N P G R S R R N M K K U D G F C
N Y Z Q M N V X L O A P T L M B F J G F
O B A P L M B L F T N Q D M O C G K I B
```

Fig. 21. Several polyphonic texts for one setting of Bazeries' cylinder

can occur if the position of the boundary between words is uncertain; for example "dark ermine" and "darker mine". The sentences

"Five fingers have I on each hand ten in all"

"Ten digits have I on each hand five and twenty on hands and feet together"

also permit of various interpretations, depending on the punctuation; only one interpretation makes logical sense. In the sentence

"Forget not to kiss thy wife"

the sense can only be derived from the context. Modern English can be just as confusing:

"British Rail hopes to have trains running normally late this afternoon"

would do little to raise the hopes of frustrated travellers, while

"The Prime Minister called for an end to violence and internment as soon as possible"

is a choice morsel for the opposition. Injectivity is often violated when there are insufficient contextual clues:

"the captive flies."

The phrase

"two thousand year old horses"

even allows for three different interpretations: *two-thousand-year-old horses, two thousand-year-old horses, two-thousand year-old horses*.

Another example where the suppression of hyphens may cause trouble is:

"a man eating fish — a man-eating fish."

There is hardly any practical requirement for polyphonic texts that are encrypted by the empty word—except, perhaps, to eliminate waffle in a text.

2.5 Character Sets

We use N to represent $|W|$, the finite size of the plaintext or cryptotext character set W. Since the case $N = 1$ conveys no information, it is a requirement that $N \geq 2$. An alphabet is a linearly ordered character set.

2.5.1 Plaintext character sets. Which ones are in use depends on the language and the epoch. In the case of Hawaiian, the character set

$$Z_{12} = \{\mathrm{a, u, i, o, e, w, h, k, l, m, n, p}\} \quad \text{is sufficient.}$$

In the Middle Ages, following the Latin tradition, 20 letters seem to have been enough for most writers, including Giovanni Battista Porta in 1563 (Fig. 23):

$$Z_{20} = \{\mathrm{a, b, \ldots, i, l, \ldots, t, v, z}\} .$$

Often /k/, /x/, and /y/ are included, or just /x/ and /y/ (Porta at other times). /w/ was long written as /vv/, so making room for /&/, as on Leon Battista Alberti's disc in 1466 (Fig. 26). By 1600 an alphabet of 24 characters had become a European standard,

$$Z_{24} = Z_{20} \cup \{\mathrm{k, w, x, y}\} ,$$

with /v/ still used for /u/. Trithemius (1508) used /w/, see Fig. 52. In a French translation of 1561 (Gabriel de Collange), the 'German' /w/ was replaced by /&/, according to Eyraud.

In the 18th century /u/ was included:

$$Z_{25}^{uw} = Z_{24} \cup \{\mathrm{u}\} .$$

But if /j/ is required (in French, for example), then /w/ must be sacrificed again (Bazeries, 1891):

$$Z_{25}^{ju} = Z_{20} \cup \{\mathrm{j, k, u, x, y}\} .$$

/j/, /k/, /w/, /x/, /y/ are very unusual in Italian, as are /k/, /w/ in French. Irish can do without /j/, /k/, /q/, /v/, /w/, /x/, /y/, /z/.

From about 1900, our present alphabet,

$$Z_{26} = Z_{24} \cup \{\mathrm{j, u}\} ,$$

was in general use. But there are exceptions even in Middle Europe. In the Second World War, the exiled Czech government used the extended character set consisting of 31 letters and 13 number symbols and other characters

$$Z_{44} = \{\mathrm{a, b, c, \check{c}, d, e, \check{e}, f, \ldots, r, \check{r}, s, \check{s}, t, \ldots, z, \check{z}}, \cdot, \text{'}, *, 0, 1, \ldots, 9\} .$$

The Italian *cifrario tascabile* from the First World War used a character set

$$Z_{36} = Z_{26} \cup \{0, 1, \ldots, 9\} .$$

The (present-day) Cyrillic alphabet has 32 letters (disregarding Ё):

Z_{32} = {А Б В Г Д Е Ж З И Й К Л М Н О П
Р С Т У Ф Х Ц Ч Ш Щ Ъ Ы Ь Э Ю Я }.

Otherwise, many different special conventions have been used to represent digits and, if necessary, punctuation marks and diacritic marks.

Spaces between words are suppressed in professional cryptography. Even in German, where the words are longer than in most languages, word spacings are commoner than /e/.

2.5.2 Technical character sets. The cryptotext character sets in use are usually determined by technical restraints; besides the alphabets mentioned above, there are other technical character sets,

Z_{256}	$= Z_2^8$	(bytes; IBM circa 1964)
Z_{32}	$= Z_2^5$	(Francis Bacon 1605, 1623, Baudot 1874)
Z_{10}	$= \{0, 1, 2, \ldots, 9\}$	(denary)
Z_6	$= \{\mathrm{A, D, F, G, V, X}\}$	(senary; these letters correspond to the easily distinguishable Morse characters ·— , —·· , ··—· , ——· , ···— , —··—)
Z_4	$= \{1, 2, 3, 4\}$	(quaternary; Alberti 1466, Caramuel 1670, Weigel 1673)
Z_3	$= \{1, 2, 3\}$	(ternary; Trithemius 1518, Wilkins 1641, Friderici 1685)
Z_2	$= \{O, L\}$	(binary; Francis Bacon 1605, 1623)

and also invented symbols, popular among amateurs (Sect. 3.1.1).

Binary[7], ternary, quinary, and denary ciphers have $W \triangleq Z_2$, $W \triangleq Z_3$, $W \triangleq Z_5$, $W \triangleq Z_{10}$ respectively.

2.5.2.1 The nine digits 1, 2, 3, 4, 5, 6, 7, 8, 9 have been used by the German *Kriegsmarine*—iterated if necessary—to encode map coordinates (Fig. 22).

1	2	3
4	5	6
7	8	9

Fig. 22.
Digit cipher in a map grid

2.5.2.2 It is fashionable to write the ciphertext in groups of five characters. This has its origins in the tariff regulations of the International Telegraph Union, which since 1875 has limited the length of a word to ten symbols (and imposed serious restrictions on the use of codes). In 1904 codes were allowed to have up to ten letters; later, telegram charges were generally based on groups of five (Whitelaw's Telegraph Cipher: 20 000 pronounceable five-letter code groups, giving 400 million ten-letter code groups).

2.5.3 In the relation $\mathbf{X} : V^* \dashrightarrow W^*$, the cryptanalyst knows neither V nor $\mathbf{X}$. However, from $\mathbf{X}(V^*)$, the set of actually occurring crypt words, he can occasionally deduce the method in use (e.g., Polybios square, Sect. 3.3.1).

2.5.4 The endomorphic case. If, as is often the case, the plaintext and the ciphertext use the same character set ($V \doteq W$, $\mathbf{X}$ endomorphic), then it is nowadays conventional in theoretical treatments to write the plaintext and

[7] Binary in the sense of biliteral, a character set of two 'bits': Bacon 1605, 1623; with explicit values a = 1, b = 2, c = 4, d = 8, etc. (abfg = 99): Napier 1617; a positional system using digits: Harriot before 1621, Caramuel 1670, Leibniz 1679.

its characters in lower-case letters, the ciphertext and its characters in small capitals. That leaves upper-case italic letters for so-called key characters. However, Alberti's disc (Fig. 26) used the opposite convention. Even in 1925, Lange-Soudart's book showed a Saint-Cyr slide (Fig. 27) with the plaintext in uppercase and the ciphertext in lowercase.

2.6 Keys

A key (French *clef, clé*, German *Schlüssel*) serves to select a step from a cryptosystem M. Keys allow one to change the encryption in accordance with previously arranged rules, for example every day, or after every message, or after every character. Frequently, keys are organized such that they allow one to produce the individual encryption steps by following simple rules. The combinatorial complexity of an encryption method is determined by the number of keys available under this method. The key technique is very varied, and will be dealt with under the individual classes of methods.

Let K denote the key character set or key vocabulary. K^* is called the key space. Let $k_j \in K$ be the j-th key used in sequence; then k_j determines a number s_j and thus the encryption step $\chi_{s_j} \in M = \{\chi_0, \chi_1, \chi_2, \ldots, \chi_{\theta-1}\}$.

2.6.1 Keys are to be changed. Repeated use of the same key is equivalent to using an encryption system with only one element. Except in the case of codes, professional cryptography makes hardly any use of such fixed encryptions. The use in diplomatic circles of the same code book for years on end is a typical case—though one can scarcely regard the diplomats of many countries as professionals in the matter of ciphers: in cities like Vienna, there has been a lively underground market for diplomatic codes at various times. The Soviet Union had a particular reputation for stealing code books. 1936, a Russian agent in Haarlem (Netherlands) used a stolen code book to decipher telegrams between the Japanese military attaché in Berlin and his government in Tokyo. At the beginning of the First World War, probably every European power possessed copies of one or more of the American diplomatic code books. In August 1941 Loris Gherardi secretly procured for the *Servizio Informazione Militare* a copy of the BLACK code used by U.S. military attachés. There is a story, told by Allen W. Dulles, of the American minister in Rumania—an ousted politician, like so many diplomats—who was unwilling to report the loss of his code book. He would wait until several messages had accumulated, then take the train to Vienna to decipher them at the embassy there. The moral is that even code books must be changed regularly, if necessary monthly.

Keys used for choosing a method must be a matter of mutual agreement. If one party is cut off, then the supply of new keys is at risk, difficult, or even impossible. In such cases use is often made of innocent sets of letters or figures, such as popular novels, statistical reports, telephone books, etc.—almost everything has been used at some time, from Jaroslav Hašek's *Good Soldier Schweik* to the 1935 Statistical Almanac of the German *Reich*. Even this system is vulnerable—if the source of keys is revealed, then a whole stream of messages becomes transparent at one blow.

2.6.2 Blocks. Following the notation of Sect. 2.3, let $\mathbf{X}$ be a finitely generated block cipher, $\mathbf{X} = [\ \chi_{s_1}, \chi_{s_2}, \chi_{s_3}, \ldots\ ]$, where $\chi_{s_j} : p_j \mapsto c_j$.

$(p_1, p_2, p_3, \ldots)$, where $p_j \in V^n$, designates the plaintext sequence;
$(c_1, c_2, c_3, \ldots)$, where $c_j \in W^m$, designates the cryptotext sequence,
$(k_1, k_2, k_3, \ldots)$, where $k_j \in K$, designates the keytext sequence.

Let k_j be a key which determines χ_{s_j} , S_j an operator standing for $\chi_{s_j}(.)$. Then we have three notations for the cryptographic equation

$$c_j = \chi_{s_j}(p_j) \qquad \text{or} \qquad c_j = \mathbf{X}(p_j, k_j) \qquad \text{or} \qquad c_j = p_j\ S_j \ .$$

Note that χ_i indicates the i-th encryption step in a numbered list of steps, while χ_{s_j} is the step used to carry out the j-th step in the encryption.

If χ_{s_j} is an *injective function*, as is usually the case, then there exists an inverse function $\chi_{s_j}^{-1}$, for which (with $\mathbf{Y} = [\ \chi_{s_1}^{-1}, \chi_{s_2}^{-1}, \chi_{s_3}^{-1}, \ldots\ ]$)

$$p_j = \chi_{s_j}^{-1}(c_j) \qquad \text{or} \qquad p_j = \mathbf{Y}(c_j, k_j) \qquad \text{or} \qquad p_j = c_j\ S_j^{-1} \ .$$

Thus $\chi_{s_j}^{-1}(\chi_{s_j}(p_j)) = p_j$ and also $\chi_{s_j}(\chi_{s_j}^{-1}(\chi_{s_j}(p_j))) = \chi_{s_j}(p_j)$.

If χ_{s_j} is also *surjective and unambiguous*, then *for all* $c_j \in W^m$ even

$$\chi_{s_j}(\chi_{s_j}^{-1}(c_j)) = c_j \ .$$

In the case of alternating traffic between two parties A and B, one of them can use a sequence of χ_{s_j} as both encryption and decryption steps, the other a sequence of $\chi_{s_j}^{-1}$ as both decryption and encryption steps.

2.6.3 Isomorphism. Let $\mathbf{X}$ again be a finitely generated block cipher. Two plaintexts $(p'_1, p'_2, p'_3, \ldots)$, $(p''_1, p''_2, p''_3, \ldots)$ such that $p'_i = p''_i\, S$, where S is a fixed substitution, are called isomorphic. Assume the same for two ciphertexts $(c'_1, c'_2, c'_3, \ldots)$, $(c''_1, c''_2, c''_3, \ldots)$ such that $c'_i = c''_i\, T$, where T is a fixed substitution. Then, for the encryption steps E'_i, E''_i, the following holds:

$$\text{If } c'_i = p'_i E'_i \text{ and } c''_i = p''_i E''_i \text{ , then } S\, E'_i = E''_i\, T \ .$$

If the encryptions E'_i and E''_i possess an inverse, then isomorphic plaintexts encrypt to isomorphic ciphertexts, and vice versa. Then

$$T = (E''_i)^{-1}\, S\, E'_i \quad \text{and} \quad S = E''_i\, T\, (E'_i)^{-1} \ .$$

If the fixed substitutions S and T possess inverses, then the keys can be transformed one into another:

$$S''_i = S S'_i T^{-1} \quad \text{and} \quad S'_i = S^{-1} S''_i T \ .$$

2.6.4 Shannon. A cryptosystem where the encryption step is uniquely determined by a pair of plaintext and cryptotext characters may be called a Shannon cryptosystem. Many customary cryptosystems have this property.

A cryptosystem where the encryption step and the decryption step coincide, and thus the crypto procedure is symmetrically determined, is called key-symmetric. In this case, every encryption step is involutory (cf. Sect. 3.2.1).

3 Encryption Steps: Simple Substitution

Among the encryption steps we find prominently two large classes: substitution and transposition. They are both special cases of the most general encryption step $V^{(n)} \dashrightarrow W^{(m)}$. We shall start by looking at several kinds of substitution and turn our attention to transposition in Chapter 6 .

A simple substitution (German *Tauschverfahren* or *Ersatzverfahren*) is a substitution with monographic encryption steps $\chi_i \in M$,

$$\chi_i : V^{(1)} \dashrightarrow W^{(m_i)} .$$

In the monoalphabetic case, an arbitrary χ_s is selected from M and encryption is done with the sequence $X = [\chi_s \,, \chi_s \,, \chi_s \,, ...]$. It is in this case sufficient to take a singleton for M .

We start with the case $m_i = 1$ for all i .

Fig. 23. Cipher disk by Giovanni Battista Porta, 1563

3.1 Case $V^{(1)} \dashrightarrow W$ (Unipartite Simple Substitutions)

The case $V^{(1)} \dashrightarrow W$ deals with a unipartite simple substitution, for short just simple substitution (French *substitution simple ordinaire*).

3.1.1 $V \longrightarrow W$ **, heterogenous encryption without homophones and nulls.** This case is primeval. For W an alphabet of strangely formed, unusual graphemes is frequently used: Examples are known from Thailand, Persia, coptic Ethiopia and elsewhere. Such marks are used by Giovanni Battista Porta (Giambattista Della Porta, 1535–1615) in his cipher disk (Fig. 23, see also Fig. 30). Charlemagne is said to have used such characters (Fig. 24) as well as the savant and mystic Hildegard von Bingen (1098–1179).

Fig. 24. Secret characters of Charlemagne

The Freemasons' cipher is to be mentioned here. It goes back to the ancient 'pigpen' cipher, and in its modern form reads

a b c d e f g h i j k l m n o p q r s t u v w x y z

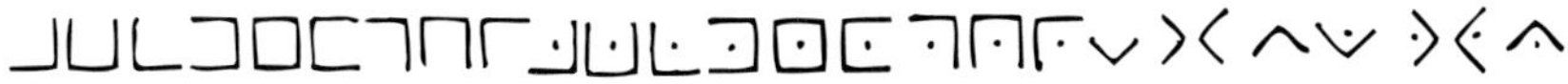

It can be memorized by the schemes

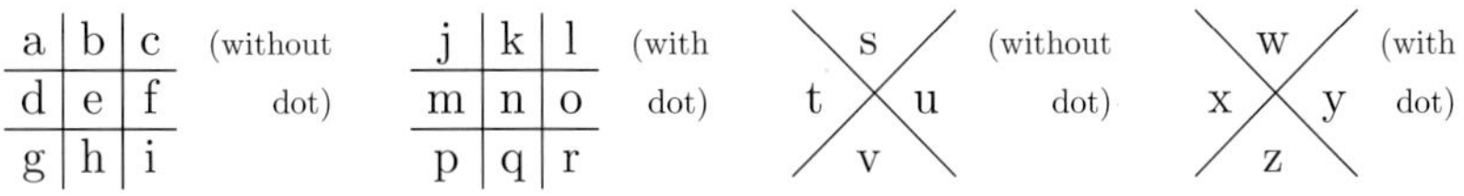

As late as 1728, when it was broken by England's Deciphering Branch, the Czar Peter the Great used (besides nomenclators) a heterogenous substitution $V \longrightarrow W$ with a bizarre cipher alphabet.

Edgar Allan Poe, famous for his literary works, used a rather trivial alphabet of common printer's types in his story "The Gold-Bug" (Sect. 15.10.1).

In this class is also the bookseller's cipher for encrypting prices and dates, a one-to-one mapping $Z_{10} \longrightarrow Z_{26}$, generated by a password ('key-phrase' cipher). An example is

1	2	3	4	5	6	7	8	9	0
M	I	L	C	H	P	R	O	B	E ,

an encryption step with the password *milchprobe* ('milk sample') used in Germany over many years for specifying the packing date of butter. Likewise, in Navy ENIGMA enciphering, sometimes figures were represented by letters

1	2	3	4	5	6	7	8	9	0
q	w	e	r	t	z	u	i	o	p .

3.1.2 $V^{(1)} \dashrightarrow W$ **, heterogenous encryption with homophones and nulls**. Homophones are found already in Muslim sources, e.g., al-Qalqashandi 1412, and in a cipher used by the Duchy of Mantua in 1401 for an exchange of letters with Simeone de Crema. The vowels—typically the more frequent characters—were given homophones, a first sign of considerations to level the character frequency. Furthermore, W was enlarged by digits. The introduction of homophones practically enforces the introduction of nulls; otherwise homophones can be recognized easily by the constant pattern of letters surrounding them in frequent words.

A method with homophones used even today is the book cipher: In some inconspicuous looking book that sender and recipient have at hand in identical copies the plaintext letters are selected one after another; the corresponding places: (page x, line y, position z) form the cipher group (x-y-z).

Choosing as the book the present volume, the word *mammal* can be encrypted as 3-5-3, 7-1-6, 6-3-11, 5-8-6, 4-3-2, 3-3-2 .

3.2 Special Case $V \longleftrightarrow V$ (Permutations)

In the case of a one-to-one mapping $V \longleftrightarrow W$ among the examples in Sect. 3.1.1 , W is called a mixed (cryptotext) alphabet of N characters (French *alphabet désordonné*, *alphabet incohérent*, German *umgeordnetes Geheimtextalphabet*), that matches a standard (plaintext) alphabet (French *alphabet ordonné*, German *Standard-Klartextalphabet*) V of N characters.

To define a substitution, it suffices to list in some way the matching pairs of plaintext characters and cryptotext characters, e.g., for $V \doteq W = Z_{26}$ (for the use of lower-case letters and small capitals see Sect. 2.5.4):

u	d	c	b	m	a	v	g	k	s	t	n	w	z	e	i	h	f	q	l	j	r	o	p	x	y
H	E	W	A	S	R	I	G	T	O	U	D	C	L	N	M	F	Y	V	B	P	K	J	Q	Z	X

For encryption, it is more convenient, of course, to have the plaintext characters ordered into a standard (plaintext) alphabet; this gives a mixed (cryptotext) alphabet:

a	b	c	d	e	f	g	h	i	j	k	l	m	n	o	p	q	r	s	t	u	v	w	x	y	z
R	A	W	E	N	Y	G	F	M	P	T	B	S	D	J	Q	V	K	O	U	H	I	C	Z	X	L

In mathematics, this ‘substitution notation’ is customary. For decryption, however, it is better to have the cryptotext characters ordered into a standard (cryptotext) alphabet; this gives a mixed (plaintext) alphabet:

b	l	w	n	d	h	g	u	v	o	r	z	i	e	s	j	p	a	m	k	t	q	c	y	f	x
A	B	C	D	E	F	G	H	I	J	K	L	M	N	O	P	Q	R	S	T	U	V	W	X	Y	Z

A new situation is given in the endomorphic case $V \doteq W$. Especially the one-to-one mapping $V \longleftrightarrow V$ is then a permutation of V . The permutation $V \longleftrightarrow V$ can be accomplished in electrical implementations by interchanging N wires (German *Umstecken*) in a wire bundle.

For permutations in particular, mathematics uses apart from the substitution notation the ‘cycle notation’

(a r k t u h f y x z l b) (c w) (d e n) (g) (i m s o j p q v)

in which the distinction between lower-case letters and small capitals has to be abandoned. For encryption, one goes in the cycle where the plaintext character is found to the cyclically next character; for decryption, to the cyclically preceding character. Cycles of length 1 (1-cycles) are often suppressed—we shall not follow this habit.

3.2.1 Self-reciprocal permutations. The most ancient sources (apart from Egypt—we shall come back to this under 'code') show a self-reciprocal ('involutory') permutation of V: in India, in the *Kāma-sūtra* of the writer *Vātsyāyana*, secret writing is mentioned as one of the sixty-four arts; *Mūladevīya* denotes the encrypting and decrypting procedure, which is a reflection ('involution'):

$$V \overset{2}{\longleftrightarrow} V : \updownarrow \begin{array}{cccccccccc} \text{a} & \text{kh} & \text{gh} & \text{c} & \text{t} & \text{ñ} & \text{n} & \text{r} & \text{l} & \text{y} \\ \text{k} & \text{g} & \text{n} & \text{ṭ} & \text{p} & \text{ṇ} & \text{m} & \text{ṣ} & \text{s} & \text{ś} \end{array}$$

(the remaining characters are left invariant, so the permutation is not properly self-reciprocal). Plaintext and cryptotext alphabets of a self-reciprocal permutation are said to be reciprocal to each other.

In the Hebrew Holy Scripture boustrophedonic substitution, called *Athbash*, was used—although not for a cryptographic purpose—which would read in the Latin alphabet $V = Z_{20}$

$$V \overset{2}{\longleftrightarrow} V : \updownarrow \begin{array}{cccccccccc} a & b & c & d & e & f & g & h & i & l \\ z & v & t & s & r & q & p & o & n & m \end{array}$$

. Such a substitution uses the reversed ('inverse') alphabet. In the case of the reflection

$$V \overset{2}{\longleftrightarrow} V : \updownarrow \begin{array}{ccccccccccc} a & b & c & d & e & f & g & h & i & l & m \\ a & z & v & t & s & r & q & p & o & n & m \end{array}$$

Charles Eyraud speaks of a complementary alphabet (French *alphabet complémentaire*), see Sect. 5.6. This permutation, however, is not properly self-reciprocal: /a/ and /m/ are left invariant.

Obvious is also a reflection with a shifted alphabet like the Hebrew *Albam*, used in 1589 by the Argentis with $V = Z_{20}$

$$V \overset{2}{\longleftrightarrow} V : \updownarrow \begin{array}{cccccccccc} a & b & c & d & e & f & g & h & i & l \\ m & n & o & p & q & r & s & t & v & z \end{array}$$

or the one used by Giovanni Battista Porta in 1563 (see Fig. 53) with $V = Z_{22}$

$$V \overset{2}{\longleftrightarrow} V : \updownarrow \begin{array}{ccccccccccc} a & b & c & d & e & f & g & h & i & l & m \\ n & o & p & q & r & s & t & v & x & y & z \end{array} .$$

The most general boustrophedonic case, showing the use of a password, is presented by the following example: ($V = Z_{26}$)

$$V \overset{2}{\longleftrightarrow} V : \updownarrow \begin{array}{ccccccccccccc} a & n & g & e & r & s & b & c & d & f & h & i & j \\ z & y & x & w & v & u & t & q & p & o & m & l & k \end{array} .$$

Reflections have, apart from the advantage of a compact notation, the property which some people have held to be of great importance that encryption and decryption steps coincide.

In the cycle notation of permutations, the last five examples would read (with cycle outsets ordered alphabetically):

(a,z) (b,v) (c,t) (d,s) (e,r) (f,q) (g,p) (h,o) (i,n) (l,m)
(a) (b,z) (c,v) (d,t) (e,s) (f,r) (g,q) (h,p) (i,o) (l,n) (m)
(a,m) (b,n) (c,o) (d,p) (e,q) (f,r) (g,s) (h,t) (i,v) (l,z)
(a,n) (b,o) (c,p) (d,q) (e,r) (f,s) (g,t) (h,v) (i,x) (l,y) (m,z)
(a,z) (b,t) (c,q) (d,p) (e,w) (f,o) (g,x) (h,m) (i,l) (j,k) (n,y) (r,v) (s,u)

Properly self-reciprocal is a permutation without 1-cycles, which means solely with 2-cycles ('swaps'). It is the target of cryptanalytic attacks (Sect. 14.1) that cease to work if some of the cycles are 1-cycles ('females').

For a binary alphabet $V = Z_2$, the sole nontrivial permutation is a reflection:
$V \overset{2}{\longleftrightarrow} V : \updownarrow \begin{matrix} O \\ L \end{matrix}$.

3.2.2 Cross-plugging. In electrical implementations, reflections are accomplished by swapping pairs of wires, simply by using double-ended connectors (Fig. 25). Such reflections were used in the ENIGMA plugboard (German *Steckerbrett*).

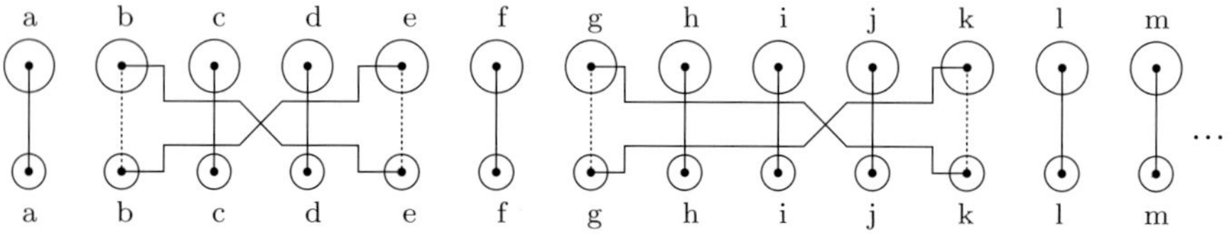

Fig. 25. Self-reciprocal permutation by cross-plugging with a pair of double-ended connectors which interrupt the direct contacts.

The number $d(k, N)$ of reflections depends on N and the number k of cinch plugs used:

$$d(k,N) = \frac{N!}{2^k \cdot (N-2k)! \cdot k!} = \binom{N}{2k} \cdot \frac{(2k)!}{2^k k!} = \binom{N}{2k} \cdot (2k-1)!! \text{ , where}$$

$$(2k-1)!! \;=\; (2k-1)\cdot(2k-3)\cdot \;\ldots\; \cdot 5 \cdot 3 \cdot 1 = \tfrac{(2k)!}{2^k k!} \;.$$

Properly self-reciprocal permutations ('genuine' reflections) require $N = 2\nu$ to be even. The number $d(\frac{N}{2}, N)$ of all genuine reflections is then (with a relative error $< 10^{-3}$ for $N \geq 6$)

$$d(\tfrac{N}{2}, N) = (N-1)!! \;=\; (N-1)\cdot(N-3)\cdot \;\ldots\; \cdot 5\cdot 3\cdot 1 = \tfrac{(2\nu)!}{\nu! 2^\nu} \;\approx\; \frac{\sqrt{(2\nu)!}}{\sqrt[4]{\pi\cdot(\nu+\frac{1}{4})}} \;.$$

The approximate value is a rather good upper limit for $(N-1)!!$.

For fixed N, however, $d(k, N)$ is maximal for $k = \lceil \nu - \sqrt{(\nu+1)/2} \rceil$:

$d(5,26) \approx 5.02 \cdot 10^9$, $d(6,26) \approx 1.00 \cdot 10^{11}$, $d(7,26) \approx 1.31 \cdot 10^{12}$,
$d(8,26) \approx 1.08 \cdot 10^{13}$, $d(9,26) \approx 5.38 \cdot 10^{13}$, $d(10,26) \approx 1.51 \cdot 10^{14}$,
$d(11,26) \approx 2.06 \cdot 10^{14}$, $d(12,26) \approx 1.03 \cdot 10^{14}$, $d(13,26) \approx 7.91 \cdot 10^{12}$,

and $d(3,10) = 3150$, $d(4,10) = 4725$, $d(5,10) = 945$.

The ENIGMA I of the *Reichswehr* of 1930 originally used six double-ended two-line connectors, the *Wehrmacht* ENIGMA, beginning October 1, 1936, five to eight, from January 1, 1939 seven to ten, and from August 19, 1939 ten double-ended two-line connectors for cross-plugging.

3.2.3 Monocyclic permutations. A compact notation describes also the monocyclic permutation, the order of which is N:

e.g., with $N = 20$ the cycle of the standard alphabet Z_{20}

$V \overset{N}{\leftarrow\!\!\!\rightarrow} V$: (a b c d e f g h i l m n o p q r s t v x)

or its third power

$V \overset{N}{\leftarrow\!\!\!\rightarrow} V$: (a d g l o r v b e h m p s x c f i n q t) ;

in substitution notation

a	b	c	d	e	f	g	h	i	l	m	n	o	p	q	r	s	t	v	x
B	C	D	E	F	G	H	I	L	M	N	O	P	Q	R	S	T	V	X	A

,

a	b	c	d	e	f	g	h	i	l	m	n	o	p	q	r	s	t	v	x
D	E	F	G	H	I	L	M	N	O	P	Q	R	S	T	V	X	A	B	C

.

The last encryption step was used by Julius Caesar (according to Suetonius), counting upwards three letters in the alphabet. His successor Augustus, inferior in several respects to Caesar, used the first encryption step (possibly he could not safely count up to three); Suetonius said he also replaced x by AA .

Every power of the cycle of the standard alphabet yields a CAESAR alphabet. We shall come back to this in Chapter 5 (CAESAR addition). But note: while the two encryption steps above are of the order twenty, the second power has only the order ten, and the tenth power has only the order two: it is a reflection as studied above. The $(N-1)$-th power is the inverse of the first power and yields the decryption step.

A monoalphabetic substitution with a CAESAR encryption step was introduced in 1915 in the Russian army after it turned out to be impossible to expect the staffs to use anything more complicated. For the Prussian Ludwig Deubner and the Austrian Hermann Pokorny, heads of the cryptanalytic services of their respective countries, it was a pleasantly simple matter to decrypt these messages.

By its very nature, a track on a disk, the rim of a washer, or a strip closed to form a ring can be used to represent a full cycle. Such gadgets have found wide use and were employed in a particular way (Sect. 7.5.3) by Thomas Jefferson and Étienne Bazeries. The q-th power of the monocyclic permutation is obtained by counting within the cycle in steps of q characters.

3.2.4 Mixed alphabets. For non-selfreciprocal and non-cyclic $V \leftarrow\!\!\!\rightarrow V$, in the most general case of a mixed alphabet (French *alphabet désordonné*, German *permutiertes Alphabet*), substitution notation is normally used:

$V \leftarrow\!\!\!\rightarrow V$:

a	b	c	d	e	f	g	h	i	j	k	l	m	n	o	p	q	r	s	t	u	v	w	x	y	z
S	E	C	U	R	I	T	Y	A	B	D	F	G	H	J	K	L	M	N	O	P	Q	V	W	X	Z

The short cycle notation is useful here, too. It shows the decomposition

$V \leftarrow\!\!\!\rightarrow V$: (a s n h y x w v q l f i) (b e r m g t o j) (c) (d u p k) (z) ,

into one 12-cycle, one 8-cycle, one 4-cycle, and two 1-cycles (cycle partition 12+8+4+1+1).

3.2.4.1 More mixed alphabets are obtained by a cyclic shift of one of the two lines in the substitution notation (shifted mixed alphabets, French *alphabet désordonné parallèle*, German *verschobenes permutiertes Alphabet*):

$V \longleftrightarrow V$:	a	b	c	d	e	f	g	h	i	j	k	l	m	n	o	p	q	r	s	t	u	v	w	x	y	z
	E	C	U	R	I	T	Y	A	B	D	F	G	H	J	K	L	M	N	O	P	Q	V	W	X	Z	S
$V \longleftrightarrow V$:	a	b	c	d	e	f	g	h	i	j	k	l	m	n	o	p	q	r	s	t	u	v	w	x	y	z
	C	U	R	I	T	Y	A	B	D	F	G	H	J	K	L	M	N	O	P	Q	V	W	X	Z	S	E

in cycle notation

(a e i b c u q m h) (d r n j) (f t p l g y z s o k) (v) (w) (x) ,

(a c r o l h b u v w x z e t q n k g) (f y s p m j) (d i) .

3.2.4.2 Iterated substitution, also called 'raising to a higher power' produces the powers of a mixed alphabet, e.g., from the substitution SECURITY... above, the second power gives

(a n y w q f) (b r g o) (c) (d p) (e m t j) (h x v l i s) (k u) (z) ,

with all cycles of even length being split in halves; in substitution notation

$V \longleftrightarrow V$:	a	b	c	d	e	f	g	h	i	j	k	l	m	n	o	p	q	r	s	t	u	v	w	x	y	z
	N	R	C	P	M	A	O	X	S	E	U	I	T	Y	B	D	F	G	H	J	K	L	Q	V	W	Z

Shifting on the one hand, raising to a power on the other do not give the same thing in general; they are two utterly different methods for producing a family of up to N (sometimes less) accompanying alphabets (Chapter 7).

3.2.5 Alphabets derived from passwords. The examples above show already the construction of an (endomorphic) simple substitution $V \longleftrightarrow V$ with the help of a password (French *mot-clef*, German *Kennwort*, *Losung*), possibly a mnemonic key or a key phrase. A classical method uses a word from V , writes its characters without repetitions and fills in alphabetic order with the characters not used. The method goes back to Giovanni Battista Argenti, around 1580. It was still a cryptologic standard even in the 20th century.[1]

This construction, however, is vulnerable: it may be easy to guess a missing part of the password (after all, the most frequent vowels /e/ and /a/ always are substituted by a letter from the password, if this has length 5 or more). A small consolation is that the password should not need much fill.

More cunning methods use therefore a reordering of the password, for example by writing it first in lines and reading it in columns (method of Charles Wheatstone, 1854, a transposition to be treated methodically in Sect. 6.2):

S	E	C	U	R	I	T	Y		a	e	i	l	o	r	u	x
A	B	D	F	G	H	J	K		b	f	j	m	p	s	v	y
L	M	N	O	P	Q	V	W		c	g	k	n	q	t	w	z
X	Z								d	h						

[1] Allowing repetitions is bad: it leads to polyphones, e.g., the 'key-phrase' cipher

a b c d e f g h i j l m n o p q r s t u v x y z
L E G O U V E R N E M E N T P R O V I S O I R E

and shortens the cryptotext character set (in our case to 13 characters, $\{b, g, j, m, z\} \mapsto$E).

This yields the alphabet

a	b	c	d	e	f	g	h	i	j	k	l	m	n	o	p	q	r	s	t	u	v	w	x	y	z
S	A	L	X	E	B	M	Z	C	D	N	U	F	O	R	G	P	I	H	Q	T	J	V	Y	K	W

or in cycle notation

(a s h z w v j d x y k n o r i c l u t q p g m f b) (e)

with the 1-cycle (e).

A further method fills also the columns of the plaintext side in the alphabetic order of the letters of the password, in the example in the order

third, second, sixth, fifth, first, seventh, fourth, eighth column

with the result

S	E	C	U	R	I	T	Y		n	d	a	u	k	h	r	x
A	B	D	F	G	H	J	K		o	e	b	v	l	i	s	y
L	M	N	O	P	Q	V	W		p	f	c	w	m	j	t	z
X	Z								q	g						

This results in the alphabet

a	b	c	d	e	f	g	h	i	j	k	l	m	n	o	p	q	r	s	t	u	v	w	x	y	z
C	D	N	E	B	M	Z	I	H	Q	R	G	P	S	A	L	X	T	J	V	U	F	O	Y	K	W

or in cycle notation

(a c n s j q x y k r t v f m p l g z w o) (b d e) (h i) (u) .

The method can also be used for the construction of cycles. The sentence *évitez les courants d'air*, "avoid drafts" (Bazeries, Sect. 7.5.3) produces the cycle

$V \overset{N}{\leftrightarrow} V$: (e v i t z l s c o u r a n d b f g h j k m p q x y)

3.2.6 Enumeration. The following table gives for $N = 26$, for $N = 10$ and for $N = 2$ a survey of the number $Z(N)$ of available alphabets $V \leftrightarrow V$:

number of permutations	$Z(N)$	$Z(26)$	$Z(10)$	$Z(2)$
total	$N\,!$	$4.03 \cdot 10^{26}$	3 628 800	2
monocyclic	$(N-1)\,!$	$1.55 \cdot 10^{25}$	362 880	1
reflections total	$\approx N \cdot (N\,!)^{\frac{1}{2}}$	$5.33 \cdot 10^{14}$	9 496	2
genuine reflections	$\approx (N\,!)^{\frac{1}{2}}$	$7.91 \cdot 10^{12}$	945	1
derived from meaningful passwords (mnemonic words)		$10^4 \ldots 10^6$		

3.2.7 Cipher disks and cipher slides. To mechanize a substitution, the fixed matching of the plaintext and the cryptotext characters, as found in the substitution notation, can be arranged on a cylinder or on a strip. Two windows allow one to see just two matching characters at any given moment. The windows can be arranged so that only the master sees the plaintext character, while the clerk only sees the cryptotext window and cannot grasp the meaning of the message (Sect. 7.5.2, Gripenstierna's machine, Fig. 54).

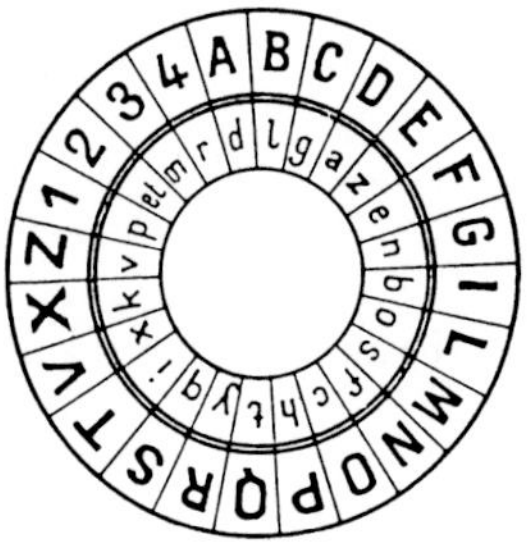

Fig. 26.
Cipher disk of Leon Battista Alberti (according to Lange-Soudart 1935)

A selection from the N accompanying shifted alphabets is obtained if one of the windows can be moved. Another possibility is to shift the plaintext alphabet with respect to the cryptotext alphabet. This leads to the use of two disks (Fig. 26) or two strips (Fig. 27). In the latter case it is necessary to repeat one of the alphabets (duplication).

a	b	c	d	e	f	g	h	i	j	k	l	m	n	o	p	q	r	s	t	u	v	w	x	y	z	a	b	c	d	e	f	g	h	i	j	k	l	m	...
						S	E	C	U	R	I	T	Y	A	B	D	F	G	H	J	K	L	M	N	O	P	Q	V	W	X	Z								

Fig. 27. Cipher slide with duplicated plaintext alphabet

Cipher disks (French *cadran*, German *Chiffrierscheibe*), mechanical tools for general substitution with shifted mixed alphabets, were described as early as 1466 by Leon Battista Alberti[2] (see Plate B). Cipher slides (French *reglette*, German *Chiffrierschieber*) were used in Elizabethan England around 1600. In the 19th century they were named *Saint-Cyr* slides after the famous French Military Academy. Cipher rods (French *bâtons*, German *Chiffrierstäbchen*) serve the same purpose.

3.2.8 Cycles with windows. Mechanizing a monocyclic permutation can also start from the cycle notation. The cycle of characters is again arranged on a cylinder or on a strip (in the latter case the first character must be duplicated). Two neighboring windows allow just two characters to be seen at any given moment, the left one of which is the plaintext character, the other one the corresponding cryptotext character.

A selection from the (up to N) accompanying powers of a mixed alphabet is obtained if the distance between the windows can be changed. In the case of a strip, it is then necessary to duplicate the whole cycle. The q-th power of the monocyclic permutation is obtained if the windows have a distance of q characters (Fig. 28 for $q = 14$).

↓ ——— 14 ——— ↑ (from H to Z)

S	E	C	U	R	I	T	Y	A	B	D	F	G	H	J	K	L	M	N	O	P	Q	V	W	X	Z	S	E	C	U	R	I	T	Y	A	B	D	F	G	...

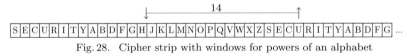

Fig. 28. Cipher strip with windows for powers of an alphabet

[2] In Alberti's illustration, differing from modern usage, capital letters are used for plaintext, small letters for cryptotext. The character /et/ presumably stands for the symbol & . The initial setting of the disk is established by lining up a key letter, say D , with a fixed character, say /a/.

3.3 Case $V^{(1)} \dashrightarrow W^m$ (Multipartite Simple Substitutions)

3.3.1 $m = 2$, **bipartite simple substitution** $V^{(1)} \dashrightarrow W^2$. Substitution by bigrams (bipartite substitution) was known in antiquity, and Polybios described a quinary ($|W| = 5$) bipartite substitution for Greek letters. In a modern form, Z_{25} is inscribed into a 5×5 checkerboard:

	1	2	3	4	5
1	a	b	c	d	e
2	f	g	h	i	k
3	l	m	n	o	p
4	q	r	s	t	u
5	v	w	x	y	z

or

	1	2	3	4	5
1	a	f	l	q	v
2	b	g	m	r	w
3	c	h	n	s	x
4	d	i	o	t	y
5	e	k	p	u	z

Decryption with the 'Polybios square' on the right hand side gives for the text semagram

33515141234333514512432411343411343442331144424333

of Sect. 1.2, Fig. 3, the plaintext

n e e d m o n e y f o r a s s a s s i n a t i o n .

While Polybios described how torches can represent the numbers 1 – 5, knock signals are used for it in more modern times. The special $Z_{25} \longrightarrow Z_5 \times Z_5$ cipher above is the ubiquitous, truly international knock cipher, used in jails from Alcatraz to Ploetzensee by criminals as much as by political prisoners. The normal speed of transmission is 8–15 words per minute.

In Czarist Russia, such a knock-cipher (with the Russian alphabet in a 6×6 square) was common and came to Western Europe with Russian anarchists as part of the 'Nihilist cipher' (Sect. 9.4.5), it was also used steganographically, see Sect. 1.2. Arthur Koestler, in *Sonnenfinsternis* , and Alexander Solzhenitsyn, in *The Gulag Archipelago*, reported on its use in the Soviet Union.

In general, a password is used, which is inscribed line by line and the remaining characters filled in. The count Honoré de Mirabeau, a French revolutionary in the 18th century, used this method in his correspondence with the Marquise Sophie de Monnier—he, too, used it steganographically and added 6 7 8 9 0 as nulls.

The ADFGVX system, invented by Fritz Nebel (1891–1967), which was installed in 1918 on the German Western Front under Quartermaster General Erich Ludendorff for wireless transmission (for the cryptotext alphabet Z_6 see Sect. 2.5.2), worked with $|W| = 6$ and checkerboards like

	A	D	F	G	V	X
A	c	o	8	x	f	4
D	m	k	3	a	z	9
F	n	w	l	0	j	d
G	5	s	i	y	h	u
V	p	1	v	b	6	r
X	e	q	7	t	2	g

Rectangular arrays are used, too. Giovanni Battista Argenti, around 1580, used the following scheme (with $W = Z_{10}$)

	0	1	2	3	4	5	6	7	8	9
1	p	i	e	t	r	o	a	b	c	d
2	f	g	h	l	m	n	q	s	u	z

with the very first application of a password.

In general, the bipartite substitution leaves ample space for homophones:

	1	2	3	4	5	6	7	8	9
9, 6, 3	a	b	c	d	e	f	g	h	i
8, 5, 2	j	k	l	m	n	o	p	q	r
7, 4, 1	s	t	u	v	w	x	y	z	

.

In this example the character 0 may serve as a null. 0, originally *nulla ziffra*, still is not taken seriously everywhere.

Preferably, homophones should smooth out the character frequencies in the cryptotext. Since the letters e t a o n i r s h in English have altogether a frequency around 70 % a good balance is reached by

	1	2	3	4	5	6	7	8	9	
4,5,6,7,8,9,0	e	t	a	o	n	i	r	s	h	71.09 %
2,3	b	c	d	f	g	j	k	l	m	19.46 %
1	p	q	u	v	w	x	y	z		9.45 %

.

Another method uses a 4-letter password and decides in this way on the outset of the cycles (**00**...24), (**25**...49), (**50**...74), (**75**...99) in defining (with $V = Z_{25}$ and $W = Z_{10}^2$) a homophonic cipher, e.g., with the password *KILO*:

	a	b	c	d	e	f	g	h	i	k	l	m	n	o	p	q	r	s	t	u	v	w	x	y	z
K	16	17	18	19	20	21	22	23	24	**00**	01	02	03	04	05	06	07	08	09	10	11	12	13	14	15
I	42	43	44	45	46	47	48	49	**25**	26	27	28	29	30	31	32	33	34	35	36	37	38	39	40	41
L	65	66	67	68	69	70	71	72	73	74	**50**	51	52	53	54	55	56	57	58	59	60	61	62	63	64
O	87	88	89	90	91	92	93	94	95	96	97	98	99	**75**	76	77	78	79	80	81	82	83	84	85	86

A denary ($|W| = 10$) bipartite cipher does not have to have homophones—the substitution does not have to be surjective and some pairs can be left unused. Such a cipher was used by the Swedish baronet Fridric Gripenstierna—possibly based on a proposal of Christopher Polhem. A funny form of a bipartite cipher with homophones was agreed upon during the development of the atom bomb by Brig. Gen. Leslie R. Groves and Lt. Col. Peer da Silva in Los Alamos (Fig. 29), to be used in telephone conversations for veiling special names and places. The point is that it takes time to look up the letters, and thus homophones are selected more at random than normally, when the encipherer is biased.

3.3.2 $m = 3$ **, tripartite simple substitution** $V^{(1)} \dashrightarrow W^3$. Substitution by trigrams (tripartite substitution) was proposed by Trithemius in the *Polygraphiæ*, with $|W| = 3$ (note that $3^3 = 27 > 26$) for steganographic reasons. Otherwise, ternary substitutions like this one are rare.

1	2	3	4	5	6	7	8	9	0	
I	P	I		O	U	O		P	N	1
W	E	U	T	E	K		L	O		2
E	U	G	N	B	T	N		S	T	3
T	A	Z	M	D		I	O	E		4
S	V	T	J		E		Y		H	5
N	A	O	L	N	S	U	G	O	E	6
	C	B	A	F	R	S		I	R	7
I	C	W	Y	R	U	A	M		N	8
M	V	T		H	P	D	I	X	Q	9
L	S	R	E	T	D	E	A	H	E	0

Fig. 29.
Bipartite cipher,
used in Los Alamos in 1944
for telephone conversations

3.3.3 $m = 5$, **quinpartite simple substitution** $V^{(1)} \dashrightarrow W^5$. Substitution by groups of five cryptotext characters (quinpartite substitution) with $|W| = 2$ was used by Francis Bacon in connection with steganographic means (note that $2^5 = 32 > 26$). Quinpartite binary encryption was resurrected in the cipher machine of Vernam in 1918 (Sect. 8.3.2) and during the Second World War in the cipher-teletype machines Siemens T52 (*Geheimschreiber*) and Lorenz SZ40/42 (*Schlüsselzusatz*), see Sect. 9.1.3 and 9.1.4 .

3.3.4 $m = 8$, **octopartite simple substitution** $V^{(1)} \dashrightarrow W^8$. Again with $|W| = 2$ (8-bit code, binary EBCDIC code, ASCII code with checkbit), this octopartite simple substitution coincides with monopartite substitution by bytes (Z_{256}) in modern computers.

3.4 The General Case $V^{(1)} \dashrightarrow W^{(m)}$, Straddling

The general case $V^{(1)} \dashrightarrow W^{(m)}$ plainly invites the use of null and homophones.

Simeone de Crema in Mantua (1401) used just homophones (with $m = 1$). With $m = 2$, apart from the use of homophones and nulls an important new thought comes into play: straddling (German *Spreizen*) of the alphabet, the mapping of V into $W^1 \cup W^2$. A cipher used at the Holy See, the papal court, devised by Matteo Argenti after 1590, shows homophones, nulls, and straddling. For an alphabet Z_{24} enriched by /et/, /con/, /non/, /che/ and with 5, 7 serving as nulls, the encryption step $Z_{24}^{(1)} \dashrightarrow Z_{10}^1 \cup Z_{10}^2$ is

a	b	c	d	e	f	g	h	i	l	m	n	o
1	86	02	20	62 82	22	06	60	3	24	26	84	9

p	q	r	s	t	v	z	et	con	non	che	ε
66	68	28	42	80 40	04	88	08	64	00	44	5 7

3.4.1 Caveat. Encryption steps with straddling are subject to the restriction that the encryption induced by them should turn out to be left-unique—this means that the hiatuses between the one-letter and the two-letter cipher elements and thus the correct decomposition are well determined. As stated in Sect. 2.4, G. B. and M. Argenti were aware of this. Their ciphers fulfill the following conditions: W is divided up into characters used for one-character cipher elements, $W' = \{1, 3, 5, 7, 9\}$ and characters used for two-character cipher elements to begin with, $W'' = \{0, 2, 4, 6, 8\}$. The Argentis made the mistake of restricting also the second character of these to W''. This exposes the straddling. Otherwise, they made some more practical recommendations: to suppress the u following the q and to suppress a duplicated letter.

The so-called spy ciphers used by the Soviet NKVD and its followers are straddling ciphers. They have been disclosed by convicted spies. By analogy with Polybios squares they are described by rectangular arrays too, e.g.,

	0	1	2	3	4	5	6	7	8	9
	s	i	o	e	r	a	t	n		
8	c	x	u	d	j	p	z	b	k	q
9	.	w	f	l	/	g	m	y	h	v

(∗)

where the first line contains the one-letter cipher elements.

With $W = Z_{10}$ 28 cipher elements are obtainable, enough for Z_{26} and two special characters, . for 'stop' and / for letter-figure swap. Because this cipher was subjected to further encryption ('closing', Sect. 9.2.1), it was tolerable to encrypt figures—after sending a letter-figure swap sign—by identical figure twins, a safeguard against transmission errors.

For the construction of this array passwords have been used, too. Dr. Per Meurling, a Swedish fellow traveler, did it 1937 as follows: He wrote down an 8-letter password (M. Delvayo was a Spanish communist) and below it the remaining alphabet; the columns were numbered backwards:

	0	9	8	7	6	5	4	3	2	1
	m	d	e	l	v	a	y	o		
1	b	c	f	g	h	i	j	k	n	p
2	q	r	s	t	u	w	x	z	.	/

.

This procedure had the disadvantage that not at all the most frequent letters obtained 1-figure ciphers. This disadvantage was also shared by the method the Swedish spy Bertil Eriksson used in 1941: He numbered the columns according to the alphabetic order of the letters occurring in the password (Sect. 3.2.5):

	6	0	8	7	5	4	9	1	2	3
3	p	a	u	s	o	m	v	e	j	k
9	b	c	d	f	g	h	i	l	n	q
	r	t	w	x	y	z				

.

The password was taken from a Swedish translation of Jaroslav Hašek's novel *Paus, som Svejk själv avbröt* .... Since encryption of the most frequent letters by 1-figure ciphers also shortens the telegraphic transmission time, the NKVD arrived in 1940 at a construction method that took this into account.

Max Clausen, wireless operator of the Russian spy Dr. Richard Sorge in Tokyo, had to memorize the sentence *"a sin to err"* (very good advice for a spy), containing the eight most frequent letters in English, 65.2 % altogether. Beginning with a password /subway/, a rectangle was started and filled with the remaining letters. Thereupon, columnwise from left to right in the order of their appearance, first, the letters from the set {a s i n t o e r} were assigned the numbers 0 ... 7; second, the remaining letters were assigned the numbers 80 ... 99:

s	u	b	w	a	y
0	82	87	91	5	97
c	d	e	f	g	h
80	83	3	92	95	98
i	j	k	l	m	n
1	84	88	93	96	7
o	p	q	r	t	v
2	85	89	4	6	99
x	z	.	/		
81	86	90	94		

In this way, the Polybios rectangle marked above by (∗) is obtained in more compact notation.

For the cyrillic alphabet, a subdivision into seven 1-figure ciphers and thirty 2-figure ciphers, altogether 37 ciphers, is suitable; it allows 5 special characters. A method that was given away by the deserted agent Reino Hayhanen, an aide to the high-ranking Russian spy Rudolf Abel, used a Russian word like СНЕГОПАД ('snowfall'), the first seven letters of which have a total frequency of 44.3 %. The rectangle was formed as usual

С	Н	Е	Г	О	П	А	.	.	.
Б	В	Д	Ж	З	И	Й	К	Л	М
Р	Т	У	Ф	Х	Ц	Ч	Ш	Щ	Ъ
Ы	Ь	Э	Ю	Я	.	.	.	.	.

and then rearranged with the help of a key that was changed from message to message and was to be found at a prearranged place within the cipher message. Finally, a closing encryption (Sect. 9.2.1) was made.

3.4.2 Russian copulation. On this occasion, it was also disclosed that the Russians used what became to be called "Russian copulation": the message was cut into two parts of roughly the same length and these parts were joined with the first after the second, burying in this way the conspicuous standard phrases at beginning and end somewhere in the middle.

Winston Churchill called Russia "a riddle wrapped in a mystery inside an enigma." This is also true for Russian cryptology.

4 Encryption Steps: Polygraphic Substitution and Coding

Simple (monographic) substitution requires a complete decomposition of the plaintext in single characters. A polygraphic substitution allows polygraphic encryption steps, i.e., encryption steps of the form $V^{(n)} \dashrightarrow W^{(m)}$ with $n > 1$.

4.1 Case $V^2 \dashrightarrow W^{(m)}$ (Digraphic Substitutions)

4.1.1 Graphemes. The oldest polygraphic encryption of this type is found in Porta's *De furtivis literarum notis* of 1563 (Fig. 30), a mapping $V^2 \longrightarrow W^1$. Porta showed great ingenuity in inventing 400 strange signs.

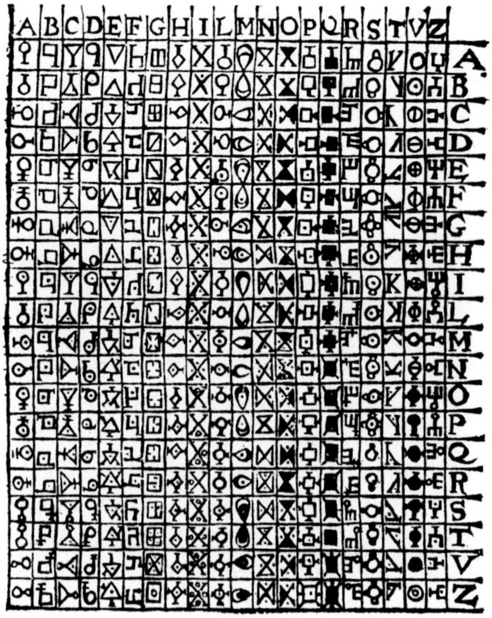

Fig. 30.
Old digraphic substitution by Giovanni Battista Porta, 1563 (Giambattista Della Porta)

4.1.2 Bipartite digraphic encryption step $V^2 \dashrightarrow V^2$. For its representation mostly a matrix is used. In case $V^2 \longleftrightarrow V^2$, it is a bigram permutation.

In the following, an example $V^2 \overset{2}{\longleftrightarrow} V^2$ of a self-reciprocal bigram permutation is given:

	a	b	c	d	e	f	g	h	i	j	k	l	m	n	o ...
a	XZ	KJ	YJ	HP	PL	EL	VB	CI	DW	XN	ZL	YP	VN	HH	CC
b	LP	QT	HE	RS	UR	CR	ZH	GV	WC	HL	YN	KT	WT	MC	KH
c	DX	MN	AO	NH	SF	GI	WL	MN	AH	GR	BZ	HS	ZU	YM	WU
d	KM	YZ	RY	FP	TR	CR	XE	JK	NY	PO	GJ	JR	PE	MO	VB
e	QU	HP	QG	JQ	YQ	OB	SA	NL	PX	OP	VS	AF	XK	XR	UQ
⋮															

The self-reciprocal character (ao ↦ CC , cc ↦AO ; ah ↦ CI , ci ↦ AH ; af ↦ EL , el ↦AF) is not superficially discernible.

Further enciphering steps $V^2 \longleftrightarrow V^2$ can be obtained again with the help of passwords, e.g., with /amerika/ and /equality/ :

	a	m	e	r	i	k	b	c	d	f	g	h	j	l	n ...
e	XZ	KJ	YJ	HP	PL	EL	VB	CI	DW	XN	ZL	YP	VN	HH	CC
q	LP	QT	HE	RS	UR	CR	ZH	GV	WC	HL	YN	KT	WT	MC	KH
u	DX	MN	AO	NH	SF	GI	WL	MN	AH	GR	BZ	HS	ZU	YM	WU
a	KM	YZ	RY	FP	TR	CR	XE	JK	NY	PO	GJ	JR	PE	MO	VB
l	QU	HP	QG	JQ	YQ	OB	SA	NL	PX	OP	VS	AF	XK	XR	UQ
⋮															

(with the effect that the self-reciprocal character disappears and the encryption work becomes more cumbersome).

Establishing a matrix needs the thorough work of a cryptologist in leveling the frequencies of the letters (Sect. 3.1.2). An imitation of the frequency distribution of the letters in the language concerned, thus feigning a transposition, is possible. The ideal result is a matrix which has in every line and in every column every letter occurring just once as first and once as second letter, e.g.,

AB	BC	CA
CC	AA	BB
BA	CB	AC

or

AC	BA	CB	DD
BD	AB	DA	CC
DB	CD	BC	AA
CA	DC	AD	BB

or

AA	BB	CC	DD	EE
BC	CD	DE	EA	AB
CE	DA	EB	AC	BD
DB	EC	AD	BE	CA
ED	AE	BA	CB	DC

Such matrices are called 'Greek-Latin squares'. Apart from the case $N = 6$ ('36-officer problem' of Euler, 1779), for all natural numbers $N > 2$, Greek-Latin squares exist, usually several.

In any case, the example given by Helen Fouché Gaines

$$\begin{array}{llll} \text{AA} & \text{BA} & \text{CA} & \text{DA} \ \ldots \\ \text{AB} & \text{BB} & \text{CB} & \text{DB} \ \ldots \\ \text{AC} & \text{BC} & \text{CC} & \text{DC} \ \ldots \\ \text{AD} & \text{BD} & \text{CD} & \text{DD} \ \ldots \\ \vdots & \vdots & \vdots & \vdots \end{array}$$

is not suitable: it results in a monographic 2-alphabetic encryption (polyalphabetic encryption, Sect. 8.2) succeeded by pairwise swapping of letters.

The following table gives for $N = 26$, for $N = 10$, and for $N = 2$ a survey of the number $Z(N)$ of available squares $V^2 \longleftrightarrow V^2$ (cf. Sect. 3.2.6):

number of squares	$Z(N)$	$Z(26)$	$Z(10)$	$Z(2)$
total	$(N^2)!$	$1.88 \cdot 10^{1621}$	$9.33 \cdot 10^{157}$	24
genuine reflections	$\approx (N^2\,!)^{\frac{1}{2}}$	$7.60 \cdot 10^{809}$	$2.72 \cdot 10^{78}$	3

K 1 Norw.	a	b	c	d	e	f	g	h	i	j	k	l	m	n	o	p	q	r	s	t	u	v	w	x	y	z	
a	ca	fn	bl	ou	ih	oo	il	bv	bw	er	rm	qm	mn	ab	zm	ns	wl	yc	zy	tr	du	wo	oa	ho	ic	pu	a
b	sk	wm	dg	ia	cw	pf	if	vd	da	xz	fo	dh	px	rr	iv	gh	mu	ae	qr	tb	og	sr	vu	qg	zt	pm	b
c	hp	no	ij	xp	ji	yf	eo	xh	zu	pl	ft	yv	qw	am	qp	lz	bg	be	lc	nw	ap	vx	rs	yi	wy	gi	c
d	ov	gg	tk	ys	hm	tx	eq	qa	iu	zo	ud	gj	lh	bn	fm	ta	ej	hi	jc	sv	vp	rd	br	rh	kt	tw	d
e	di	wz	qo	pz	ag	wk	fl	uo	ll	oe	ph	jq	gl	vy	lf	af	vt	cj	vq	yz	rz	fc	ps	pq	ro	aq	e
f	cu	rf	nt	xr	ya	tg	xj	db	sc	hg	zr	hs	em	xv	vr	ul	wn	sh	ku	my	va	ad	fg	zp	ut	lb	f
g	sx	hd	vk	st	lk	xf	gn	lv	yr	yd	xg	kr	hc	xl	xw	pa	au	eb	gb	li	id	rj	tz	xq	wd	rn	g
h	bq	oy	sb	mw	qx	zd	ar	po	on	rx	sj	om	as	mb	vs	ke	yy	xy	uj	hb	rc	jg	co	fq	jr	pe	h
i	cb	sl	ri	cf	qt	ek	un	kl	nx	to	hk	ew	yo	wp	kj	kh	su	xi	jo	of	dt	ml	zi	bk	qq	gu	i
j	vv	tf	fi	mp	ky	hl	qc	iq	na	gd	up	tq	hq	xs	xb	wt	ez	mm	hj	vg	eh	dc	qe	ti	uk	cg	j
k	uv	bt	bf	ux	kz	zw	ex	nh	ac	av	tt	aw	ye	dw	dy	nv	wf	dn	sf	eg	lg	wc	kx	ur	pc	od	k
l	ir	ea	kn	le	jb	nu	at	hu	zl	fw	ce	ka	jv	bm	ev	ak	cp	gm	yn	cd	kd	ue	xm	ig	fy	ht	l
m	mv	el	yg	ny	bu	cq	fk	wq	pk	oo	ms	sz	rl	pr	qi	te	qn	kf	gs	uc	kv	kc	dl	kp	cl	lp	m
n	je	sq	gz	ts	dk	vo	xo	ge	mj	qv	mi	dp	vf	rb	yj	bj	mg	vl	qs	uw	rq	pb	mh	lt	oz	qk	n
o	vc	gk	al	vz	np	vm	by	cm	re	wv	uz	yt	ww	gp	js	en	tv	jn	bo	tm	sp	or	fj	ub	ck	td	o
p	hr	ah	ik	xn	mo	zk	ds	in	dz	ym	ci	qu	dv	df	nk	yk	pt	iz	ef	ws	es	ip	fz	ss	jk	ct	p
q	ec	xc	jj	vb	vh	ot	pg	ib	ty	ch	pd	qz	qf	fd	oh	sa	bc	zj	ba	fp	nq	wa	ie	vi	oq	lw	q
r	wi	uq	ln	ja	gq	lo	rp	sd	ko	iy	si	mc	uu	io	yh	ru	xx	qy	fr	hy	ob	ox	nl	uh	fh	ga	r
s	zg	nf	sy	jw	nn	kq	vn	ld	go	mt	pn	jf	he	um	ua	za	xt	bb	op	qh	gf	yl	md	os	ju	ei	s
t	yw	wg	mx	ol	sw	se	rv	yp	us	rk	dx	zs	bz	dj	cn	mf	hx	de	it	ai	ug	mk	ql	cs	ix	pi	t
u	gy	fa	ow	gr	vw	bh	ly	kw	ry	mz	pj	sg	jz	gt	dd	nd	et	az	tp	jh	cx	iw	la	zq	rw	lm	u
v	gv	bi	oi	ii	zb	lj	hz	zh	nb	ks	cy	yq	jx	dq	ma	hf	wr	lq	jp	ng	gw	jl	rg	tl	lr	wh	v
w	aj	gx	nr	qb	uf	ok	rt	xu	bp	wb	qd	jt	mr	aa	pv	yu	nj	xd	eu	mq	hw	nz	ze	km	uy	tn	w
x	kb	yx	ui	pw	we	xk	fe	vj	gc	pp	ep	hh	zn	ha	zf	ax	do	py	nm	xe	ff	so	tc	sm	fb	fx	x
y	fs	ay	ni	wj	wu	fu	ed	an	fv	xa	cv	cz	bs	ve	th	cc	bx	ra	cr	im	ne	hn	zv	oj	yb	tj	y
z	kg	bd	wx	zz	zx	lu	jy	sn	zc	tu	is	ao	dr	ki	ls	ey	qj	ee	lx	hv	nc	dm	jd	me	jm	kk	z
	a	b	c	d	e	f	g	h	i	j	k	l	m	n	o	p	q	r	s	t	u	v	w	x	y	z	Pnr. **0033**

Fig. 31a. Bipartite digraphic encryption of the RSHA call signals in Norway

A classical example is given by Fig. 31a, a $V^2 \longleftrightarrow V^2$ step for the encryption of call signals of the RSHA radio network in Norway. Ten such tables were fabricated in *Amt* VI (headed up to 1941 by Jost, then Schellenberg), the Foreign Intelligence section of Himmler's R.S.H.A. (*Reichssicherheitshauptamt*), possibly by Andreas Figl (1873–1967). The R.S.H.A. got hold of the Austrian Colonel Andreas Figl, former head of the Austrian *Dechiffrierdienst* (*'Chiffrengruppe'*), together with useful documents, in 1938, when Austria was occupied (*Anschluss*). The importance of this booty was "discovered" by the young Austrian SS-Sturmbannführer Dr. Wilhelm Höttl (1915–1999), who later became deputy head of the Vienna desk of group *VI E*[1]. Figl, held until 1941 in "custody" by the SS, worked as an "advisor" in Berlin-Wannsee.

Höttl also helped *Amt VI* of the *R.S.H.A.*, employing from mid-1944 onward a group of Hungarian army cryptologists at Budapest, headed by Major Bibo who in 1944 succeeded in penetrating Allen Dulles' Bern-Washington traffic.

Figl, a very shrewd cryptanalyst, was Captain when in 1911 he built up the cryptanalytic bureau of the *k.u.k. Armee*, in the best tradition of the Viennese court. In 1915 Major Figl solved Italian cryptograms, and in 1926 he had the rank of Colonel, when he wrote a good textbook, *Systeme des Chiffrierens* (243 pages with 45 supplements, Graz 1926). A planned second volume, *Systeme des Dechiffrierens*, was in 1926 disallowed to be printed; a copy, but not the original of the manuscript now seems to be accessible.

Self-reciprocal bipartite digraphic encryption was used since May 1, 1937 for the superencryption of the indicators (*Spruchschlüssel*) that preceded the wireless messages encrypted by the *Marine* ENIGMA. There was a choice between ten such tables with names like FLUSS (Fig. 31b), BACH, STROM, TEICH, UFER that had been in use; they were known to the British who had seized (U-110, Mai 1941, Gedania, June 1941; VP 5904, January 1942; U 505, June 1944) or possibly stolen them, later also reconstructed them. The indicator, such as /psq/, was to be chosen at random; it was doubled to /psqpsq/ and ENIGMA-enciphered with a given ground setting (*Grundstellung*), say /iaf/ ; the cipher, say SWQRAF, was split into two parts:

$$\begin{array}{cccc} \mathrm{S} & \mathrm{W} & \mathrm{Q} & * \\ * & \mathrm{R} & \mathrm{A} & \mathrm{F} \end{array} \quad \text{and filled with nulls ('padding letters'):} \quad \begin{array}{cccc} \mathrm{S} & \mathrm{W} & \mathrm{Q} & \mathrm{X} \\ \mathrm{P} & \mathrm{R} & \mathrm{A} & \mathrm{F} \end{array} .$$

The digraphic encipherment was done with vertical pairs, with, say

$$\begin{array}{c} \mathrm{S} \\ \mathrm{P} \end{array} \leftrightarrow \begin{array}{c} \mathrm{Q} \\ \mathrm{A} \end{array} , \quad \begin{array}{c} \mathrm{W} \\ \mathrm{R} \end{array} \leftrightarrow \begin{array}{c} \mathrm{F} \\ \mathrm{P} \end{array} , \quad \begin{array}{c} \mathrm{Q} \\ \mathrm{A} \end{array} \leftrightarrow \begin{array}{c} \mathrm{C} \\ \mathrm{D} \end{array} , \quad \begin{array}{c} \mathrm{X} \\ \mathrm{F} \end{array} \leftrightarrow \begin{array}{c} \mathrm{S} \\ \mathrm{Z} \end{array} ,$$

the result was $\begin{array}{cccc} \mathrm{Q} & \mathrm{F} & \mathrm{C} & \mathrm{S} \\ \mathrm{A} & \mathrm{P} & \mathrm{D} & \mathrm{Z} \end{array}$. The encrypted indicator QFCSAPDZ was transmitted. On the receiving side, the procedure was applied backwards: first the (self-reciprocal) bipartite digraphic substitution and the removal of the nulls, then the (self-reciprocal) ENIGMA-enciphering with the given ground setting /iaf/ ; the result had to have the pattern *123123*, the first part of which was the individual wheel setting for the message. In this way, a key was negotiated between the two parties.

The procedure seemed complicated enough to lull those who invented it into a sense of security. For the British, the obstacles nevertheless were surmountable. A conjecture Turing had before the end of 1939 was confirmed when the German Boat Polares was seized in 1940; in 1941 the British then succeeded in reconstructing the bigram tables after a few ones had been 'pinched'.

---

[1] After the Second World War, Höttl played an unsuccessful role in the Austrian right-wing party *Wahlpartei der Unabhängigen*; he was arrogant ("I was Hitler's Master Spy") and also wrote books (*The Secret Front*, 1954 and under the pseudonym Walter Hagen *The Paper Weapon*, 1955).

Geheim! Prüfnr. 516

Kennwort: Fluß

Doppelbuchstabentauschtafel für Kenngruppen — Tafel B

AD	AE	AJ	AK	AL	AM	BC	BD	BE	CE	CF	CG	DG
AA = RN	BA = IK	CA = KJ	DA = PK	EA = TC	FA = XP	GA = NE	HA = JR	IA = NN	JA = WE	KA = EI	LA = EU	MA = RG
B = KW	B = RT	B = PO	B = EZ	B = JX	B = OI	B = JO	B = NO	B = VF	B = OY	B = GW	B = RH	B = IP
C = FM	C = EY	C = JV	C = AW	C = OM	C = IU	C = BK	C = GY	C = DN	C = NQ	C = IM	C = VO	C = WW
D = YE	D = AK	D = BM	D = JM	D = MJ	D = RB	D = FL	D = TB	D = FW	D = KK	D = SE	D = YA	D = TA
E = NR	E = OW	E = MZ	E = WB	E = NY	E = PA	E = ZT	E = ZI	E = RP	E = TN	E = AG	E = CV	E = BQ
F = UC	F = WQ	F = EK	F = XY	F = AS	F = DZ	F = SA	F = QY	F = EO	F = VS	F = JH	F = SC	F = KV
G = KE	G = QA	G = KT	G = ZA	G = PU	G = NV	G = LR	G = OA	G = WS	G = FR	G = PN	G = JU	G = NS
H = XU	H = ZZ	H = AZ	H = BS	H = WO	H = ZK	H = TP	H = CU	H = NU	H = KF	H = DT	H = ZQ	H = VK
I = PC	I = OG	I = ND	I = MT	I = KA	I = QR	I = MW	I = QS	I = TM	I = PM	I = LV	I = RX	I = XC
J = JP	J = HQ	J = TQ	J = OE	J = GZ	J = LN	J = AU	J = IS	J = XO	J = SV	J = CA	J = WZ	J = ED
K = BD	K = GC	K = GX	K = FP	K = CF	K = EL	K = QN	K = PG	K = BA	K = IT	K = JD	K = EM	K = ZF
L = QI	L = PR	L = RE	L = RI	L = FK	L = GD	L = WH	L = KR	L = MS	L = UP	L = TO	L = OK	L = DR
M = HT	M = CD	M = WA	M = VV	M = LK	M = AC	M = PB	M = SF	M = KC	M = DD	M = BW	M = TR	M = SU
N = MR	N = NL	N = OS	N = IC	N = TY	N = CP	N = OX	N = SZ	N = QZ	N = PX	N = UX	N = FJ	N = LO
O = BZ	O = US	O = DY	O = YJ	O = IF	O = VE	O = JT	O = FY	O = YV	O = GB	O = QC	O = MN	O = NX
P = XI	P = SX	P = FN	P = NF	P = NC	P = DK	P = RY	P = MX	P = MB	P = AJ	P = VJ	P = BT	P = FZ
Q = OZ	Q = ME	Q = QF	Q = GU	Q = WV	Q = PY	Q = IZ	Q = BJ	Q = OV	Q = XH	Q = RS	Q = IV	Q = OJ
R = UK	R = YN	R = XJ	R = ML	R = KS	R = JG	R = CY	R = OP	R = SH	R = HA	R = HL	R = GG	R = AN
S = EF	S = DH	S = ZB	S = QG	S = QW	S = UE	S = RF	S = RJ.	S = HJ	S = YZ	S = ER	S = NW	S = IL
T = IY	T = LP	T = SW	T = KH	T = XD	T = SR	T = XV	T = AM	T = JK	T = GO	T = CG	T = UF	T = DI
U = GJ	U = XK	U = HH	U = WH	U = LA	U = WX	U = DQ	U = UQ	U = FC	U = LG	U = XZ	U = XW	U = BY
V = QU	V = TI	V = LE	V = HW	V = RL	V = TL	V = UM	V = LZ	V = LQ	V = CC	V = MF	V = KI	V = UT
W = DC	W = KM	W = VP	W = SO	W = SK	W = ID	W = KB	W = DV	W = PH	W = QL	W = AB	W = PW	W = GI
X = UV	X = VY	X = UG	X = NT	X = UZ	X = YS	X = CK	X = WJ	X = UD	X = EB	X = ZY	X = PP	X = HP
Y = SG	Y = MU	Y = GR	Y = CO	Y = BC	Y = HO	Y = HC	Y = VN	Y = AT	Y = TU	Y = NZ	Y = QD	Y = VB
Z = CH	Z = AO	Z = YI	Z = FF	Z = DB	Z = MP	Z = EJ	Z = YD	Z = GQ	Z = UW	Z = WP	Z = HV	Z = CE

Fortsetzung s. Rückseite

Kennwort: Fluß

Tafel B

DH	DJ	DR	DS	DT	EG	EH	EJ	EK	EO	ER	ES	ET
NA = TZ	OA = HG	PA = FE	QA = BG	RA = QH	SA = GF	TA = MD	UA = QX	VA = ON	WA = CM	XA = TX	YA = LD	ZA = DG
B = QV	B = ZX	B = GM	B = ZD	B = FD	B = OT	B = HD	B = SD	B = MY	B = DE	B = UL	B = VG	B = CS
C = EP	C = TH	C = AI	C = KO	C = PL	C = LF	C = EA	C = AF	C = ZO	C = QJ	C = MI	C = WL	C = SI
D = CI	D = XS	D = NH	D = LY	D = OQ	D = UB	D = ZN	D = IX	D = SY	D = PV	D = ET	D = HZ	D = QB
E = GA	E = DJ	E = QT	E = TJ	E = CL	E = KD	E = YX	E = FS	E = FO	E = JA	E = WM	E = AD	E = TT
F = DP	F = QM	F = XX	F = CQ	F = GS	F = HM	F = RO	F = LT	F = IB	F = VT	F = ZL	F = OR	F = MK
G = XM	G = BI	G = HK	G = DS	G = MA	G = AY	G = WK	G = CX	G = YB	G = ZM	G = SS	G = VQ	G = RM
H = PD	H = NP	H = IW	H = RA	H = LB	H = IR	H = OC	H = ZJ	H = RK	H = GL	H = JQ	H = QQ	H = VL
I = TW	I = FB	I = ZR	I = AL	I = DL	I = ZC	I = BV	I = ST	I = XR	I = YR	I = AP	I = CZ	I = HE
J = VR	J = MQ	J = TS	J = WC	J = HS	J = PQ	J = QE	J = NM	J = KP	J = HX	J = CR	J = DO	J = UH
K = YU	K = LL	K = DA	K = SN	K = VH	K = EW	K = XN	K = AR	K = MH	K = TG	K = BU	K = WR	K = FH
L = BN	L = RZ	L = RC	L = JW	L = EV	L = VX	L = FV	L = XB	L = ZH	L = YC	L = ZP	L = SQ	L = XF
M = UJ	M = EC	M = JI	M = OF	M = ZG	M = ZV	M = II	M = GV	M = ZU	M = XE	M = NG	M = VW	M = WG
N = IA	N = VA	N = KG	N = GK	N = AA	N = QK	N = JE	N = YY	N = HY	N = DU	N = TK	N = BR	N = TD
O = HB	O = UU	O = CB	O = VZ	O = TF	O = DW	O = KL	O = TV	O = LC	O = EH	O = IJ	O = PZ	O = VC
P = OH	P = HR	P = LX	P = XT	P = IE	P = RV	P = GH	P = JL	P = CW	P = KZ	P = FA	P = ZS	P = XL
Q = JC	Q = RD	Q = SJ	Q = YH	Q = UR	Q = YL	Q = CJ	Q = HU	Q = YG	Q = BF	Q = YT	Q = RW	Q = LH
R = AE	R = YF	R = BL	R = FI	R = WU	R = FT	R = LM	R = RQ	R = NJ	R = YK	R = VI	R = WI	R = PI
S = MG	S = CN	S = UY	S = HI	S = KQ	S = XG	S = PJ	S = BO	S = JF	S = IG	S = OD	S = FX	S = YP
T = DX	T = SB	T = WY	T = PE	T = BB	T = UI	T = ZE	T = MV	T = WF	T = OU	T = QP	T = XQ	T = GE
U = IH	U = WT	U = EG	U = AV	U = ZW	U = MM	U = JY	U = OO	U = YW	U = RR	U = AH	U = NK	U = VM
V = FG	V = IQ	V = WD	V = NB	V = SP	V = JJ	V = UO	V = AX	V = DM	V = EQ	V = GT	V = IO	V = SM
W = LS	W = BE	W = LW	W = ES	W = YQ	W = CT	W = NI	W = JZ	W = YM	W = MC	W = LU	W = VU	W = RU
X = MO	X = GN	X = JN	X = UA	X = LI	X = BP	X = XA	X = KN	X = SL	X = FU	X = PF	X = TE	X = OB
Y = EE	Y = JB	Y = FQ	Y = HF	Y = GP	Y = VD	Y = EN	Y = PS	Y = BX	Y = PT	Y = DF	Y = UN	Y = KX
Z = KY	Z = AQ	Z = YO	Z = IN	Z = OL	Z = HN	Z = NA	Z = EX	Z = QO	Z = LJ	Z = KU	Z = JS	Z = BH

Fig. 31b. Self-reciprocal bigram table FLUSS of the *Kriegsmarine*

The key negotiation was and is up to the present days a particular weakness of cryptology.

The British should have been warned. Nevertheless, the British Merchant Navy used bipartite digraphic substitution for the superencryption of their

Verschlüsselungstafel.

	0	1	2	3	4	5	6	7	8	9
0	23	48	60	05	78	35	58	64	29	52
1	20	77	33	59	21	70	02	40	63	08
2	11	49	01	69	47	41	79	74	22	42
3	32	76	39	18	75	30	09	51	80	65
4	61	19	43	81	06	56	73	62	10	28
5	85	50	24	88	31	84	27	90	55	57
5	03	91	96	53	68	16	44	89	15	87
7	97	25	71	04	95	34	14	37	93	38
8	26	72	54	92	13	83	45	00	66	67
9	86	12	98	36	99	46	82	17	94	07

Entschlüsselungstafel.

	0	1	2	3	4	5	6	7	8	9
0	87	22	16	60	73	03	44	99	19	36
1	48	20	91	84	76	68	65	97	33	41
2	10	14	28	00	52	71	80	56	49	08
3	35	54	30	12	75	05	93	77	79	32
4	17	25	29	42	66	86	95	24	01	21
5	51	37	09	63	82	58	45	59	06	13
6	02	40	47	18	07	39	88	89	64	23
7	15	72	81	46	27	34	31	11	04	26
8	38	43	96	85	55	50	90	69	53	67
9	57	61	83	78	98	74	62	70	92	94

Fig. 32. Bipartite digraphic substitution (*Geheimklappe*) for the superencryption of numeral codes

BAMS code. The codebook fell into German hands in 1940, when the German raider *Atlantis* seized the vessel *City of Bagdad* in the Indian Ocean. The *B-Dienst* of the German *Kriegsmarine* was successful until 1943 in stripping off the superencipherment of allied ships' radio signals.

For the superencryption of numeral codes a permutation $Z_{10}^2 \longleftrightarrow Z_{10}^2$, as specified by the *Geheimklappe*, suffices. This was a bipartite digraphic substitution introduced in March 1918 by the Germans for tactical communications on the Western Front, with one table for enciphering and one table for deciphering (Fig. 32). Towards the end of the First World War, this bipartite digraphic substitution was changed every day.

4.1.3 Tripartite digraphic substitution $V^2 \longrightarrow W^3$. It is sometimes used, e.g., the denary tripartite digraphic substitution ($V = Z_{26}$, $W = Z_{10}$):

	a	b	c	d	e	...
a	148	287	089	623	094	
b	243	127	500	321	601	
c	044	237	174	520	441	
d	143	537	188	257	347	
⋮	⋮	⋮	⋮	⋮	⋮	

Cryptanalytically, $V^2 \dashrightarrow W^{(n)}$ falls for arbitrary n into one and the same class and can be interpreted as a $|V|$-fold homophonic simple substitution of the odd-numbered letters plus a $|V|$-fold homophonic simple substitution of the even-numbered letters. Correspondingly, it is trivial to break the encryption, if, as in the example $V^2 \longleftrightarrow V^2$ of Helen Fouché Gaines, a standard cipher table is used, provided there is enough material available. Eyraud points out that in particular the method of cutting the message into halves, writing them in two lines, and using digraphic substitution for vertical pairs, is a *complication illusoire*.

4.2 Special Cases of Playfair and Delastelle: Tomographic Methods

4.2.1 Playfair cipher. In 1854, Charles Wheatstone invented a special bipartite digraphic substitution (Fig. 33); his friend Lyon Playfair, Baron of St. Andrews, recommended it to high-ranking government and military persons. The system may have been used for the first time in the Crimean War and was reportedly used in the Boer War; the name of Playfair remained attached to it. The military appreciated it as a field cipher because it needed neither tables nor apparatus. The British Army adopted it around the turn of the century and continued to keep it secret. Nevertheless, in the First World War, by mid-1915, the Germans could solve it routinely.

The PLAYFAIR encryption step goes as follows: From a password, a permuted alphabet Z_{25} (say, omitting the J of Z_{26}) is inscribed into a 5×5 square (French *damier*):[2]

P	A	L	M	E		T	O	N	R	S
R	S	T	O	N		D	F	G	B	C
B	C	D	F	G	or	K	Q	U	H	I
H	I	K	Q	U		X	Y	Z	V	W
V	W	X	Y	Z		L	M	E	P	A

and this is thought to be closed like a torus, such that the two examples mean the same. Now, if the two letters of a bigram stand in one and the same line (or column), each is replaced by the letter to its right (or beneath it, respectively); e.g., both squares yield consistently

$$\text{am} \mapsto \text{LE} \qquad \text{dl} \mapsto \text{KT} \quad .$$

Otherwise, the first letter is replaced by the letter in the same line, but in the column of the second letter; likewise the second letter is replaced by the letter in the same line, but in the column of the first letter ("crossing step", French *substitution orthogonale et diagonale*). Thus

$$\text{ag} \mapsto \text{EC} \qquad \text{ho} \mapsto \text{QR} \quad .$$

The step is undefined if the bigram is a doubled letter[3] or if the final letter is unpaired. This situation is avoided by inserting x :

ba ll oo n is replaced by ba lx lo on ; le ss se ve n by le sx sx se ve nx .

This is a dangerous weakness. Notwithstanding, the PLAYFAIR step fascinates by its relative simplicity. But because of the torus symmetry its combinatory complexity is even less than that of a simple substitution.

In the third case above, the PLAYFAIR step can be interpreted as a composition of mappings: a mapping of the plaintext bigram into a pair of line-

---

[2] Wheatstone actually used alphabets that were better mixed (Sect. 3.2.5), and rectangular matrices. These important safety measures were soon dropped.

[3] The advice of Matteo Argenti, to suppress duplicated letters altogether (Sect. 3.4.1), was probably forgotten.

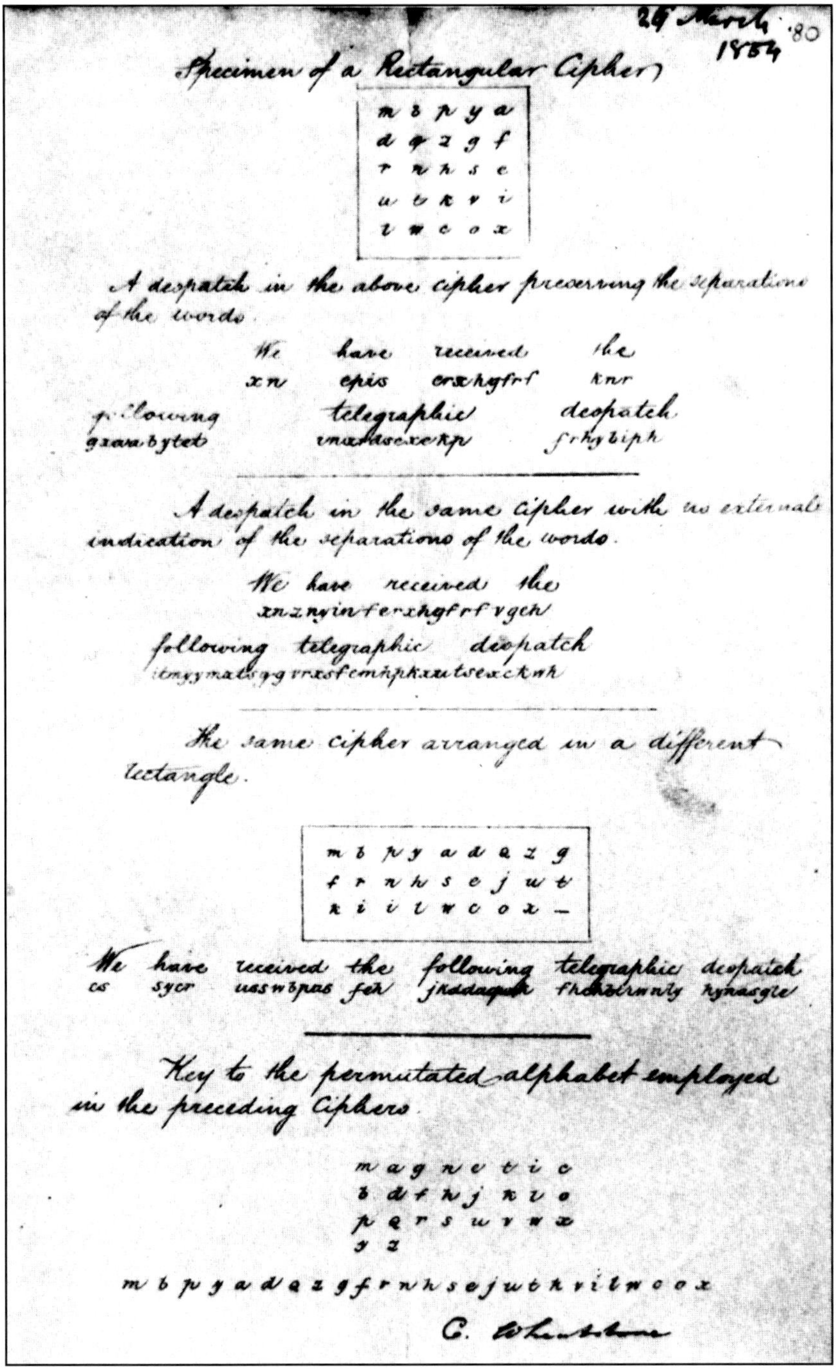

26 March 1854 '80

Specimen of a Rectangular Cipher

m b p y a
d q z g f
r n h s e
u t k v i
l w c o x

A despatch in the above cipher preserving the separations of the words.

We have received the
xn epis erxhgfrf knr
following telegraphic despatch
gxambytet tnxzdsexekp frhgbiph

A despatch in the same cipher with no external indication of the separations of the words.

We have received the
xnznyinferxhgfrfvgen
following telegraphic despatch
itnyymatsygvrxsfemnhpkxxutsexcknh

The same cipher arranged in a different rectangle.

m b p y a d q z g
f r n h s e j u t
k i v l w c o x –

We have received the following telegraphic despatch
cs sycr uaswbpas fen jkddaqu fhehoirnnly hynasgle

Key to the permutated alphabet employed in the preceding Ciphers.

m a g n e t i c
b d f h j k l o
p q r s u v w x
y z

m b p y a d q z g f r n h s e j u t k v i l w c o x

C. Wheatstone

Fig. 33. Description of the PLAYFAIR cipher, signed by its inventor Charles Wheatstone, March 26, 1854 and-column coordinates, a permutation of the column coordinates, and re-translation into a bigram (similar to a spoonerism, Sect. 6.1.2):

	1	2	3	4	5
1	P	A	L	M	E
2	R	S	T	O	N
3	B	C	D	F	G
4	H	I	K	Q	U
5	V	W	X	Y	Z

a	g
12	35
15	32
E	C

Such compositions of encryptions, that amount to a decomposition and re-combination, are called tomographic or fractionating methods ('*chiffres à damiers*', Auguste L. A. Collon, 1899); we shall study them in Sect. 9.4.4.

4.2.2 Modified PLAYFAIR. Such a step was used by the German Army and by the SD (*Sicherheitsdienst*, the Nazi party and government political police), starting mid-1941, as a *Handschlüssel*—and regularly broken until the fall of 1944 by the British at Bletchley Park under Colonel John H. Tiltman (1900–1984), who was since 1924 head of the Military Section. It was named double casket (German *Doppelkas[set]tenverfahren*), also double PLAYFAIR, two-table-PLAYFAIR, because it used (say, omitting the J) two different 5×5 squares, e.g.,

A	Y	K	I	H		Y	X	U	H	A
L	B	M	N	P		T	R	K	B	I
Q	R	C	O	G		P	M	C	G	S
Z	X	V	D	S		F	D	L	Q	V
F	W	U	T	E		E	N	O	W	Z

They were not constructed from passwords, but were formed "at random" and then distributed. As in the original PLAYFAIR, bipartite digraphic steps like

$$\text{ah} \mapsto \text{AY}\,, \quad \text{nr} \mapsto \text{KP}\,, \quad \text{nb} \mapsto \text{IP}$$

occur if plaintext letters stand in the same line (closed like a cylinder), in all other cases a "crossing step" is used:

$$\text{xe} \mapsto \text{FW}\,, \quad \text{or} \mapsto \text{MN}\,, \quad \text{bx} \mapsto \text{RY}\,.$$

Moreover, the plaintext was cut into groups of predetermined length, so for example the message

anxobergruppenfuehrerxvonxdemxbachxkiewxbittexdreixtausendxschuss
xpatronenxschickenstopx

—with /x/ for the space, which was not suppressed (Sect. 2.4.2)—was cut into groups of, say, 17 letters and each group encrypted in the following way:

a	n	x	o	b	e	r	g	r	u	p	p	e	n	f	u	e
h	r	e	r	x	v	o	n	x	d	e	m	x	b	a	c	h
A	K	F	M	R	Z	C	M	M	N	T	R	N	I	Z	O	W
Y	P	W	N	Y	S	W	E	Y	V	E	G	H	P	A	C	H

This procedure was applied once again: ay ↦ XY, kp ↦ YC, fw ↦ ZW, … so that the final cryptotext, read off in groups of five, amounts to

XYYCZ WRUPY VQGUT UTKID … .

The modified PLAYFAIR is, like the classical one, a little bit cumbersome and error-prone; this results in frequent queries and brings the danger of compromising encryption security (see Chapter 11). The violations on the German side helped the British as much as the Prussian predilection to be "both methodical and courteous" and to indulge in titles and other formalities.

4.2.3 Delastelle cipher. A tomographic method in the purest form ("while searching for a method of digraphic encipherment that did not require cumbersome 26×26 enciphering tables", Kahn) was published in 1901 by Félix Marie Delastelle (1840–1902), author of the *Traité Élementaire de Cryptographie* (Gauthier-Villars, Paris 1902): This was a one-to-one bipartite simple (i.e., monographic) substitution (very much like a Polybios square), then a transposition over four places, finally the same bipartite simple substitution in reverse, e.g.,

	1	2	3	4	5
1	B	O	R	D	E
2	A	U	X	C	F
3	G	H	I	J	K
4	L	M	N	P	Q
5	S	T	V	Y	Z

o n
12 ╳ 43
14 23
D X

or

o n
1 4 D
2 3 X

The encryption step is self-reciprocal and results in a bipartite digraphic substitution, similar to the one in Sect. 4.1.2. For the reverse translation another, conjugated bipartite simple substitution can be used; then the self-reciprocal character disappears.

A warning of a *complication illusoire* is appropriate: a mere gliding by one place, a *Kulissenverfahren* (Rohrbach 1948) does not give the wanted effect:

... a b s a l o m ...
3 2 1 1 1 5 1 2 1 4 1 1 2 4 2 3
H B E O D B C X

The cryptanalyst does not need to reconstruct the square: it suffices to interpret the encryption step as a $V \dashrightarrow V^2$ with homophones:

$$a \mapsto \left.\begin{matrix} O \\ U \\ H \\ M \\ T \end{matrix}\right\} \times \left\{\begin{matrix} B \\ O \\ R \\ D \\ E \end{matrix}\right. , \quad b \mapsto \left.\begin{matrix} B \\ A \\ G \\ L \\ S \end{matrix}\right\} \times \left\{\begin{matrix} B \\ O \\ R \\ D \\ E \end{matrix}\right. , \quad s \mapsto \left.\begin{matrix} E \\ F \\ K \\ Q \\ Z \end{matrix}\right\} \times \left\{\begin{matrix} B \\ O \\ R \\ D \\ E \end{matrix}\right. , \quad \ldots$$

under the side condition of overlaying:

$$\text{a b s a l o m} \mapsto \text{HB}\smile\text{BE}\smile\text{EO}\smile\text{OD}\smile\text{DB}\smile\text{BC}\smile\text{CX} .$$

This opens an unexpected line of attack.

An earlier example of a numeral tomographic method is found in a 1876 publication by the Danish engineer Alexis Koehl (Sect. 9.4.6). It is related to a method Pliny Earle Chase invented in 1859 (Sect. 9.5.4).

4.3 Case $V^3 \dashrightarrow W^{(m)}$ (Trigraphic Substitutions)

4.3.1 Gioppi. Trigraphic substitution in full generality soon leads into technical difficulties. Paper is not three-dimensional, unfortunately, so the listing of trigrams is more cumbersome, and $26^3 = 17\,576$ trigrams is a respectable number—a booklet of 26 pages may be needed. Trigraphic substitutions are hard to mechanize with simple means—although the use even of small hand-held computers can help a lot. Special substitutions by trigrams à la PLAYFAIR have not been very successful; the count Luigi Gioppi di Türkheim in Milano published such a system in 1897. William Friedman dealt around 1920 with trigram substitutions as well, and that makes them somewhat interesting. In the very special case of linear substition (Chapter 5) have trigraphic substitutions found application by Jack Levine (1958, 1963).

4.3.2 Henkels. A cipher machine,which mechanically performs a quadrupartite tetragraphic substitution, was patented in 1922 for a certain Henkels.

4.4 The General Case $V^{(n)} \dashrightarrow W^{(m)}$: Codes

Instead of being able to encrypt $26^3 = 17\,576$ trigrams, it may be better to be able to encrypt several hundred, thousand, or tens of thousands of frequently occurring multigrams of different length; this means that the encryption step operates on a subset C of $V^{(n)}$ (with a rather big n); with the proviso that every plaintext $x \in V^*$ can be decomposed into elements of C :

$$x = x_1 \star x_2 \star x_3 \star \ldots \star x_k \quad \text{(for some } k \in \mathbb{N} \text{ and suitable } x_j \in C \subseteq V^{(n)}) \ .$$

This can be guaranteed by the following 'single letter condition': $C \supseteq V$.

Following Kahn, such an encryption is called a code, if the choice of C is determined linguistically: frequent diphthongs, syllables, prefixes, endings, words, phrases are listed in a code book together with their code groups.

The single letter condition guarantees that even queer, curious, strange words, including those from biology and chemistry, or names of places, rivers, mountains, and proper names can be encrypted. Of course, it does not mean that every word should be resolved into individual letters and every sentence should be split into words—on the contrary, the longer the code entry that is found, the better; the best resolution is one that needs a minimum of code book entries. In full generality, the optimum may not be determined uniquely, but this indeterminism does no harm.

To maintain coding discipline is difficult. In 1918, the Commanding Officer of the *American Expeditionary Force* (A.E.F.) in France had reasons to admonish the staff, that *boche*, to be spelled with five codegroups, should be replaced by *German*, which needs one codegroup; and that the eighteen codegroups needed for *almost before the crack of dawn* were better replaced by the two codegroups for *day break*. Use of codes requires high education, since good encoding is a question of intellect; bad encoding helps the unauthorized decryptor to break the code. Thus, codes should be disallowed if the right people are not available. In the First World War, a Lieutenant Jäger from the staff of the German 5th Army did a great service for the foe when he signed his well-meaning orders to maintain signal security regularly with his name, which, unfortunately, was missing in the codebook and had to be spelled letter by letter every time. He "was beloved by his adversaries because he kept them up with code changes," writes Kahn. In 1918, Jäger endangered both the superencipherment with the *Geheimklappe* and the new codebook, the *Schlüsselheft*.

Coding discipline on the American side in the First World War was even worse, according to the G.2 A.6 Chief, Major Frank Moorman, who felt responsible for it; this is explained by a "well-known American disregard for regulations—especially ones as persnickety as these" (Kahn).

In the Second World War things improved slightly. Cryptographic control officers were assigned to each headquarters. Still, there were the diplomats. The anti-hero is Roosevelt's diplomat Robert Murphy (1894–1978), who insisted, for prestige reasons, on always using a diplomatic code; the stereotyped beginnings "For Murphy" or "From Murphy" helped Rohrbach's group at the German *Auswärtiges Amt* to break the code. Fräulein Asta Friedrichs, who took part in this activity, said after the war, as she was detained in Marburg and saw him drive by one day: "*Ich wollte ihn anhalten und ihm die Hand schütteln, – so viel hatte er für uns getan.*" [I wanted to stop him and shake his hand—he'd done so much for us.]

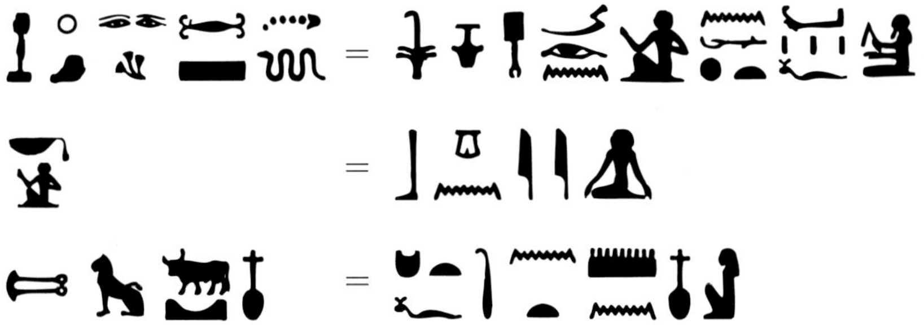

Fig. 34. Hieroglyphic inscriptions: unusual forms (left) and ordinary hieroglyphs (right)

4.4.1 Nomenclators. The oldest codings, seen from the Occident, are Chinese ideograms—although not perceived as such by the Chinese. Indeed the lack of cryptologic achievements in the ancient high cultures of China has been explained by the fact that written messages were anyhow understandable only for a few. The Egyptian hieroglyphs, however, were based—2000 years B.C.—on the principle of the rebus and on acrophony. The graphic of a 'rer' (pig) supplies the character for the letter /r/, the graphic of a 'wr', meaning a swallow as well as big, supplies the character for the letter /wr/; special marks (determinatives) clarify, if necessary, the difference. Hieroglyph writing is to a large extent coded writing—if necessary, a word can even be decomposed into characters for single consonants. But the secrecy aspect is missing. However, it is also missing if diplomats little by little get to know a code by heart and are able to give an impromptu speech in this code, as the American consul in Shanghai did in GRAY at his retirement dinner speech in the early 1920s (Sect. 4.4.7).

Where in Egypt inscriptions (Fig. 34) with unusual graphemes are found side by side with plaintext, secrecy may not be intended primarily; the pompous epigraph on a tombstone should impress and conjure arcane magical powers.

In the Occident, around 1380 the first mixtures of monographic substitutions and polygraphic elements are found, initially for some very frequent words only, among them *et* (see Fig. 26), *con, non, che* (Sect. 3.4). As soon as these collections became somewhat more voluminous, they were called nomenclators. An early example is shown in Fig. 35.

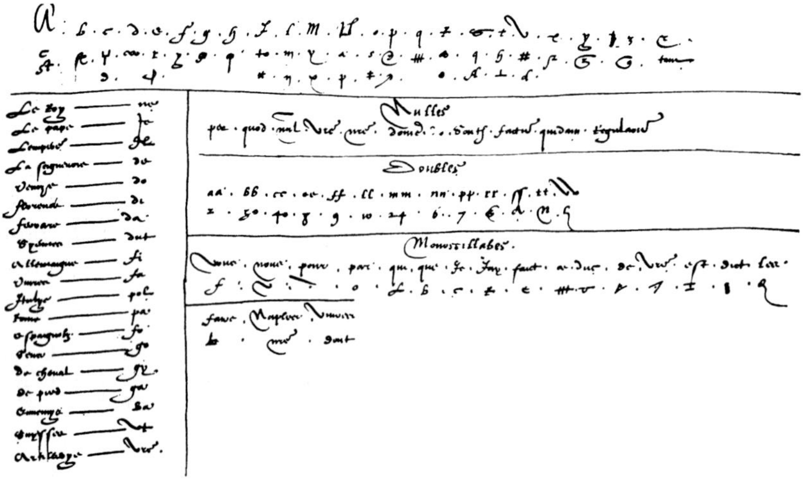

Fig. 35. An early nomenclator from Florence, 1554

Nomenclators kept their great importance during the whole Renaissance time. Charles I used a nomenclator with homophones, reconstructed in 1860 by Wheatstone: /a/ was represented by 12 ... 17 , /b/ by 18 ... 19 , and so on, /france/ by 9476 . By 1600, some nomenclators ran to several hundred entries, and numeral ones had three-digit code groups.

Nomenclators were also solved. Philip II of Spain gave his envoy Juan de Moreo a nomenclator containing about 400 codegroups. Viète worked from October 28, 1589 till March 15, 1590 to break the riddle, then he gave the complete solution to his king Henri IV. King Philip, who found out that a cipher he had thought unbreakable was compromised, complained to the pope that Henri had used black magic. The pope was better informed: his own cryptologist Giovanni Battista Argenti had also served him well, and Philip was held up to ridicule.

In the game of political intrigues, codes played a continuing role, from the sentence of Mary Stuart in 1587 to the conviction of the French anarchists who had been accused in the trial of Saint-Etienne in 1892 on the basis of secret messages Bazeries had solved.

In the 17th century, not only the Italian principalities, but every one of the great European courts had their Black Chamber, *Cabinet Noir*, *Geheimkabinett*. The statesmen had important cryptologists as aides and confidants: Louis XIV had Antoine Rossignol, the Czarina had Christian Goldbach, Charles II had John Wallis, and Maria Theresia had Baron Ignaz de Koch. These people were paid well and knew their importance. Kahn writes:

> "Though Wallis entreated Nottingham not to publicize his solutions for fear France would again change her ciphers, as she had done nine or ten times before (probably under the expert Rossignol tutelage), word of his prowess somehow spread. The King of Prussia gave him a gold chain for solving a cryptogram, and the Elector of Brandenburg a medal for reading 200 or 300 sheets of cipher. The Elector of Hanover, not wanting to depend on a foreign cryptanalyst, got Wallis' fellow intellectual, Baron Gottfried von Leibnitz, to importune him with lucrative offers to instruct several young men in the art. When Wallis put off Leibnitz' query as to how he did these amazing things by saying that there was no fixed method, Leibnitz quickly acknowledged it and, hinting that Wallis and the art might die together, pressed his request that he instruct some younger people in it. Wallis finally had to say bluntly that he would be glad to serve the elector if need be, but he could not send his skill abroad without the king's leave."

C. Goldbach (1690–1764), from 1742 privy councillor in the Russian Foreign Office, deciphered a letter of the French ambassador with unpleasant remarks about the reigning czarina Jelisaweta Petrowna, the daughter of Peter the Great. Sometimes a fiasco happens, like the one de Koch experienced, when a letter to the Duke of Modena was sealed by mistake with the signet of the Duke of Parma. Nevertheless, the Austrian emperors' *Kaiserliche Geheime Kabinetts-Kanzlei* continued into the 19th century and was able to read among others Napoleon's and Talleyrand's correspondence.

In the New World, too, cryptology assumed importance and fame. George Washington had the help of two agents, Sam Woodhull and Robert Townsend, with the cover names CULPER SR. and CULPER JR. In 1779 they used a nomenclator with about 800 entries, compiled by Major Benjamin Tallmadge. Jefferson, too, concocted a nomenclator for James Madison in 1785 (Fig. 36).

4.4.2 Two-part books. Around 1630, Antoine Rossignol (1600–1682) made a big step forward. Apart from the earliest examples, all nomenclators were up to then order-preserving mappings of the lexicographically ordered plaintext elements onto lexicographically ordered literal codegroups or mathematically ordered numeral codegroups. This allowed the user to get away in practice with only one codebook which could be inspected for plaintext as well as for cryptotext in the order the entries were printed.

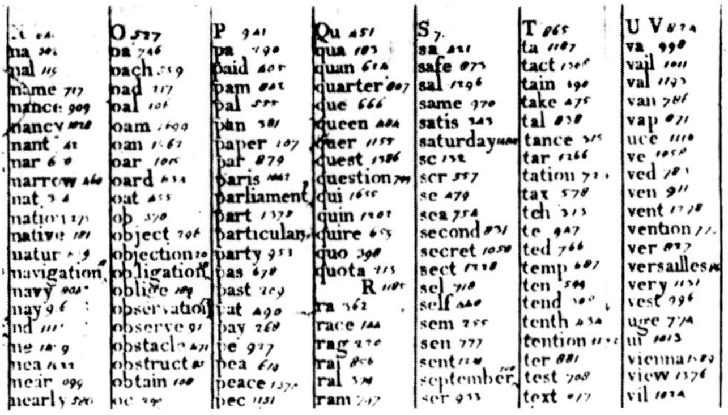

Fig. 36. Nomenclator, 1785, made by Jefferson for use with Madison and Monroe

This system, however, had a great disadvantage: as soon as the plaintext equivalent for one codegroup was known, all codegroups standing lower could only have plaintext equivalents standing lower as well. With a few established codegroup-plaintext relations, a good deal of the fine-structure of the code can be obtained.

Astonishingly, even during the First World War this outdated system of a one-part nomenclator, one-part code (French *dictionnaire à table unique*, German *einteiliges Satzbuch*) was still in practical use with the GREEN code of the U.S. State Department, which had up to $20^3 \cdot 6^2 = 288\,000$ codegroups of type $\mathcal{CVCVC}$ ($\mathcal{C}$ stands for a consonant, $\mathcal{V}$ for a vowel): FYTIG, MIHAK, PEDEK.

A *Signalbuch* of the German *Kaiserliche Marine* that was captured from the cruiser *Magdeburg* in August 1914, was a one-part code, showing for example

63940	OAT	Ohnmacht, -ig
63941	OAU	Ohr, Ohren-
63942	OAÜ	Okkupation, Okkupations, -ieren
63943	OAV	Ökonomie, -isch
63944	OAW	Oktant

Rossignol introduced thoroughly mixed codes. This, however, means that for decryption a second, lexicographically or mathematically ordered listing of the codegroups is needed, a two-part nomenclator, two-part code (French *dictionnaire à table double, dictionnaire à deux tables*, German *zweiteiliges Satzbuch*). Jefferson's nomenclator provides an example. A modern example with homophones from the military genre is

⋮	⋮	⋮	⋮
flap	XYMAS	RATPA	ship
	TIBAL	RATPE	quite
flapjack	UPTON	RATPI	enough
flapper	UPABS	RATPO	happy
flare	OHPAP	RATPU	loxodromic
⋮	⋮	⋮	⋮

where for a frequent plaintext word like /army/ five homophones are available: TORMA, RAFEM, LABAR, ROMUF, IBEXO .

Some codebooks provided digit groups (numeral codes) as well as letter groups (literal codes). An example of a German *Satzbuch* from the year 1944 could have read:

a	0809	XCL	b	1479	MLA
Abend	8435	PUV	Bad	1918	TID
aber	7463	NAS	bald	1492	LGD
acht	6397	DXL	⋮	⋮	⋮
Achtung	1735	APS	z	2467	VBH
an	7958	EVG	⋮	⋮	⋮
auf	6734	UNO	zyklotron	5116	JLD
⋮	⋮	⋮			

The *Kurzsignalheft* (short signal book) of the *Kriegsmarine* (since summer 1941) was a caption code containing codegroups for stereotyped commands:

AAAA	*Beabsichtige gemeldete Feindstreitkräfte anzugreifen*
AAEE	*Beabsichtige Durchführung Unternehmung wie vorgesehen*
AAFF	*Beabsichtige Durchführung Unternehmung mit vollem Einsatz*
AAGG	*Beabsichtige Durchführung Unternehmung unter Vermeidung vollen Einsatzes*

The *Wetterkurzschlüssel* (short weather cipher) of the *Kriegsmarine* coded air temperatures by a polyphonic single letter code (X was missing!):

A $\hat{=}$ +28° B $\hat{=}$ +27° C $\hat{=}$ +26° D $\hat{=}$ +25° ... W $\hat{=}$ +6° Y $\hat{=}$ +5° Z $\hat{=}$ +4°
A $\hat{=}$ +3° B $\hat{=}$ +2° C $\hat{=}$ +1° D $\hat{=}$ 0° E $\hat{=}$ −1° F $\hat{=}$ −2° ... Z $\hat{=}$ −21°

In a similar way, water temperature, atmospheric pressure, humidity, wind direction, wind velocity, visibility, degree of cloudiness, geographic latitude, and geographic longitude had to be coded in a prescribed order; a weather report consisted of a single short word. This seemed to be very economical and also made direction-finding reconnaissance difficult, but it was cryptologically utterly stupid: the superencrypted weather reports the U-boats were ordered to broadcast regularly were for the enemy's cryptanalysis of the superencipherment almost as good as plaintext.

4.4.3 Modern codes. Around 1700, the nomenclators had 2000–3000 entries, and they kept growing, although the two-part books needed more space. Modern codes with homophones and polyphones are found in Figs. 37, 38.

The Black Chambers were dissolved in Europe in the mid-19th century (1844 in England, 1848 in Vienna and Paris); this ended the surreptitious, clandestine opening of diplomatic and other mail. The Age of Enlightenment had its victory. The industrial revolution brought about the telegraph and as a consequence commercial codebooks with the main use of condensing telegrams and thus lowering transmission time.

In 1845, Francis O. J. Smith published a code—*The Secret Corresponding Vocabulary. Adapted for Use to Morse's Electro-Mechanic Telegraph*—even before the Morse alphabet was introduced. Smith had 50 000 codegroups, and only 67 sentences. His codegroups were built up from digits (numeral code)

Shershel

– 51648 C...**Shershel**
– 07510 B...**Shetland** Islands
– 18855 B....Shetland Mainland
– 43026 C...**Shetlands**
– 53038 A...**Shiant** Islands
– 04216 C...**Shield**—*for*
– 35998 C...**Shielday**
– 43144 B...**Shielded**
– 35732 B....Shielded by
– 10726 B....Shielded from
– 53124 C...**Shielding**
– 06656 B...**Shields**—*for*—*of*
– 17848 B....Shields, North
– 41802 A....Shields, South
– 28814 C...**Shift**-s

A 10569 B, B 53472 C, C 03917 A } Ship is
– 35613 A....Ship is **not**
– 50968 C....Ship is **not** to
– 06679 A....Ship is **not** to be
– 18641 C....Ship is now—*at*
– 42583 C....Ship is to
– 10247 A....Ship is to be
– 53180 C....Ship must
– 07006 A....Ship must be
A 51738 B, B 41759 C, C 10994 C } Ship of

– **07700** B...Spontaneous-ly
– 07701 B...Sow-s-ing
– 07703 B...Rodd
– 07704 C...Vacate-s
– 07705 B...To what
– 07707 A...What time—*is*—*are*
A 07708 C...Hornet, H.M.S.
B 07708 A...Referring
C 07708 B...Wednesday
– 07709 A...Send-s mails for
– **07710** C...Worth
– 07712 B...Riddled by (with)
A 07713 A...Smoke-s—*from*—*of*
B 07713 B...Will be
C 07713 C...13th April
– 07714 A...Tsu Sima

– **07750** A...Dummy group
– 07751 A...Recurrences—*of*
– 07752 B...Report when she
– 07754 A...Rush-es-ing
– 07755 C...Purpose of
– 07756 C...Withdrawn from
– 07758 B...Sheep
A 07759 C...12th April
B 07759 A...Was no-t
C 07759 B...In convoy
– **07760** C...She could
– 07761 A...That every
– 07763 A...Sulen Isles
A 07764 C...Begins
B 07764 B...Spell word of **13 letters**
C 07764 A...Acknowledge

Fig. 37. SA Cipher of the British Admiralty (1918).
One page of the homophonic encoding part and one page of the polyphonic decoding part

and were intended to allow (Sect. 9.2) superencryption. Later, a transition to codes with codegroups built from letters (literal code), mainly groups of five, took place; the number of sentences went into the hundreds, the number of code groups up to 100 000. Because of the great volume, one-part codes were used again, and this even in diplomatic services and in military staffs, although secrecy was vital there. However, this did give elaborate genuine ciphers more and more cryptological importance.

Little by little, hundreds of commercial codes came into existence; among the earliest were one by Henry Rogers and one by John Wills (both in 1847). According to Friedman, in 1860, "a man named Buell published in Buffalo his *Mercantile Cipher for Condensing Telegrams*". In 1874, eight years after completion of the transatlantic cable, the widely used ABC Code by William Clausen-Thue, a five-letter code, appeared. Other five-letter codes were compiled by Bolton (*Dictionnaire pour la Correspondance anglais*), by Krohn in Berlin (1873), and by Walter in Winterthur (1877). A four-letter code (*Chiffrier-Wörterbuch*) was published by Katscher in Leipzig (1889) and a three-letter code (*Dictionnaire télégraphique, économique et secret*) by Mamert-Gallian in Paris (1874). In the U.S.A., famous codes are named after John Charles Hartfield (1877, continued since 1890 by his son John William Hartfield) and Henry Harvey (1878). The codebook by Benjamin Franklin Lieber, with 75 800 codegroups, was also translated into French and German. Even seven-letter codes found use, such as the *Ingenieur-Code* (in

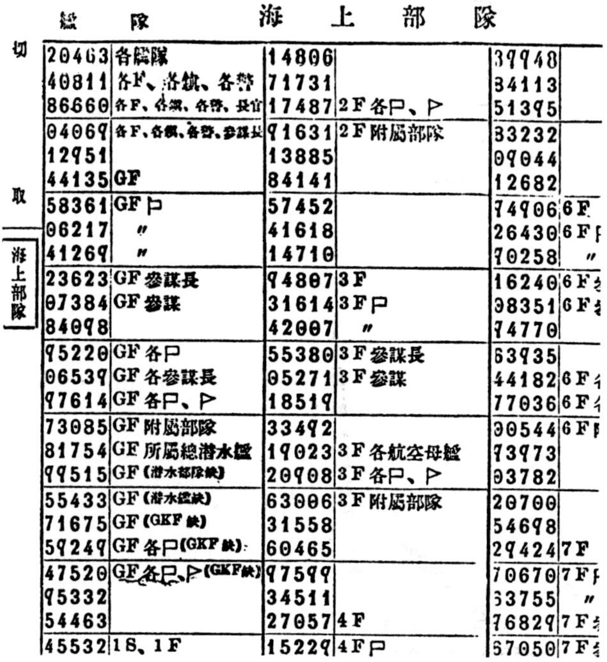

総 隊 海 上 部 隊

切 取 海上部隊

20463	各艦隊	14806		39948	
40811	各F、各戦、各警	71731		34113	
86660	各F、各戦、各警、長官	17487	2F 各P、▷	51395	
04069	各F、各戦、各警、参謀長	91631	2F 附属部隊	33232	
12951		13885		09044	
44135	GF	84141		12682	
58361	GF P	57452		74906	6F
06217	〃	41618		26430	6F
41269	〃	14710		90258	〃
23623	GF 参謀長	94807	3F	16240	6F
07384	GF 参謀	31614	3F P	98351	6F
84098		42007	〃	94770	
95220	GF 各P	55380	3F 参謀長	63935	
06539	GF 各参謀長	05271	3F 参謀	44182	6F
97614	GF 各P、▷	18519		77036	6F
73085	GF 附属部隊	33492		90544	6F
81754	GF 所属総潜水艦	19023	3F 各航空母艦	93973	
99515	GF (潜水部隊缺)	20908	3F 各P、▷	93782	
55433	GF (潜水艦缺)	63006	3F 附属部隊	20700	
71675	GF (GKF 缺)	31558		54698	
59249	GF 各P (GKF 缺)	60465		29424	7F
47520	GF 各P、▷ (GKF 缺)	97599		70670	7F
95332		34511		63755	〃
54463		27057	4F	76829	7F
45532	1S、1F	15229	4F P	67050	7F

Fig. 38. Sample from the encoding part of a Japanese Navy code (1943)

German) by Galland. The aim was mainly to reduce the cost of telegraphic transmission (*Mercantile Cypher for Condensing Telegrams*, by Buell 1860). This was particularly important for transatlantic traffic.

In Europe, numeral codes were preferred which allowed a simple additive superencryption. An epoch-making prototype of a four-digit code was the *Dictionnaire abréviatif chiffré* by F. J. Sittler in Paris (1868), besides the *Dictionnaire pour la Correspondance télégraphique secrète* by Brunswick in Paris (1868) and the *Dictionnaire chiffré* by Nilac. Bazeries (1893) as well as de Viaris produced codes; other four-digit codes were the *Dizionario per corrispondenze in cifra* by Baravelli in Torino (1896), *Chiffrier-Wörterbuch* by Friedmann in Berlin, and *Chiffrierbuch* by Steiner & Stern in Vienna (1892).

4.4.4 Telegraph codes. The tariff policy of the International Telegraph Union (Sect. 2.5.2.2) led in 1890 to the widespread use of five-digit codes. Brachet in Paris published such a code in 1850 (*Dictionnaire chiffré*), others are *Diccionario para la correspondencia secreta* by Vaz Subtil in Lisbon (1871), *Wörterbuch* by Niethe in Berlin (1877), and *Dictionnaire pour la Correspondance secrète* by N. C. Louis in Paris (1881). Among the later ones were the *Dictionnaire chiffré Diplomatique et Commercial* by Airenti and the *Telescand Code* in France, *Diccionario Cryptographico* in Lisbon (1892), *Nuovo Cifrario* by Mengarini in Rome (1898), *Cifrario per la corrispendenza* *segreta* by Cicero in Rome (1899), *Slater's Code* by Slater in London (1906), and *Clave telegrafica* by Darhan in Madrid (1912).

4.4.5 Commercial codes. In the 20th century, many code books offered numeral and literal codegroups as well. Until recently frequently used codes include *Bentley's Code* (since 1922), *ABC Code 6th edition* (since 1925), *Peterson's Code 3rd edition* by Ernest F. Peterson, *Acme Code* by William J. Mitchel, *Rudolf Mosse Code* (since 1922), *Lombard Code*, and *AZ Code*. The largest codebook ever in general use was compiled by Cyrus Tibbals for the *Western Union Code*; it contained 379 300 entries, while the *ABC Code* had only 103 000 .

In the Second World War, the Allies used the BAMS code ("Broadcasting for Allied Merchant Ships"), which was widely compromised, as a basis of superencryption. Plaintext would have been no worse.

For long years, an *Internationaler Hotel-Telegraphenschlüssel für Zimmerbestellung* was reproduced in German calendar notebooks, with codegroups

ALBA	for *"1 Zimmer mit 1 Bett"*,	ARAB	for *"1 Zimmer mit 2 Betten"*,
ABEC	for *"1 Zimmer mit 3 Betten"*,	BELAB	for *"2 Zimmer mit je 1 Bett"*,
BIRAC	for *"2 Zimmer mit 3 Betten"*,	BANAD	for *"2 Zimmer mit 4 Betten"*,
CIROC	for *"3 Zimmer mit 3 Betten"*,	CARID	for *"3 Zimmer mit 4 Betten"*,
CALDE	for *"3 Zimmer mit 5 Betten"*		and so on.

Some codes have been translated into foreign languages. The Marconi Code, by James C. H. Macbeth, is truly multilingual (nine languages in four volumes), making true a dream of Athanasius Kircher (Fig. 39).

4.4.6 Error-detecting and -correcting codes. In 1880, J. C. Hartfield introduced for checking purposes the 'two-character differential' of the codegroups (for a 27-character alphabet Z_{27}, from $27^5 = 14\,348\,907$ codegroups of a five-character code, there remain $27^4 = 531\,441$ ones). Around 1925, W. J. Mitchel introduced also a check against transposition of adjacent characters (leading to reverses like in LABED and ALBED), which reduced the number of usable codegroups, in the example to 440 051 . Mitchel's idea of the 'adjacent-letter restriction' spread rapidly. Both the one-part code of the Japanese Navy, dubbed by OP-20-G JN-25A (1. 6. 1939, broken Sept. 1940) and the two-part code JN-25B (1. 12. 1940, broken March 1942) used five-digit codegroups, divisible by 3 . These codes were forerunners of the error-detecting and error-correcting codes introduced by Richard W. Hamming in 1950; today this checking principle is everywhere present in the bar codes of the European Article Number (weights alternatingly 1 and 3, divisible by 10) or in the ISBN system (weights $10, 9, 8, \ldots, 2, 1$ in this order, divisible by 11).

4.4.7 Shortlived codes. In contrast to commercial codebooks, which are (for not too low a price) generally accessible and therefore should have a lifetime as long as possible, diplomatic and military codes should be subjected to "planned obsolescence" (Sect. 2.1.1), and correspondingly should be changed as often as possible. It therefore seems hopeless to list the ones used in this

M. N. O. P. R. S. T. U. Y. Z.
0 1 2 3 4 5 6 7 8 9

19140	UVVIM	slackness.	relâchement.	flojedad, descuido.
19141	UVVON	Slag(s).	Scorie(s).	Escoria(s).
19142	UVWEO	Slander(s).	Diffame(r), diffamation.	Calumnia(r), calumnia(s).
19143	UVWUP	slandered.	diffamé.	calumniado.
19144	UVWYR	slandering.	diffamant.	calumniando.
19145	UVYBS	slanderous.	diffamatoire.	calumnioso.
19146	UVYCT	Slate(s).	Ardoise(s).	Pizarra(s).
19147	UVYDU	Sleeper(s).	Traverse(s) (chemins de fer).	Traviesa(s), durmiente(s) (f.c.).
19148	UVYFY	Sleeve-valve.	Soupape à manchon.	Válvula de manguito.
19149	UVYMZ	Slide(s).	Glisse(r), glissière(s). tiroir de distribution.	Resbala(r), corredera(s). válvula de distribución, de corredera.
19150	UVYUM	slide-valve.	glissant, à coulisse.	resbalando, resbalamiento, deslizamiento.
19151	UVYVN	sliding.		
19152	UVYWO	sliding scale.	échelle mobile.	escala móvil.
19153	UVYZP	Slight.	Léger, peu important.	Ligero, leve.
19154	UVZUR	slightest.	le (la) moindre.	lo más ligero, leve.
19155	UVZYS	not the slightest.	pas le (la) moindre.	no lo más ligero, mínimo.

M. N. O. P. R. S. T. U. Y. Z.
0 1 2 3 4 5 6 7 8 9

19140	UVVIM	slackness.	Schlaffheit, Geschäftsstille.	slapheid, stilte.
19141	UVVON	Slag(s).	Schlacke(n).	Schuim, slak(ken), metaalschuim(-slakken).
19142	UVWEO	Slander(s).	Verleumden(-e,-t), Verleumdung(en).	Belasteren, belaster(t), laster.
19143	UVWUP	slandered.	verleumdet.	belasterd.
19144	UVWYR	slandering.	verleumdend.	belasterend.
19145	UVYBS	slanderous.	verleumderisch.	lasterlijk.
19146	UVYCT	Slate(s).	Schiefer, Schiefertafel(n).	Lei(en).
19147	UVYDU	Sleeper(s).	(Bahn-)Schwelle(n).	Dwarsligger(s).
19148	UVYFY	Sleeve-valve.	Muffenventil.	Mofklep.
19149	UVYMZ	Slide(s).	Gleiten(-e, -et), Gleitbahn(en), Gleitführung(en).	Glijden(-t); schuif (schuiven), leibaan (leibanen), windklep(pen).
19150	UVYUM	slide-valve.	Schieberventil, Ventilschieber.	schuif, stoomschuif, schuifklep.
19151	UVYVN	sliding.	gleitend.	verschuifbaar, glijdend.
19152	UVYWO	sliding scale.	Gleitskala.	kalibermaat, proportioneele schaal.
19153	UVYZP	Slight.	Gering, von geringer Wichtigkeit.	Gering, onbeduidend.
19154	UVZUR	slightest.	geringst.	geringste.
19155	UVZYS	not the slightest.	nicht das (der, die) geringste.	niet de (het) geringste.

Fig. 39. The Marconi Code: Corresponding pages from the English-French-Spanish and the English-German-Dutch edition

century, although often parsimony and laziness have prevented sufficiently frequent change. U.S.-American diplomatic codes, which were far too long in use, are RED and BLUE (before 1914), both five-digit codes, and GREEN (from about 1914 until about 1919), likewise with five digits. Around 1920 GRAY, mentioned in Sect. 4.4.1, arrived, and was still in use under Franklin Delano Roosevelt in 1941. Roosevelt sent Dec. 6, 1941, a note to *Cordell Hull*: *"Dear Cordell—Shoot this to Grew* [the ambassador in Tokio]*—I think can go in gray code—saves time—I don't mind if it gets picked up—FDR"*. Under the security-conscious Roosevelt, in the mid-1930s, the two-part code BROWN was introduced; after 1939 the distrustful statesman nevertheless preferred encryption systems from the Navy Department "for matters of utmost secrecy," as he put it. The further diplomatic codes A-1, B-1, C-1, D-1 did not change Roosevelt's harsh opinion about the security risk of the State Department codes. The BLACK code appeared about 1940.

Sometimes the code producers ran into surprises: When the *American Expeditionary Force* (A.E.F.) in 1917 got involved in France in the First World War, it turned out that the *War Department Telegraph Code*, issued 1915, was both unsafe and inadequate for tactical use. In great haste, the *Code Compilation Subsection* of MI-8 in July 1917 started work on a suitable code; it was ready after one year, on July 1, 1918. The *Military Intelligence Code No. 5* was a one-part code, though with two-character difference of the codegroups of type $\mathcal{VCVCV}$, $\mathcal{VCCVC}$ or $\mathcal{CVCCV}$. Although rather soon a better, two-part code (*Military Intelligence Code No. 9*) was available, *No. 5* stayed valid until September 1, 1934, with the classification "SECRET", then under the short name SIGCOT it was demoted to "CONFIDENTIAL". Likewise, *No. 9*, which was taken out of use around 1923, was reactivated April 1, 1933, demoted to "CONFIDENTIAL", with the short name SIGSYG for the encryption and SIGPIK for the decryption part. Lack of money was responsible for this, but not even a superencipherment was prescribed, which would have needed no investment.

4.4.8 Trench codes. On the lower military level, the combat level, codes had sometimes a better, sometimes a worse reputation. The *kaiserliche Heer* changed in 1917 from the hitherto used turning grilles (Sect. 6.1.4) to codes. For radio traffic in the 3-km combat zone, in March 1917 a simple digraphic substitution, a *Befehlstafel* was introduced. Already in 1916 the French issued a three-letter code, the *carnet réduit*, with names like *olive* und *urbain*. The codegroups were ordered with headings like infantry, artillery, numbers, clock times, common words, place names, cover names and so on, so it was a caption code.

In March 1918, the *kaiserliche Heer* made the transition step—foreseen by the Allies—to a superencrypted, but still one-part code, a three-digit code. The superencryption was extended to the first two digits only and was done with the *Geheimklappe* (Sect. 4.1.2), which was changed frequently. The fixed third digit allowed the occurrence of patterns and thus helped cryptanalysis.

For higher cryptanalytic security requirements, outside the 3-km combat zone, the Germans introduced in June 1917 a two-part three-letter code (*Satzbuch*). No superencryption was intended; the cryptanalytic security of this cryptosystem was from the beginning based on planned obsolescence (something like 14 days). The code contained a great number of homophones (KXL, ROQ, UDZ for *Anschluß fehlt*) and nulls. It was called KRU code by the Allies, because all codegroups started with one of the letters K, R, U; or also *Fritz* code. Later, codegroups beginning with S were added and furthermore some beginning with A (KRUSA code); finally the 26 letters of the alphabet were supplemented by the mutated vowels Ä, Ö, Ü (this code was—not very systematically—called KRUSÄ code).

The armistice in November 1918 ended this unpeaceful epoch of the 'trench codes'. But trench codes were not forgotten. In a manual of the U.S. War Department (1944) can be found: "Cipher machines cannot, as a rule, be carried forward of the larger headquarters, such as Division. Hence, code methods may predominate in the lower echelons and troop formations."

This can be interpreted as the view that—in military radio traffic—codes without superencryption can only be tolerated at the lowest level of security. For this reason, Friedman introduced polyalphabetic enciphering with a handy device, the M-94, as a field cipher in the U.S. Army. During the Second

Fig. 40. Hagelin M-209 in combat situation

World War, even this was no longer considered sufficient in the U.S.A. Boris Hagelin—who had travelled in May 1940 at the last minute from Sweden across Germany to Genoa and then boarded the *Conte di Savoia* to the United States, with two of his C-36 (Sect. 8.5.2) in his luggage—impressed Friedman and the U.S. Signal Corps with his improved version C-38. Hagelin had to wait for a full year while his machine was thoroughly tested. In June 1941, the decision was made for a mechanical machine at the low-echelon level. Figure 40 shows a soldier, rifle slung on back, enciphering with a Hagelin M-209 cipher machine at the message center of the command post of the 3rd Division, U.S. Infantry in Hyopchong, Korea, on October 1, 1951. Indeed, Boris Hagelin had quite early considered the use of mechanical cipher machines in the front line. For the Hagelin C-35, the base-plate was formed in such a way that the machine, when used at the front line, could be strapped onto the knee of the operator. As Hagelin writes, the operator could walk, if necessary, with the machine fixed to his knee—whether he liked it, is left open. For the French constabulary, Hagelin even designed in the 1950s a pocket cipher machine of roughly the same power (CD-55, CD-57).

In the meantime, the struggle between codebooks and cipher machines has become obsolete—microelectronics collapses the differences and opens completely new avenues. One of these was prepared by the special case of linear substitutions, to be discussed in the next chapter.

5 Encryption Steps: Linear Substitution

Although Hill's cipher system itself
saw almost no practical use, it had
a great impact upon cryptology.
David Kahn 1967

A linear (geometrically 'affine') substitution is a special polygraphic substitution. The injective encryption step of a polygraphic block encryption

$$\chi : V^n \dashrightarrow W^m$$

with relatively large n and m is restricted in a particular way.

The finite character sets V and W are now interpreted *essentially* as being linearly ordered, with a first character $\alpha(V)$ and a last character $\omega(V)$. The ordered character set is called a standard alphabet in the proper sense.

In this order there is for each character x except the last character exactly one *next* character succ x; for the last character the next character is defined as the first one, succ $\omega(V) = \alpha(V)$. Thus, the mapping succ defines uniquely an inverse mapping pred; the cyclically closed standard alphabet is a finite non-branching (i.e., linear) cyclic quasiordering.

In $V = Z_{|V|}$ and $W = Z_{|W|}$ an addition can be defined recursively:
For $a,\ b \in V$ or $a,\ b \in W$ holds

$$a + b \;=\; \text{succ } a + \text{ pred } b \;,$$
$$a + \alpha \;=\; a \;.$$

This means that the sets $Z_{|V|}$ and $Z_{|W|}$ are mapped uniquely and order-preservingly on $\mathbb{Z}_{|V|}$ and $\mathbb{Z}_{|W|}$, where $\mathbb{Z}_N$ denotes the group of residual classes *modulo* the natural number N of the group of integers $\mathbb{Z}$, the elements of which are represented by the cycle of natural numbers $(0, 1, \dots N-1)$. Addition in V and W corresponds to addition of the residual classes. Commonly, the alphabet $\{\alpha , \dots , \omega\}$ is identified with the cycle numbers ('cyclotomic numbers') $\{0 , \dots , N-1\}$, where $N = |V|$ or $N = |W|$.[1]

Addition in V or W is now carried over to V^n and W^m componentwise. Moreover we identify $V = W = \mathbb{Z}_N$, $V^n = \mathbb{Z}_N^n$, $W^m = \mathbb{Z}_N^m$.

---

[1] For $Z_{26} \leftrightarrow \mathbb{Z}_{26}$, the identification is as follows ('algebraic alphabet'):

a	b	c	d	e	f	g	h	i	j	k	l	m	n	o	p	q	r	s	t	u	v	w	x	y	z
0	1	2	3	4	5	6	7	8	9	10	11	12	13	14	15	16	17	18	19	20	21	22	23	24	25

With these definitions, a mapping $\varphi : \mathbb{Z}_N^n \dashrightarrow \mathbb{Z}_N^m$ is said to be additive if and only if

$$\forall\, x,\ y \in \mathbb{Z}_N^n : \ \varphi(x+y) \ = \ \varphi(x) + \varphi(y)\ ,$$

in words: "The image of the sum is the sum of the images." Consequently

$$\forall\, x \in \mathbb{Z}_N^n \ : \ \varphi(x+x+\ldots+x) \ = \ \varphi(x)+\varphi(x)+\ldots+\varphi(x)\ ,$$

in words: "The image of a multiple is the multiple of the images."

Indeed, $\mathbb{Z}_N$ is a ring and $\mathbb{Z}_N^n$ a vector space with the origin $\mathbf{o} = (0\,0\ldots 0)$; φ is a linear mapping of the vector space $\mathbb{Z}_N^n$ in the vector space $\mathbb{Z}_N^m$. If $N = p$ is prime, and only then, $\mathbb{Z}_N$ is even a field, the Galois field $\mathbb{F}(p)$. However, in the sequel we shall not require the primality of N.

Notationally, we use a square matrix T over $\mathbb{Z}_N$ for the representation of $\varphi : \varphi(x) = xT$ with the inverse $\varphi^{-1}(y) = yT^{-1}$.

A linear substitution $\chi : \mathbb{Z}_N^n \longrightarrow \mathbb{Z}_N^m$ is defined as the sum of a homogeneous part, a linear mapping φ represented by a matrix $T \in \mathbb{Z}_N^{n,m}$ and a translation of the origins, represented by a vector $t \in \mathbb{Z}_N^m$:

$$\chi(x) = xT + t\ .$$

If T is the identity, there is the special case $\chi(x) = x + t$ of a translation (polygraphic CAESAR addition).

If a linear mapping φ is injective, then it is regular, i.e., it has a unique inverse φ^{-1} on its image. If in the sequel we assume the endomorphic case with equal width $m = n$, then a regular linear mapping is a one-to-one mapping.

Example:

Given over $\mathbb{Z}_{26}$ a square matrix T and a vector t,

$$T \ = \ \begin{pmatrix} 15 & 2 & 7 \\ 8 & 10 & 23 \\ 0 & 2 & 8 \end{pmatrix} \qquad t \ = \ (\,17 \quad 4 \quad 20\,)$$

The 3×3 matrix T and the 3-component vector t define a tripartite trigraphic substitution. The image of the trigram /mai/ $\hat{=}$ $(\,12 \quad 0 \quad 8\,)$ is obtained by calculation *modulo* 26:

$$(\,12 \quad 0 \quad 8\,)\begin{pmatrix} 15 & 2 & 7 \\ 8 & 10 & 23 \\ 0 & 2 & 8 \end{pmatrix} + (\,17 \quad 4 \quad 20\,) \overset{26}{\simeq}$$

$$(\,24 \quad 14 \quad 18\,) + (\,17 \quad 4 \quad 20\,) \overset{26}{\simeq} (\,15 \quad 18 \quad 12\,) \ \hat{=} \ /\mathrm{psm}/\ .$$

But the trigram /ecg/ $\hat{=}$ $(\,4 \quad 2 \quad 6\,)$ has the same image /psm/ , since

$$(\,4 \quad 2 \quad 6\,)\begin{pmatrix} 15 & 2 & 7 \\ 8 & 10 & 23 \\ 0 & 2 & 8 \end{pmatrix} \overset{26}{\simeq} (\,24 \quad 14 \quad 18\,)\ .$$

Encryption by the given matrix T is therefore not injective, in fact T is not regular and does not have an inverse. The vector $(8\ \ 24\ \ 2)$ annihilates T.

5.1 Self-reciprocal Linear Substitutions

The question is obvious: When is an endomorphic linear substitution χ self-reciprocal? The condition is that $\chi(x) = xT + t = \chi^{-1}(x)$, thus

$$x = \chi(\chi(x)) = (xT + t)T + t = xT^2 + tT + t \text{ , from which}$$

$$T^2 = I \quad \text{and} \quad tT + t = \mathbf{o}$$

follows. This is to say that the matrix T of the homogenous part is self-reciprocal, whence only 1 or $N-1$ can be eigenvalues of T, and the translation vector t either is zero or eigenvector of T for the eigenvalue $N-1$.

In particular, χ can be a reflection with a reflecting plane having as its normal the vector v :

$$\chi(x) = x + (1 - \gamma(x))\, v \quad \text{with} \quad v \neq \mathbf{o}$$

where the linear functional γ fulfils the condition $\gamma(v) = 2$ (in the case $N = 2$ the condition $\gamma(v) = \mathbf{o}$). Then $\chi(v) = \mathbf{o}$, and $\chi(\mathbf{o}) = v$. A simple calculation confirms $\chi^2(x) = x$:

$$\begin{aligned}\chi(\chi(x)) &= \chi(x) + [1 - \gamma(\chi(x))]\, v \\ &= x + (1 - \gamma(x))v + [1 - \gamma(x) - (1 - \gamma(x))\gamma(v)]\, v \\ &= x + (1 - \gamma(x))v + [\ (1 - \gamma(x)) - 2(1 - \gamma(x))\ ]\, v \quad = x \ .\end{aligned}$$

Example: $\mathbb{Z}_2^2 \to \mathbb{Z}_2^2 \quad (N = 2\ , n = 2)$

$$\chi((x_1\ x_2)) = (x_1\ x_2) + (1{-}x_1{-}x_2)\ (1 \quad 1) = (1 - x_2\ 1 - x_1) =$$

$$(x_1\ x_2)\begin{pmatrix} 0 & -1 \\ -1 & 0 \end{pmatrix} + (1 \quad 1)\ .$$

$$\begin{aligned}\chi((0\,0)) &= (1\,1) \\ \chi((0\,1)) &= (0\,1) \\ \chi((1\,0)) &= (1\,0) \\ \chi((1\,1)) &= (0\,0)\end{aligned}$$

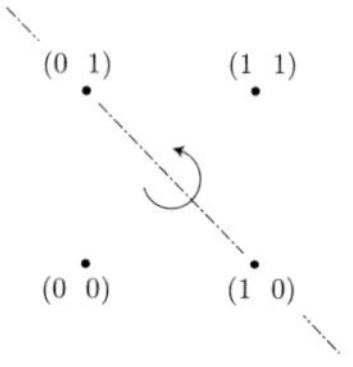

The plaintext 0 0 1 0 0 1 1 1 0 1 1 0 1 1 0 0 0 1 yields the cryptotext 1 1 1 0 0 1 0 0 0 1 1 0 0 0 1 1 0 1 and vice versa.

5.2 Homogeneous Linear Substitutions

5.2.1 Hill. The special case of homogeneous linear substitution, $t = \mathbf{o}$, was studied by Hill as a cryptographic instrument (HILL encryption step).

How do we obtain T together with its inverse T^{-1}? Over $\mathbb{Z}$, start with a (square) matrix of determinant $+1$ and its inverse, e.g., with $n = 4$:

$$T = \begin{pmatrix} 8 & 6 & 9 & 5 \\ 6 & 9 & 5 & 10 \\ 5 & 8 & 4 & 9 \\ 10 & 6 & 11 & 4 \end{pmatrix} \qquad T^{-1} = \begin{pmatrix} -3 & 20 & -21 & 1 \\ 2 & -41 & 44 & 1 \\ 2 & -6 & 6 & -1 \\ -1 & 28 & -30 & -1 \end{pmatrix}.$$

For numerical work it is recommended to use small negative numbers as representatives of the residual classes, thus over $\mathbb{Z}_{26}$ besides

$$T^{-1} \stackrel{26}{\simeq} \begin{pmatrix} 23 & 20 & 5 & 1 \\ 2 & 11 & 18 & 1 \\ 2 & 20 & 6 & 25 \\ 25 & 2 & 22 & 25 \end{pmatrix} \quad \text{also} \quad T^{-1} \stackrel{26}{\simeq} \begin{pmatrix} -3 & -6 & 5 & 1 \\ 2 & 11 & -8 & 1 \\ 2 & -6 & 6 & -1 \\ -1 & 2 & -4 & -1 \end{pmatrix} .$$

Example: The image of the tetragram /ende/ $\hat{=}$ $(4 \ \ 13 \ \ 3 \ \ 4)$ is calculated *modulo* 26 to yield /jhbl/ $\hat{=}$ $(9 \ \ 7 \ \ 1 \ \ 11)$:

$$(4 \ \ 13 \ \ 3 \ \ 4) \begin{pmatrix} 8 & 6 & 9 & 5 \\ 6 & 9 & 5 & 10 \\ 5 & 8 & 4 & 9 \\ 10 & 6 & 11 & 4 \end{pmatrix} \stackrel{26}{\simeq} (9 \ \ 7 \ \ 1 \ \ 11) ,$$

$$(9 \ \ 7 \ \ 1 \ \ 11) \begin{pmatrix} 23 & 20 & 5 & 1 \\ 2 & 11 & 18 & 1 \\ 2 & 20 & 6 & 25 \\ 25 & 2 & 22 & 25 \end{pmatrix} \stackrel{26}{\simeq} (4 \ \ 13 \ \ 3 \ \ 4) .$$

5.2.2 Inhomogenous case. With $t = (3 \ \ 8 \ \ 5 \ \ 20)$ and T as above an inhomogeneous linear substitution χ is obtained

$$\chi((x_1 \ x_2 \ x_3 \ x_4)) = (x_1 \ x_2 \ x_3 \ x_4) \begin{pmatrix} 8 & 6 & 9 & 5 \\ 6 & 9 & 5 & 10 \\ 5 & 8 & 4 & 9 \\ 10 & 6 & 11 & 4 \end{pmatrix} + (3 \ \ 8 \ \ 5 \ \ 20)$$

with the inverse substitution

$$\chi^{-1}((y_1 \ y_2 \ y_3 \ y_4)) = (y_1 \ y_2 \ y_3 \ y_4) \begin{pmatrix} 23 & 20 & 5 & 1 \\ 2 & 11 & 18 & 1 \\ 2 & 20 & 6 & 25 \\ 25 & 2 & 22 & 25 \end{pmatrix} + (3 \ \ 24 \ \ 21 \ \ 14) .$$

5.2.3 Enumeration. The number of regular $n \times n$ matrices over $\mathbb{Z}_N$ depends on the primality of N. A well-known result is (L. E. Dickson, *Linear groups*. Leipzig 1901):

Theorem. Let $N = p$, p prime. The number $g(p, n)$ of regular matrices from $\mathbb{Z}_p^{n,n}$ is equal to the number of bases of the vector space $\mathbb{Z}_p^{n,n}$, i.e.,

$$g(p, n) = (p^n - 1)\,(p^n - p)\,(p^n - p^2) \,.\,.\, (p^n - p^{n-1}) \ .$$

The number of different matrices altogether is p^{n^2} . Thus

$$g(p, n) = p^{n^2} \cdot \rho(p, n) , \qquad \text{where}$$

$$\rho(p, n) = \prod_{k=1}^{n} (1 - (\frac{1}{p})^k) \ .$$

For the binary case $N{=}2$: $g(2,1) = 1$, $g(2,2) = 2^1 \cdot 3$, $g(2,3) = 2^3 \cdot 3 \cdot 7$, $g(2,4) = 2^6 \cdot 3 \cdot 7 \cdot 15$, $g(2,5) = 2^{10} \cdot 3 \cdot 7 \cdot 15 \cdot 31$, $g(2,6) = 2^{15} \cdot 3 \cdot 7 \cdot 15 \cdot 31 \cdot 63$.

For $\rho(p,\ n)$, a limit as n goes to infinity can be given (Euler 1760):

$$\lim_{n\to\infty} \rho(p,\ n) = h\ (\frac{1}{p})\ , \qquad \text{where}$$

$$\begin{aligned} h(x) &= 1 + \sum_{k=1}^{\infty} (-1)^k\ [x^{(3k^2-k)/2} + x^{(3k^2+k)/2}] \\ &= 1 - x - x^2 + x^5 + x^7 - x^{12} - x^{15} + x^{22} + x^{26} \ldots\ . \end{aligned}$$

$h(x)$ is a 'lacunary' series, connected with theta series and elliptic functions. For details see R. Remmert, Funktionentheorie I, Springer, Berlin 1984, p. 263. $h(\frac{1}{p})$ provides for larger n a rather good estimate for $\rho(p, n)$; some values for primes p are

$N = p$	$h(\frac{1}{p})$	$\ln h(\frac{1}{p})$
2	0.28879	−1.24206
3	0.56013	−0.57959
5	0.76033	−0.27400
7	0.83680	−0.17817
11	0.90083	−0.10444
13	0.91716	−0.08647
17	0.93772	−0.06430
19	0.94460	−0.05699

For $p > 10$, $1 - \frac{1}{p} - \frac{1}{p^2}$ gives already five correct figures for $h(\frac{1}{p})$; $1/(\frac{3}{2} - p)$ approximates $\ln h(\frac{1}{p})$ with a relative error less than $\frac{1}{p^2}$.

For powers of a prime, the situation is more complex.

Theorem (Manfred Broy 1981)

Let $N = p^s$ and $A \in \mathbb{Z}_N^{n,n}$. Then there exist $A_i \in \mathbb{Z}_p^{n,n}$, $0 \leq i < s$ such that A can be uniquely represented in the form $A = \sum_{i=0}^{s-1} A_i\ p^i$. A is regular if and only if A_0 is regular.

From this theorem,

$$g(p^s,\ n) = g(p,\ n) \cdot (p^{s-1})^{n^2} = (p^s)^{n^2} \cdot \rho(p,\ n) = N^{n^2} \cdot \rho(p,\ n)\ .$$

Finally, for the general case $N = p_1^{s_1} \cdot p_2^{s_2}\ \ldots\ p_k^{s_k}$, the number of regular matrices is

$$g(N,\ n) = N^{n^2} \cdot \rho(p_1\ , n) \cdot \rho(p_2\ , n)\ \cdot\ \ldots\ \rho(p_k\ , n)\ .$$

N^{n^2} is a small number compared with the number $(N^n)!$ of *all* n-partite n-graphic substitutions: for $N = 25$, $n = 4$ we have $(N^n)! \approx 10^{2\,184\,284}$, compared to $N^{n^2} = 2.33 \cdot 10^{22}$ and $g(N,\ n) = 1.77 \cdot 10^{22}$. This comes close to the number $6.20 \cdot 10^{23}$ of simple cyclic permutations for $N = 25$.

For $N = 25$: $g(25, 1) = 20$, $g(25, 2) = 300\,000$, $g(25, 3) = 2\,906\,250\,000\,000$;
for $N = 26$: $g(26, 1) = 12$, $g(26, 2) = 157\,248$, $g(26, 3) = 1\,634\,038\,189\,056$.

5.2.4 Construction of pairs of matrices. The construction of a regular square matrix is most simply done as a product of a lower and an upper triangular regular matrix. This means that the diagonal elements should be invertible (Sect. 5.5, Table 1); most simply 1s are chosen. Moreover, the transpose of the lower triangular matrix can be chosen for the upper triangular matrix, which produces a symmetric matrix. Inversion of the triangular matrices with the elimination method leads to the inverse matrix.

Choosing 1s in the diagonal, it is also possible, although not always preferable, to do all the computations first in $\mathbb{Z}$ and then to pass over to the residual classes.

Example:

$$\begin{pmatrix} 1 & & \\ 3 & 1 & \\ 5 & 2 & 1 \end{pmatrix} \begin{pmatrix} 1 & 3 & 5 \\ & 1 & 2 \\ & & 1 \end{pmatrix} = \begin{pmatrix} 1 & 3 & 5 \\ 3 & 10 & 17 \\ 5 & 17 & 30 \end{pmatrix} ;$$

$$\begin{pmatrix} 1 & & \\ 3 & 1 & \\ 5 & 2 & 1 \end{pmatrix}^{-1} = \begin{pmatrix} 1 & & \\ -3 & 1 & \\ 1 & -2 & 1 \end{pmatrix} , \begin{pmatrix} 1 & 3 & 5 \\ & 1 & 2 \\ & & 1 \end{pmatrix}^{-1} = \begin{pmatrix} 1 & -3 & 1 \\ & 1 & -2 \\ & & 1 \end{pmatrix} ,$$

$$\begin{pmatrix} 1 & -3 & 1 \\ & 1 & -2 \\ & & 1 \end{pmatrix} \begin{pmatrix} 1 & & \\ -3 & 1 & \\ 1 & -2 & 1 \end{pmatrix} = \begin{pmatrix} 11 & -5 & 1 \\ -5 & 5 & -2 \\ 1 & -2 & 1 \end{pmatrix} .$$

For $\mathbb{Z}_{26}$, $\mathbb{Z}_{25}$ respectively,

$$\begin{pmatrix} 1 & 3 & 5 \\ 3 & 10 & 17 \\ 5 & 17 & 4 \end{pmatrix} , \begin{pmatrix} 11 & 21 & 1 \\ 21 & 5 & 24 \\ 1 & 24 & 1 \end{pmatrix} \quad \text{and} \quad \begin{pmatrix} 1 & 3 & 5 \\ 3 & 10 & 17 \\ 5 & 17 & 5 \end{pmatrix} , \begin{pmatrix} 11 & 20 & 1 \\ 20 & 5 & 23 \\ 1 & 23 & 1 \end{pmatrix}$$

are pairs of (symmetric) mutually inverse matrices; for $\mathbb{Z}_{10}$, $\mathbb{Z}_2$,

$$\begin{pmatrix} 1 & 3 & 5 \\ 3 & 0 & 7 \\ 5 & 7 & 0 \end{pmatrix} , \begin{pmatrix} 1 & 5 & 1 \\ 5 & 5 & 8 \\ 1 & 8 & 1 \end{pmatrix} \quad \text{and} \quad \begin{pmatrix} 1 & 1 & 1 \\ 1 & 0 & 1 \\ 1 & 1 & 0 \end{pmatrix} , \begin{pmatrix} 1 & 1 & 1 \\ 1 & 1 & 0 \\ 1 & 0 & 1 \end{pmatrix} .$$

If for given n and N the lower triangular matrix L and the upper triangular matrix U (with 1s in the diagonal) are chosen arbitrarily and for D an arbitrary diagonal matrix with invertible elements is taken, one obtains up to reordering of rows and columns all pairs of mutually reciprocal matrices $L\,D\,U$ and $U^{-1}\,D^{-1}\,L^{-1}$.

5.2.5 The construction of a self-reciprocal matrix is scarcely more difficult: If, for given n and N, $(X \, , X^{-1})$ is a pair of mutually reciprocal matrices and J is a self-reciprocal diagonal matrix, with elements $+1$ or -1 (more generally, self-reciprocal in $\mathbb{Z}_N$), then $X\,J\,X^{-1}$ is self-reciprocal.

Example:

$$\begin{pmatrix} 1 & 3 & 5 \\ 3 & 10 & 17 \\ 5 & 17 & 30 \end{pmatrix} \begin{pmatrix} 1 & & \\ & -1 & \\ & & 1 \end{pmatrix} \begin{pmatrix} 11 & -5 & 1 \\ -5 & 5 & -2 \\ 1 & -2 & 1 \end{pmatrix} = \begin{pmatrix} 31 & -30 & 12 \\ 100 & -99 & 40 \\ 170 & -170 & 69 \end{pmatrix} .$$

For $\mathbb{Z}_{26}$, $\mathbb{Z}_{25}$, $\mathbb{Z}_{10}$, $\mathbb{Z}_2$ the following self-reciprocal matrices are obtained:

$$\begin{pmatrix} 5 & 22 & 12 \\ 22 & 5 & 14 \\ 14 & 12 & 17 \end{pmatrix} , \begin{pmatrix} 6 & 20 & 12 \\ 0 & 1 & 15 \\ 20 & 5 & 19 \end{pmatrix} , \begin{pmatrix} 1 & 0 & 2 \\ 0 & 1 & 0 \\ 0 & 0 & 9 \end{pmatrix} , \begin{pmatrix} 1 & 0 & 0 \\ 0 & 1 & 0 \\ 0 & 0 & 1 \end{pmatrix} .$$

Over $\mathbb{Z}_2$ the identity is the only self-reciprocal diagonal matrix.

No simple expression is known for the number of self-reciprocal $n{\times}n$ matrices over $\mathbb{Z}_N$.

5.3 Binary Linear Substitutions

For $\mathbb{Z}_2$, i.e., for binary words (of length n_0) of plaintext and cryptotext, the technical execution of linear substitutions is particularly simple. The arithmetic *modulo* 2 can be translated into Boolean algebra and implemented in parallel with binary circuits of a width n_0, for not too big n_0 , say up to 64. Comparing the case $\mathbb{Z}_{2^s}^{n_0 \times n_0}$ $(n = n_0, N = 2^s)$ with the case $\mathbb{Z}_2^{s \cdot n_0 \times s \cdot n_0}$ $(n = s \cdot n_0, N = 2)$ which is obtained by decomposing the 2^s characters of $\mathbb{Z}_{2^s}^{n_0 \times n_0}$ in binary words of length s, gives the following conclusion: the number of all regular linear substitutions is $2^{s \cdot n_0^2} \cdot \rho(2, n_0) = K \cdot \rho(2, n_0)$ in the first case, $2^{s^2 \cdot n_0^2} \cdot \rho(2, s \cdot n_0) = K^s \cdot \rho(2, s \cdot n_0)$ in the second case. The coarse structure of $\mathbb{Z}_{2^s}$ decreases the power within the class of linear substitutions.

5.4 General Linear Substitutions

Including the N^n translations, there are roughly N^{n^2+n} linear substitutions altogether. Self-reciprocal homogenous linear substitutions (with $N = 26$) were proposed in 1929 by Lester S. Hill[2] (a precursor was F. J. Buck in 1772, L. J. d'Auriol used in 1867 a bipartite digraphic cipher $V^2 \to V^2$ which possibly is a special linear substitution.) Hill's ideas were taken up in 1941 by A. A. Albert in a wave of both patriotic and mathematical enthusiasm, in particular at a meeting of the American Mathematical Society. By then, Hill's ideas had already had their impact on W. F. Friedman in the U.S.A. and

[2] Lester S. Hill was assistant professor of mathematics at Hunter College in New York. He received his Ph.D. in 1926 at Yale, aged 35, having been a college teacher for a while. The paper was published in *The American Mathematical Monthly* under the title *Cryptography in an Algebraic Alphabet* (Vol. **36**, p. 306–312, June–July 1929), with a follow-up *Concerning certain linear transformation apparatus of cryptography* (Vol. **37**, p. 135–154, March 1931). Hill received U.S. patent 1 845 947 on his apparatus, Feb. 16, 1932. He was until 1960 professor at Hunter College and died January 9, 1961.

on W. Kunze in the German *Auswärtiges Amt.*[3] The importance of Hill's invention stems from the fact that since then the value of mathematical methods in cryptology has been unchallenged. Consequently in the 1930s mathematicians entered the cipher bureaus: Solomon Kullback (1907–1994), Abraham Sinkov (b. 1907), Werner Kunze, the Dutch statistician Maurits de Vries and many more whose names remain unpublished.

Lester S. Hill designed a machine for linear substitutions ($n = 6$), U.S. Patent 1 845 947. Such a purely mechanical device with geared wheels was rather slow, therefore in the Second World War Hill's machines were only used for superencrypting three-letter code groups of radio call signs—which was, compared to hand computation, quite a saving.

5.5 Decomposed Linear Substitutions

As a further special case, linear substitution contains a certain poly*alphabetic* enciphering. This occurs if T decomposes in a direct sum, $T = T_1 \oplus T_2 \oplus \cdots \oplus T_r$, i.e., its matrix has blockdiagonal form,

$$T = \left(\begin{array}{c|c|c|c} T_1 & 0 & \dots & 0 \\ \hline 0 & T_2 & \dots & 0 \\ \hline \vdots & \vdots & \dots & 0 \\ \hline 0 & 0 & 0 & T_r \end{array}\right)$$

where T_i is $n_i \times n_i$. In this case, each T_i, together with the corresponding part t_i of $t = t_1 \oplus t_2 \oplus \cdots \oplus t_r$, still is a polygraphic substitution, an enciphering of n_i-grams. Provided these r substitutions are pairwise different, the encryption step is an r-fold polyalphabetic linear polygraphic step. In other words, a whole period of a periodic polyalphabetic encryption is comprised in the single matrix. More about this in Sect. 7.4.1 .

An important extreme case has $n_i = 1$, $r = n$. Then T is a diagonal matrix and to every line there corresponds a simple linear substitution, a very special unipartite monographic substitution $T_i : V^1 \to V^1$.

Let us study this substitution—a permutation—more closely: It reads $\chi(x) = h{\cdot}x{+}t$ and is certainly regular for $h = 1$, yielding $\chi(x) = x{+}t$. Thus, a simple linear substitution with $h = 1$ is a monographic CAESAR addition $\chi(x) = x + t$ with the inverse $\chi^{-1}(x) = x - t$ (for $t \neq 0$ a proper one).

[3] Dr. Werner Kunze, b. about 1890, studied mathematics, physics and philosophy in Heidelberg, was with the cavalry in the First World War, and in January 1918 started work on cryptology in the *Auswärtiges Amt*. In 1923, he solved a superencrypted French diplomatic code, in 1936 ORANGE and later RED, two Japanese rotor-cipher machines. Kunze was presumably the first professional mathematician to serve in a modern cryptanalytic bureau. Kunze was, like Mauborgne, a passable violin player and Oliver Strachey was known to be a good musician, while Painvin was an excellent cellist. Lambros D. Callimahos, at N.S.A., was a famous flutist.

N=2	1									$\mathcal{M}_2 = \mathcal{C}_1$
N=3	1	**2**								$\mathcal{M}_3 = \mathcal{C}_2$
N=4	1	**3**								$\mathcal{M}_4 = \mathcal{C}_2$
N=5	1	**2** 3	4							$\mathcal{M}_5 = \mathcal{C}_4$
N=6	1	**5**								$\mathcal{M}_5 = \mathcal{C}_2$
N=7	1	2 4	**3** 5	6						$\mathcal{M}_7 = \mathcal{C}_6$
N=8	1	**3**	5	**7**						$\mathcal{M}_8 = \mathcal{C}_2 \times \mathcal{C}_2$
N=9	1	**2** 5	4 7	8						$\mathcal{M}_9 = \mathcal{C}_6$
N=10	1	**3** 7	9							$\mathcal{M}_{10} = \mathcal{C}_4$
N=11	1	**2** 6	3 4	5 9	7 8	10				$\mathcal{M}_{11} = \mathcal{C}_{10}$
N=12	1	**5**	7	**11**						$\mathcal{M}_{12} = \mathcal{C}_2 \times \mathcal{C}_2$
N=13	1	**2** 7	3 9	4 10	5 8	6 11	12			$\mathcal{M}_{13} = \mathcal{C}_{12}$
N=14	1	**3** 5	9 11	13						$\mathcal{M}_{14} = \mathcal{C}_6$
N=15	1	**2** 8	4	7 13	11	**14**				$\mathcal{M}_{15} = \mathcal{C}_4 \times \mathcal{C}_2$
N=16	1	**3** 11	5 13	7	9	**15**				$\mathcal{M}_{16} = \mathcal{C}_4 \times \mathcal{C}_2$
N=17	1	2 9	**3** 6	4 13	5 7	8 15	10 12	11 14	16	$\mathcal{M}_{17} = \mathcal{C}_{16}$

Table 1a. Reciprocal pairs in $\mathbb{Z}_N$ for N from 2 to 17
(Bold-faced figures: Generating elements of the multiplicative group $\mathcal{M}_N$)

N=18	1 **5**/11 7/13 17	$\mathcal{M}_{18} = \mathcal{C}_6$
N=19	1 **2**/10 3/13 4/5 6/16 7/11 8/12 9/17 14/15 18	$\mathcal{M}_{19} = \mathcal{C}_{18}$
N=20	1 **3**/7 9 11 13/17 **19**	$\mathcal{M}_{20} = \mathcal{C}_4 \times \mathcal{C}_2$
N=21	1 **2**/11 4/16 5/17 8 10/19 13 **20**	$\mathcal{M}_{21} = \mathcal{C}_6 \times \mathcal{C}_2$
N=22	1 3/15 5/9 **7**/19 13/17 21	$\mathcal{M}_{22} = \mathcal{C}_{10}$
N=23	1 2/12 3/8 4/6 **5**/14 7/10 9/18 11/21 13/16 15/20 17/19 22	$\mathcal{M}_{23} = \mathcal{C}_{22}$
N=24	1 **5** 7 11 **13** 17 19 **23**	$\mathcal{M}_{24} = \mathcal{C}_2 \times \mathcal{C}_2 \times \mathcal{C}_2$
N=25	1 **2**/13 3/17 4/19 6/21 7/18 8/22 9/14 11/16 12/23 24	$\mathcal{M}_{25} = \mathcal{C}_{20}$
N=26	1 3/9 5/21 **7**/15 11/19 17/23 25	$\mathcal{M}_{26} = \mathcal{C}_{12}$
N=27	1 **2**/14 4/7 5/11 8/17 10/19 13/25 16/22 20/23 26	$\mathcal{M}_{27} = \mathcal{C}_{18}$
N=28	1 3/19 **5**/17 9/25 11/23 13 15 **27**	$\mathcal{M}_{28} = \mathcal{C}_6 \times \mathcal{C}_2$
N=29	1 **2**/15 3/10 4/22 5/6 7/25 8/11 9/13 12/17 14/27 16/20 18/21 19/26 23/24 28	$\mathcal{M}_{29} = \mathcal{C}_{28}$
N=30	1 **7**/13 11 17/23 19 **29**	$\mathcal{M}_{30} = \mathcal{C}_4 \times \mathcal{C}_2$
N=31	1 2/16 **3**/21 4/8 5/25 6/26 7/9 10/28 11/17 12/13 14/20 15/29 18/19 22/24 23/27 30	$\mathcal{M}_{31} = \mathcal{C}_{30}$
N=32	1 **3**/11 5/13 7/23 9/25 15 17 19/27 21/29 **31**	$\mathcal{M}_{32} = \mathcal{C}_8 \times \mathcal{C}_2$
N=33	1 2/17 4/25 **5**/20 7/19 8/29 10 13/28 14/26 16/31 23 **32**	$\mathcal{M}_{33} = \mathcal{C}_{10} \times \mathcal{C}_2$

Table 1b. Reciprocal pairs in $\mathbb{Z}_N$ for N from 18 to 33
(Bold-faced figures: Generating elements of the multiplicative group $\mathcal{M}_N$)

For $\chi(x) = x + i$ we write from now on also $\chi(x) = \rho^i(x)$, where $\rho(x) = x + 1$.

Quite generally:

A simple linear substitution $\chi(x) = h \cdot x + t$ is regular, $\chi^{-1}(x) = h^{-1} \cdot (x - t)$ if and only if h is relatively prime to N .

Table 1 gives for some values of N reciprocal pairs of h and h^{-1} , including certain self-reciprocal h yielding involutory permutations.

5.6 Decimated Alphabets

The homogeneous case $t = 0$ has the trivial cases

$h = h^{-1} = 1$ (unchanged alphabet) and

$h = h^{-1} = N - 1$ (complementary alphabet)

and otherwise the decimated alphabets (French *alphabets chevauchants*, German *dezimierte Alphabete*), studied by Eyraud: alphabets whose representants are the h-folds of the integers *modulo* N—provided h and N are relatively prime. Thus, the alphabets are obtained by going in steps of h ('symbolic multiplication', 'decimation by h').

Examples for $N = 8$:

$$h = 1: \begin{pmatrix} a & b & c & d & e & f & g & h \\ a & b & c & d & e & f & g & h \end{pmatrix} = (a)\,(b)\,(c)\,(d)\,(e)\,(f)\,(g)\,(h)$$

$$h = 7: \begin{pmatrix} a & b & c & d & e & f & g & h \\ a & h & g & f & e & d & c & b \end{pmatrix} = (a)\,(bh)\,(cg)\,(df)\,(e)$$

$$h = 3: \begin{pmatrix} a & b & c & d & e & f & g & h \\ a & d & g & b & e & h & c & f \end{pmatrix} = (a)\,(bd)\,(cg)\,(e)\,(fh)$$

$$h = 5: \begin{pmatrix} a & b & c & d & e & f & g & h \\ a & f & c & h & e & b & g & d \end{pmatrix} = (a)\,(bf)\,(c)\,(dh)\,(e)\,(g) \quad .$$

There is a distinction between the complementary alphabet with

$$\chi(x) = (N-1) \cdot x \quad \overset{N}{\simeq} \quad N - x \quad \overset{N}{\simeq} \quad -x$$

and the reversed alphabet, originating from the inhomogeneous case with

$$\chi(x) = (N-1) \cdot (x+1) \overset{N}{\simeq} (N-1) - x \overset{N}{\simeq} -x - 1 \ .$$

The number $g(N, 1)$ of *regular* homogeneous simple linear substitutions coincides with the Euler totient function $\varphi(N)$, the number of numbers from $1, 2, \ldots, N-1$ relatively prime to N.

For $N = p_1^{s_1} \cdot p_2^{s_2} \cdot \ldots \cdot p_k^{s_k}$,

$$\begin{aligned} \varphi(N) &= (p_1 - 1) \cdot p^{s_1 - 1} \cdot (p_2 - 1) \cdot p^{s_2 - 1} \cdot \ \ldots \ \cdot (p_k - 1) \cdot p^{s_k - 1} \\ &= N \cdot (1 - \tfrac{1}{p_1}) \cdot (1 - \tfrac{1}{p_2}) \cdot \ \ldots \ \cdot (1 - \tfrac{1}{p_s}) \ . \end{aligned}$$

5.7 Linear Substitutions with Decimal and Binary Numbers

Note the following difference: $\mathbb{Z}_N^n$ belongs to V^n, while $\mathbb{Z}_{N^n}$ belongs to V; one obtains $\mathbb{Z}_{N^n}$ if V^n is ordered lexicographically.

5.7.1 Case $N = 10$ $(\mathbb{Z}_{10^n})$: The decimated alphabets (Sect. 5.6) are particularly interesting for amateurs encrypting n-digit decimal numbers with a pocket calculator, since they allow use of multiplication besides addition.

Even with a mechanical adder it is easy to calculate in $\mathbb{Z}_{10^n}$. (For the transition to calculation in $\mathbb{Z}_{10}^n$, it is only necessary to dismantle the carry device, see Sect. 8.3.3).

Example $n = 2$ $(\mathbb{Z}_{100})$: It suffices to know the reciprocals *modulo* 100 of the primes up to 97 (excluding 2 and 5):

h =	3	7	11	13	17	19	23	29	31	37	41	43	47	53	59	61	67	71	73	79	83	89	97
h^{-1} =	67	43	91	77	53	79	87	69	71	73	61	7	83	17	39	41	3	31	37	19	47	9	33

Note that the last figure of the reciprocal is determined by the reciprocal *modulo* 10 of the last figure, see Table 1, $N = 10$.
This observation suggests for larger values of n a stepwise procedure: In every step, just one new figure is suitably chosen.

Example $n = 5$ $(\mathbb{Z}_{10^5})$: The reciprocal *modulo* 10^5 of the number $h = 32\,413$ is $h^{-1} = 3\,477$ according to the following algorithm:

$$\begin{array}{rlllrcl}
3: & & & 3 \cdot & 7 & = & 2 \cdot 10 + 1 \\
13: & 2 + 1 \cdot 7 + 3 \cdot x \overset{10}{\simeq} 0 & x=7 & 13 \cdot & 77 & = & 10 \cdot 10^2 + 1 \\
413: & 10 + 4 \cdot 7 + 3 \cdot x \overset{10}{\simeq} 0 & x=4 & 413 \cdot & 477 & = & 197 \cdot 10^3 + 1 \\
2413: & 197 + 2 \cdot 7 + 3 \cdot x \overset{10}{\simeq} 0 & x=3 & 2413 \cdot & 3477 & = & 839 \cdot 10^4 + 1 \\
32413: & 839 + 3 \cdot 7 + 3 \cdot x \overset{10}{\simeq} 0 & x=0 & 32413 \cdot & 03477 & = & 1127 \cdot 10^5 + 1
\end{array}$$

The costs for the determination of the reciprocal of a n-figure number are proportional to n^2.

5.7.2 Case $N = 2$ $(\mathbb{Z}_{2^n})$: For professional work the binary number system is preferable. The cases $n = 8, 16, 32$, or even 64 fit directly the internal arithmetical architecture of microprocessors. The algorithm for the determination of a reciprocal *modulo* 2^n is completely analogous the decimal one above, moreover it is—like the classical division algorithm for binary numbers—simpler than that for decimal numbers.

For example, the number $LOOO\ OOOO\ OOLL\ OLLL = 32\,823$
is reciprocal *modulo* 2^{16} to $OOLL\ OLOL\ LOOO\ OLLL = 13\,703$

Indeed, $32823 \cdot 13703 = 449773569 = 1 + 6863 \cdot 2^{16}$.

5.7.3 Turing, Already in the fall of 1937, two years before he became seriously involved with cryptology, Alan Turing had thoughts about encryption by multiplication in the binary number system. This may have occurred incidentally to other mathematicians, too. Turing, however, designed a relay multiplication circuit for this purpose and built a few stages, supported by the Princeton physicist Malcolm MacPhail. Circumstances prompted Turing to drop this project after he returned in July 1938 from his Princeton stay, but he was well prepared to take up mechanical cryptanalysis on September 4, 1939, one day after the outbreak of the war, when he entered Bletchley Park. This was a Victorian country mansion in Buckinghamshire, halfway between Oxford and Cambridge, and was the place where the Government Code and Cypher School had been evacuated in August 1939. G.C. & C.S. invited Turing already in summer 1938, after his return, to a course in cryptology "just in case". He passed another one around Christmas, and he met quite regularly with the chief cryptologist Dillwyn Knox, who had struggled hard to solve Italian and later Spanish messages encrypted by the commercial ENIGMA without plugboard[4] (and who died in 1943). Gordon Welchman, too, had been recruited into intelligence work before the war broke out.

The first mathematician recruited by the Government Code and Cypher School was Peter Twinn, an Oxford graduate who entered service in February 1939. He was told later that there had been some doubts about the wisdom of recruiting a mathematician "as they were regarded as strange fellows notoriously unpractical" (Andrew). In fact, some other early Bletchleyites like Alan Turing, Gordon Welchman, and Dennis Babbage had at least some skill at chess, not to speak of the chess masters Stuart Milner-Barry, Harry Golombek, and Hugh Alexander, all recruited with the help of Welchman.

Great Britain was well prepared for the war that was brewing; from Oxford and Cambridge the best people, if they didn't want to become fighter pilots, were recruited for Bletchley Park. G.C. & C.S., a branch of the Foreign Office, had started in mid-1938 to be alarmed. Neither the United States nor Germany had made such painstaking preparations in the recruitment of scientists. In Great Britain, it took longer than in Germany or in the U.S.A. to recognize the importance of mathematics for cryptanalysis, but with Turing and Welchman at hand the arrears were made up completely. Exploiting the talents of unconventional and eccentric personalities enabled the Foreign Office to establish the ablest team of cryptanalysts in British history.

In France, whose cryptology like that of the British earlier was extremely language oriented, the opportunity passed by in 1940.

---

[4] It may not be wise to believe the story Frederick W. Winterbotham started and Cave Brown told in his book, that Knox and Turing travelled in the middle of 1938 to Warsaw, to meet there, arranged by the Polish Secret Service, a Pole with the pseudonym Richard Lewinski, who allegedly had worked at the firm Heimsoeth & Rincke in Berlin as mathematician and engineer and had offered to procure a copy of the ENIGMA. Marian Rejewski, in 1982, called this "a fable". However, Harry Hinsley reports that already in 1938 the Polish Secret Service had contacted the G.C. & C.S. and Knox re ENIGMA. This first contact, however, was not flourishing; Knox called the Polish 'stupid and ignorant'.

6 Encryption Steps: Transposition

En un mot, les méthodes de transposition
sont une salade des lettres du texte clair.
[In one word, the transposition methods
give a nice mess of cleartext letters.]
Bazeries

An extreme special case, not discussed at all in Chapter 5, requires that the matrix of the homogeneous linear substitution has only zeroes and ones as its elements. This is for $N > 2$ a severe restriction. Requiring moreover that in every row and in every column, one occurs just once and thus the elements are zero otherwise, leads to permutation matrices effectuating a mere permutation of the vector space basis.

The transposition (German *Würfelverfahren* or *Versatzverfahren*) is a polygraphic substitution $V^n \to V^n$, an *encoding* of most special kind

$$(x_1\,,\ x_2\,,\ \ldots\ x_n) \mapsto (x_{\pi(1)}\,,\ x_{\pi(2)}\,,\ \ldots\ x_{\pi(n)})$$

where $\pi \in \gamma_k$ is a permutation of $\{1 \ldots n\}$, γ_k denoting the full group of $n!$ permutations.

A transposition is *not* a permutation of alphabet characters, but a permutation of places. Its use for anagrams (bolivia – lobivia) is primeval, in particular for the construction of pseudonyms (Améry – Mayer).

6.1 Simplest Methods

The simple classes of methods use one or a few encryption steps repeated over and over with a not-too-big n ('complete-unit transposition').

6.1.1 Crab. Simplest is back slang or crab (German *Krebs*) : The message is reversed word for word or in toto: LIRPA OCCABOT KOOL (Sect. 1.5).

This 'reversed writing' comprises ananymes like REMARQUE for Kramer, and AVE for Eva. Crab is also known in music, for example in the crab canon.

Palindromes are words or sentences which are invariant under a crab:

Madam	été	Reittier	summus
Able was I ere I saw Elba		Ein Neger mit Gazelle zagt im Regen nie	
Esope reste ici et se repose		in girum imus nocte et consumimur igni	

Every language has its palindromes. Some more examples in English are:

Red rum & murder. A man, a plan, a canal: Panama. Ma is as selfless as I am.
Was it a cat I saw? *(Henry E. Dudeney)* Madam, I'm Adam. *(Sam Loyd)*
Lewd did I live, & evil I did dwel. *(John Taylor)* Draw pupil's lip upward.
Doc note, I dissent; a fast never prevents a fatness; I diet on cod. *(Peter Hilton)*

6.1.2 Spoonerism. A harmless non-cryptographic use of syllable transposition ($n = 4$) is found in the spoonerism (German *Schüttelreim*):

they hung flags — they flung hags   dear old Queen — queer old Dean
wasted the term — tasted the worm   missed the history — hissed the mystery

The transposition $\pi : \pi(1,2,3,4) = (3,2,1,4)$ found in spoonerisms is used cryptographically in Medical Greek, according to Kahn a mild epidemic disease of London medical students: POKE A SMIPE stands for *smoke a pipe*.

1	4	53	18	55	6	43	20
52	17	2	5	38	19	56	7
3	64	15	54	31	42	21	44
16	51	28	39	34	37	8	57
63	14	35	32	41	30	45	22
50	27	40	29	36	33	58	9
13	62	25	48	11	60	23	46
26	49	12	61	24	47	10	59

Fig. 41. Route for knight's tour transposition ($n = 64$)

6.1.3 Route transcription. Then there is route transcription ('tramp', German *Würfel*): The plaintext is written in l rows of a fixed length k and read out in some prescribed way. Thus, a $k \times l$ rectangle is used for the encryption step, $n = k \times l$. Frequently, a square is used, then $n = k^2$. For a 2×2 square, the spoonerism is included as an 'overcrossing' route.

The cryptotext can be read out in columns (row-column transcription):

```
i c h b i n
d e r d o k        I D T E C E O N H R R B B D E A I O I R N K S T
t o r e i s
e n b a r t
```

This method we have met already for the construction of an alphabet using a mnemonic password. Variants read out along the diagonals:

E T N D O B I E R A C R E R H D I T B O S I K N

or boustrophedonically[1], alternatingly down and up the columns (every second column in a crab):

I D T E N O E C H R R B A E D B I O I R T S K N

or even in a spiral:

T S K N I B H C I D T E N B A R I O D R E O R E .

[1] Greek *bustropheidon*, German *furchenwendig*, turning like oxen in plowing.

A more complex route is given by a knight's tour (German *Rösselsprungwürfel*) (Fig. 41); its decryption, if the start is known or can be guessed, is not very difficult, as is familiar from knight's tour puzzles.

Instead of rectangles, other geometric patterns have been used from time to time, primarily triangles, also crosses of varying forms and other arrays (Figs. 42, 43). There are no limits to fantasy. But these simple transposition methods are quite open to cryptanalysis.

```
a       e       i       n       r       v       z
  b   d   f   h   k   m   o   q   s   u   w   y
    c       g       l       p       t       x
```

A E I N R V Z B D F H K M O Q S U W Y C G L P T X

Fig. 42. 'Rail fence' (Smith) transposition, $n = 25$

```
  b     f     k     o     s     w
a   c e   g i   l n   p r   t v   x
  d     h     m     q     u     z
```

B F K O S W A C E G I L N P R T V X D H M Q U Z

Fig. 43. *'Croix Grecque'* (Muller), 'Four winds' (Nichols) transposition, $n = 24$

6.1.4 Grilles. Convenient as tools for transposition and, if the pattern is irregular enough, more secure than the routing methods are grilles, also called trellis ciphers (French *grille*, German *Raster*). Generally, a set of prefabricated grilles is needed. An important practical simplification appears in the turning grille, which brings the different windows of one and the same grille into action by rotation. It was described in 1885 by Jules Verne (1828–1905) in the story *Mathias Sandorff*. Grilles were used in the 18th century, for example in 1745 in the administration of the Dutch *Stadthouder* William IV.

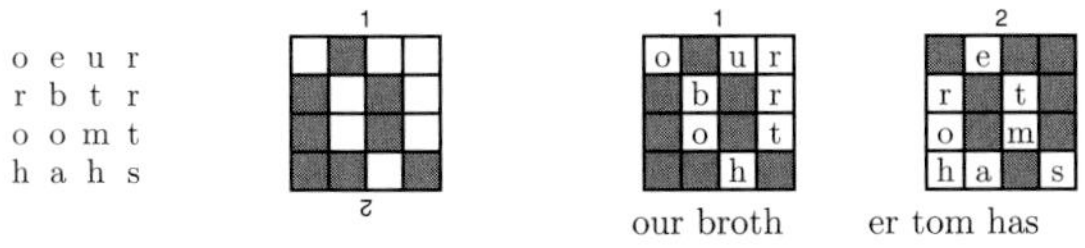

Fig. 44. Turning grille with two positions

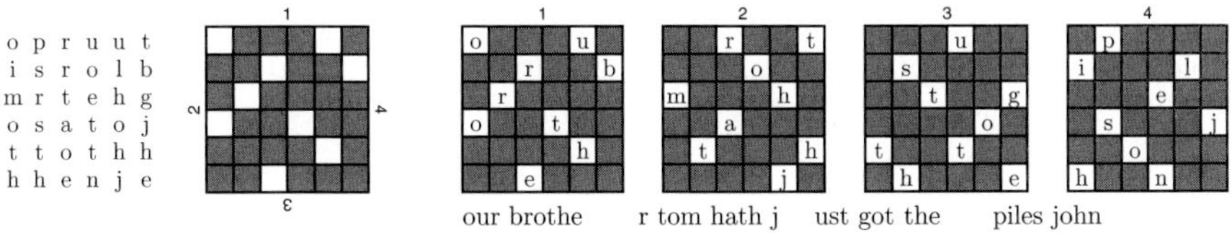

Fig. 45. Turning grille with four positions

The mathematician C. F. Hindenburg studied turning grilles more systematically in 1796, followed by Moritz von Prasse 1799, Johann Ludwig Klüber

1809. There are grilles with two positions (Fig. 44) and, preferably, with four positions (Fig. 45), which are often called Fleissner grilles[2] in ignorance of their historical origin.

The construction of turning grilles is simple enough: A quadrant of a square checkerboard (with an even number 2ν of rows and columns) is marked with the numbers $1 \dots \nu$, all numberings produced by rotation are superimposed, then for each number, a position of the rotated grille is selected and the corresponding window is cut. The turning grille of Fig. 45 is obtained like this:

$$\begin{array}{|ccc|ccc}
\hline
\mathbf{1} & 2 & 3 & 7 & \mathbf{4} & 1 \\
4 & 5 & \mathbf{6} & 8 & 5 & \mathbf{2} \\
7 & \mathbf{8} & 9 & 9 & 6 & 3 \\
\hline
\mathbf{3} & 6 & 9 & \mathbf{9} & 8 & 7 \\
2 & 5 & 8 & 6 & \mathbf{5} & 4 \\
1 & 4 & \mathbf{7} & 3 & 2 & 1
\end{array} \quad .$$

This technique allows the production of turning grilles according to a key pattern that lists the grille positions in the consecutive construction steps, for example in the following suggestive way:

┘ └ ┐ └ ┌ ┘ ┐ ┘ ┌ .

The power of the class of turning grilles (for given $n = 4\nu^2$) can thus be determined: For a turning grille with two positions $2^{n/2}$ possibilities, for a turning grille with four positions $4^{n/4} = 2^{n/2}$ possibilities. For $n = 36$ as above there exist $\approx 2.62 \cdot 10^5$ Fleissner grilles, the number of all permutations is $36! \approx 3.72 \cdot 10^{41}$.

Some of the military powers developed in the late 19th century a liking for route transcription and turning grilles on the tactical combat level. In the First World War, the German *Heer* early in 1917 suddenly introduced turning grilles with denotations like *ANNA* (5×5), *BERTA* (6×6), *CLARA* (7×7), *DORA* (8×8), *EMIL* (9×9) and *FRANZ* (10×10). After four months this was discontinued—to the distress of the French, who had easily broken the encryption.

Route transcription and turning grilles result in a transposition of the message and nothing else. A precursor, the Cardano grille, gives *no* transposition, instead it introduces besides the characters visible through the window a larger set of nulls (Sect. 1.6). Grilles of any sort should normally not be used by serious cryptographers, but one cannot be sure that amateurs would not use them. Moreover, transposition in connection with substitution is to be taken quite seriously.

---

2 Eduard Baron Fleißner von Wostrowitz (1825–1888), Austrian Colonel, "Neue Patronen-Geheimschrift" (*Handbuch der Kryptographie*, Wien 1881). The word 'Patrone', mlat. 'father form', specimen form, was used in the textile industry for a drawing of the weaving *pattern* on checkered paper. In Jaroslav Hašek's novel *The Good Soldier Schwejk* a "Handbuch der militärischen Kryptographie von Oberleutnant Fleissner" is mentioned, and other details pointing to a certain familiarity of Hašek with cryptography.

6.2 Columnar Transpositions

Serious use of transpositions requires rather large values of n, coming close to the length of the whole message, in connection with the use of passwords for selecting the transposition from a rather powerful set of encryption steps. In German, such a password is called a *Losung*.

6.2.1 Passwords. They are already used in the simple columnar transposition (French *transposition simple à clef*): The plaintext is written in rows of the chosen length k, the resulting columns are reordered according to a permutation $\pi \in \gamma_k$ (*Losung*), and the cryptotext is read out *column by column*:

$$\underbrace{\text{e s w a r s c h o n d u n k e l}} \qquad \pi:\ 2\ 1\ 4\ 3$$

$$\left.\begin{array}{ll} \mathit{2\ 1\ 4\ 3} & \mathit{1\ 2\ 3\ 4} \\ \text{e s w a} & \text{S E A W} \\ \text{r s c h} & \text{S R H C} \\ \text{o n d u} & \text{N O U D} \\ \text{n k e l} & \text{K N L E} \end{array}\right\} \quad \text{S S N K E R O N A H U L W C D E}\ .$$

For cryptanalysis equivalent is obviously the block transposition or 'complete-unit transposition' (Gaines), French *variante de Richelieu* (Eyraud), German *Gruppen-Transposition, Umstellung*, which is as above except that the cryptotext is read out *row by row*:

$$\underbrace{\text{e s w a r s c h o n d u n k e l}} \qquad \pi:\ 2\ 1\ 4\ 3$$

$$\left.\begin{array}{ll} \mathit{2\ 1\ 4\ 3} & \mathit{1\ 2\ 3\ 4} \\ \text{e s w a} & \text{S E A W} \\ \text{r s c h} & \text{S R H C} \\ \text{o n d u} & \text{N O U D} \\ \text{n k e l} & \text{K N L E} \end{array}\right\} \quad \text{S E A W S R H C N O U D K N L E}\ .$$

This encryption can also be interpreted in the following way: the plaintext is divided into blocks of k elements and each block is permuted according to π —this means a repeated monoalphabetic polygraphic encryption step of *blocks* of width k: block transposition is a complete-unit transposition.

Simple columnar transposition is carried out with pencil and paper more easily and with less risk of error than block transposition. But although it spreads over the whole plaintext, it offers no more cryptanalytic security than complete-unit transposition—it is an example of a *complication illusoire*.

6.2.2 Rectangular schemes. For route transcription and simple columnar transposition, decryption is made easier if the plaintext fits a rectangle or square of l rows. To accomplish this, nulls are frequently used. If they are not chosen with great care, unauthorized decryption is also much facilitated—e.g., if the plaintext is filled up with q q q ... q q . By no means is this filling necessary; the length of the last row is determined by the division rest.

Since simple columnar transposition and block transposition—even with incomplete rectangles—can be solved easily, more complicated transposition methods are worthwhile. They all can be understood to be composite methods (see Sect. 9.1.1).

6.2.3 Two-step methods. An additional permutation is introduced in the mixed-rows columnar transposition (French *transposition double*, Givierge, Eyraud): The plaintext is written in rows of the chosen length k *according to some permutation* π_1, the resulting columns are reordered according to a permutation π_2, and the cryptotext is read out column by column,

e s w a r s c h o n d u n k e l π_1: *2 4 1 3* π_2: *2 1 4 3*

			2 1 4 3	*1 2 3 4*	
1	e s w a	*2*	r s c h	S R H C	S K S N R N E O H L A U C E W D .
2	r s c h	*4*	n k e l	K N L E	
3	o n d u	*1*	e s w a	S E A W	
4	n k e l	*3*	o n d u	N O U D	

Correspondingly, the mixed-rows block transposition is handled as above, but the cryptotext is read out *row by row*:

e s w a r s c h o n d u n k e l π_1: *2 4 1 3* π_2: *2 1 4 3*

			2 1 4 3	*1 2 3 4*	
1	e s w a	*2*	r s c h	S R H C	S R H C K N L E S E A W N O U D .
2	r s c h	*4*	n k e l	K N L E	
3	o n d u	*1*	e s w a	S E A W	
4	n k e l	*3*	o n d u	N O U D	

The same effect can be attained by using π_1 afterwards:

e s w a r s c h o n d u n k e l π_2: *2 1 4 3* π_1: *2 4 1 3*

2 1 4 3	*1 2 3 4*				
e s w a	S E A W	*1*	S R H C	*2*	S R H C K N L E S E A W N O U D .
r s c h	S R H C	*2*	K N L E	*4*	
o n d u	N O U D	*3*	S E A W	*1*	
n k e l	K N L E	*4*	N O U D	*3*	

Mixed-rows columnar or block transposition with a square can use the same permutation both times, $\pi_2 = \pi_1$. Taking $\pi_2 = \pi_1^{-1}$ gives a method attributed by Kerckhoffs in 1883 to the Russian *Nihilists* ('Nihilist transposition').

For a mathematical treatment of these transpositions, we assume that plaintext and ciphertext are represented by a rectangular (or square) matrix X of l rows, each with k elements.

Permutation of the rows then means multiplication from the left by an $l \times l$ permutation matrix π_1,

$$X \mapsto \pi_1 X \ .$$

while permutation of the columns means multiplication from the right by a $k \times k$ permutation matrix π_2,

$$X \mapsto X \pi_2 \ .$$

Row-column transcription means a *matrix* transposition (mirroring at the diagonal),

$$X \mapsto X^T \, .$$

Now, block transposition is just $X \mapsto X\pi_2$, while columnar transposition is described by a column permutation, followed by a row-column transcription:

$$X \mapsto (X\pi_2)^T$$

therefore, since a permutation matrix is orthogonal, $\pi_2^T = \pi_2^{-1}$, also as a row permutation with π_2^{-1} of the transposed matrix,

$$X \mapsto \pi_2^{-1} X^T \, .$$

A mixed-rows block transposition is just

$$X \mapsto (\pi_1 X)\pi_2 = \pi_1 (X\pi_2) \, ,$$

while a mixed-rows columnar transposition reads

$$X \mapsto ((\pi_1 X)\pi_2)^T \, .$$

or in the variant form

$$X \mapsto \pi_2^{-1} X^T \pi_1^{-1} \, .$$

6.2.4 Ubchi. The double columnar transposition (French *double transposition*[3], German *doppelte Spaltentransposition, Doppelwürfelverfahren*) uses simple columnar transposition twice. This would in principle mean using two different passwords, which for square matrices is not always done.

Double columnar transposition can be interpreted as

$$X \mapsto ((X\pi)^T \pi')^T = (\pi')^{-1} X\pi \, ,$$

which is indistinguishable from a mixed-rows block transposition. Both are mappings by a cross-product $\pi_i \times \pi_k$. For the Nihilist transposition the mapping is a similarity transformation, the same is true for the U.S. Army Double Transposition in case $l = k$ (see below). All these methods require essentially the same cryptanalytic techniques as simple columnar transposition.

Double columnar transposition with one password was used in the U.S. Army for quite a while ('U.S. Army Double Transposition'). It was also used unsuspectingly by the German *Kaiserliches Heer*—the French under Major, later Colonel and even General, François Cartier called it *ubchi*, since during German prewar manoeuvres, drill messages were marked *übchi*, short for *übungschiffrierung*. The French learned to break the encryption from this material and read the serious material until November 18, 1914.

Strangely, the *Deutsche Wehrmacht* had not learnt from this and returned to its sins: From the outbreak of World War II until July 1, 1941 and again from June 1, 1942 double column transposition with a password that was changed every day served as an emergency cipher for the *Heer* (*Handschlüsselverfahren*), used from regiments downwards, and for the *Kriegsmarine* (*Reserve-Handverfahren, Notschlüssel*). This time, the British read along too. The use of two different passwords, or even triple columnar transposition, would not have helped: the relevant method of "multiple anagramming" was quite general. More about cryptanalysis of columnar transposition in Chapter 21.

[3] Note the difference in French: *transposition double* (6.2.3), *double transposition* (6.2.4).

Double columnar transposition (with *one* password) was the preferred cryptographic system of the Dutch Resistance and the French Maquis. It was also used by the British espionage and sabotage organisation *Special Operations Executive* (S.O.E.), founded by *Churchill* in 1942. This was finally confirmed in 1998 by *Leo Marks*, former head of the S.O.E. code department.

A true complication for the unauthorized decryption of simple transpositions is the introduction of irregularly distributed positions which are left blank (PA-K2 system, Japan 1941; *Rasterschlüssel 44*, German Armed Forces, introduced in March 1944). The U.S.A. broke the PA-K2 encryption routinely, although often with considerable delay. The *Rasterschlüssel 44* would have been no exception, had it not been introduced very late in the war.

6.2.5 Construction of permutations. Many schemes can be imagined for the derivation of permutations to be used in the columnar transpositions from a mnemonic password. One frequently mentioned in the literature goes as follows.

Each password letter is given the number of its alphabetic ranking:

M	*A*	*C*	*B*	*E*	*T*	*H*
6	*1*	*3*	*2*	*4*	*7*	*5*

This is simple enough if the password has no repeated letters. If it has, a slightly corrected scheme ranks repeated letters consecutively:

A	*M*	*B*	*A*	*S*	*S*	*A*	*D*	*E*	*D*	*A*	*L*	*L*	*E*	*M*	*A*	*G*	*N*	*E*
1	*15*	*6*	*2*	*18*	*19*	*3*	*7*	*9*	*8*	*4*	*13*	*14*	*10*	*16*	*5*	*12*	*17*	*11*

6.3 Anagrams

Transposition leaves invariant the bag[4] of characters of a plaintext. An anagram poses the problem of reconstructing from the bag the plaintext. If anagrams could be solved systematically, then all transposition encryption would be broken.

6.3.1 Origins. Anagrams have a rich history. Huyghens gave the following

$$\mathrm{a}^7\ \mathrm{c}^5\ \mathrm{d}^1\ \mathrm{e}^5\ \mathrm{g}^1\ \mathrm{h}^1\ \mathrm{i}^7\ \mathrm{l}^4\ \mathrm{m}^2\ \mathrm{n}^9\ \ \mathrm{o}^4\ \mathrm{p}^2\ \mathrm{q}^1\ \mathrm{r}^2\ \mathrm{s}^1\ \mathrm{t}^5\ \mathrm{u}^5\ ,$$

which allows the interpretation *'annulo cingitur tenui plano, nusquam cohaerente, ad eclipticam inclinato'* ([Saturn] is girdled by a thin flat ring, nowhere touching, inclined to the ecliptic).

Newton wrote to Leibniz

$$\mathrm{a}^7\ \mathrm{c}^2\ \mathrm{d}^2\ \mathrm{e}^{14}\ \mathrm{f}^2\ \mathrm{i}^7\ \mathrm{l}^3\ \mathrm{m}^1\ \mathrm{n}^8\ \ \mathrm{o}^4\ \mathrm{q}^3\ \mathrm{r}^2\ \mathrm{s}^4\ \mathrm{t}^8\ \mathrm{v}^{12}\ \mathrm{x}^1\ ,$$

which *could* have meant *'data aequatione quodcumque fluentes quantitates involvente, fluxiones invenire et vice versa'* (from a given equation with an arbitrary number of *fluentes* to find the *fluxiones* and vice versa).

---

[4] i.e., repeated elements counted one by one. The statement is also trivially valid for sets.

Anagrams were a pastime for scientists in the 17th century, and this may be reflected in the liking amateurs have for transposition methods even nowadays.

Galilei wrote to Kepler a masked anagram:

HAEC IMMATURA A ME IAM FRUSTRA LEGUNTUR O. Y.

(These unripe things are now read by me in vain); it was to mean

'cynthiae figuras aemulatur mater amorum' (The mother of love [= Venus] imitates the phases of Cynthia [= Moon]).

A modern example is ASTRONOMERS, which can be read as *moon starers*, but also *no more stars*.

King Ludwig II of Bavaria (the insane builder of Neuschwanstein) wrote the (not very pensive) masked anagram MEICOST ETTAL, to be read *l'état c'est moi.*

The pharmaceutical industry makes use of anagrams, too: The trademark KLINOMYCIN® (Lederle) denotes the agent *Minocyclin.* This is only one example of wordplay in sales promotion.

In experimental lyrics, anagram poems are found like the following one by Francesco Gagliardi

Glück und Sommer weinen Waden, Röhricht neu,
Rad und Röcke suchen Note: Glühweinwimmern.
Randenhügel, Wut und Nock: wie Öre schimmern.
Wandertürme, Gnom in Köchern wund, eil scheu.

of type $a^1\ c^2\ d^2\ e^5\ g^1\ h^2\ i^2\ k^1\ l^1\ m^2\ n^5\ o^1\ r^3\ s^1\ t^1\ u^2\ w^2$ ö^1 ü^1 .

Among British intellectuals, anagrams are still popular today (Fig. 46). They are also the subject of riddles in German weeklies:

IRI BRÄTER, GENF	Briefträgerin
FRANK PEKL, REGEN	Krankenpfleger
PEER ASTIL, MELK	Kapellmeister
INGO DILMUR, PEINE	Diplomingenieur
EMIL REST, GERA	Lagermeister
KARL SORDORT, PEINE	Personaldirektor
GUDRUN SCHRILL, HERNE	Grundschullehrerin

6.3.2 Uniqueness. The question arises, whether from a heap of letters more than one meaningful message can be constructed. Jonathan Swift already answered this question when he pointed out in his satire *Gulliver's Travels*, that a malicious political enemy could interpret a harmless sentence like

OUR BROTHER TOM HATH JUST GOT THE PILES

by transposition (*'Anagrammatick Method'*) as the conspirative message

Resist, — a Plot is brought home — The Tour.

admonition	domination	alarmingly	marginally
algorithms	logarithms	alienators	senatorial
ancestries	resistance	antagonist	stagnation
auctioning	cautioning	australian	saturnalia
broadsides	sideboards	catalogued	coagulated
catalogues	coagulates	certifying	rectifying
collapsing	scalloping	compressed	decompress
configures	refocusing	conserving	conversing
contenting	contingent	coordinate	decoration
countering	recounting	creativity	reactivity
dealership	leadership	decimating	medicating
decimation	medication	deductions	discounted
denominate	emendation	denotation	detonation
denouncers	uncensored	deposition	positioned
descriptor	predictors	directions	discretion
discoverer	rediscover	earthiness	heartiness
egocentric	geocentric	enduringly	underlying
enervating	venerating	enervation	veneration
excitation	intoxicate	filtration	flirtation
harmonicas	maraschino	impregnate	permeating
impression	permission	impressive	permissive
indiscreet	iridescent	introduces	reductions
mouldering	remoulding	nectarines	transience
ownerships	shipowners	percussion	supersonic
persistent	prettiness	persisting	springiest
pertaining	repainting	petitioner	repetition
platitudes	stipulated	positional	spoliation
procedures	reproduces	profounder	underproof

Fig. 46. Ten-letter word anagrams (by Hugh Casement)

Indeed, experience shows, and is supported by Shannon's theory, that there is no length for which an anagram must have a unique decryption.

Historically, it should be added that an early first form of transposition is found in the ancient Greek *skytale* ($\sigma\kappa\upsilon\tau\acute{\alpha}\lambda\varepsilon$) that is known from the fifth century B.C.: a staff of wood, around which a strip of papyrus is wrapped. The secret message is written on the papyrus down the length of the staff.

After the decline of classical culture and the collapse of the Roman empire, the first encryption by transposition is found, according to Bernhard Bischoff, in the mediæval handwriting of bored monks: here and there crab, vertical writing, and play on words, often rather perfunctory.

Transposition lost its importance with the surge of mechanical cipher machines at the beginning of the 20th century, since it is hard for a mechanical device to store a great number of letters. Things have changed since then. Semiconductor technology now offers enough storage to encrypt effectively with transposition, and tiny chips provide millions of bits, with very short access time, for the price of a bus ticket. The 21st century will see transposition regain its true importance.

7 Polyalphabetic Encryption: Families of Alphabets

Monoalphabetic encryption uses some encryption step (possibly a polygraphic one) over and over. All the encryption steps treated in Chapters 3–6 can be used monoalphabetically—this was tacitly assumed in the examples.

Genuine polyalphabetic encryption requires that the set M of available encryption steps has at least two elements, i.e., that the cryptosystem M has at least the cardinality $\theta = 2$. For the frequent case $\theta = N$, where $N = |V|$, the French literature speaks of a *chiffre carré*.

The individual encryption steps can be of quite different nature, for example the cryptosystem M could consist of one or more simple substitutions and one or more transpositions of some width. This could drive an unauthorized professional decryptor crazy, since customarily all encryption steps within the same cryptosystem should belong to the same narrow class—for example, all substitutions, or all linear substitutions of the same width, or all transpositions. Frequently it is even required that all steps have equal encryption width, and block encryption may be wanted for technical reasons.

The main problem is to characterize in a simple way many different encryption steps or, as one says, to generate many different alphabets. Surprisingly, the imagination of inventors has so far left open many possibilities.

7.1 Iterated Substitutions

A natural idea is to build a cryptosystem by systematically deriving from one encryption step (primary alphabet, German *Referenzalphabet*) other encryption steps. This we have seen for simple substitution in Sect. 3.2.4, where families of derived alphabets were obtained by taking all shifts and raising to all powers.

We shall see that both these families are constructed using the concept of iterated substitution S^i, defined by $p\,S^{j+1} = (p\,S)\,S^j$ for a necessarily endomorphic ($V = W$) substitution S. In fact, iterated substitution is predominant in generating accompanying alphabets.

7.1.1 We concentrate our attention on the endomorphic case $V \doteq Q$, $W \doteq Q$. Let $S : Q^n \longleftrightarrow Q^n$. Then $S^i : Q^n \longleftrightarrow Q^n$ and we have

(a^o) $\{ S^i : i \in \mathbb{N} \}$, the *group* of powers of a mixed alphabet S.

With some substitutions $P_1 : V \longleftrightarrow Q^n$ and $P_2 : Q^n \longleftrightarrow W$, there is the set

(a) $\{ P_1 S^i P_2 : i \in \mathbb{N} \}$ where $P_1 S^i P_2 : V \longleftrightarrow W$

with the special cases (′) $V = Q^n, P_1 = \text{id}$ and (″) $W = Q^n, P_2 = \text{id}$.

Furthermore, with some additional substitution $R : Q^n \longleftrightarrow Q^n$, we have

(b^o) $\{ S^{-i} R S^i : i \in \mathbb{N} \}$, the *group* of S-similarities of R.

Again, with some substitutions P_1, P_2 as above, there is the set

(b) $\{ P_1 S^{-i} R S^i P_2 : i \in \mathbb{N} \}$ where $P_1 S^{-i} R S^i P_2 : V \longleftrightarrow W$

The families are in any case finite, since $Q^n \longleftrightarrow Q^n$ with finite $|Q| = N$ contains not more than $(N^n)!$ different permutations.

For S of the order $h \leq (N^n)!$, i.e., $S^h = \text{id}$ and $S^i \neq \text{id}$ for $i < h$, the powers produce h different alphabets. Note, that $h > N^n$ is possible: For $N = 5$, $q = 1$, the substitution (in cycle notation) $(ab)(cde)$ is of the order 6. It may happen that h is rather small: For a self-reciprocal S, there is apart from id no other power of S.

7.1.2 It is by no means necessary, but it may be advantageous, to choose for S a cyclic permutation σ. Such a permutation is of the order N^n. There are $(N^n - 1)!$ different cyclic permutations. The N^n powers of σ can be mechanized in this case, as already mentioned in Sect. 3.2.8 (Fig. 28). If N is of the order of magnitude 25, $n = 2$ will rarely be surpassed.

7.2 Shifted and Rotated Alphabets

Once a standard alphabet in Q is distinguished, there is also a standard alphabet fixed in Q^n by lexicographic ordering. The cycle belonging to this ordering (Sect. 3.2.3) and the corresponding substitution of the standard alphabet are in the following denoted by ρ. The P_i and R above are then functioning as primary alphabets.

7.2.1 With ρ^i for S^i in Sect. 7.1.1, we have the powered cycles

(a^o) $\{ \rho^i : i \in \mathbb{N} \} = \{ \rho^i \rho : i \in \mathbb{N} \} = \{ \rho \rho^i : i \in \mathbb{N} \}$,

the group of shifted standard alphabets (French *alphabets normalement parallèle*, German *verschobene Standardalphabete*). Particular cases of (a):

(a′) $\{ \rho^i P : i \in \mathbb{N} \}$ is the set of horizontally shifted (mixed) P-alphabets (French *alphabets désordonné et parallèle*),

(a″) $\{ P \rho^i : i \in \mathbb{N} \}$ is the set of vertically continued (mixed) P-alphabets (French *alphabets désordonné et étendu verticalement*).

For the general case (Eyraud: *alphabets non-normalement parallèles*)

(a*) $\{ P_1 \rho^i P_2 : i \in \mathbb{N}$ see Sect. 8.2.3 (and Sect. 19.5.3).

The designations will become clear by a look at the tables for the families of substitutions: With $V = Q = W = Z_{26}$ and $N = 26$, for the primary alphabet P generated by the mnemonic password NEWYORKCITY,

a	b	c	d	e	f	g	h	i	j	k	l	m	n	o	p	q	r	s	t	u	v	w	x	y	z
N	E	W	Y	O	R	K	C	I	T	A	B	D	F	G	H	J	L	M	P	Q	S	U	V	X	Z

the set $\{ \rho^i P : i \in \mathbb{N}\}$ has the following table (in the form of a *tabula recta*, i.e., with identical letters along the diagonals)

i	a	b	c	d	e	f	g	h	i	j	k	l	m	n	o	p	q	r	s	t	u	v	w	x	y	z
0	N	E	W	Y	O	R	K	C	I	T	A	B	D	F	G	H	J	L	M	P	Q	S	U	V	X	Z
1	E	W	Y	O	R	K	C	I	T	A	B	D	F	G	H	J	L	M	P	Q	S	U	V	X	Z	N
2	W	Y	O	R	K	C	I	T	A	B	D	F	G	H	J	L	M	P	Q	S	U	V	X	Z	N	E
3	Y	O	R	K	C	I	T	A	B	D	F	G	H	J	L	M	P	Q	S	U	V	X	Z	N	E	W
4	O	R	K	C	I	T	A	B	D	F	G	H	J	L	M	P	Q	S	U	V	X	Z	N	E	W	Y
5	R	K	C	I	T	A	B	D	F	G	H	J	L	M	P	Q	S	U	V	X	Z	N	E	W	Y	O
⋮	⋮	⋮	⋮	⋮	⋮	⋮	⋮	⋮	⋮	⋮	⋮	⋮	⋮	⋮	⋮	⋮	⋮	⋮	⋮	⋮	⋮	⋮	⋮	⋮	⋮	⋮
25	Z	N	E	W	Y	O	R	K	C	I	T	A	B	D	F	G	H	J	L	M	P	Q	S	U	V	X

while the set $\{ P\rho^i : i \in \mathbb{N}\}$ has the table

i	a	b	c	d	e	f	g	h	i	j	k	l	m	n	o	p	q	r	s	t	u	v	w	x	y	z
0	N	E	W	Y	O	R	K	C	I	T	A	B	D	F	G	H	J	L	M	P	Q	S	U	V	X	Z
1	O	F	X	Z	P	S	L	D	J	U	B	C	E	G	H	I	K	M	N	Q	R	T	V	W	Y	A
2	P	G	Y	A	Q	T	M	E	K	V	C	D	F	H	I	J	L	N	O	R	S	U	W	X	Z	B
3	Q	H	Z	B	R	U	N	F	L	W	D	E	G	I	J	K	M	O	P	S	T	V	X	Y	A	C
4	R	I	A	C	S	V	O	G	M	X	E	F	H	J	K	L	N	P	Q	T	U	W	Y	Z	B	D
5	S	J	B	D	T	W	P	H	N	Y	F	G	I	K	L	M	O	Q	R	U	V	X	Z	A	C	E
⋮	⋮	⋮	⋮	⋮	⋮	⋮	⋮	⋮	⋮	⋮	⋮	⋮	⋮	⋮	⋮	⋮	⋮	⋮	⋮	⋮	⋮	⋮	⋮	⋮	⋮	⋮
24	L	C	U	W	M	P	I	A	G	R	Y	Z	B	D	E	F	H	J	K	N	O	Q	S	T	V	X
25	M	D	V	X	N	Q	J	B	H	S	Z	A	C	E	F	G	I	K	L	O	P	R	T	U	W	Y

The *horizontally shifted P-alphabets* show the primary alphabet P in every line, shifted from line to line by one position to the left; the *vertically continued P-alphabets* show the primary alphabet P in the first line only, and it is continued vertically in the *standard* order.

7.2.2 Furthermore in Sect. 7.1.1 for the S-similarities of R, we have

(b°) $\{ \rho^{-i} R \rho^i : i \in \mathbb{N} \}$,

the group of R-rotated (the designation will be motivated in Section 7.3) standard alphabets.

For the particular case $P_1 = P, P_2 = P^{-1}$ of (b), there is

(b*) $\{ P \rho^{-i} R \rho^i P^{-1} : i \in \mathbb{N} \}$

the set of R-rotated (mixed) P-alphabets.

Taking now for R the same primary alphabet NEWYORKCITY as above, the set $\{ \rho^{-i} R \rho^{i} : i \in \mathbb{N} \}$ has the table

i	a	b	c	d	e	f	g	h	i	j	k	l	m	n	o	p	q	r	s	t	u	v	w	x	y	z
0	N	E	W	Y	O	R	K	C	I	T	A	B	D	F	G	H	J	L	M	P	Q	S	U	V	X	Z
1	A	O	F	X	Z	P	S	L	D	J	U	B	C	E	G	H	I	K	M	N	Q	R	T	V	W	Y
2	Z	B	P	G	Y	A	Q	T	M	E	K	V	C	D	F	H	I	J	L	N	O	R	S	U	W	X
3	Y	A	C	Q	H	Z	B	R	U	N	F	L	W	D	E	G	I	J	K	M	O	P	S	T	V	X
4	Y	Z	B	D	R	I	A	C	S	V	O	G	M	X	E	F	H	J	K	L	N	P	Q	T	U	W
5	X	Z	A	C	E	S	J	B	D	T	W	P	H	N	Y	F	G	I	K	L	M	O	Q	R	U	V
⋮	⋮	⋮	⋮	⋮	⋮	⋮	⋮	⋮	⋮	⋮	⋮	⋮	⋮	⋮	⋮	⋮	⋮	⋮	⋮	⋮	⋮	⋮	⋮	⋮	⋮	⋮
21	M	F	X	D	O	V	W	Y	A	B	C	E	G	H	K	L	N	P	Q	S	U	I	Z	R	T	J
22	K	N	G	Y	E	P	W	X	Z	B	C	D	F	H	I	L	M	O	Q	R	T	V	J	A	S	U
23	V	L	O	H	Z	F	Q	X	Y	A	C	D	E	G	I	J	M	N	P	R	S	U	W	K	B	T
24	U	W	M	P	I	A	G	R	Y	Z	B	D	E	F	H	J	K	N	O	Q	S	T	V	X	L	C
25	D	V	X	N	Q	J	B	H	S	Z	A	C	E	F	G	I	K	L	O	P	R	T	U	W	Y	M

The *R-rotated standard alphabets* show the primary alphabet R in the first line only, and it is continued *along the diagonals* in the *standard* order.

7.2.3 The family of shifted primary alphabets can be mechanized, as mentioned in Sect. 3.2.7, by an Alberti cipher disk or a cipher slide. We shall speak of ALBERTI encryption steps in case of horizontally shifted P-alphabets, of ROTOR encryption steps in case of R-rotated standard alphabets.

7.2.4 The cycle decomposition of the accompanying alphabets is interesting.

Example : For $Q = \begin{pmatrix} \text{a b c d e} \\ \text{b a d e c} \end{pmatrix}$ and $\rho = (\text{a b c d e}) = \begin{pmatrix} \text{a b c d e} \\ \text{b c d e a} \end{pmatrix}$,

one obtains

$$\rho Q = \begin{pmatrix} \text{a b c d e} \\ \text{b c d e a} \end{pmatrix} \begin{pmatrix} \text{b c d e a} \\ \text{a d e c b} \end{pmatrix} = \begin{pmatrix} \text{a b c d e} \\ \text{c b e a d} \end{pmatrix}$$

$$Q \rho = \begin{pmatrix} \text{a b c d e} \\ \text{b a d e c} \end{pmatrix} \begin{pmatrix} \text{b a d e c} \\ \text{c b e a d} \end{pmatrix} = \begin{pmatrix} \text{a b c d e} \\ \text{c b e a d} \end{pmatrix}$$

$$\rho^{-1} Q \rho = \begin{pmatrix} \text{a b c d e} \\ \text{e a b c d} \end{pmatrix} \begin{pmatrix} \text{e a b c d} \\ \text{d c b e a} \end{pmatrix} = \begin{pmatrix} \text{a b c d e} \\ \text{d c b e a} \end{pmatrix} .$$

In the substitution notation the alphabets are in this example

i	a b c d e
0	B A D E C
1	A D E C B
2	D E C B A
3	E C B A D
4	C B A D E

i	a b c d e
0	B A D E C
1	C B E A D
2	D C A B E
3	E D B C A
4	A E C D B

i	a b c d e
0	B A D E C
1	D C B E A
2	B E D C A
3	B C A E D
4	E C D B A

In cycle notation one obtains

for the set of horizontally shifted P-alphabets
{ (a b)(c d e) , (a)(b d c e) , (a d b e)(c) , (a e d)(b c) , (a c)(b)(d)(e)}
for the set of vertically continued P-alphabets
{ (a b)(c d e) , (a c e d)(b) , (a d b c)(e) , (a e)(b d c) , (a)(b e)(c)(d)}
for the set of R-rotated standard alphabets
{ (a b)(c d e) , (b c)(d e a) , (c d)(e a b) , (d e)(a b c) , (e a)(b c d) } .

From the theory of groups it is known that a similarity transformation $\rho^{-i}Q\rho^{i}$ leaves the length of the cycles of a permutation Q invariant. All substitutions from the family of R-rotated alphabets have the same cycle decomposition. This has been called 'The Main Theorem of Rotor Encryption'. For the (horizontally or vertically) shifted P-alphabets, this is not the case.

In our case, the partition belonging to the cycle decomposition is $3+2$. In the example of Sect. 7.2.2 , the partition is 10+8+6+1+1 , belonging to the cycle decomposition (a n f r l b e o g k) (c w u q j t p h) (d y x v s m) (i) (z) .

7.2.5 The number of different alphabets among the accompanying ones is exactly N^n for (a), for (b) it is between 1 and N^n depending on R . It is 1 if P is the identity; it is N^n , if $\rho^j P \neq P\rho^j$ for $j = 1 , 2 , \ldots , N^n - 1$. For small values of N^n , there are only few 'rotors' R fulfilling this condition.

For $N^n = 4$ and $\rho =$ (a b c d) , there are only four maximal 'rotor' families:
{ (a b), (b c), (c d), (d a) } , { (a c b d), (b d c a), (c a d b), (d b a c) } ;
{ (a c b), (b d c), (c a d), (d b a) } , { (a b c), (b c d), (c d a), (d a b) } ;
for $N^n = 3$ and $\rho =$ (a b c) only one : { (a b), (b c), (c a) } .
For $N^n = 2$, there is no 'rotor' family of two members.

7.3 Rotor Crypto Machines

With the introduction of electric typewriters, electromechanical ciphering machines came to the fore. For a realization of a fixed substitution P with electric contacts, a switchboard may serve, with N entry sockets for the plaintext characters and N exit sockets for the cryptotext characters, internally connected by N wires, Fig. 47 (a).

To obtain a realization by electric contacts for a family $\{P\rho^i\}$ of shifted P-alphabets, sliding contacts are put behind the exit sockets of the switchboard, or rather the switchboard slides before a contact row, as shown in Fig. 47 (b). In any case, flexible wires are needed, which leads to a problem of mechanical breakage.

This can be avoided if a movable contact row is attached on the entry side as well as on the exit side of the switchboard, coupling both rigidly. Sliding the switchboard does it equally well, and now flexible wires are no longer needed. This is shown in Fig. 47 (c). Duplication of the contacts is not necessary any

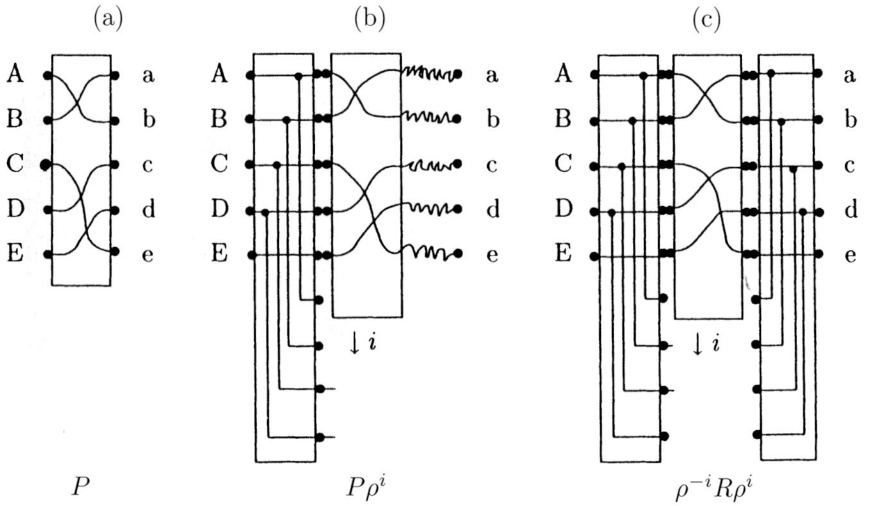

Fig. 47. Fixed substitution, shift and rotation realized by electrical contacts

longer, if a movable switching drum (German *Walze*), a rotor, is used. In this way, a realization of the family $\{\rho^{-i}R\rho^{i}\}$ is obtained, which gives it the name R-rotated (standard) alphabets.

Using slip-rings with a switching drum allows the realization of shifted P-alphabets $\{P\rho^{i}\}$. This has been called more recently a half-rotor (Fig. 48).

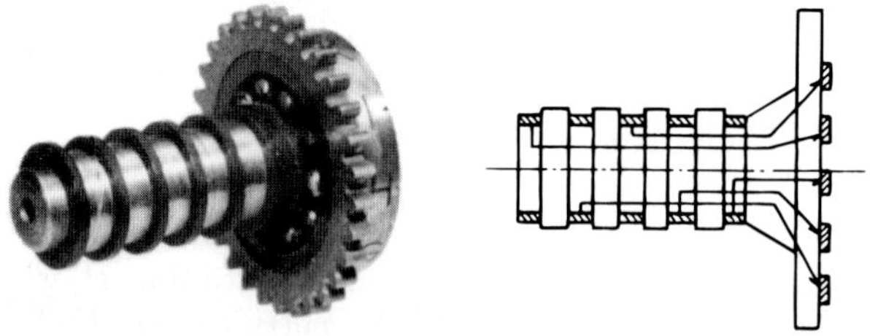

Fig. 48. 'Half-rotor' of *Arvid Damm* (1919)

7.3.1 Origins. The idea of the electric contact-rotor originated before 1920 independently in four places. According to Kahn, material from a patent hearings testimony (U.S. Patent Office Interference 77 716) shows that the American Edward Hugh Hebern (1869–1952), who had in 1915 connected two electric typewriters monoalphabetically by 26 wires, made in 1917 first drawings of a rotor to change the connection mechanically and thus to have 26 alphabets available. He only filed for a U.S. patent in 1921 and received one (No. 1 510 441) at last in 1924. Thus, three other patents were filed earlier: by the German Arthur Scherbius (patent filed February 23, 1918, German Patent 416 219), and then, almost in a dead-heat, by the Dutchman Hugo

Alexander Koch (1870–1928), patent filed October 7, 1919, and the Swede Arvid Gerhard Damm, patent filed October 10, 1919 (half-rotors).

Edward Hugh Hebern

Arvid Gerhard Damm

Arthur Scherbius

None of these inventors found fortune or happiness. Hebern was treated very badly by the U.S. Navy in 1934 and later by the U.S. Government; in 1941 he lost a patent interference case against International Business Machines. He had little income when he died from a heart attack at the age of 82. Koch died already in 1928, after Scherbius' company had bought his patents. Scherbius (October 30, 1878 – May 13, 1929) suffered a fatal accident; his company's name *Chiffriermaschinen Aktiengesellschaft* in Berlin W 35, Steglitzerstraße 2 was changed later to Heimsoeth & Rinke, and lasted until 1945. Damm was an *homme galant* and died in 1928, and his company was taken over by Boris Hagelin (July 2, 1892 – September 7, 1983), who abandoned the half-rotor in 1935 and in 1939 renamed the company in *Aktiebolaget Ingenjörsfirman Cryptoteknik*. Damm is actually out of place here in that he used his 5-pin half-rotors in pairs for a Polybios-type cipher. In the Scherbius patents, also 10-pin rotors, suited for the encryption of numeral codes, are used for demonstration purposes.

In his patent application of 1918, Scherbius discussed multiple rotors, up to ten, used sucessively. Likewise, Hebern used five rotors (two of which had a fixed position), and Scherbius used in the first commercial models ENIGMA A and ENIGMA B of 1923 four rotors (German *Durchgangsräder*), see Plate K. In the last case there is the family $\{R_{(i_1,i_2,i_3,i_4)}\}$ with

$$R_{(i_1,i_2,i_3,i_4)} = \rho^{-i_1} R_N \rho^{i_1-i_2} R_M \rho^{i_2-i_3} R_L \rho^{i_3-i_4} R_K \rho^{i_4} \quad ,$$

which has under suitable choice of R_K, R_L, R_M, R_N $26^4 = 456\,976$ members.

7.3.2 Reflector. A few years later, when the commercial ENIGMA C of 1926 showed up, Scherbius' colleague Willi Korn (patent filed March 21, 1926, German Patent 452 194) had the seemingly clever idea of introducing a reflector (German *Umkehrwalze*, in Bletchley Park frequently misspelled *Umkerwaltz* and pronounced 'Uncle Walter'). With now only three, but exchangeable rotors R_L , R_M , R_N and a properly self-reciprocal substitution U (which requires N to be even) we have (Fig. 49) the family $\{P_{(i_1,i_2,i_3)}\}$, where

$$P_{(i_1,i_2,i_3)} = S_{(i_1,i_2,i_3)}\ U\ S^{-1}_{(i_1,i_2,i_3)} \quad \text{and}$$
$$S_{(i_1,i_2,i_3)} = \rho^{-i_1}\ R_N\ \rho^{i_1-i_2}\ R_M\ \rho^{i_2-i_3}\ R_L\ \rho^{i_3} \quad .$$

All members of this family are now properly self-reciprocal substitutions. It was thought to be an advantage that encryption and decryption coincided and no longer a switch was needed. The rotor solution, however, had the consequence that no letter could be encrypted as itself. This would turn out in the end to be a great security risk (Sects. 11.2.4, 14.5.1, 19.7.2). Likewise, the fact that the electric current went through six rotors was by some people wrongly interpreted as additional cryptanalytical security.

What happened to Korn later is not known. The last patent he filed in Germany was submitted March 5, 1930, after which there are in Berlin no traces.

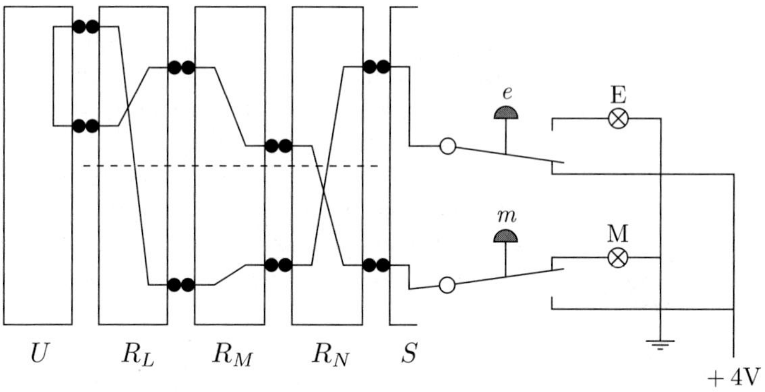

Fig. 49. Electric current in a 3-rotor ENIGMA for push-button e and lamp M.

The price of an ENIGMA was at that time about 600 Reichsmark ($ 140). Figure 49 shows the electric current for the plaintext character e and the corresponding cryptotext character M, S denotes the stator. In the ENIGMA C the reflector U could be inserted in two fixed positions, but in the commercial ENIGMA D of 1927 it could be turned like the three rotors; from the outside, it looked like another, fourth rotor, the reflecting rotor U.

Thus we have the family $\{P_{(i_1,i_2,i_3,i_4)}\}$ of proper involutions, where

$$P_{(i_1,i_2,i_3,i_4)} = S_{(i_1,i_2,i_3,i_4)}\ U\ S^{-1}_{(i_1,i_2,i_3,i_4)} \quad \text{and}$$
$$S_{(i_1,i_2,i_3,i_4)} = \rho^{-i_1}\ R_N\ \rho^{i_1-i_2}\ R_M\ \rho^{i_2-i_3}\ R_L\ \rho^{i_3-i_4} \quad .$$

The commercial ENIGMA D (Welchman: 'Glowlamp Machine') was for the first time furnished with reserve rotors; the machine was widely used and went to Sweden, the Netherlands, England, Japan, Italy, Spain, U.S.A. and was bought legally by the Polish *Biuro Szyfrów*. Its successor ENIGMA K was delivered 1938–1940 to the Swiss Army (U.S. codename INDIGO). In the later models for the *Reichswehr*, the reflecting rotor was fixed again ($i_4 = 0$).

Fig. 50a. 3-rotor *Wehrmacht* ENIGMA (1937)

Fig. 50b. The three removable rotors of the Wehrmacht ENIGMA (1937)

On July 15, 1928, the Polish Cipher Bureau for the first time picked up ENIGMA-enciphered radio signals from the *Reichswehr*. The German *Reichsmarine* had started experiments in 1925 with a 28-contact 3-rotor-ENIGMA (*Funkschlüssel C*, 1926) with an alphabetically ordered keyboard comprising additional characters Ä, Ö, Ü (X bypassed). The reflector could now be inserted in four fixed positions, denoted by $\alpha, \beta, \gamma, \delta$. In 1933, minor modifications were made to the Funkschlüssel C; a version including Ä, Ü was in test use. In the *Reichswehr* models ENIGMA G (1928) and ENIGMA I (1930) there were again only 26 contacts and a keyboard similar to that of the standard German typewriter (up to the position of the letter P); the connections to the contacts of the stator were in alphabetic order. The reflector had one fixed position. There was also a ENIGMA II with a typewriter.

The ENIGMA I ('One') of the *Heer* (introduced June 1, 1930), which became later the common *Wehrmacht* ENIGMA, was protected by a plugboard that provided for an additional (unnecessarily self-reciprocal) substitution T, called the *Steckerverbindung* (cross-plugging). This results in the

ENIGMA equation $\quad c_i = p_i\ T\ S_i\ U\ S_i^{-1}\ T^{-1} \quad (= \ p_i\ T\ U_i\ T^{-1}\ )$

between plaintext characters p_i and cryptotext characters c_i ; there is an isomorphism (Sect. 2.6.3) between $c_i\ TS_i$ and $p_i\ TS_i$, since

$$c_i\ TS_i = p_i\ TS_i\ U\ .$$

7.3.3 The *Wehrmacht* version. In 1934, *Reichsmarine* and *Heer* agreed on a common version, under pressure from Colonel Erich Fellgiebel (1886–1944), later (from 1939) Major-General and Chief, OKW Signal Communications.

The three rotors of the Navy ENIGMA (*Funkschlüssel M*, introduced in October 1934) could be selected respectively from a set of five (1934), seven (1938), or eight (1939) rotors (moreover, they could be permuted). They were marked with the roman numerals I ...VIII. Before December 15, 1938, the Army released only three of the five rotors provided for. The Air Force, too, introduced in August 1935 the *Wehrmacht* ENIGMA (Fig. 50a) for its new *Luftnachrichtentruppe*. The three removable rotors are shown in Fig. 50b.

The railroad company (*Deutsche Reichsbahn*), the post office (*Deutsche Reichspost*), and the police used less secure older models without *Steckerbrett* (plugboard), although for example messages concerning railroad transports in Russia were liable to give many clues to the enemy.

The reflector marked "A" of the *Wehrmacht* ENIGMA was replaced on November 1, 1937 by *Umkehrwalze* "B". In 1941, "C" appeared. In 1944, a 'pluggable' reflector "D", capable of rewiring, was first observed January 2, 1944 in Norway traffic. The wiring was changed thrice a month (Fig. 51b).

A special device, the *Uhr* box (Plate M) was introduced in 1944 to replace the steckering of the *Wehrmacht* ENIGMA plugboard by a non-reciprocal substitution, which also could be changed easily by turning the knob (presumably every hour). Despite the extra security it added, it was not widely used.

The rotors could be inserted in arbitrary order into the ENIGMA. Until the end of 1935, this wheel order (German *Walzenlage*) and the cross-plugging (German *Steckerverbindung*) were fixed for three months. Beginning January 1, 1936, they changed every month; from October 1, 1936, every day. Later, during the Second World War, they were changed every eight hours. The question is, why not earlier?

VIII. Beispiel.

17. Gültiger Tagesschlüssel:
(Ausschnitt aus der für die Verschlüsselung des Klartextes in Betracht kommenden Schlüsseltafel, z. B. »............ Maschinenschlüssel für Monat Mai«)

Datum	Walzenlage	Ringstellung	Grundstellung
4.	I III II	16 11 13	01 12 22

Steckerverbindung	Kenngruppen-Einsatzstelle ...... Gruppe	Kenngruppen
CO DI FR HU JW LS TX	2	adq nuz opw vxz

Fig. 51a. From the *Wehrmacht* ENIGMA operating manual, dated June 8, 1937. Setting of ring position and rotor order was in the Navy the prerogative of an officer

Geheime Kommandosache! Jeder einzelne Tagesschlüssel ist geheim! Mitnahme im Flugzeug verboten! Nº 000082

Luftwaffen-Maschinen-Schlüssel Nr. 2744

Achtung! Schlüsselmittel dürfen nicht unversehrt in Feindeshand fallen. Bei Gefahr restlos und frühzeitig vernichten.

	Monats-Tag	Walzenlage			Ringstellung			Steckerverbindungen an der Umkehrwalze	Steckerverbindungen am Steckerbrett 1	2	3	4	5	6	7	8	9	10	Zusatzsteckerverbindungen 1 1500	2 2300	Kenngruppen			
2744	31	III	V	IV	17	11	04	HK GL NQ SV UX TZ RW AD BP CO EP IM	TW	BI	UY	GP	CK	JQ	DL	RV	EM	AH	NS	FO	kim	pwh	sbx	csw
2744	30	I	IV	V	08	17	21		LS	DH	MT	EO	AP	UZ	FQ	WY	BK	GR	CI	JN	uaq	omn	ume	duf
2744	29	V	II	III	11	14	05		DO	JW	CN	IV	PZ	BM	HU	AL	FR	KX	EQ	GT	don	cqo	xum	bpg
2744	28	II	IV	V	02	20	16		NT	HK	BW	EP	LQ	AU	OY	FJ	CX	GI	DZ	MR	lui	pyg	sby	dtq
2744	27	III	V	IV	18	13	22		HM	GV	KZ	AI	DQ	NR	ES	BL	OU	FT	CP	JY	cmy	fqr	scl	bur
2744	26	I	III	II	24	10	01		GW	AQ	MO	FV	PS	DI	RU	JZ	BN	EH	KT	CL	kbj	yaq	udm	cnz
2744	25	IV	I	III	04	25	23		LT	DR	QX	AG	IN	EU	BJ	KP	FW	CM	SZ	HO	kqz	yar	vdb	coa
2744	24	V	III	I	09	19	06		GL	MY	CR	HN	JX	DT	AF	PU	IQ	BO	EW	KS	cmz	aoj	zod	auh
2744	23	IV	I	V	15	03	19		IT	DV	HQ	AJ	MU	EX	KO	CS	FY	LN	BP	GZ	kra	yas	xun	cob
2744	22	I	V	III	12	26	07		EY	JL	AK	NV	FZ	CT	HP	MX	BQ	GS	DW	IO	jdm	uhf	xuo	bph
2744	21	III	IV	II	15	09	12	AG IR BH CS DZ EW FK LX MP OU NT QV	JP	DY	QS	HL	AE	NW	CU	IK	FX	BR	MV	GO	jpf	aok	iys	btx
2744	20	IV	II	I	02	22	05		HT	NP	AM	DX	GJ	KQ	BS	OV	ER	CW	IU	FL	boy	wac	uow	cse
2744	19	V	I	II	08	19	17		GM	OX	BT	QU	DP	HJ	FK	SW	AN	EL	CY	IR	xjc	wad	unj	ctd
2744	18	III	IV	I	11	21	01		KW	IP	DM	SV	JR	CX	EN	AZ	QT	BU	FH	GY	kpn	rzi	vcm	bpo
2744	17	I	V	II	18	23	14		BV	HW	AR	NX	DS	PT	CZ	FI	LY	EJ	GK	MQ	kdx	crq	vcn	cod
2744	16	III	IV	V	16	04	07		LU	CV	FM	KR	BY	GN	QW	DJ	PS	AO	EI	HX	lgx	jri	uob	aur
2744	15	V	III	IV	24	13	10		HZ	NQ	AD	TV	IX	KM	BG	LO	CE	RY	JU	FP	wpt	vhy	zoe	aus
2744	14	I	IV	II	06	20	25		FN	UY	CJ	IW	LP	AS	DK	GQ	MO	BZ	ET	HR	wog	hxi	zxi	bpi
2744	13	III	II	I	03	26	18		KR	IZ	AT	NV	BH	MP	CG	OY	ES	DF	UW	LQ	lqv	iqb	zsy	coe
2744	12	II	IV	III	04	11	15		DT	JV	HS	CI	AY	KU	EN	FQ	LR	BW	MP	GO	zic	myt	zof	dtr
2744	11	V	I	IV	16	07	02		JS	PW	AV	QX	DN	IZ	KM	CO	EG	FL	HY	BR	inf	zbm	krz	dug
2744	10	IV	III	II	20	12	14	HL KN FM EI AC BG DS OW PZ QX RU TV	FS	CQ	JO	PR	AW	HV	EZ	KN	DU	GT	IL	BY	ink	acu	zxj	cnu
2744	9	III	II	V	06	18	10		HK	TZ	MX	LW	GQ	AD	NY	BE	CS	JP	RV	IO	efm	pmi	snw	cof
2744	8	V	I	III	01	21	17		GU	SW	BF	RX	EV	OT	LQ	CH	IP	KY	JM	NZ	imy	rjw	tjm	cog
2744	7	II	V	I	25	08	23		CX	AZ	DV	KT	HU	LW	GP	EY	MR	FQ	IN	OS	inv	rkc	snx	bpj
2744	6	IV	II	V	13	26	03		DV	LP	NQ	GZ	OS	FK	EW	MR	IT	HX	UY	BJ	yvu	hsb	swq	aut
2744	5	III	I	II	24	19	22		SY	EK	NZ	OR	CG	JM	QU	PV	BI	LW	TX	DF	seu	iqe	swr	auv
2744	4	II	IV	I	17	05	09		BD	GV	AX	KP	EM	FN	CW	RU	HO	JT	IL	QS	zfj	hxj	zxk	dpt
2744	3	V	II	IV	20	16	11		JT	NW	DU	EO	KV	BY	FS	HQ	IM	LX	GP	CR	clx	zbn	zxa	buk
2744	2	II	III	V	14	03	19		RW	OQ	GI	AZ	EJ	MS	CU	DH	PY	BF	LV	TX	ljs	jre	zpq	coh
2744	1	III	I	IV	18	24	15		NP	JV	LY	IX	KQ	AO	DZ	CR	FT	EM	GS	HW	plf	dgw	tjn	cnv

Diese Zusatzsteckerverbindungen ersetzen zu den angegebenen Zeiten die Steckerverbindungen 1 und 9

Fig. 51b. Cipher document *Nr. 2744* of the *Luftwaffe* ENIGMA, presumably for August or September 1944, showing a column "Steckerverbindungen an der Umkehrwalze"

The Navy, always suspicious that the ENIGMA could be compromised, introduced on February 1, 1942 for the key net TRITON a new version *Funkschlüssel M4* (Plate I) with a fourth rotor marked β and therefore called *Griechenwalze*. The additional rotor could be turned, but was not moved during encryption. The 4-rotor ENIGMA was first used only by the U-boats in the Atlantic. By July 1, 1943, an additional rotor γ came into use. To ensure compatibility of the new 4-rotor ENIGMA with the old 3-rotor ENIGMA, the old reflector "B" or "C" was split into a fixed thin reflecting disk "B dünn" or "C dünn" and the turnable additional rotor β or γ, respectively. Rohwer presumed in 1978 a further 'Greek rotor' α; Deavours and Kruh (1985) followed him. To this David Kahn: "No α rotor was ever recovered. In fact, splitting the reflector "A", which had disappeared in 1937, did not make sense.

Another invention Paul Bernstein made already for the ENIGMA A did not develop into a cryptological fiasco: The ring which allowed the rotor position (alphabet ring, German *Sperr-Ring*) to be read was like a tyre mounted round the rim of the rotor, and its position with respect to the core, the ring setting (German *Ringstellung*), could be fixed with a pin.

Figure 51a shows a section from the ENIGMA operating manual, with daily changing rotor order, ring settings, and cross-pluggings. The *Grundstellung* (ground setting) characterizes the situation of the three rotors when enciphering is started. The *Kenngruppen* had no cryptological meaning. Figure 51b shows a cipher document of the *Luftwaffe* ENIGMA from 1944, indicating that in 1944 the pluggable reflector was changed every 10 days and that some *Steckerverbindungen* were changed every 8 hours.

The *Heer* used ENIGMAs from regiments upwards. An estimated total of 200 000 (Johnson) is certainly too high, and a low estimate from the Polish side is that 20000 ENIGMAs were in use in until 1938, 40000 at the outbreak of the war, which would suggest nearly 100 000 altogether. After the Second World War, the victorious countries sold captured ENIGMA machines, at that time still widely thought to be secure, to developing countries.

7.3.4 TYPEX. In England, too, rotor machines were built in the Second World War: TYPEX was quite an improved ENIGMA (instead of the plugboard there was an entrance substitution performed by two fixed rotors which was not self-reciprocal, Sect. 22.2.7)). In the U.S.A., under the early influence of William Friedman (1891–1969) and on the basis of the Hebern development, there was in the early 1930s a more independent line of rotor machines, leading in 1933 to the M-134-T2, then to the M-134-A (SIGMYC), and in 1936 to the M-134-C (SIGABA) of the Army, named CSP889 (ECM Mark II) by the Navy. The Germans obviously did not succeed in breaking the SIGABA, which had been made watertight by Frank Rowlett (1908–1998), an aide of Friedman since April 1930.

An interesting postwar variant of the ENIGMA with seven rotors and a fixed reflector was built and marketed by the Italian company Ottico Meccanica

Italiana (OMI) in Rome. The Swiss army used from 1946 on an ENIGMA variant NEMA with four active rotors, built by Zellweger A.G., Uster.

7.3.5 The ENIGMA substitutions

7.3.5.1 The substitutions performed by the stator, the three rotors, and the reflecting rotor of the ENIGMA D are (Cipher A. Deavours, Louis Kruh):

entry	a b c d e f g h i j k l m n o p q r s t u v w x y z
stator	J W U L C M N O H P Q Z Y X I R A D K E G V B T S F
exit rotor 1	L P G S Z M H A E O Q K V X R F Y B U T N I C J D W
exit rotor 2	S L V G B T F X J Q O H E W I R Z Y A M K P C N D U
exit rotor 3	C J G D P S H K T U R A W Z X F M Y N Q O B V L I E
exit reflector	I M E T C G F R A Y S Q B Z X W L H K D V U P O J N

7.3.5.2 The substitutions performed by the stator, the eight rotors, and the reflecting rotors of the *Wehrmacht* ENIGMA are (Ralph Erskine, Frode Weierud, Heinz Ulbricht):

entry	a b c d e f g h i j k l m n o p q r s t u v w x y z
stator	A B C D E F G H I J K L M N O P Q R S T U V W X Y Z
exit rotor I	E K M F L G D Q V Z N T O W Y H X U S P A I B R C J
exit rotor II	A J D K S I R U X B L H W T M C Q G Z N P Y F V O E
exit rotor III	B D F H J L C P R T X V Z N Y E I W G A K M U S Q O
exit rotor IV	E S O V P Z J A Y Q U I R H X L N F T G K D C M W B
exit rotor V	V Z B R G I T Y U P S D N H L X A W M J Q O F E C K
exit rotor VI	J P G V O U M F Y Q B E N H Z R D K A S X L I C T W
exit rotor VII	N Z J H G R C X M Y S W B O U F A I V L P E K Q D T
exit rotor VIII	F K Q H T L X O C B J S P D Z R A M E W N I U Y G V
exit reflector A	E J M Z A L Y X V B W F C R Q U O N T S P I K H G D
exit reflector B	Y R U H Q S L D P X N G O K M I E B F Z C W V J A T
exit reflector C	F V P J I A O Y E D R Z X W G C T K U Q S B N M H L
exit rotor Beta	L E Y J V C N I X W P B Q M D R T A K Z G F U H O S
exit reflector B dünn	E N K Q A U Y W J I C O P B L M D X Z V F T H R G S
exit rotor Gamma	F S O K A N U E R H M B T I Y C W L Q P Z X V G J D
exit reflector C dünn	R D O B J N T K V E H M L F C W Z A X G Y I P S U Q

Note: Beta, followed by B dünn, followed by Beta^{-1}, equals B,
e.g. $\text{Beta}(a) = \text{L}$, $\text{B dünn}(\text{L}) = \text{O}$, $\text{Beta}^{-1}(\text{O}) = y$; thus $\text{B}(a) = y$.

7.3.5.3 The rotors of the ENIGMA D and of the *Wehrmacht* ENIGMA show the following cycle partitions of the substitutions, allowing easy identification:

Stator	13+7+3+2+1	Rotor I	10+4+4+3+2+2+1
Rotor 1	25+1	Rotor II	8+7+3+2+2+2+1+1
Rotor 2	17+7+2	Rotor III	17+8+1
Rotor 3	19+6+1	Rotor IV	22+2+2
		Rotor V	11+9+6
		Rotor VI	14+8+4
		Rotor VII	26
		Rotor VIII	17+3+3+3

7.3.5.4 Rotor I of the *Wehrmacht* ENIGMA gives the following rotated alphabets (with ring setting A for $i = 0$):

ring setting	i	a	b	c	d	e	f	g	h	i	j	k	l	m	n	o	p	q	r	s	t	u	v	w	x	y	z
A	0	E	K	M	F	L	G	D	Q	V	Z	N	T	O	W	Y	H	X	U	S	P	A	I	B	R	C	J
Z	1	K	F	L	N	G	M	H	E	R	W	A	O	U	P	X	Z	I	Y	V	T	Q	B	J	C	S	D
Y	2	E	L	G	M	O	H	N	I	F	S	X	B	P	V	Q	Y	A	J	Z	W	U	R	C	K	D	T
X	3	U	F	M	H	N	P	I	O	J	G	T	Y	C	Q	W	R	Z	B	K	A	X	V	S	D	L	E
W	4	F	V	G	N	I	O	Q	J	P	K	H	U	Z	D	R	X	S	A	C	L	B	Y	W	T	E	M
V	5	N	G	W	H	O	J	P	R	K	Q	L	I	V	A	E	S	Y	T	B	D	M	C	Z	X	U	F
⋮	⋮	⋮	⋮	⋮	⋮	⋮	⋮	⋮	⋮	⋮	⋮	⋮	⋮	⋮	⋮	⋮	⋮	⋮	⋮	⋮	⋮	⋮	⋮	⋮	⋮	⋮	⋮

7.4 Shifted Standard Alphabets: Vigenère and Beaufort

Referring to Sect. 7.2.1, choosing in (a′) $P=\rho$, gives $\{ \rho^i \rho \; : \; i \in \mathbb{N}\}$, the set of horizontally shifted standard alphabets, which coincides, see (a°), with the set $\{\rho^i : i \in \mathbb{N} \}$ of powers of the standard alphabet and likewise, see (a″), with the set of vertically continued standard alphabets. This case was treated by the Benedictine abbot Johannes Heidenberg from Trittenheim on the Moselle river, latinized Trithemius (1462–1516), in the fifth book of his *Polygraphiae* in a standard regular table ('*tabula recta*', Fig. 52, French *table régulière*). For its mechanization, an Alberti disk can be used, which carries also on the inner ring the standard alphabet; as well as a disk with the cycle ρ of the standard alphabet and a movable window, since powers are to be formed. The literature speaks of VIGENÈRE encryption steps. More correctly, this trivial case should be named after Trithemius. The secondary literature of the 19th century was unjust to Vigenère inasfar as only shifted *standard* alphabets were connected with his name, while his proposal actually was not limited to this: Vigenère wrote in the table heading of a *tabula recta* a mixed alphabet; obviously this was equivalent to Alberti's disk.

The set $\{\rho^i : i \in \mathbb{N} \}$ is implemented electrically by a half-rotor (Fig. 48).

7.4.1 VIGENÈRE. Evidently, a VIGENÈRE encryption step is a linear substitution: The cycle ρ defines the linear cyclic quasiordering of V^n and thus an addition *modulo* N^n (Chapter 5); the set of alphabets $\{ \rho^i : i \in \mathbb{N}\}$ corresponds to the addition of shift numbers A_i (French: *nombre de décalage*):

$$\{ A_i : \; i \in \mathbb{Z}_{N^n} \} \quad \text{with} \quad A_i : A_i(x) \overset{N^n}{\simeq} x + i \; .$$

The decryption step $\qquad A_i^{-1} : A_i^{-1}(y) \overset{N^n}{\simeq} y - i$

amounts to a subtraction of the shift number. The case $n=1$ is predominant. Such 'cryptographic equations' were used about 1846 by Charles Babbage (British Museum, Add. Ms. 37205, Folio 59), but he did not publish them.

VIGENÈRE encryption comes sometimes in disguise: In 1913, when Wilson was President, the U.S. State Department and the U.S. Army introduced a VIGENÈRE variant named 'Larrabee', using twenty-six cards, each one

Recta transpositionis tabula.

a	b	c	d	e	f	g	h	i	k	l	m	n	o	p	q	r	s	t	u	x	y	z	w
b	c	d	e	f	g	h	i	k	l	m	n	o	p	q	r	s	t	u	x	y	z	w	a
c	d	e	f	g	h	i	k	l	m	n	o	p	q	r	s	t	u	x	y	z	w	a	b
d	e	f	g	h	i	k	l	m	n	o	p	q	r	s	t	u	x	y	z	w	a	b	c
e	f	g	h	i	k	l	m	n	o	p	q	r	s	t	u	x	y	z	w	a	b	c	d
f	g	h	i	k	l	m	n	o	p	q	r	s	t	u	x	y	z	w	a	b	c	d	e
g	h	i	k	l	m	n	o	p	q	r	s	t	u	x	y	z	w	a	b	c	d	e	f
h	i	k	l	m	n	o	p	q	r	s	t	u	x	y	z	w	a	b	c	d	e	f	g
i	k	l	m	n	o	p	q	r	s	t	u	x	y	z	w	a	b	c	d	e	f	g	h
k	l	m	n	o	p	q	r	s	t	u	x	y	z	w	a	b	c	d	e	f	g	h	i
l	m	n	o	p	q	r	s	t	u	x	y	z	w	a	b	c	d	e	f	g	h	i	k
m	n	o	p	q	r	s	t	u	x	y	z	w	a	b	c	d	e	f	g	h	i	k	l
n	o	p	q	r	s	t	u	x	y	z	w	a	b	c	d	e	f	g	h	i	k	l	m
o	p	q	r	s	t	u	x	y	z	w	a	b	c	d	e	f	g	h	i	k	l	m	n
p	q	r	s	t	u	x	y	z	w	a	b	c	d	e	f	g	h	i	k	l	m	n	o
q	r	s	t	u	x	y	z	w	a	b	c	d	e	f	g	h	i	k	l	m	n	o	p
r	s	t	u	x	y	z	w	a	b	c	d	e	f	g	h	i	k	l	m	n	o	p	q
s	t	u	x	y	z	w	a	b	c	d	e	f	g	h	i	k	l	m	n	o	p	q	r
t	u	x	y	z	w	a	b	c	d	e	f	g	h	i	k	l	m	n	o	p	q	r	s
u	x	y	z	w	a	b	c	d	e	f	g	h	i	k	l	m	n	o	p	q	r	s	t
x	y	z	w	a	b	c	d	e	f	g	h	i	k	l	m	n	o	p	q	r	s	t	u
y	z	w	a	b	c	d	e	f	g	h	i	k	l	m	n	o	p	q	r	s	t	u	x
z	w	a	b	c	d	e	f	g	h	i	k	l	m	n	o	p	q	r	s	t	u	x	y
w	a	b	c	d	e	f	g	h	i	k	l	m	n	o	p	q	r	s	t	u	x	y	z

Fig. 52. '*tabula recta*' of Trithemius
(Original in the State Library Munich)

showing the standard plaintext alphabet Z_{26} and the cryptotext alphabet obtained by adding the shift number.

7.4.2 EYRAUD. Another family of accompanying encryption steps, involving multiplication, is suggested in Sect. 5.6 (EYRAUD encryption steps)

$$\{\, C_q : q \in \mathbb{Z}_{N^n} \ \wedge\ gcd(q,\ N^n) = 1 \,\} \quad \text{with} \qquad C_q : C_q(x) \overset{N^n}{\simeq} q \cdot x\ .$$

The decryption step, using q' such that $q{\cdot}q' \overset{N^n}{\simeq} 1$, is $C_q^{-1} : C_q^{-1}(x) \overset{N^n}{\simeq} q'{\cdot}x$.

This is the family of decimated alphabets (French *alphabets chevauchants*, Eyraud), the properly decimated alphabets are those with $|q| \neq 1$.

The most general linear substitution in the ring $\mathbb{Z}_{N^n}$ is a composition of VIGENÈRE and EYRAUD encryption steps:

$$T_{q,i} :\ T_{q,i}(x) \overset{N^n}{\simeq} q \cdot x + i \ =\ A_i(C_q(x))\ .^{1}$$

---

1 Eyraud has described a method producing this most general family: After two cycles of cleartext characters and of cryptotext characters are given, for every permissible pair q, q' the following is done: The one cycle is written in lines of q, the other one in lines of q' elements, in both cases the cycles are periodically continued. An initial pairing is assumed—this amounts to determining the i. By reading rowwise in the cleartext block

7.4.3 BEAUFORT. The case of fixed $q = q' = N^n - 1$ yields the family

$$\{ B_i : i \in \mathbb{Z}_{N^n} \} \quad \text{with} \quad B_i : \; B_i(x) \overset{N^n}{\simeq} i - x \; .$$

The literature calls this BEAUFORT encryption, although it was studied already by Giovanni Sestri in 1710 and rediscovered in 1857 by Admiral Sir Francis Beaufort (1774–1857), who is famous for the scale of wind speed. The decryption step

$$B_i^{-1} : \; B_i^{-1}(y) \overset{N^n}{\simeq} i - y$$

coincides with the encryption step, so the BEAUFORT encryption step is self-reciprocal, but not properly:

$$x \text{ is fixpoint of } B_i \text{, if and only if } \; 2 \cdot x \overset{N^n}{\simeq} i \; .$$

In the classical cases of VIGENÈRE and BEAUFORT encryption steps, naturally $n = 1$. Incidentally, it was de Viaris[2], a man oriented towards mathematics, who in 1888 published the interpretation of the VIGENÈRE and BEAUFORT encryption steps as addition and subtraction *modulo* N, after Kerckhoffs in 1883 (without knowing of the studies of Babbage) had shown the mathematical relations between VIGENÈRE and BEAUFORT. Before this time, and partly even later, the processes had been explained by the slides which were in practical use (*Saint-Cyr* slide).

---

and columnwise in the cryptotext block, the encryption step is obtained by juxtaposition; reading columnwise in the cleartext block and rowwise in the cryptotext block gives the decryption step.

Example: $n = 1$, $N^n = 26$, $q = 3$, $q' = 9$; Initial pairing: **c** – **P** .

Cleartext cycle e v i t z l s **c** o u r a n d b f g h j k m p q w x y
Cryptotext cycle S E C U R I T Y A B D F G H J K L M N O **P** Q V W X Z

e v i	S E C U R I T Y A	e v i	S E C U R I T Y A
t z l	B D F G H J K L M	t z l	B D F G H J K L M
s **c** o	N O **P** Q V W X Z S	s **c** o	N O **P** Q V W X Z S
u r a	E C **U** R I T Y A B	u **r** a	E C U R I T Y A B
n d b	D F **G** H J K L M N	n **d** b	D F G H J K L M N
f g h	O P **Q** V W X Z S E	f **g** h	O P Q V W X Z S E
j k m	C U **R** I T Y A B D	j **k** m	C U R I T Y A B D
p q w	F G **H** J K L M N O	p **q** w	F G H J K L M N O
x y e	P Q **V** W X Z S E C	x **y** e	P Q V W X Z S E C
v i t	U R **I** T Y A B D F	v **i** t	U R I T Y A B D F
z l s	G H **J** K L M N O P	z **l** s	G H J K L M N O P
⋮	⋮	⋮	⋮

Encryption step c o u r a n d b f g h j k m p q w x y e v i t z l s
P U G Q R H V I J W T K X Y L Z A M S B N E D O C F

Decryption step P Q V W X Z S E C U R I T Y A B D F G H J K L M N O
c r d g k q y i l o a b h m w e t s u n f j p x v z

in cycle notation (a r q z o u g w)(b i e)(c p l)(d v n h t)(f j k x m y s) .

[2] Marquis Gaëtan Henri Léon de Viaris, 1847–1901, French cryptologist. De Viaris invented in about 1885 one of the first printing cipher machines—according to Kahn, the very first were invented presumably before 1874 by Émile Vinay and Joseph Gaussin.

With any fixed slide position, a VIGENÈRE encryption step turns into a CAESAR encryption step, while a BEAUFORT encryption step turns into a CAESAR encryption step for the reversed alphabet.

7.4.4 BACKWARDS VIGENÈRE, BACKWARDS BEAUFORT. In the English literature the backwards VIGENÈRE

$$E_i: \quad E_i(x) \overset{N^n}{\simeq} -i+x = -(i-x)$$

is also called 'Variant Beaufort', in French 'variante à l'allemande'. It was proposed in 1858 by Lewis Carroll and described by de Viaris.

The trivial self-reciprocal backwards BEAUFORT

$$F_i: \quad F_i(x) \overset{N^n}{\simeq} -i-x,$$

also described by de Viaris, was rediscovered in 1972 by Ole Immanuel Franksen.

A 'variant', attributed by Caspar Schott in the *Schola steganographia* (1665) to Count Gronsfeld, is nothing more than a literal VIGENÈRE, using only ten alphabets, which were designated by the figures 0 ... 9 . Cryptographically this brings nothing but disadvantages. Jules Verne describes it in his novel *The Giant Raft*, 1881. In 1892, a group of French anarchists used it, and the cipher was broken by Étienne Bazeries.

7.4.5 PORTA. A family of eleven self-reciprocal substitutions ($V = Z_{22}$) was used as early as 1563 by Giovanni Battista Porta (Fig. 53). These alphabets are called shifted self-reciprocal alphabets, they are designated homophonically by 22 key letters arranged in pairs. A similar arrangement with ten alphabets ($V = Z_{20}$) was used in 1589 by G. B. and M. Argenti (Fig. 69). For an alphabet with an even number $N = 2\nu$ of characters, there are ν such properly self-reciprocal alphabets; we shall call them PORTA encryption steps.

7.5 Unrelated Alphabets

Porta robbed himself of the fame of being the inventor of a general polyalphabetic substitution based on a number θ ($\theta \le (N^n)!$) of 'mutually unrelated' mixed alphabets, that is to say of alphabets which are not related one to another in any such a simple algebraic way as shift or similarity transformation (Kahn: "The order of the letters in the tableau may be arranged arbitrarily, provided no letter is omitted").

7.5.1 PERMUTE. Although Porta described in his *De furtivis literarum notis* this case of multiple mixed alphabets (French *alphabets indépendants*, German *unabhängige Alphabete*), he did not illustrate it except with shifted self-reciprocal alphabets like the ones in Fig. 53. Eyraud is inclined to give this glory solely to the Frenchman Vigenère. Likewise, Luigi Sacco, author of the excellent *Manuale di crittografia* (3rd ed., Rome 1947), favored

LITERAE SCRIPTI.

LITERAE CLARIS.

AB	a	b	c	d	e	f	g	h	i	l	m
	n	o	p	q	r	ſ	t	u	x	y	z
CD	a	b	c	d	e	f	g	h	i	l	m
	z	n	o	p	q	r	ſ	t	u	x	y
EF	a	b	c	d	e	f	g	h	i	l	m
	y	z	n	o	p	q	r	ſ	t	u	x
GH	a	b	c	d	e	f	g	h	i	l	m
	x	y	z	n	o	p	q	r	ſ	t	u
IL	a	b	c	d	e	f	g	h	i	l	m
	u	x	y	z	n	o	p	q	r	ſ	t
MN	a	b	c	d	e	f	g	h	i	l	m
	t	u	x	y	z	n	o	p	q	r	ſ
OP	a	b	c	d	e	f	g	h	i	l	m
	ſ	t	u	x	y	z	n	o	p	q	r
QR	a	b	c	d	e	f	g	h	i	l	m
	r	ſ	t	u	x	y	z	n	o	p	q
ST	a	b	c	d	e	f	g	h	i	l	m
	q	r	ſ	t	u	x	y	z	n	o	p
VX	a	b	c	d	e	f	g	h	i	l	m
	p	q	r	ſ	t	u	x	y	z	n	o
YZ	a	b	c	d	e	f	g	h	i	l	m
	o	p	q	r	ſ	t	u	x	y	z	n

Fig. 53. Eleven self-reciprocal alphabets for polyalphabetic encryption (Porta 1563)

Italy (Kahn: "trying to prove that everything was Italian first"). Charles J. Mendelsohn, who was beyond favoritism, praised Porta as "the outstanding cryptographer of the Renaissance." When we are dealing with most general permutations, we shall speak of a family of PERMUTE encryption steps.

A table for general polyalphabetic substitution with unrelated alphabets could look like this (note the construction principle from a password, namely

passwords serve to select a method from a class of methods and keys especially to select encryptions e ...):

	c	h	a	p	t	e	r	l	v	n	b	d	f	g	i	j	k	m	o	q	s	u	w	x	y	z
C	**P**	**A**	**S**	**W**	**O**	**R**	**D**	**E**	**V**	**T**	B	C	F	G	H	I	J	K	L	M	N	Q	U	X	Y	Z
R	N	P	Q	R	U	V	W	X	Y	Z	**O**	**S**	**E**	**L**	**C**	**T**	**A**	**M**	**H**	**D**	B	F	G	I	J	K
Y	**L**	**S**	**E**	**T**	B	D	G	H	I	J	K	N	P	Q	U	V	W	X	Y	Z	**F**	**R**	**O**	**M**	**A**	**C**
P	V	W	X	Z	**H**	**O**	**D**	**S**	**A**	**N**	**K**	**E**	**Y**	**P**	B	C	F	G	I	J	L	M	Q	R	T	U
T	F	G	H	J	K	M	P	Q	R	U	V	W	X	Z	**E**	**C**	**I**	**A**	**L**	**Y**	**T**	**O**	**S**	**N**	B	D
O	**Y**	**P**	**T**	**I**	**O**	**N**	**S**	**E**	A	B	D	F	G	H	J	K	L	M	Q	U	V	W	X	Z	**C**	**R**
⋮	⋮	⋮	⋮	⋮	⋮	⋮	⋮	⋮	⋮	⋮	⋮	⋮	⋮	⋮	⋮	⋮	⋮	⋮	⋮	⋮	⋮	⋮	⋮	⋮	⋮	⋮

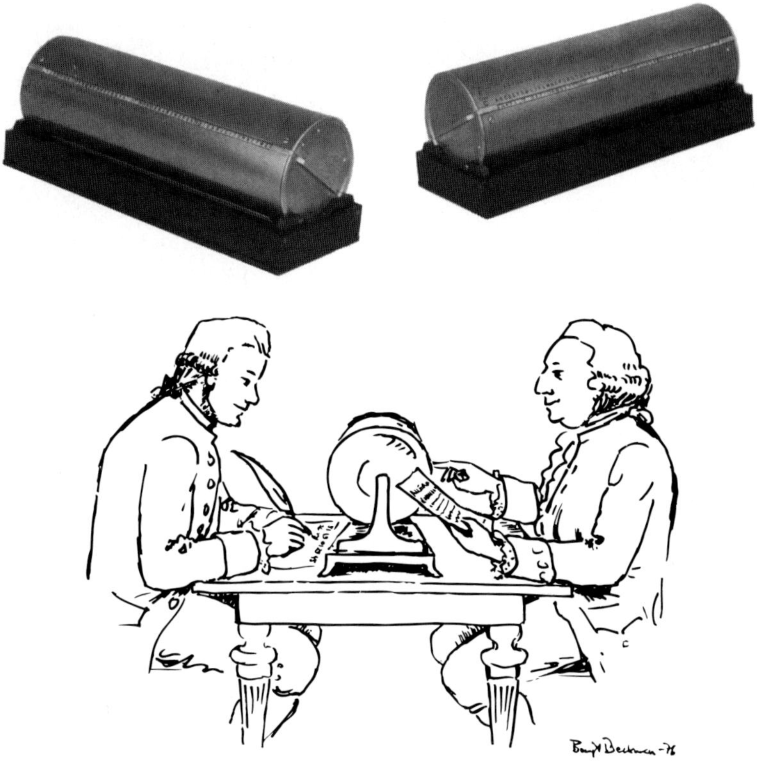

Fig. 54. Ciphering device of Baron Fredrik Gripenstierna (Crypto AG)

7.5.2 Gripenstierna. General polyalphabetic substitution with unrelated alphabets is present in the little known ciphering device of 1786, built by the Swedish baron Fredrik Gripenstierna (1728–1804) for King Gustav III of Sweden (Fig. 54), reconstructed by Crypto AG, Zug (Switzerland) from documents discovered by Sven Wäsström in the State Archive Stockholm. The device had 57 disks, each one for a different (fixed) bipartite simple substitution $Z_{26} \rightarrow Z_{10}^2$. Considered as a polygraphic substitution with a width of 57, permutation of the disks gave the family the immense number of $57! \approx 4.05 \cdot 10^{76}$ alphabets. Even if used with unchanged order of the disks for a message of several hundred characters or for several such messages, the cryptosystem was far better than anything else around at that time.

Likewise, in 1799, the Roman Catholic priest Johann Baptist Andres (1770–1823) described the use of a table with 26 unrelated mixed alphabets, to be selected periodically according to a key.

In 1915, the Swedish inventor Arvid Damm conceived a device somewhat along these lines, using a number of exchangeable bands with unrelated mixed alphabets in an arrangement on a drum parallel to its axis (A-21: Fig. 55). Next to the bands was a straight edge for the plaintext alphabet; after each

Fig. 55. A-21 (1915) by Arvid Damm (A.B. Cryptograph, Stockholm)

step the drum was moved one step. The straight edge for the plaintext alphabet could be brought into two positions, which were changed with a relatively short period. This cryptosystem was far inferior to Gripenstierna's.

Polyalphabetic encryption with unrelated alphabets was used in the First World War by a German radio station for messages to a sabotage group in North Africa ("für GOD" system), as well as by the U.S. Air Force in the Second World War for air-ground traffic (SYKO)—and on both occasions it was insecure. SYKO consisted of thirty self-reciprocal alphabets, printed on cards—the same old "Larrabee" idea (Sect. 7.4.1). The alphabets were used in some cyclic order—the encipherer using an indicator to define the beginning—for a whole day. That was far too long and was the reason for the weakness of an otherwise good system.

1	a	b	c	d	e	f	g	h	i	j	k	l	m	n	o	p	q	r	s	t	u	v	x	y	z
2	b	c	d	f	g	h	j	k	l	m	n	p	q	r	s	t	v	x	z	a	e	i	o	u	y
3	a	e	b	c	d	f	g	h	i	o	j	k	l	m	n	p	u	y	q	r	s	t	v	x	z
4	z	y	x	v	u	t	s	r	q	p	o	n	m	l	k	j	i	h	g	f	e	d	c	b	a
5	y	u	z	x	v	t	s	r	o	i	q	p	n	m	l	k	e	a	j	h	g	f	d	c	b
6	z	x	v	t	s	r	q	p	n	m	l	k	j	h	g	f	d	c	b	y	u	o	i	e	a
7	a	l	o	n	s	e	f	t	d	p	r	i	j	u	g	v	b	c	h	k	m	q	x	y	z
8	b	i	e	n	h	u	r	x	l	s	p	a	v	d	t	o	y	m	c	f	g	j	k	q	z
9	c	h	a	r	y	b	d	e	t	s	l	f	g	i	j	k	m	n	o	p	q	u	v	x	z
10	d	i	e	u	p	r	o	t	g	l	a	f	n	c	b	h	j	k	m	q	s	v	x	y	z
11	e	v	i	t	z	l	s	c	o	u	r	a	n	d	b	f	g	h	j	k	m	p	q	x	y
12	f	o	r	m	e	z	l	s	a	i	c	u	x	b	d	g	h	j	k	n	p	q	t	v	y
13	g	l	o	i	r	e	m	t	d	n	s	a	u	x	b	c	f	h	j	k	p	q	v	y	z
14	h	o	n	e	u	r	t	p	a	i	b	c	d	f	g	j	k	l	m	q	s	v	x	y	z
15	i	n	s	t	r	u	e	z	l	a	j	b	c	d	f	g	h	k	m	o	p	q	v	x	y
16	j	a	i	m	e	l	o	g	n	f	r	t	h	u	b	c	d	k	p	q	s	v	x	y	z
17	k	y	r	i	e	l	s	o	n	a	b	c	d	f	g	h	j	m	p	q	t	u	v	x	z
18	l	h	o	m	e	p	r	s	t	d	i	u	a	b	c	f	g	j	k	n	q	v	x	y	z
19	m	o	n	t	e	z	a	c	h	v	l	b	d	f	g	i	j	k	p	q	r	s	u	x	y
20	n	o	u	s	t	e	l	a	c	f	b	d	g	h	i	j	k	m	p	q	r	v	x	y	z

Fig. 56. The twenty cycles of Bazeries

7.5.3 MULTIPLEX. Restricted to full cyclic permutations, polyalphabetic encryption with unrelated alphabets has found classical use in theform of a special ciphering device, the cylinder used by Jefferson (between 1790 and 1800) and reinvented in 1891 by Bazeries. The cyclic substitutions are represented one by one as a cycle at the rim of a thin cylinder (French *rondelle*). Jefferson ordered 36 such cylinders (each one with a mixed Z_{26}) into a long cylinder; Bazeries used 20 cylinders (each one with a mixed Z_{25}), as shown by Fig. 19. Friedman called these families of unrelated cycles 'multiplex systems.'

Thomas Jefferson (1743–1826)

Fourteen of the twenty cycles Bazeries used—they are found in Fig. 56—originated from whimsical dicta, passwords like

Allons enfants de la patrie, le jour de gloire est arrivé
Bienheureux les pauvres d'esprit, le royaume des Cieux
Charybde et Scilla
Dieu protège la France
Évitez les courants d'air
Formez les faisceaux
Gloire immortelle de nos aïeux
Honneur et Patrie
Instruisez la jeunesse
J'aime l'oignon frit à l'huile
Kyrie eleison
L'homme propose et Dieu dispose
Montez à cheval
Nous tenons la clef

Bazeries was not successful in convincing the French *état-major général* to accept his invention—de Viaris (Sect. 14.3.1) succeeded in showing how to break messages encrypted with the cylinder if the alphabets were known, a realistic assumption in the military combat situation (Sect. 11.2.3). Apparently, Bazeries did not know that Jefferson long before had had the same idea, and most likely—he died in 1931, at the age of 85—he did not learn of the late vindication of his proposal in 1922 by the U.S. Army. A device with thirteen cylinders was proposed in 1900 by the Italian Colonel Oliver Ducros.

The cylinders of Jefferson and Bazeries allowed the encrypted text to be read off (Fig. 21) not only in the next, but in an arbitrarily chosen i-th line (the "i-th generatrix"). Thus, the encryption was polyphonic. The authorized decryptor, after having set the cryptotext, simply looked for a line that struck the eye. For unauthorized decryption, this complication was less harmful than one would naively think (Sect. 14.3.1).

Fig. 57. Cryptographer assembling M-94

Normally, the order of the cylinders was left unchanged for a whole message, even for several messages, or for a predetermined period, such as a day.

Instead of cylinders, strips with a duplication of the alphabet can be used. Such a ciphering device was proposed in 1893 by the Frenchman Arthur J. Hermann. It was propagated in 1914 by Captain, later Colonel, Parker Hitt, referring to Bazeries. He also had no success at first. Meanwhile, in 1917, Russell Willson, a naval lieutenant, also invented a strip device, the NCB (Navy Code Box), which was used in the U.S. Navy at least until 1935. But the U.S. Army turned after all to the Jefferson cylinder; the famous M-94, introduced in 1922 under the influence of Friedman after substantial improvements had been made in the alphabets by Colonel Mauborgne[3], then head of the Signal Corps' research and engineering division. It had 25 thin aluminum cylinders the size of a silver dollar, turning on a spindle 110 mm long. The M-94 (Plate D, Fig. 57) was declared obsolete in 1943, when sufficient M-209s (Sect. 4.4.8) were available.

Gilbert Vernam

Parker Hitt

Joseph O. Mauborgne

[3] It was the same Mauborgne who had in 1918 improved Vernam's bitwise encryption through the introduction of endless and senseless keys ('one-time keys'), see Sect. 8.8.2.

In 1934, M-138, a strip version, was adopted; one hundred strips were available and thirty were used at a time. The improved M-138-A from 1938 served military officers and diplomats. It was thought to be so secure from unauthorized decryption that a radio signal from Roosevelt to Churchill, immediately after the Atlantic Conference in August 1941, was sent via M-138-A . This was very much to the misgivings of the distrustful Roosevelt, who was cryptologically experienced—Churchill was irresponsible as far as encryption security goes. While the Japanese seemingly were not able to break the American strip cipher, the Germans did: Hans Rohrbach in 1944 (Sect. 14.3.6) broke it without having access to the alphabets. (His success did not last long, for soon afterwards the U.S.A. made a change to the SIGTOT Vernam-type machines.) Plate E shows the version M-138-T4 with 25 strips used at a time.

The U.S. State Department system 'O-2' that Rohrbach had defeated used from fifty available strips thirty at a time, namely two groups of fifteen strips each. Here was a risk: The total number of available strips should be considerably larger than the period, i.e., the number of strips used at a time.

By the way, the U.S. Navy used as a successor to the Cipher Box the ciphering device CSP 642, also with thirty strips. The Japanese seized some of these and took great pains to break messages, without success—presumably they had not studied the methods of de Viaris and Friedman (Sect. 14.3).

For cylinder and strip cipher devices, we shall follow Friedman and denote the encryption steps as MULTIPLEX encryption steps. They are special, i.e., fully cyclic, PERMUTE encryption steps.

7.5.4 The Latin square requirement. In the special case $\theta \leq N$ it can be required that the N permuted alphabets of N characters each, written row by row, have the following property: in no column does a character occur more than once (Eyraud: '*alphabets réellement non-parallèles*'). In the case $\theta = N$, the alphabets form a 'Latin square' in the sense that also in every column each character occurs just once. This requirement (with the insufficient justification that it allows the table to be turned around) was already mentioned in the *Geheimschreibekunst* of Johann Baptist Andres in 1799 (see Sect. 7.5.2) who also gave an example with his table. The *tabula recta* trivially fulfills the requirement, but its alphabets are not unrelated.

The requirement can be postulated for the permuted alphabets belonging to the cycles of a multiplex system (Sect. 7.5.3) with the effect of preventing the de Viaris attack (Sect. 14.3.1) . Cycles derived from mnemonic passwords are unlikely to qualify, in fact the permuted alphabets belonging to the cycles of Bazeries (Table 2a) show a peculiar effect: most columns have one or two letters occurring frequently. It is clear that many lacking letters help to make a break. The alphabets belonging to the cycles of Bazeries—irrespective of how they are supplemented by five more—cannot give a Latin square.

Usually, a standard alphabet of N characters is chosen for the first row and for the first column of a Latin square. For $N = 2$ and $N = 3$ there are only

	a	b	c	d	e	f	g	h	i	j	k	l	m	n	o	p	q	r	s	t	u	v	x	y	z
1	b	c	d	e	f	g	h	i	j	k	l	m	n	o	p	q	r	s	t	u	v	x	y	z	a
2	e	c	d	f	i	g	h	j	o	k	l	m	n	p	u	q	r	s	t	v	y	x	z	b	a
3	e	c	d	f	b	g	h	i	o	k	l	m	n	p	j	u	r	s	t	v	y	x	z	q	a
4	z	a	b	c	d	e	f	g	h	i	j	k	l	m	n	o	p	q	r	s	t	u	v	x	y
5	j	y	b	c	a	d	f	g	q	h	e	k	l	m	i	n	p	o	r	s	z	t	v	u	x
6	z	y	b	c	a	d	f	g	e	h	j	k	l	m	i	n	p	q	r	s	o	t	v	u	x
7	l	c	h	p	f	t	v	k	j	u	m	o	q	s	n	r	x	i	e	d	g	b	y	z	a
8	v	i	f	t	n	g	j	u	e	k	q	s	c	h	y	a	z	x	p	o	r	d	l	m	b
9	r	d	h	e	t	g	i	a	j	k	m	f	n	o	p	q	u	y	l	s	v	x	z	b	c
10	f	h	b	i	u	n	l	j	e	k	m	a	q	c	t	r	s	o	v	g	p	x	y	z	d
11	n	f	o	b	v	g	h	j	t	k	m	s	p	d	u	q	x	a	c	z	r	i	y	e	l
12	i	d	u	g	z	o	h	j	c	k	n	s	e	p	r	q	t	m	a	v	x	y	b	f	l
13	u	c	f	n	m	h	l	j	r	k	p	o	t	s	i	q	v	e	a	d	x	y	b	z	g
14	i	c	d	f	u	g	j	o	b	k	l	m	q	e	n	a	s	t	v	p	r	x	y	z	h
15	j	c	d	f	z	g	h	k	n	b	m	a	o	s	p	q	v	u	t	r	e	x	y	i	l
16	i	c	d	k	l	r	n	u	m	a	p	o	e	f	g	q	s	t	v	h	b	x	y	z	j
17	b	c	d	f	l	g	h	j	e	m	y	s	p	a	n	q	t	i	o	u	v	x	z	r	k
18	b	c	f	i	p	g	j	o	u	k	n	h	e	q	m	r	v	s	t	d	a	x	y	z	l
19	c	d	h	f	z	g	i	v	j	k	p	b	o	t	n	q	r	s	u	e	x	l	y	m	a
20	c	d	f	g	l	b	h	i	j	k	m	a	p	o	u	q	r	v	t	e	s	x	y	z	n

Table 2a. The 20 permuted alphabets corresponding to the cycles of Bazeries

the trivial solutions of a *tabula recta*. For $N = 4$ there are, apart from the *tabula recta*, three more of these 'reduced' Latin squares:

a	b	c	d		a	b	c	d		a	b	c	d		a	b	c	d
b	c	d	a		b	d	a	c		b	a	d	c		b	a	d	c
c	d	a	b		c	a	d	b		c	d	b	a		c	d	a	b
d	a	b	c		d	c	b	a		d	c	a	b		d	c	b	a

The numbers grow fast: 56 for $N = 5$, 9 408 for $N = 6$, 16 942 080 for $N = 7$, 535 281 401 856 for $N = 8$. For $N = 9$ there are 377 597 570 964 258 816 reduced Latin squares, as calculated by S. E. Bammel and J. Rothstein in 1975, while the total number of all reduced square sets of nine alphabets with nine characters amounts to $(8!)^8 \approx 6.98 \cdot 10^{36}$.

Two examples of Latin squares with $N = 10$, with $Z_{10} = \{0, 1, 2, \ldots, 9\}$ are:

0	1	2	3	4	5	6	7	8	9		0	1	2	3	4	5	6	7	8	9
1	5	7	2	8	9	0	3	4	6		1	4	3	2	0	9	8	5	6	7
2	4	6	1	3	8	9	0	5	7		2	6	5	4	3	0	9	8	7	1
3	0	5	7	2	4	8	9	6	1		3	8	7	6	5	4	0	9	1	2
4	9	0	6	1	3	5	8	7	2		4	9	8	1	7	6	5	0	2	3
5	8	9	0	7	2	4	6	1	3		5	0	9	8	2	1	7	6	3	4
6	7	8	9	0	1	3	5	2	4		6	7	0	9	8	3	2	1	4	5
7	6	1	8	9	0	2	4	3	5		7	2	1	0	9	8	4	3	5	6
8	3	4	5	6	7	1	2	9	0		8	3	4	5	6	7	1	2	9	0
9	2	3	4	5	6	7	1	0	8		9	5	6	7	1	2	3	4	0	8

Table 2b shows the alphabets belonging to the 25 cycles of the M-94; they almost form a Latin square for $N = 26$. Why Mauborgne provided for the three exceptions in alphabet 16 is unknown.

Note also that the rotated alphabets (7.2.2)—in contrast to the shifted alphabets (7.2.1)—usually do not form a Latin square.

Simple arithmetic formulas for the number $l(N)$ of reduced Latin squares of N rows and columns have not been given so far. Erdős (1913–1996) and Kaplanski conjectured in 1946 that asymptotically

$$l(N) \asymp N \cdot (N!)^{N-2} / e^{N \cdot (N-1)/2} \qquad (\, l(9) < 1.73 \cdot 10^{24} \,).$$

For $N \leq 9$, empirically a pretty good upper bound is

$$l(N) \leq \sqrt{((N-1)!)^{N-1}} \qquad (\, l(9) < 2.64 \cdot 10^{18} \,).$$

A very crude lower bound (Heise) is

$$l(N) \geq 2! \cdot 3! \cdot 4! \cdot \ldots \cdot (N-2)! \qquad (\, l(9) > 1.25 \cdot 10^{11} \,).$$

Note that $l(9) \approx 3.78 \cdot 10^{17}$. For $l(26)$, the above bounds give $10^{243} < l(26) < 10^{499}$. A value closer to the upper bound is to be expected.

	a	b	c	d	e	f	g	h	i	j	k	l	m	n	o	p	q	r	s	t	u	v	w	x	y	z
1	b	c	e	j	i	v	d	t	g	f	z	r	h	a	l	w	k	x	p	q	y	u	n	s	m	o
2	c	a	d	e	h	i	z	f	j	k	t	m	o	p	u	q	x	w	b	l	v	y	s	r	g	n
3	d	g	z	k	p	y	e	s	n	u	o	a	j	x	m	h	r	t	c	v	b	w	l	f	q	i
4	e	i	b	c	d	g	j	l	f	h	m	k	r	w	q	t	v	u	a	n	o	p	y	z	x	s
5	f	r	y	o	m	n	a	c	t	b	d	w	z	q	p	i	u	h	l	j	k	x	e	g	s	v
6	g	j	i	y	t	k	p	w	x	s	v	u	e	d	c	o	f	n	q	a	r	m	b	l	z	h
7	h	n	f	u	z	m	s	x	k	e	p	c	q	i	g	v	t	o	y	w	l	r	a	j	d	b
8	i	w	v	x	r	z	t	p	h	o	c	q	g	s	b	j	e	y	u	d	m	f	k	a	n	l
9	j	x	r	s	f	h	y	g	v	d	q	p	b	l	i	m	o	a	k	z	n	t	c	w	u	e
10	k	d	a	f	l	j	h	o	c	g	e	b	t	m	n	r	s	q	v	p	x	z	i	y	w	u
11	l	e	g	i	j	b	k	u	z	a	r	t	s	o	h	n	p	f	x	m	w	q	d	v	c	y
12	m	y	u	v	w	l	c	q	s	t	x	h	n	f	a	z	g	d	r	b	j	e	o	i	p	k
13	n	m	j	h	a	e	x	b	l	i	g	d	k	c	r	f	y	p	w	s	z	o	q	u	v	t
14	o	l	t	w	g	a	n	z	u	v	j	e	f	y	d	k	h	s	m	x	q	i	p	b	r	c
15	p	v	x	r	n	q	u	i	y	z	s	j	a	t	w	b	d	l	g	c	e	h	f	o	k	m
16	q	t	s	**e**	o	p	i	d	m	n	f	x	w	u	k	y	j	v	h	g	**b**	l	z	c	**a**	r
17	r	k	w	p	u	t	q	e	b	x	l	n	y	v	f	c	i	m	z	h	s	a	g	d	o	j
18	s	o	n	m	q	u	v	a	w	r	y	g	c	e	z	l	b	k	d	f	i	j	x	h	t	p
19	t	s	m	z	k	x	w	v	r	y	u	f	i	g	j	d	a	b	e	o	p	c	h	n	l	q
20	u	p	k	g	s	c	f	j	o	w	a	y	d	h	v	e	l	z	n	r	t	b	m	q	i	x
21	v	f	l	q	y	s	o	r	p	m	h	z	u	k	x	a	c	g	j	i	d	n	t	e	b	w
22	w	h	o	l	b	d	m	k	e	q	n	i	x	r	t	u	z	j	f	y	c	s	v	p	a	g
23	x	z	p	t	v	o	b	m	q	c	w	s	l	j	y	g	n	e	i	u	f	d	r	k	h	a
24	y	q	h	a	c	r	l	n	d	p	b	o	v	z	s	x	w	i	t	e	g	k	u	m	j	f
25	z	u	q	n	x	w	r	y	a	l	i	v	p	b	e	s	m	c	o	k	h	g	j	t	f	d

Table 2b. Almost Latin square, belonging to Mauborgne's alphabets for the M-94. A cyclic permutation of the three boldface letters in line 16 establishes a correct Latin square.

8 Polyalphabetic Encryption: Keys

No message is safe in cipher unless the key phrase
is comparable in length with the message itself.
Parker Hitt 1914

8.1 Early Methods with Periodic Keys

8.1.1 Alberti. The earliest attempts at polyalphabetic encryption can be found in the work *De cifris* of Leon Battista Alberti (1404–1472), an essay of 25 pages he wrote in 1466 or 1467 for his friend Leonardo Dato, the papal secretary. The Latin original is reproduced in Aloys Meister, *Die Geheimschrift im Dienste der Päpstlichen Kurie*, Paderborn, Schöningh, 1906, pp. 125–141 (Italian translation *Trattati in cifra*, manuscript, about 1470).

Alberti was not only an architect, painter, music composer, and organ player, but also a great Renaissance scholar. He knew how to break a simple substitution cipher, and so he had thoughts on how to avoid this. He proposed to change the substitution alphabet after every three or four words, "introducing a new meaning of the cipher letters." For this purpose he invented a device, the turnable cipher disk (Fig. 26), which made quite a number of derived substitution alphabets available. Three or four words is on average 18 letters. Thus, Alberti unconsciously stayed below Shannon's unicity distance (Sect. 12.6) for simple substitution. This was great progress compared to the then common use of homophones: While in a homophonic simple substitution $Z_{25} \dashrightarrow Z_{10}^2$ the bigrams 89, 43, 57, and 64 could mean the letter /a/ , now every bigram could mean /a/ . Of course, encryptor and decryptor had identical disks.

The secondary literature does not agree completely on how with Alberti's disk the change of alphabet was accomplished. Sacco, an Italian cryptologist and Eyraud, a French one, give the following explanation: A particular letter, say /b/, is agreed upon as indicator (French *index*). The encryptor inserts in front of every part of the text, that is to be encrypted with a new alphabet, an arbitrary one of the four figures 1 ... 4. Whenever a figure has been enciphered, the new position of the turnable disk is determined by juxtaposing the indicator with the cipher equivalent of this figure. The decryptor knows that when a figure appears in the decrypted plaintext, his next step is to turn the disk and set the cryptotext letter against the indicator. By the way, this procedure is the first instance of covert key communication, so important with modern encryption machines.

Kahn sees the main use of the figures on Alberti's disk as a superencryption of the quaternary code introduced by Alberti as well; a code of 336 groups of two, three and four of the figures 1 ... 4, to be interspersed within literal text. Alberti mentioned also ordering the code for encryption in words and for decryption in groups. Whether this early two-part code was thoroughly mixed, as it was by Rossignol for increased security, is unclear.

By introducing polyalphabetic encryption *and* superencrypted code, Alberti may be called the father of modern cryptology, without disrespect for the architect of the Palazzo Pitti, the churches Sant' Andrea at Mantua, Santa Maria Novella at Florence, and the Tempio Malatestiana at Rimini. These works established his fame, but his cryptologic relevance was forgotten for a long time.

8.1.2 Trithemius. While Alberti changed the alphabet after every three or four words, Trithemius proposed already in Vol. V of his *Polygraphiæ* (1508–1518) to proceed after every letter to the next alphabet. He did so, however, according to a regular, periodic progression—just line by line of his *tabula recta*. Thus, his method was with respect to this far inferior to that of Alberti.

On the other hand, he used all available alphabets before an alphabet was used a second time. Following Kahn, this is called a "progressive key" crypto system—not to be confused with the expression "running key" (Sect. 2.3.3) introduced by Friedman. Modern cipher machines display a special liking for progressive encryption, but use many more than two dozen alphabets. More about this in Sect. 8.4.3.

Trithemius' encryption was a fixed crypto system with a period of 24; it can be considered a monoalphabetic polygraphic encryption of width 24. One would not believe that such a method would be used professionally in the 20th century. Actually, it was even used with period 3 (followed by a simple columnar transposition) in 1914 by the Germans on the Western front. The French called it ABC—in today's language it is a periodic VIGENÈRE encryption with special key $A\,B\,C$—and liked it, of course (Sect. 2.1.1).

Porta then showed in 1602 how a Trithemius encryption can be attacked: If the plaintext contains three alphabetically consecutive letters like *pon* in *pondus*, they give a triple letter in the cryptotext.

Neither Alberti nor Trithemius used key words in connection with polyalphabetic encryption. Giovanni Battista Belaso was the first (1553) to denote the encryption steps consecutively with letters A, B, ..., Z and to use a keytext to select the alphabets—either by turning the cipher disk or by choosing a line in a table. His keytext consisted of rather long phrases like *OPTARE MELIORA* and *VIRTUTI OMNIA PARENT*, to be repeated if necessary. Such a repeated key leads to a periodic polyalphabetic (monographic) encryption (Sect. 2.3.3). Still using the standard alphabet only, the combinatorial complexity Z of this method is N^d for keys with d letters, for the polygraphic case of width n it is $Z = (N^n)^d = N^{n \cdot d}$.

8.2 'Double Key'

8.2.1 Porta. Ten years after Belaso, in 1563, his use of a key was combined with Alberti's use of a mixed alphabet and other ones derived from it by shifts. This decisive step was made by 28-year-old Giovanni Battista Porta (1535–1615) in his *De furtivis literarum notis*, a book Kahn calls "extraordinary, with freshness, charm, and ability to instruct."

Since the password determining the mixed alphabet was already called the key, it became customary to speak of a 'double cipher' (in the French terminology, this is alive even today, *substitution à double clef*), but this is now obsolete. Kahn writes: "Givierge was even then [1920s] calling polyalphabetic systems by the almost obfuscatory 'double substitution' which tells absolutely nothing at all about the system." Givierge speaks of *clef principale* for the 'actual key'. The password is sometimes called 'second key'.

The combinatorial complexity of this method for keys with d letters and an arbitrary mixed alphabet is for simple (monographic) encryption $N!\cdot N^{d-1}$, for the polygraphic case of width n it is $(N^n)!\cdot(N^n)^{d-1}$.

Instead of a disk, a table can be used, of course. It would read for the case of Alberti's disk in Fig. 26 (with $\{\rho^{-i}P : i \in \mathbb{N}\}$):

	a	b	c	d	e	f	g	i	l	m	n	o	p	q	r	s	t	v	x	z	1	2	3	4
0	D	L	G	A	Z	E	N	B	O	S	F	C	H	T	Y	Q	I	X	K	V	P	&	M	R
1	R	D	L	G	A	Z	E	N	B	O	S	F	C	H	T	Y	Q	I	X	K	V	P	&	M
2	M	R	D	L	G	A	Z	E	N	B	O	S	F	C	H	T	Y	Q	I	X	K	V	P	&
3	&	M	R	D	L	G	A	Z	E	N	B	O	S	F	C	H	T	Y	Q	I	X	K	V	P
⋮	⋮	⋮	⋮	⋮	⋮	⋮	⋮	⋮	⋮	⋮	⋮	⋮	⋮	⋮	⋮	⋮	⋮	⋮	⋮	⋮	⋮	⋮	⋮	⋮

The decryption steps are obviously obtained, if the mixed alphabet of the cryptotext characters in the line $i = 0$ is put atop a *tabula recta* of the cleartext characters, in our example (with $\{P^{-1}\rho^{i} : i \in \mathbb{N}\}$):

	D	L	G	A	Z	E	N	B	O	S	F	C	H	T	Y	Q	I	X	K	V	P	&	M	R
0	a	b	c	d	e	f	g	i	l	m	n	o	p	q	r	s	t	v	x	z	1	2	3	4
1	b	c	d	e	f	g	i	l	m	n	o	p	q	r	s	t	v	x	z	1	2	3	4	a
2	c	d	e	f	g	i	l	m	n	o	p	q	r	s	t	v	x	z	1	2	3	4	a	b
3	d	e	f	g	i	l	m	n	o	p	q	r	s	t	v	x	z	1	2	3	4	a	b	c
⋮	⋮	⋮	⋮	⋮	⋮	⋮	⋮	⋮	⋮	⋮	⋮	⋮	⋮	⋮	⋮	⋮	⋮	⋮	⋮	⋮	⋮	⋮	⋮	⋮

8.2.2 Vigenère. It is just this connection of a *tabula recta* with an arbitrary substitution (cf. Sect. 7.4) that was proposed in 1585 by Vigenère. Proposing the use of keys as well, he obtained the full power of ALBERTI encryption steps. He also recognized how important it was to choose quite long key words for making cryptanalysis difficult.

Blaise de Vigenère was born April 5, 1523 in Saint-Pourçain, "halfway between Paris and Marseilles", writes Kahn with American liberality. He went to the Diet of Worms as a very young secretary, and subsequent travels through Europe in diplomatic missions widened his experience, then for the rest of his life he served the Duke of Nevers. He read Trithemius, Belaso, Cardano and Porta and had access to Alberti's manuscript. In 1570, at the age of 47, he concentrated fully on writing; until his death in 1596 he wrote on everything on earth, even a *Traicté des Comètes*. He wrote his *Traicté des Chiffres* at age 62 in 1585, "despite the distraction of a year-old baby daughter," as Kahn writes. In 1570 Vigenère had married Marie Varé, who was many years younger. The cryptologic book had more than 600 pages and contained a lot more than cryptography—Japanese ideograms, alchemy, magic, kabbalah, recipes for making gold, but also a reliable, precise reflection of the status of cryptology at that time. Discussing polyalphabetic encryption, he followed Alberti and Trithemius in the use of alphabets obtained by shifts, and marked the rows by key characters as Belaso and Porta had done for their self-reciprocal alphabet. Altogether, he gave the picture of polyalphabetic simple substitution its modern form.

8.2.3 A 'treble key' (French *triple clef*) is obtained, if *two* primary alphabets are combined with a keyed iterated substitution, for example if for given P_1, P_2 the case (a) in Sect. 7.2.1 is taken into account, i.e., the set of alphabets $\{P_1 \rho^i P_2 \; : \; i \in \mathbb{N}\}$ (Sect. 19.5.3). Vigenère had gone into this case by denoting the VIGENÈRE encryption steps with a mixed alphabet of key letters.

8.3 Vernam Encryption

Modern communication channels work in a binary alphabet $Z_2 = \{O, L\}$ or $\mathbb{Z}_2 = \{0, 1\}$. Encrypting the symbols of the International Teletype Alphabet CCITT No. 2 can be seen as a polygraphic binary encoding with $N=2$ and $n=5$; for the encryption of bytes, i.e., of 8-bit characters, which often serve in today's computers as basic units, we have the case $N = 2$ and $n = 8$ of binary octograms, for blocks of 8 bytes $N=2$ and $n=64$.

Restricted to VIGENÈRE encryption steps, there are 32 for $\mathbb{Z}_2^5$ and 256 for $\mathbb{Z}_2^8$; their execution as addition *modulo* 32 or *modulo* 256 requires a cyclic adder with a width of 5 bits or 8 bits. A suitable binary circuitry (with $n = 5$) is shown in Fig. 58. Larger microprocessors today allow even 64-bit addition and can directly encrypt byte octograms.

8.3.1 Bitwise encryption. On the other hand, a VIGENÈRE encryption step can be executed bit by bit. This extreme case of a bitwise encryption will become particularly important later. If a bitwise binary encryption $\mathbb{Z}_2 \rightarrowtail \mathbb{Z}_2$ is a mapping, it is a permutation of the two characters 0 and 1 ;

the identity $O : \begin{array}{ccc} 0 & \mapsto & 0 \\ 1 & \mapsto & 1 \end{array}$ and the reflection $L : \begin{array}{ccc} 0 & \mapsto & 1 \\ 1 & \mapsto & 0 \end{array}$

are the only encryption steps (VERNAM encryption steps). They coincide with the VIGENÈRE and BEAUFORT steps $+0$ and $+1$. The encryption is necessarily self-reciprocal, but not properly self-reciprocal. $|M| = 2$ is the smallest integer that allows a polyalphabetic encryption. Thus, the key of a

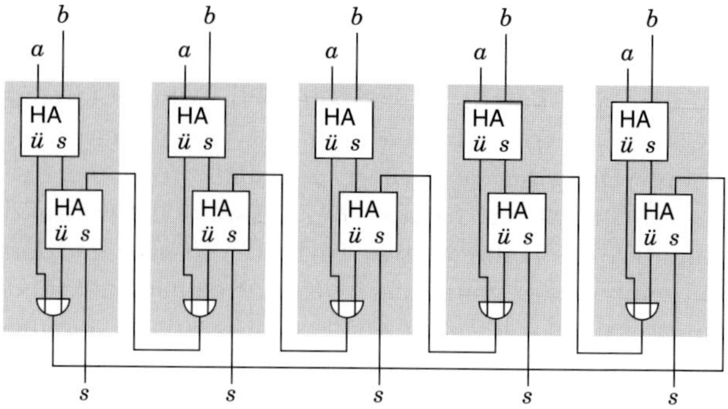

Fig. 58.
Addition circuit built from half-adders **HA**

VERNAM encryption is generated by a finite (O,L) word that is periodically repeated, or it is an infinite (O,L) sequence like $O\ L\ L\ O\ L\ O\ O\ L\ L\ O \ldots$.

Since in $\mathbb{Z}_2$ the identity O can be performed by the addition of 0 and the reflection L by the addition of 1, the encryption $\mathbb{Z}_2 \to \mathbb{Z}_2$ is a linear transformation. Addition in $\mathbb{Z}_2$, addition *modulo* 2, frequently denoted by $\oplus$, coincides with the Boolean operation $\not\leftrightarrow$ (Exclusive Or), which is also called 'non-carrying binary addition', since it is the sum output of a half-adder.

8.3.2 Vernam. The idea of realizing these two encryption steps by electric contacts came in 1917 (before Lester S. Hill) to a young employee of AT & T in New York, Gilbert S. Vernam (1890–1960).

Vernam constructed for a commercial teletypewriter a binary VIGENÈRE encryption supplement. The key was punched on normal 5-channel teletype tape that could be linked to form a rather long loop. By double encrytion with short loops of 999 and 1000 characters, Lyman F. Morehouse, in Vernam's team, obtained a key that was 999000 characters long and, more important, was 'senseless'. Vernam applied on September 13, 1918 for a U.S. patent and obtained it in 1919 under the number 1 310 719. On the commercial level, it was not successful; codes were more in demand. But the idea was adopted in professional diplomatic and military cryptology, among others in the Siemens double punch-tape reader with 'mixer' and later in the U.S. Army SIGTOT machine.

8.3.3 Mutilated carry. Transition from a VIGENÈRE encryption step in $\mathbb{Z}_{2^n}$ (performed by addition of $(a_1\ a_2\ a_3\ \ldots\ a_n)$ in the binary system) to n polyalphabetic VERNAM encryption steps (that is, VIGENÈRE encryption steps in $\mathbb{Z}_2$ with addition of a_1, of a_2, $\ldots$, of a_n) amounts to dismantling the carry part of the electronic binary addition circuitry.

The same goes for VIGENÈRE encryption steps in $\mathbb{Z}_{10^n}$, performed by addition *modulo* 10^n of numbers from $\{0, \ldots\ 10^n - 1\}$. For a mechanical desk calculator, transition to n VIGENÈRE encryption steps in $\mathbb{Z}_{10}$ amounts to dismantling the mechanical carry device (Sect. 5.7); such a mutilated desk calculator performs polyalphabetic encryption over its full working width.

8.4 Quasi-nonperiodic Keys

8.4.1 Polyalphabetic encryption considered arduous. Despite the cryptanalytic security it offered when used properly, periodic polyalphabetic encryption with long keys found it difficult to win against the nomenclators. It was first used in exceptional cases only: in the Papal Curia in 1590, where it was broken by Chorrin, a decryptor of Henri IV; and by the Cardinal de Retz in 1654 for communications with the Prince of Condé (Louis II of Bourbon) before the outbreak of the Huguenot War, when it was broken by Guy Joly who guessed the key word, which was the preferred method. In 1791 Marie Antoinette used polyalphabetic encryption in her exchange of amatory and conspiratorial letters (Sect. 2.1.1). Her lover from 1783, the Swedish Count Axel von Fersen, concocted a Porta-like assembly of 23 self-reciprocal alphabets (Fig. 59). Axel von Fersen was cautious and did not use obvious mnemonic words but *mots vides* like DEPUIS, VOTRE. It was not to be blamed on cryptanalysis that the escape of Louis XVI and Marie Antoinette ended at the bridge of Varennes, for none of their messages had been decrypted.

A	(ab)	(cd)	(ef)	(gh)	(i k)	(lm)	(no)	(pq)	(rs)	(tu)	(xy)	(z&)
B	(bk)	(du)	(ei)	(f l)	(gn)	(ho)	(my)	(ps)	(qx)	(rt)	(ac)	(&z)
C	(l r)	(ad)	(bg)	(cz)	(s&)	(ek)	(fm)	(ht)	(ix)	(np)	(oq)	(uy)
⋮	⋮	⋮	⋮	⋮	⋮	⋮	⋮	⋮	⋮	⋮	⋮	⋮

Fig. 59. Marie Antoinette's polyalphabetic self-reciprocal encryption

Before it was mechanized, polyalphabetic substitution had a reputation for being cumbersome and prone to error. William Blair wrote in an 1819 encyclopædia article: "polyalphabetic substitution requires too much time and by the least mistake in writing is so confounded ... ."

The same complaint is found in a 17th-century Brussels book, *Traitté de l'art de dechiffrer*: "... takes too long to encipher them, dropping of a single ciphertext letter garbles the message from that point on ... ."

8.4.2 Polyalphabetic encryption considered safe. However, polyalphabetic substitution was also reputed to be unbreakable. Matteo Argenti wrote, "The key cipher is the noblest and the greatest in the world, the most secure and faithful that never was there a man who could find it out."

Until the 19th century, the only genuine break occurred when trivial alphabets with shifts were used, words of the cleartext could be guessed, and a short key could be reconstructed—not to speak of guessing the key word, as Porta and the Argentis succeeded in doing. This changed only with the rise of systematic solution in the middle of the 19th century.

To exclude these possibilities of attack, it is advisable (Parker Hitt) to take the period length of the keytext to be considerably larger than the whole plaintext (quasi-nonperiodic key), or to use a nonperiodic key (Sect. 8.7).

8.4.3 Progressive encryption. But if quasi-nonperiodic encryption is envisaged, it is advisable for the sake of security to use many more alphabets than one usually has key characters. These alphabets should also be selected less regularly than mnemonic keywords provide for (Kahn: "irregular sequence of alphabets"). Moreover, if a great number of alphabets are available, it may be worth using progressive encryption in the following sense.

Progressive encryption is a periodic polyalphabetic encryption that uses no alphabet again before all other alphabets have been used. Thus, the period d of a progressive encryption coincides with the power θ of the set of encryption steps. A quasi-nonperiodic encryption results when the message is shorter than θ .

Progressive encryption was already proposed by Trithemius with his *tabula recta* (Sects. 7.4, 8.1.2), although with 24 shifted standard alphabets it did not provide much security. Progressive encryption is systemic with the cylinder and strip devices, where each alphabet is only available in *one* copy. Progressive encryption was also favored in the mechanical or electromechanical encryption machines of the first half of the 20th century. In the following, cascading stepwise movement of a set of rotors is typically progressive.

8.4.4 'Regular' rotor movement. Although nothing prevented rotors from having many contacts (the half-rotor of the Japanese *angō kikai taipu A*, Fig. 66, had 60), it seemed natural in case of Z_N to have N contacts and thus only N alphabets. To achieve a high period for a progressive encryption, the weak solution Hebern and Scherbius found independently was to step several rotors successively as in a counter ('regular' rotor movement). With the four rotors, as in the ENIGMA A ($N=26$), the period d is equal to or at least (in 'almost progressive encryption') not much smaller than $\theta = 26^4 = 456\,976$. This is an impressive number, which means the period does not need to be exhausted for a message the length of a typical novel. With five rotors, θ is 12 million, which is many more letters than there are in the entire Bible.

8.5 Machines that Generate Their Own Key Sequences

Crypto machines of some comfort frequently have a double function: They perform polyalphabetic encryptions, and they generate their own key sequence for the selection of these encryptions. If keytext generation is included, it is the crucial issue of mechanization.

8.5.1 Wheatstone. Trithemius used the shifted (standard) alphabets straightforwardly one by one, and this was still done (with shifted mixed alphabets) in the *Cryptograph* of Wheatstone, 1867 (Plate C). This meant the use of a fixed key. The use of keys by Belaso allowed the experienced encryptor enough irregularity in the selection of the alphabets.

8.5.2 Attempts at irregularity. Keytext generators with such a long period that normally for *one* message the full period by far is not exhausted, offer an 'irregularity in the selection of the alphabets', caused by the encipherer in choosing the starting point of the keytext cycle. *Arthur Scherbius* therefore provided in his basic patent application filed February 23, 1918 (German Patent 416219) in the first instance the regular, cyclometric rotor movement ("like that of counters") only as *one* possibility.

Arvid Gerhard Damm, one of the inventors of the rotor principle, made in his Swedish patent application of October 10, 1919 a rather weak attempt at irregularity: four gears ('key wheels'), one for each rotor, move each half-rotor after each encryption step a varying number of positions. This 'irregularity' was not very deep, and more likely to impress a naive person was the period $d = 17 \cdot 19 \cdot 21 \cdot 23$ of the rotor movement; at more than 150 000 it was about one third of $\theta = 26^4 = 456\,976$. It was almost progressive encryption.

In a later application filed September 26, 1920 (German Patent 425 147), Scherbius mentioned gears ('key wheels') with irregularly dispersed cams. For the ENIGMA A of 1923 with four rotors, the rotor movement (patented for Paul Bernstein, filed March 26, 1924, German Patent 429 122) was somewhat irregular insofar as the four gears had gaps: one wheel with 11 positions had 5 teeth and 6 gaps, one wheel with 15 positions had 9 teeth and 6 gaps, one wheel with 17 positions had 11 teeth and 6 gaps, one wheel with 19 positions had 11 teeth and 8 gaps. Thus, a period of $d = 11 \cdot 15 \cdot 17 \cdot 19$, that is more than 50 000, was obtained for the rotor movement, which was only about one ninth of θ, but certainly again providing almost progressive encryption.

Irregular movement by means of gears with varying numbers of teeth and gaps was also used by the cipher machine (Plate F) of Alexander von Kryha (patent filed January 16, 1925, German Patent 434 642), but with a period of between 260 and 520 the machine was cryptologically very weak.

When Boris Hagelin, who took over Damm's company *Aktiebolaget Cryptograph*, in 1935 replaced the half-rotors by a 'bar drum', also called 'lug cage' (German *Stangenkorb*), for performing BEAUFORT encryption steps, he nevertheless continued to use irregular movement produced by 'step figures' of the key wheels. In the machines C-35/C-36 (Fig. 60a, Plate G), the number of key wheels was increased to five to give a period of $17 \cdot 19 \cdot 21 \cdot 23 \cdot 25 = 3900225$. In a later model, improved on the advice of Yves Gyldén, six key wheels were used, for a period of $17 \cdot 19 \cdot 21 \cdot 23 \cdot 25 \cdot 26$, i.e., more than 100 million (Plate H).[1] Hagelin got from France an order for 5000 machines, to be fabricated under license by Ericsson-Colombes. In the Second World War, 140 000 machines were built in the U.S.A. under license by the typewriter company L. C. Smith & Corona, named M-209 by the U.S. Army, CSP 1500 by the U.S. Navy. An unsafe C-38m was used by the Italian Navy in the Mediterranean Sea. Later Hagelins ('hags') had a period of $29 \cdot 31 \cdot 37 \cdot 41 \cdot 43 \cdot 47$, i.e., over 2 billion. BC-543 (Fig. 60b), a printing version operated electrically from a keyboard, was used in the U.S.A. for medium-level communications and was copied as C-41 late in the war by the German

[1] For details see Arto Salomaa, *Public-Key Cryptography*, Springer, Berlin 1990, pp. 44 ff.

typewriter company Wanderer Werke. After the war, Hagelin further improved his machines. In 1952, the Hagelin-Crypto CX-52 entered the market, using six out of twelve available key wheels (H 54 built in licence by the Dr. Hell Co., Kiel).

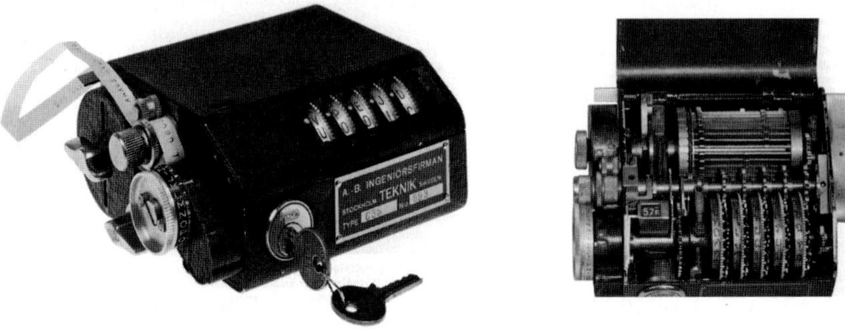

Fig. 60a. C-35 constructed by B. C. W. Hagelin (A. B. Cryptoteknik, Stockholm)

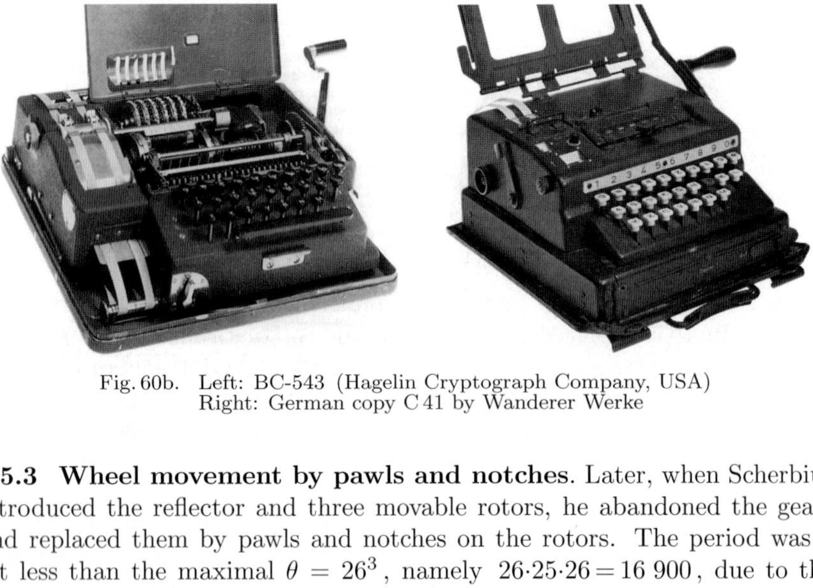

Fig. 60b. Left: BC-543 (Hagelin Cryptograph Company, USA)
Right: German copy C 41 by Wanderer Werke

8.5.3 Wheel movement by pawls and notches. Later, when Scherbius introduced the reflector and three movable rotors, he abandoned the gears and replaced them by pawls and notches on the rotors. The period was a bit less than the maximal $\theta = 26^3$, namely $26{\cdot}25{\cdot}26 = 16\,900$, due to the construction of the cam mechanism. Correspondingly, the maximal length of a message was limited to 180 characters (from Jan. 13, 1940: 250 characters).

In the ENIGMA I and in the *Wehrmacht* ENIGMA the 'regular' movement of the rotors was accomplished by using *one* notch at the *alphabet ring* of each rotor. The 'fast' (rightmost) rotor R_N moved at each encryption step. It caused for each full turn one step of the 'medium' (middle) rotor R_M, which again for each full turn caused one step of the 'slow' (leftmost) rotor R_L. This in fact meant regular, cyclometric rotor movement. (The 'Greek' rotors β and γ that were introduced later could not step.)

To be at least a little bit irregular, notches were cut for the new *Wehrmacht* ENIGMA rotors I – V at different positions of the alphabet ring:[2]

Rotor	I	II	III	IV	V
Alphabet letter	Y	M	D	R	H

But this was only a *complication illusoire*. Even worse, it was "a complication that defeats itself," as Kahn said ironically: If all rotors had the notches cut at the same letter, the cryptanalysts would not have been able to find out which rotor was used as the fast rotor by finding out (for known rotors) what letter caused the turnover. The *Kriegsmarine* seemingly found out about this and cut the notches on the new rotors VI and VII (1938) and VIII (1939) at the same letters. Moreover, these rotors had *two* notches: one at the letter H, one at U (Plate K). Unlike in the commercial ENIGMA, the notches were on the alphabet ring. Thus, the movement depended on the ring setting.

Although by using two notches, the period was halved and the danger of a superimposition (Sect. 19.1) increased (as a countermeasure, the permissible length of any one message had been so drastically limited), this change made cryptanalysis more difficult: "We would have had great trouble if each wheel had had two or three turnover positions instead of one" (Welchman). Three notches, by the way, would not have shortened the period, since 3 is relatively prime to 26. Was this overlooked by the OKW specialists?

Fig. 61. Rotors of the Abwehr-ENIGMA, seen from two sides, left with 2×26 teeth, right with pairs of teeth used as notches

A very special variant of the ENIGMA for the counterespionage and espionage service (*Abwehr*) under Admiral Wilhelm Canaris had—strangely—no plugboard, but like the D model a *rotating* reflector. A difference of the *Abwehr* ENIGMA from the D model was that it used coupled gears instead of pawls and notches (following Korn's 1928 German patent); allowing, in connection with a revolution counter, forward and backward moving by means of a crank. It had new rotors; but as in the *Wehrmacht* ENIGMA, turnover positions fixed to the alphabet ring. The three rotors had 11, 15, and 17 (not 19,

[2] The turnover has happened when (with a difference of 19 letters) the letters R, F, W, K, A are in the window. At Bletchley Park there was a corresponding, rather silly mnemonic verse *Royal Flags Wave Kings Above*.

as Twinn said) turnover positions. This gave Dillwyn Knox quite a headache; but he succeeded in autumn 1941 in the cryptanalysis. Knox developed on the *Abwehr* ENIGMA a special terminology: some particular simultaneous movements of R_N, R_M, R_L and the reflector he called 'crab' and 'lobster'.

8.5.4 Typex. Plate L shows the British TYPEX (Type-X), developed under the supervision of a government commission and ready in 1935, after nine years (but not commercially available), was in some respects similar to the 3-rotor ENIGMA; one difference was that among its five rotors the two next to the entry were not movable. The light-bulb output of the ENIGMA was replaced by a tape printer. In this respect, TYPEX was cryptanalytically equivalent to a 3-rotor ENIGMA with a non-self-reciprocal plugboard. Essential differences existed in the rotor movement: It was regular, too, but basically multi-notched. The notched rim was rigidly fixed to the rotor rim as in the commercial ENIGMA and the rotor core was a 'wiring slug' sitting in a receptacle carrying the rotor rim and the alphabet. The wiring slugs could be inserted in two orientations, P or P^{-1}. In a typical version, five slugs could be selected out of ten. There were rims with five, seven and nine notches; in the last case the notches were arranged so that a turnover occurred when in the window one of the letters B, G, J, M, O, R, T, V, X was shown. All rotors when used together had identical notchings.

The rotors of the Italian OMI Cryptograph-CR (Sect. 7.3.4) could be assembled from a receptacle, that contained the notches, and a pair of rotor cores; in this way, one could speak of 14 rotors, whose movement was coupled in pairs, or of 7 rotors with a choice from $\binom{14}{2} = 91$ rotors (Fig. 62).

Fig. 62.
Left:
TYPEX Mark II
Right:
Rotor machine of the Ottica Meccanica Italiana

Besides ENIGMAs, TYPEXs from the Second World War surplus went after 1945 into many small countries. Some were in use until 1975.

From the late 1940s until the early 1980s, the North Atlantic Treaty Organization (NATO) used rotor machines KL-7 developed in the U.S.A. for multinational communications. (The American SIGABA was considered too good to be shared). The KL-7 (Figure 63) had seven cipher rotors; it was in its mechanical aspects faintly similar to the British TYPEX, using interchangeable coding cylinders and rings. Plastic slip rings which controlled the rotor movement could be permuted among the coding cylinders. Each rotor had 36 contacts, providing for letters and digits. The KL-7 was one of the last rotor machines ever produced. The security of these machines had to be very good, since they were available even to some non-NATO countries and were sure to fall into the other side's hands. This shows that cryptologists in the 1960s had fully accepted Shannon's maxim (Sect. 11.2.3) that a cryptosystem must be safe even if the device is in the hands of the enemy. Indeed, already in 1962 the U.S. Officer *Joseph G. Helmich* sold to the sowjets technical informations about rotors and key lists; he was arrested in 1982 by the FBI. Use of the KL-7 ended in 1985 after the espionage case Walker—at that time it was outdated anyway.

The Russian counterpart, a 10-rotor machine, was named FIALKA ('violet').

Fig. 63. Rotor machine KL-7 (cover name ADONIS)

8.5.5 Hebern. Early in his career, the great William Friedman studied the rotor machines of Hebern, who was in contact with the U.S. Navy, and in 1925 also gave an evaluation. His test of the Hebern machine was a *chef-d'œuvre*. He was given ten messages of about 300 characters, all encrypted with the same rotor arrangement, and the initial setting of the rotors. In two weeks of labor, he found the solution, including the reconstruction of the wiring of at least some rotors. The resulting report was finally declassified in 1996; obviously his *index of coincidence* (Sect. 16.1) is involved.

The Navy under Laurance F. Safford and the Army under Friedman, with Sinkov, Rowlett, and Kullback, spent long years hunting for improvements to the Hebern machine. In 1932, Hebern finally designed a satisfactory machine, the HCM with five rotors. It had reasonably irregular rotor movement, but still Friedman's group was not satisfied.

In about 1935, Friedman himself designed for the U.S. Army Signal Corps a 4-rotor machine based on the ENIGMA, Converter M-325 (which was even built in 1944 and dubbed SIGFOY but, because of some practical drawbacks, was not generally introduced). Then, again under pressure from the Navy, the 'Electric Cipher Machine' ECM Mark I appeared, a 5-rotor-machine with rotor movement controlled by pinwheels, which finally fulfilled even the highest requirements. But Frank Rowlett succeeded in inventing further improvements, leading to the ECM Mark II, often simply called ECM, with the U.S. Army also M-134-C and SIGABA, with the Navy CSP 889 (Fig. 64). The SIGABA had 15 rotors, with 5 cipher rotors and 5 rotors for irregular movement sitting in a basket, another five were equivalent to a plugboard. In the 1940s, it proved to be the securest machine in the (Western) world. It was also the most expensive. The system was in use until 1959. CCM, the 'Combined Cipher Machine', also named CSP-1700, was a hybrid machine for connections with SIGABA and TYPEX. From the Hebern machine, with further improvements from the professional cryptologists, the ECM Mark III was developed after 1934, and was still used in the Second World War.

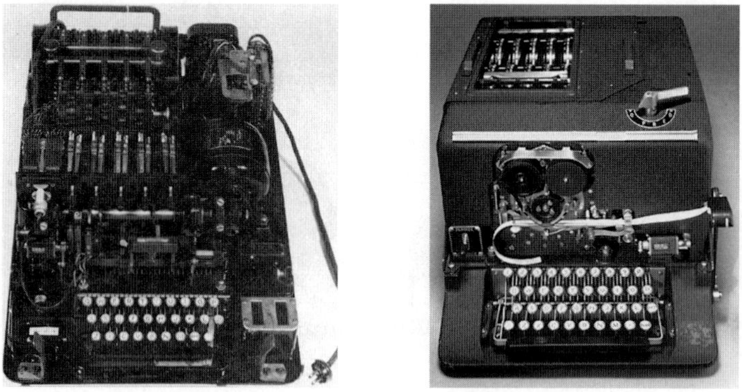

Fig. 64. Rotor machines ECM Mark II (M-134-C SIGABA, CSP 889)

8.5.6 Yardley. Japan, on the way to becoming an East Asian great power, could not get away without diplomacy after the First World War and thus needed cryptology. The diplomats used code books, like everyone else. The government had advisors; the Polish captain Jan Kowalewski taught it the simplest security measures such as Russian copulation (Sect. 3.4.2). Between 1919 and the spring of 1920, the Japanese introduced eleven codebooks, among them voluminous ones with 25 000 code groups. The Japanese radio signals naturally attracted the interest of the *Black Chamber* the U.S. State and War Departments were jointly running. Supported by the section MI-8 of the U.S. War Department, the Black Chamber was organized after 1918 by Herbert Osborne Yardley (1889–1958). Officially, it was ancillary to the Military Intelligence Division. It was housed in New York City under strict shielding; after the office was broken into, from 1925 it used the cover of a code compiling company, which indeed compiled and sold the *Universal Trade Code*. Yardley and his people were rather industrious and diligent; in the summer of 1921 they decrypted a telegram from the Japanese ambassador in London to his Foreign Ministry, containing delicate information about the International Maritime Disarmament Conference that was just in preparation and revealing expansionist Japanese dreams in the Far East. By 1929, the Black Chamber had decrypted 45 000 telegrams from all parts of the globe.

On March 4, 1929, Herbert C. Hoover acceded to office as the 31st president of the United States of America, and suddenly all that changed. Hoover's *naïveté* had the effect that he and Secretary of State Henry L. Stimson no longer wanted the disreputable services of the decryptors, the Black Chamber was without hesitation dissolved, effective October 31, 1929. The working material went to the Signal Corps of the Army, directed by Friedman. Yardley had to find another position; at the height of the Depression this was extremely difficult. He was forced to earn money and decided in his bitterness and distress to write a book, a startling *exposé* with the title *The American Black Chamber* (Indianapolis, 1931). Yardley was a superb storyteller and the book was an immediate success. But he provoked the anger and scorn of his government. In defense, he accused the State Department of grossly neglecting the interests of the U.S.A. in using "sixteenth-century codes," and stated that it had no right to bring moral pressure to bear on him. More serious were the objections from his professional colleagues; they knew better than Stimson that in view of the possibility of war the national interest not only disallowed a violation of state secrets, but also called for continuity of cryptological competence.

The Yardley case had a legislative sequel. The 73rd Congress of the U.S.A. debated in 1933 a controversial bill introduced by the Roosevelt administration, making it punishable to publish or furnish without authorization matter which was obtained while in the process of transmission between any foreign government and its diplomatic mission in the United States. The freedom of the press was infringed and Public Law 37, the *Lex Yardley*, went into Section 952 of Title 18 of the United States Code, but no criminal prosecution ensued against Yardley.

Yardley had enumerated nineteen countries whose diplomatic codes had been compromised, among them eleven South-American countries, Liberia and China—not surprising anyone—but also England, France, Germany, Spain and the Soviet Union, where at least officially nobody could express moral disgust (it was said that in the 1920s every larger European country was in the possession of one or more American code books)—and Japan.

The book became a tremendous success, not least due to the public stir the affair created. It sold 17 931 copies in the U.S.A. and a further 5 480 in Great Britain, which were unheard-of numbers for a cryptology book. Translations followed into French, Swedish, Chinese—and Japanese. A sensational 33 931 copies were sold in Japan, showing that Yardley had touched a nerve in the Japanese soul.

There, a member of the House of Peers used quite impolite and harsh words; throwing blame on his own Foreign Ministry, he spoke of a "breach of faith committed by the United States Government"; the foreign minister and former Japanese ambassador to the United States spoke of "dishonor". Yardley was slandered. And still he had rendered Japan the greatest service he could by stimulating a radical improvement of her cryptanalytic security. As a result, the Japanese multiplied their efforts on mechanizing encryption.

In 1938, Yardley was hired by Chiang Kai-shek and subsequently broke Japanese columnar transpositions. In 1940 he returned; he went in June 1941 to Canada to work on spy ciphers, but was replaced after six months, under Anglo-American pressure, by the approved and reliable Oliver Strachey.

Fig. 65. Japanese ENIGMA imitation, GREEN machine

8.5.7 Green, red and purple. Following a familiar pattern, the Japanese studied the machines of other countries, in particular those that were accessible through the patent literature: the ENIGMA, the machines of Damm and Hagelin, and those of Hebern. For machines with the Latin alphabet, the common Hepburn transliteration of *kana* into the Latin alphabet was used. The Japanese imitation of the ENIGMA D, denoted GREEN by American cryptanalysts, was a strange construction with four vertically mounted rotors (Fig. 65) and did not achieve great importance. Next, the half-rotors of Damm showed up in the *angō kikai taipu A* (Cipher Machine A), called RED in American jargon. Apart from a fixed permutation by a plugboard, it had a half-rotor with 26 slip-rings (Fig. 66). The wiring permuted the six vowels onto them and therefore also the 20 consonants and thus needed (two times) 60 exit contacts, since 60 is the least common multiple of 6 and 20. The reason for this cryptologically rather disadvantageous separation may have been in the tariff regulations of the international telegraph union, requiring 'pronouncable' words.

The Japanese RED machine was a very poor cryptosystem, not much better than Kryha's machine. Rotor movement was accomplished by a gear with 47 positions, with 4, 5 or 6 gaps. Cryptologically *angō kikai taipu A* performed two separate ALBERTI encryptions of the vowel and the consonant group. It is not at all surprising that the RED machine was attacked in 1935 by Kullback and Rowlett of the U.S. Army and in 1936 completely reconstructed by Agnes Driscoll (actually with two half-rotors, one with six, one with 20 slip-rings, Fig. 67). In the spring of 1936, Kunze at *Pers Z* of the German *Auswärtiges Amt* directed his interest to a variant, working with the *kana* alphabet (called ORANGE in American jargon). Jack S. Holtwick from the U.S. Navy had a similar goal. They both succeeded, Kunze in August 1938.

Fig. 66. Half-rotor with 26 slip-rings in the Japanese machine *angō kikai taipu A*

Fig. 67. American reconstruction RED of the *angō kikai taipu A* with two half-rotors

In 1937, Japan started the development of an encryption machine that was much more secure. It replaced the RED machine in the diplomatic service and was put in operation in 1939—the first messages picked up went in March 1939 from Warsaw to Tokyo. The *angō kikai taipu B* (Cipher Machine B, also 97-*shiki obun injiki*, alphabetic typewriter '97), called PURPLE in American jargon, included a new feature, used for the first time by the Japanese, namely stepping switches (uniselectors), known from telephone exchanges. The separation into two groups of six and 20 characters was kept, although later the six characters no longer had to be vowels. It turned out that the number of available alphabets was cut to 25, and the mapping was quite irregular and determined by the internal wiring. To find this needed the concentrated, month-long work of a whole group of people, not only William Friedman and Frank Rowlett, but also Robert O. Ferner, Albert W. Small, Samuel Snyder, Genevieve Feinstein (née Grotjan), and Mary Jo Dunning. They first found the mapping of the 6-vowel group, and they had indications that the number of alphabets was 25. But as for the 20-consonants group, they were in the dark, and no one could identify a known electromechanical encryption step that would produce the observed effects. When the situation seemed almost hopeless, in midsummer 1940, a newly arrived recruit from MIT, Leo Rosen, was initiated—and he hit upon the idea that the Japanese may have used stepping switches (Fig. 68). That gave the work a fresh impulse, and the mystery was soon solved: there were three banks of stepping switches, and the wiring connections could be established. In due time, following an important discovery Genevieve Grotjan had made, a working reconstruction was built, and in August 1940, after 18 months of work, the first complete PURPLE solution was achieved. In January 1941, the British in Bletchley Park were sent a copy. The RED machine had paved the way, and many weaknesses in the encryption discipline of the Japanese gave clues, hints and cribs, but it was a victory of the U.S. Army cryptanalytic bureau "that has not been duplicated elsewhere ... the British cryptanalytic service and the German cryptanalytic service were baffled in their attempts" (Friedman). They spoke of MAGIC. Soon the British had their victory, too—over the German ENIGMA; they called it ULTRA. However, David Kahn and Otto Leiberich reported, that the Russians and the Germans also solved PURPLE.

Fig. 68. Stepping switch bank of the Japanese PURPLE machine

Once the PURPLE machine was understood, it proved to possess only mediocre security, comparable to RED. It seems that the Japanese underrated the cleverness of the Americans; also they believed that their language would protect them and would not be understood fully elsewhere. The follow-up machines they built used stepping switches too and were only slightly more complicated: one, called CORAL in the American jargon, gave up the separation into 20+6; it was finally broken by OP-20-GY in March 1944. Another one, called JADE, was unique because it printed in *kana* symbols. Otherwise it had only minor added complications and was broken in due course.

The Japanese also had a very transparent systematics in their daily plugboard arrangements, and the bad habit of sending changes to the keying as encrypted messages—thus keeping the foe, once he had broken in, always up to date. Even the 'key to the keys' was discovered by Frank Raven in 1941.

8.6 Off-Line Forming of Key Sequences

8.6.1 Matrix powers. For VIGENÈRE and BEAUFORT encryption, 'irregular' key sequences of cycle numbers from $\mathbb{Z}_N$ are required. A much favored method uses successive powers *modulo* N of a regular $k \times k$ matrix T, sufficiently different from the identity. Since the number of such matrices (Sect. 5.2.3) is less than N^{k^2}, some power T^r must give identity for the first time. The number $r = r(T, N)$ is called order of the matrix T in $\mathbb{Z}_N$.

For example, the matrix $A = \begin{pmatrix} 0 & 1 \\ 1 & 1 \end{pmatrix}$ with $k = 2$ has the following order (see also 9.4.2):

$N=$	2	3	4	5	6	7	8	9	10	11	12	13	16	20	23	24	25	26	32	48	64	80	160
$r=$	3	8	6	20	24	16	12	24	60	10	24	28	24	60	48	24	100	84	48	24	96	120	240

Picking up a suitable i-j-element of the matrix powers produces a sequence of cycle numbers with the period $r(A, N)$ such that no smaller period exists.

A particularly convenient form of a matrix T is a $k \times k$ 'companion matrix' of the form

$$A = \begin{pmatrix} 0 & 0 & 0 & \dots & 0 & \alpha_k \\ 1 & 0 & 0 & \dots & 0 & \alpha_{k-1} \\ 0 & 1 & 0 & \dots & 0 & \alpha_{k-2} \\ & & \vdots & & & \\ 0 & 0 & 0 & \dots & 0 & \alpha_2 \\ 0 & 0 & 0 & \dots & 1 & \alpha_1 \end{pmatrix}$$

The 1-k-element of the powers of this matrix is then the last element of the iterated vector $t_i = t_0 A^i = t_{i-1} A$, if the initial vector is $t_0 = (1\ 0\ 0\ \dots\ 0\ 0)$.

To produce these iterated vectors, a shift register with k positions is used. Shift registers in connection with a companion matrix are also called linear shift registers. Using a basis analysis, they allow an easy break (Sect. 20.3). Non-linear shift registers are preferable: they form the next element of the sequence by some arbitrary function of the last k elements.

Simple steps for achieving non-linearity, like reversing the order of the vector components after each step, may be dangerous: non-linearity may lead to very short periods (Selmer 1993, Brynielsson 1993).

8.6.2 Bit sequences. For the binary case $N = 2$ of a VERNAM encryption, the key sequences are $(0,1)$-sequences, where 0 stands for the identity O and 1 for the reflection L. For example, the matrix $(k = 3)$

$$A = \begin{pmatrix} 0 & 0 & 1 \\ 1 & 0 & 1 \\ 0 & 1 & 0 \end{pmatrix}$$

modulo 2 yields the sequence (0 1 0 1 1 1 0 0 1 0 1 1 1 0 0 1 0 1 1 1 0 ...) with the period $7 = 2^3-1$. Since there are 2^k different k-bit vectors, and the zero vector is invariant, it is obvious, that the maximal reachable period is 2^k-1. One can show (Oystein 1948):

If the polynomial $x^k - \alpha_1 x^{k-1} - \alpha_2 x^{k-2} - \ldots - \alpha_k$ over the field $\mathbb{Z}_2 = \mathbb{F}(2)$ is irreducible, then every vector sequence iterated with A has a period, which is a divisor of 2^k-1.

For $k = 31$, the polynomial $x^{31} + x^{13} + 1$ is irreducible, the corresponding period $2^{31}-1$ amounts to more than 2 billion.

If 2^k-1 is prime (2^k-1 is then called a Mersenne prime[3]), then there are only the periods 2^k-1 and 1; the sequence (0 0 0 0 ...) has the period 1.

For $N{=}2$, i.e., in $\mathbb{Z}_2$, there are only the VIGENÈRE and BEAUFORT steps $+\mathbf{O}$ and $+\mathbf{L}$, i.e., the VERNAM steps $O \mathrel{\hat{=}} 0$ and $L \mathrel{\hat{=}} 1$. Polyalphabetic *binary* encryption needs a particularly long period and a good mechanism for the generation of an irregular $(0,1)$ sequence.

8.6.3 In principle, from every polyalphabetic set of block encryption steps χ_i with uniform encryption width m, which can be rather large, a finite sequence (Sect. 2.3) $X = (\chi_{i_1}, \chi_{i_2}, \cdots, \chi_{i_s})$ can be formed and X can be iterated on an initial key $u = (u_1, u_2, \ldots, u_s)$; the progressive sequence

$$u\ ,\ X(u)\ ,\ X^2(u)\ ,\ X^3(u)\ ,\ \ldots$$

is in fact periodic, but mostly with a very large period[4] such that it may be usable as a quasi-nonperiodic key sequence. For example, in Sect. 9.5.2, X will be defined with the help of the h-th power *modulo* a prime p,

$$X(u) = u^h \bmod p,\quad X^s(u) = u^{(h^s)} \bmod p\,.$$

8.7 Nonperiodic Keys

A nonperiodic encryption (Sect. 2.3.3) requires $\theta \geq 2$ and a nonperiodic sequence $(\chi_{i_1}, \chi_{i_2}, \chi_{i_3} \cdots)$ of polyalphabetic encryption steps. It is characterized by the index sequence $(i_1, i_2, i_3, \ldots)$ with $0 \leq i_\mu < \theta$, or by

---

[3] This is so for $k = 2, 3, 5, 7, 13, 17, 19, 31, 61, 89, 107, 127$. The further primes $2^{521}-1$, $2^{607}-1$, $2^{1279}-1$, $2^{2203}-1$, $2^{2281}-1$ were discovered in 1952 by Ralph M. Robinson using the SWAC. Fifteen more followed, then $2^{859433}-1$ (1994), $2^{1257787}-1$ (1996), $2^{1398269}-1$ (1996), $2^{2976221}-1$ (1997), $2^{3021377}-1$ (1998), $2^{6972593}-1$ (1999).

[4] Following Robert Floyd, with considerable computational effort, but with minimal storage requirement, the period of X can be determined in the following way: Let $a_0 = u$, $b_0 = u$ and $a_{i+1} = X(a_i)$, $b_{i+1} = X^2(b_i)$. As soon as $a_n = b_n$, there is $X^n(u) = X^{2n}(u)$ and n is the period.

the proper fraction $0.i_1 i_2 i_3 \ldots$ in a number system with the base $\theta \geq 2$. Thus, there exists for every irrational real number and for every $\theta \geq 2$ a nonperiodic encryption.

8.7.1 Delusions. For $\theta = 2$, a nonperiodic VERNAM encryption, such as one with the *infinite* index sequence (the 'running key')

$$(L\ L\ O\ L\ O\ O\ O\ L\ O\ O\ O\ O\ O\ O\ O\ L \ldots)$$

i.e., $i_\mu = \begin{cases} L & \text{if } \mu = 2^k \text{ for some } k\,, \\ O & \text{otherwise}\,, \end{cases}$

gives no advantage compared with a periodic encryption—it is even worse. But even a nonperiodic encryption with the key sequence (Axel Thue, 1904; Marston Morse, 1921) of the 'Mephisto-Polka', as used by Max Euwe in 1929,

$$(1\ 0\ 0\ 1\ 0\ 1\ 1\ 0\ 0\ 1\ 1\ 0\ 1\ 0\ 0\ 1\ 0\ 1\ 1\ 0 \ldots)$$

has a quite transparent law of key formation, allowing a recursive calculation. And the fractal sequence of (0-1)-words

$$\begin{aligned} a_0 &\mathrel{\hat{=}} (0) \\ a_1 &\mathrel{\hat{=}} (1) \\ a_2 &\mathrel{\hat{=}} (0\ 1) \\ a_3 &\mathrel{\hat{=}} (1\ 0\ 1) \\ a_4 &\mathrel{\hat{=}} (0\ 1\ 1\ 0\ 1) \\ a_5 &\mathrel{\hat{=}} (1\ 0\ 1\ 0\ 1\ 1\ 0\ 1) \\ a_6 &\mathrel{\hat{=}} (0\ 1\ 1\ 0\ 1\ 1\ 0\ 1\ 0\ 1\ 1\ 0\ 1) \\ a_7 &\mathrel{\hat{=}} (1\ 0\ 1\ 0\ 1\ 1\ 0\ 1\ 0\ 1\ 1\ 0\ 1\ 1\ 0\ 1\ 0\ 1\ 1\ 0\ 1) \\ &\vdots \qquad\qquad \vdots \end{aligned}$$

defined by the Lindenmayer term replacement system (Aristide Lindenmayer, 1968)

$$\begin{cases} 0 \rightarrow 1 \\ 1 \rightarrow 0\ 1 \end{cases}$$

also has a transparent law of key formation: for $i \geq 2$ it is $a_i = a_{i-2} \circ a_{i-1}$.
Obviously, nonperiodic sequences can be quite 'regular'. How easily can a nonperiodic index sequence be obtained, that is 'irregular' and nevertheless known to both the encryptor and the authorized decryptor?

The idea of taking as a key a text from a widespread book is reinvented mainly by amateurs. According to Shannon's rule "the enemy knows the system being used" this leads to a fixed key, with all the dangers already mentioned in Sect. 2.6.1. For meaningful key texts in a common language, a systematic zig-zag approach for breaking the encryption exists (Sect. 14.4) for Shannon cryptosystems with known alphabets (Sect. 2.6.4).

8.7.2 Autokeys. It is not surprising that the prospects for deriving a nonperiodic key from the plaintext were discussed very early. The decisive step was made by Geronimo (Girolamo) Cardano (1501–1576). After Belaso had introduced polyalphabetic substitution with keys, Cardano used the plaintext in his book *De Subtilitate* in 1550, starting the key over from the beginning with each new plaintext word:

s	i	c	e	r	g	o	e	l	e	m	e	n	t	i	s	
S	I	C	S	I	C	E	S	I	C	E	R	G	O	E	L	.
N	T	F	Z	C	L	T	Z	V	H	R	Y	V	I	P	E	

The alphabet is $Z_{20} \cup \{x, y\}$, the encryption linear polyalphabetic with

a	b	c	d	e	f	g	h	i	l	m	n	o	p	q	r	s	t	v	x	y	z
1	2	3	4	5	6	7	8	9	10	11	12	13	14	15	16	17	18	19	20	21	22

The idea of an autokey (French *autoclave*, *autochiffrant*) was conceived with the best intentions, was even fascinating; but Cardano presumably never tried it. The encryption is polyphonic, s and S as well as f and F yield N ; i and *I* as well as x and *X* yield T ; c and *C* as well as p and *P* yield F , etc. The unauthorized decryptor has no more work to find the right combination among 2^k ones (if the first word has k letters) than the authorized decryptor. Belaso tried to remedy the defect by encrypting the first word according to Trithemius, then for each following word the first letter of the previous word and the letters following it were used as keys:

s	i	c	e	r	g	o	e	l	e	m	e	n	t	i	s	
A	B	C	S	T	V	X	E	F	G	H	I	L	M	N	O	.
T	M	E	Z	N	D	M	L	R	N	V	P	Z	G	Y	H	

But this was still a fixed method. Then Blaise de Vigenère had the brilliant idea of introducing a short, freely selected priming key: He chose at will the first letter of the key and took as further key characters either those of the plaintext or those of the cryptotext ('autokey'):

a	u	n	o	m	d	e	l	e	t	e	r	n	e	l
<u>*D*</u>	A	U	N	O	M	D	E	L	E	T	E	R	N	E
X	I	A	H	G	U	P	T	M	L	S	H	I	X	T

a	u	n	o	m	d	e	l	e	t	e	r	n	e	l
<u>*D*</u>	X	H	E	E	C	O	U	M	X	G	N	A	B	Q
X	H	E	E	C	O	U	M	X	G	N	A	B	Q	O

In this case, the polyalphabetic encryption over Z_{20} (Fig. 69) was a self-reciprocal PORTA encryption and not *à la* VIGENÈRE.

The second kind, however, is useless: the key is completely exposed, and the whole message (except for the first character) can be decrypted at once (Shannon 1949).

Security is better with the first kind: It is a recurrent method, only knowing the first key character helps. But the two dozen or so possibilities are quickly tested. A remedy is to use a priming key of d letters instead of only one letter. The combinatorial complexity is nevertheless the same as that of a periodic encryption with a key of length d. For sufficiently large d testing is

A *B*	↕	a	b	c	d	e	f	g	h	i	l
		m	n	o	p	q	r	s	t	u	x
C *D*	↕	a	b	c	d	e	f	g	h	i	l
		x	m	n	o	p	q	r	s	t	u
E *F*	↕	a	b	c	d	e	f	g	h	i	l
		u	x	m	n	o	p	q	r	s	t
G *H*	↕	a	b	c	d	e	f	g	h	i	l
		t	u	x	m	n	o	p	q	r	s
I *L*	↕	a	b	c	d	e	f	g	h	i	l
		s	t	u	x	m	n	o	p	q	r
M *N*	↕	a	b	c	d	e	f	g	h	i	l
		r	s	t	u	x	m	n	o	p	q
O *P*	↕	a	b	c	d	e	f	g	h	i	l
		q	r	s	t	u	x	m	n	o	p
Q *R*	↕	a	b	c	d	e	f	g	h	i	l
		p	q	r	s	t	u	x	m	n	o
S *T*	↕	a	b	c	d	e	f	g	h	i	l
		o	p	q	r	s	t	u	x	m	n
U *X*	↕	a	b	c	d	e	f	g	h	i	l
		n	o	p	q	r	s	t	u	x	m

Fig. 69. PORTA encryption for Z_{20} by G. B. and M. Argenti

no longer feasible, but if the same priming key is used repeatedly for different messages, superimposition (Sect. 19.1) may help break it. Thus, the priming key should be comparable in length with the message—but then an autokey continuation no longer makes sense.

A further disadvantage is the spreading of encryption errors—a general weakness of all autokey methods.

Babbage reinvented the autokey—this time even with a mixed alphabet—and, although he first thought it to be unbreakable, also gave solutions in particular cases. Much later, in 1949, Shannon remarked that recurrent VIGENÈRE encryption is equivalent to VIGENÈRE encryption of period 2. If the plaintext is divided into groups $a_1\ a_2\ a_3\ \ldots$ of length d and if D is the priming key, then the following identities (mod N) hold for the cryptotext $C_1\ C_2\ C_3\ \ldots$:

$$C_1 = a_1 + D, \qquad C_i = a_i + a_{i-1} \quad (i = 2,\ 3,\ \ldots)$$

and thus the recurrent identities

$$\begin{aligned} C_1 &= a_1 + D \\ C_2 - C_1 &= a_2 - D \\ C_3 - C_2 + C_1 &= a_3 + D \\ C_4 - C_3 + C_2 - C_1 &= a_4 - D \quad \text{and so on, thus the sequence} \end{aligned}$$

$$C_1 \, , \; C_2 - C_1 \, , \; C_3 - C_2 + C_1 \, , \; C_4 - C_3 + C_2 - C_1 \; , \; \ldots$$

can be treated like a polygraphic VIGENÈRE of period 2, i.e., like two alternating polygraphic CAESAR additions. Even the use of a mixed alphabet doesn't change this. An analogous result holds for recurrent BEAUFORT.

8.7.3 Klartextfunktion. The idea of influencing the keying procedure of encryption machines in some hidden way by the plaintext shows up again in the patent literature around 1920 ('influence letter', in the patent application of October 10, 1919, by Arvid Gerhard Damm, Swedish Patent 52279, U.S. Patent 1 502 376). Thus, with the cipher teletype machines T 52d and T 52e of Siemens and SZ 42 of Lorenz, the (irregular) movement of the encryption elements could be further obfuscated ("*mit Klartextfunktion*") and the encryption was practically nonperiodic. However, in the case of noisy transmission channels this frequently led to an 'out-of-phase' problem with the encryption; the *Klartextfunktion* was therefore, very much to the relief of the British decryptors, used only for a few months towards the end of 1944.

8.7.4 Stream cipher. A recurrent encryption of the kind $c_i = f(p_i, p_{i-1})$ Cardano and Vigenère used is a special case of the modern stream cipher (German *Stromchiffrierung*) $c_i = \mathbf{X}(p_i, k_i)$, a nonperiodic encryption where the infinite key k_i is generated by a finite automaton G as key generator $k_i = G(k_{i-1}, p_{i-1})$, with k_1 as priming key. The hidden complexity lies in G.

8.8 Individual, One-Time Keys

8.8.1 Vernam. Given the fact that recurrent encryption is not much better than quasi-nonperiodic encryption, it is still possible that in a secure cryptosystem sender and receiver be equipped with a theoretically unlimited supply of secret keys, each one being genuinely irregular, with no meaning and holding no information, being random and used only one time, an individual key (German jargon *i-Wurm*). Vernam seems to have evolved this idea incidentally in 1918, but it spread fast between the two World Wars; early traces can be found in the U.S.A., in the Soviet Union, and in Germany.

8.8.2 Endless and senseless. Major Joseph O. Mauborgne, later Major General and Chief Signal Officer, U.S. Army (1937–1941), took heed in 1918 of Parker Hitt's 1914 admonition—"no message is safe in [the Larrabee] cipher unless the key phrase is comparable in length with the message itself"—and introduced in connection with the VERNAM encryption steps (Sect. 8.3) the concept of a one-time key (*one-time tape, one-time pad*, OTP), thus welding the epithet *endless* to Morehouse's (see Sect. 8.3.2) *senseless*.

In Germany, Kunze, Schauffler, and Langlotz in 1921 proposed blocks with 50 sheets, each one containing 240 digits (in 48 groups of fives) for the superencryption of numeral codes. From 1926, the cipher system of the German *Auswärtiges Amt* was superencryption (see Sect. 9.2.1) by one-time pads.

The Soviets, too, changed to the use of individual keys in 1926, very much to the distress of Fetterlein, the specialist for the Soviet Union in the British M.I.1(b). The Soviets kept a liking for individual keys; Plate O shows a matchbook-sized sheet found with a Russian spy.

By their very nature, one-time keys should be destroyed immediately after use. With Vernam type machines, shredding the tape can be done mechanically. A great practical difficulty is to provide enough key material for heavy traffic, in particular in unstable situations on the battlefield. These difficulties are more manageable for military headquarters, at diplomatic posts, or in a strictly two-way spy correspondence—and in such situations one-time keys are frequently used, provided the key supply cannot be cut off.

8.8.3 Bad habits. The sequences of letters or digits of an individual key should not show any regularity, should be random. Good stochastic sources are expensive. Kahn made the following remark on Russian individual keys:

"Interestingly, some pads seem to be produced by typists and not by machines. They show strike-overs and erasures—neither likely to be made by machines. More significant are statistical analyses of the digits. One such pad, for example, has seven times as many groups in which digits in the 1-to-5 group alternate with digits in the 6-to-0 group, like 18293, as a purely random arrangement would have. This suggests that the typist is striking alternately with her left hand (which would type the 1-to-5 group on a Continental machine) and her right hand (which would type the 6-to-0 group). Again, instead of just half the groups beginning with a low number, which would be expected in a random selection, three quarters of them do, possibly because the typist is spacing with her right hand, then starting a new group with her left. Fewer doubles and triples appear than chance expects. Possibly the girls, ordered to type at random, sensed that some doublets and triplets would occur in a random text but, misled by their conspicuousness, minimized them. Despite these anomalies, however, the digits still show far too little pattern to make cryptanalysis possible."

8.8.4 Holocryptic encryptions. If the individual key comes from a stochastic source emitting all characters independently and with equal probability, then common sense says that the plaintext encrypted with this keytext is an 'unbreakable' cryptotext, is holocryptic. (The expression was used by Pliny Earle Chase as early as 1859.) What this intuitively means, seems to be clear at first sight; it is also worth observing that in this book all cryptanalytic methods assume preconditions that are violated for holocryptic encryptions. But this is no proof; in fact the problem is to give a precise formulation of 'holocryptic', necessarily one of stochastic nature. The most intelligible one so far was given in 1974 by Gregory J. Chaitin, based on the work of A. N. Kolmogorov. Following him and Claus-Peter Schnorr (1970), we require that for the infinite index sequence of a nonperiodic encryption to be rightfully called holocryptic the following holds:

For every finite subsequence there does not exist a shorter algorithmic characterization than the listing of the subsequence—no subsequence can be condensed into a shorter algorithmic description.

As a consequence, no sequence generated by a machine, i.e., by a fixed algorithm, is holocryptic. Algorithms in this context are to be understood in the universal sense of the Church thesis. Thus, no digit sequences are suitable that characterize rational or computable irrational numbers (Sect. 8.7). Numbers like $\sqrt{2}$, $\sqrt{5}$, $\sqrt{17}$ are not suited anyhow, for they can be guessed too easily.

The set of non-computable real numbers is still very large. It is not known whether every non-computable real number defines a holocryptic encryption.

8.8.5 Fabrication of holocryptic key sequences. Physical effects, used today for the generation of 'true' random keys are based on the superimposition of incommensurable oscillations or on chaotic nonlinear systems. It seems that they are more reliable than the noise effects of vacuum tubes and Zener diodes used around 1950, or Geiger counter recordings. Vacuum tube noise was used in 1943 for the production of individual keys for the British ROCKEX system, a VERNAM encryption that served the highly sensitive traffic of the British with the U.S.A.—about one million words per day, or in more modern terms, the content of four 3.5″·2HD floppy disks per day.

8.8.6 Practical use. The U.S. State Department started in 1944 to use SIGTOT, an Army VERNAM crypto system with one-time keys for its most secret diplomatic messages. The Army also used M-134-A (SIGMYC), a five-rotor machine whose rotors were moved by a one-time 5-channel tape. The VERNAM system was in January 1943 replaced by a rotor system, too, the M-228 (SIGCUM) developed by Friedman. After a few days of practical use, Rowlett found a weakness of the system, which was therefore temporarily withdrawn and replaced in April 1943 by an improved version.

Just a handful of postwar designs were true one-time key systems, among which may be mentioned the Mi-544 from Standard Elektrik Lorenz (Germany), and the Hagelin T-52 and T-55 from Crypto AG (Switzerland). The Russians called their one-time tape machine AGAT ('agate').

8.8.7 Misuse. The practical use of one-time keys raises its own philosophical problems. Erich Hüttenhain has reported that in the *Auswärtiges Amt*, according to the security regulations, each one-time key sheet in a block of one hundred should have existed only in one original and one copy. In fact, nine copies were made and distributed, with permuted ordering, to five diplomatic missions.

The "Venona breaks" of Richard Hallock, Cecil James Phillips (1925–1998), Genevieve Feinstein, and Lucille Campbell into the highest Soviet cryptosystems (starting in November 1944) were also achieved on account of occasional re-issue of the same one-time pads. Phillips found out in summer 1944, that the first 5-digit cipher group is the key indicator. This break later broke the necks of the Soviet spies Julius and Ethel Rosenberg and revealed finally Harold 'Kim' Philby, Guy Burgess and Donald Duart MacLean, Klaus Fuchs, Harry Gold, David Greenglass, Harry C. White, and Pierre Cot as spies. On the other hand, the Soviets were warned in 1946 by William Weisband and in August 1949 by Philby, which may have caused the Soviets to stop using the duplicate OTPs after 1949.

A clear violation of the idea of a holocryptic encryption is the fabrication of key sequences by a machine. If then a cryptotext-cryptotext compromise happens between such a system and, say, a system using additives periodically and if the latter system is duly broken, then the one-time pad with the alleged stochastic key lies open. Provided there is enough material, the machine that generated the keying sequence can be reconstructed. This happened indeed for the German diplomatic cipher dubbed FLORADORA by the British (see Sect. 9.2.1): The *i-Wurm* the German *Auswärtiges Amt* had used showed a regularity and Bletchley Park could even find out what machine was involved—according to P. W. Filby, who took part in the break, Nigel de Grey said a Mr. Lorenz had offered such a machine in 1932 to the Foreign Office. Erich Langlotz on the German side did not yet know of the Chaitin doctrine. And tests can only disprove, but cannot prove randomness.

8.9 Key Negotiation and Key Management

8.9.1 The single characters of a key serve for the formation or selection (see Sect. 2.6) of an encryption step in an encryption system. Such a system can be monoalphabetic or polyalphabetic: in any case an encryption step should never be used a second time, if scrupulous cipher security is required.

In the monoalphabetic case, an encryption satisfying this requirement must be polygraphic with a width that can cover a whole message. This would be a great practical inconvenience. Thus, polyalphabetic encryptions with a lesser width come under consideration, in particular monographic ones. Moreover, the strict requirement never to use an encryption step a second time, may be weakened to the requirement of an individual one-time key (Sect. 8.8),—i.e., a key never used again as a whole—which shows a complete lack of any regularity in the sequence of encryption steps, since this already guarantees in the sense of Chaitin and Kolmogoroff that the encryption cannot be broken.

Although before 1930 in the U.S.A., Germany, the Soviet Union, and elsewhere individual one-time keys were already highly appreciated for very special tasks, their practical drawbacks led to a widespread tendency to accept weaker, only relative, encryption security.

8.9.2 It cannot be emphasized strongly enough (see Sect. 2.6.1) that the key negotiation between two partners is a particular weakness of every cryptological system. To master a frequently large distance safely depends (see below) on the reliability of the messenger, which is difficult to guarantee, as well as on their availability.

Therefore, there have been many attempts in the history of cryptology to cover the key negotiation itself by cryptological remedies; possibly even by steganographic measures.

Though it may look promising to perform the key negotiation for some cryptological system within this system itself—the more if one is strongly convinced of the unbreakability of such a system—, just this should be avoided by all means, since a break into the material that is serving for key negotiation may then compromise the whole system. At least it is necessary, as the German Navy did later in the war by using bigram tables, to submit the key negotiation to some additional enciphering in a different kind of system.

The idea of encrypted key negotiation by a message key indicating the starting position of some mechanical key generator was latent for quite some while and was not only propagated, e.g., for the commercial ENIGMA of 1923, but also accepted for the 3-rotor ENIGMA of the *Reichswehr* and of the *Wehrmacht*. The key negotiation was then the entry point for the break the young Polish mathematicians succeeded with in 1932 against the German ENIGMA enciphered traffic. The German side—except for the Navy—had strongly underrated the capabilities of their adversaries and had not considered it necessary to make the procedure of key negotiation more complicated; always with the excuse, not to burden more than necessary the capacity of the signals traffic and the capabilities of the cipher clerks.

Extravagant methods to bypass such a vulnerable key negotiation are feasible, for example by using two encryptions $\mathbf{X}^{(1)}$, $\mathbf{X}^{(2)}$ that commute:

$$\chi_i^{(1)} \chi_i^{(2)} x = \chi_i^{(2)} \chi_i^{(1)} x \ ,$$

say two VIGENÈRE or VERNAM encryptions. In this case, the sender encrypts his plain message with $\mathbf{X}^{(1)}$ according to a key $k^{(1)}$ chosen at random by him; the recipient applies $\mathbf{X}^{(2)}$ according to a key $k^{(2)}$ chosen at random by him *and sends this new cipher back to the sender*. This one interprets it because of the commutativity of $\mathbf{X}^{(1)}$ and $\mathbf{X}^{(2)}$ as a message he has encrypted, which he can decrypt by means of his key $k^{(1)}$. *The partly decrypted message he sends now to the recipient*, who in turn interprets it as a message he has encrypted, which he can decrypt by means of his key $k^{(2)}$. Thus, he obtains the original plain text. The disadvantage of this method is the need for a threefold transmission. If the message is short, this can be tolerated. Therefore the method would be good for transmitting vital information like passwords or a key to be used subsequently by some different encryption method. Since after all neither the plain message nor one of the keys are transmitted openly, the method seems to be safe. However, the devil is lurking already, as the following simple example with two VIGENÈRE encryptions over $\mathbb{Z}_{26}$ shows:

Sender A chooses key $A\ Q\ S\ I\ D$, which is not known to the recipient.
Recipient B chooses key $P\ Z\ H\ A\ F$, which is not known to the sender.

The plaintext /image/ is encrypted by the sender with $A\ Q\ S\ I\ D$:

$$\begin{array}{rl} & \text{i m a g e} \\ + & A\ Q\ S\ I\ D \\ \hline & \text{I C S O H} \end{array}$$

I C S O H is sent to the recipient, who encrypts it with $P\ Z\ H\ A\ F$:

$$\begin{array}{rl} & \text{I C S O H} \\ + & P\ Z\ H\ A\ F \\ \hline & \text{X B Z O M} \end{array}$$

X B Z O M is sent back to the sender, who decrypts it with the help of $A\ Q\ S\ I\ D$:

$$\begin{array}{rl} & \text{X B Z O M} \\ - & A\ Q\ S\ I\ D \\ \hline & \text{X L H G J} \end{array}$$

X L H G J is finally sent back to the recipient, who decrypts it with the help of $P\ Z\ H\ A\ F$: and thus obtains the message /image/ .

$$\begin{array}{rl} & \text{X L H G J} \\ - & P\ Z\ H\ A\ F \\ \hline & \text{i m a g e} \end{array}$$

Over the open transmission line the two signals X B Z O M and I C S O H are sent, whose difference exposes the key of B :

$$\begin{array}{rl} & \text{X B Z O M} \\ - & \text{I C S O H} \\ \hline & P\ Z\ H\ A\ F \end{array}$$

(likewise, X B Z O M and X L H G J expose the key of A).

This means that by decrypting X L H G J with the help of this key $P\ Z\ H\ A\ F$ the plaintext /image/ is compromised:

$$\begin{array}{rl} & \text{X L H G J} \\ - & P\ Z\ H\ A\ F \\ \hline & \text{i m a g e} \end{array}$$

The reason for this possibility of a break is that the keys form a group with respect to the composition of encryptions (see Sect. 9.1) and moreover one that is typical for the encryption method—the cyclic group of order 26. The encryption steps can be expected to be known.

A safeguard against the break is only given if at least one of the two decryption processes is made so difficult that it is practically intractable. This amounts to using an encryption method where the knowledge of an encryption key does not suffice to derive the decryption key efficiently. Such a thought was expressed in 1970 by *James H. Ellis* (†1997), as was disclosed in 1998 by the British *Communication-Electronics Security Group*. But if so, then the recipient B might as well publicly announce the key to be applied for messages that B should be able to decrypt. Moreover, the first and second steps of the method can be omitted. This produces the idea of an asymmetric encryption method, *published* in this form for the first time in 1976 by *Whitfield Diffie* and *Martin E. Hellman*—see more in Sect. 10.1.2 .

8.9.3 As soon as a communications network includes a large number of nodes and links, "key handling" is to be extended to "key management". The secure distribution of keys becomes the most difficult task of a key management scheme. Keys in transit must be protected from interception. Keys can be distributed on a secure path manually by couriers (preferred by diplomats and the military) or by registered mail (formerly preferred by commercial users) while telegraph, telephone, telefax, and the Internet are dubious. Normally the older channels cannot be utilized for the transmission of the secret message itself because they are too slow and, in most cases, too expensive. Frequently, they cannot safely carry the full load of messages.

Moreover, the safe insertion of keys into a crypto system, with tamperproof key carriers and "emergency clear" devices, belongs just as much to good key management as certification of the quality of keys.

Key management schemes that include key registration and allotment run a risk, which can be reduced steganographically by a special abbreviation nomenclature.

Following this line of thought and guided by practical requirenments, key hierarchies with different security levels (master key systems, use of primary, secondary, and tertiary keys) come under consideration. To give an example, the primary key may be machine-generated, but since the machine itself may fall into the hands of an enemy, a secondary key may be used, valid only for a rather short message, say of not more than 250 characters, and protected mildly by a system different from the main system, but which is also not unbreakable. Therefore a tertiary key transmitted by safe means is used that may hold for a longer period—say one day.[5] For example, the Key Exchange Algorithm (KEA) developed by N.S.A., using a key length of 1024 bit, declassified in June 1998 by the U.S. Department of Defense, is protected by the intractability of computing the discrete logarithm, see Sect. 10.2.4.2.

Such hierarchical systems render the task of key management even more complex. Moreover, they run the risk of a step-by-step attack: Compromise the key generator, compromise the secondary key, compromise the complete system. This is particularly dangerous if the key negotiation for the secondary key is done within the primary system: a one-time break may lead only too easily to a permanent break.

All the rules of keymanagement hold also for individual, one-time keys. They trivially comply with Hitts admonition that the keytext length be equal to or greater than the plaintext length. But this excludes the genuine unbreakable systems in many practical cases. They are increasingly being superseded by quasi-nonperiodic keys, in reality periodic keys of extremely long, guaranteed period and certified as having passed strict stochastic tests (pseudorandom keys). Frequently, complicated number-theoretical investigations are involved in the certification of these keys. It is not only publicly offered, commercial cryptosystems that are subject to this trend. But progress in storing techniques may stop this, for lightweight memory disks (CD-ROM) with a density of gigabytes per decagram give individual, one-time random keys a new chance to be used in high-level diplomatic, strategic military, and commercial links where there is a real need for absolute unbreakability.

---

[5] For the example of the *Wehrmacht* ENIGMA, according to the procedure that held from July 8, 1937 until September 15, 1938: the primary key is machine-generated, a secondary message key (indicator) determines the starting position of the rotors of each message, a tertiary *Tagesschlüssel* (Fig. 51a) comprises wheel order, ring setting, basic wheel setting ("*Grundstellung*"), and steckering. However, the primary and the secondary cipher systems were identical; and the tertiary key was transmitted by courier.

9 Composition of Classes of Methods

Let us recall that an encryption $\mathbf{X} : \; V^* \dashrightarrow W^*$ is usually finitely generated by a cryptosystem M. Let M^* denote the set of all encryptions defined in this way by M. An encryption method S is a subset of M^*. M^d indicates the subset of periodic encryptions with key sequences of period d, M^∞ the subset of encryptions with non-computable key sequences.

A composition of two encryptions by serial connection of their encryption steps requires that the cryptotext space of the first method coincides with the plaintext space of the second method.

Amateurs are inclined to believe that the composition of two classes of methods offers more resistance to unauthorized decryption than either of the two alone. That is not necessarily so. The second method can even partly or completely counterbalance the effect of the first. To give an example, let S be a simple substitution, generated, as usual, by a password, say the following, which could well come from Bazeries: BASEDOW'S DISEASE IS CURABLE .

The substitution is then

a	b	c	d	e	f	g	h	i	j	k	l	m	n	o	p	q	r	s	t	u	v	w	x	y	z
B	A	S	E	D	O	W	I	C	U	R	L	F	G	H	J	K	M	N	P	Q	T	V	X	Y	Z

In fact, it has four 1-cycles, two 2-cycles, and one 18-cycle. Applied twice it results in

a	b	c	d	e	f	g	h	i	j	k	l	m	n	o	p	q	r	s	t	u	v	w	x	y	z
A	B	N	D	E	H	V	C	S	Q	M	L	O	W	I	U	R	F	G	J	K	P	T	X	Y	Z

where eight letters, included the frequent vowels e a, are invariant.

9.1 Group Property

Some crypto systems M with $V \doteqdot W$ have the property that the composition of two encryption steps from M does not lead outside M. One says that such a crypto system forms a group. Examples are the group $\mathcal{P}_{26}$ of all simple substitution steps over Z_{26}, the group $\mathcal{P}_{24}$ of all transpositions of width 24.

For other crypto systems with $V \doteqdot W$, this is not necessarily so: the set of monocyclic simple substitution steps does not form a group, since the group identity is not monocyclic. The examples in Sect. 7.2.4 show that the set of ALBERTI encryption steps and the set of ROTOR encryption steps for some primary alphabets do not form a group. The composition of such steps increases the combinatorial complexity. This justifies the use of three and four rotors in the ENIGMA. The group property would be detrimental.

9.1.1 Key groups. However, if a cryptosystem M forms a group, then the composition of two $\chi_s, \chi_t \in M$ is some $\chi_\iota \in M$, where ι is uniquely determined by s and t: $\iota = s \bullet t$. Thus $\chi_s(\chi_t(p)) = \chi_{s \bullet t}(p)$; $.\bullet.$ is the group composition of the key characters, which form a key group ('key space').

A crypto system with a key group has been called 'pure' by Shannon, 'closed under composition' by Salomaa.

9.1.2 Composition of methods. An encryption method may also be a group with respect to composition of its encryptions, such as the group of all linear substitutions of a given width n, the group of all polyalphabetic (monographic) substitutions of a given period d, or the group of all block transpositions of a given width n.

The composition of two encryption methods ('product encryption') leads in general, however, to a new encryption method, although often to a related one: the composition of two general or linear polyalphabetic encryption methods with the periods d_1 and d_2 is a general or linear polyalphabetic encryption method with the period $lcm(d_1, d_2)$; analogously for block transposition of width n_1 and n_2. For substitutions, this was already pointed out by Babbage in 1854. Here, too, the combinatorial complexity is increased.

Sometimes, the composition of two encryption methods is commutative, like the composition of the group of all simple substitutions with the group of all block transpositions of a given width n. If two encryption methods, each one being a group, commute, then the product encryptions also form a group (Shannon: "The product of two pure ciphers which commute is pure.")

9.1.3 T52. The encryption steps of the cipher teletype machines made by Siemens worked over $\mathbb{Z}_2^5$ and used a composition of pentagraphic substitutions (VERNAM steps operating on the 5-bit code groups) and transpositions of the five bits (permutation of their positions)—in group-theoretic terms a subset of the hyper-octahedral group of order $2^5 \cdot 5! = 3840$. They were based on a patent applied for by August Jipp and Ehrhard Rossberg on July 18, 1930. The models T 52a and T 52b were used by the *Kriegsmarine* from 1931; the model T 52c was first used by the *Luftwaffe*, and by mid-1941 was used generally by the *Wehrmacht* (*Geheimschreiber*, British code-name *sturgeon*). It is estimated that about 1000 machines were built over the years.

Encryption and decryption were controlled by ten cipher wheels w_s, each one operating a binary switch i_s with $i_s = 0$ or $i_s = 1$, $s = 1...10$. Five wheels $w_1...w_5$ performed on $\mathbb{Z}_2^5$ 32 VERNAM substitutions, five more $w_6...w_{10}$ performed transpositions generated by 2-cycles. In the T 52a this was the set $\{(12)^{i_6}(23)^{i_7}(34)^{i_8}(45)^{i_9}(51)^{i_{10}}\}$; since $(12)(23)(34)(45) = (23)(34)(45)(51) = (54321)$ and $(23)(34)(45) = (12)(23)(34)(45)(51) = (5432)$, the number of different ones among these transpositions is 30. Altogether, the ten wheels generated 960 alphabets. In the T52c, developed under Herbert Wüsteney (1899–1988), a message key could be easily changed. In the T52e, due to new circuitry, only 16 different substitutions and 15 different transpositions occurred, reducing the number of alphabets used for one message to 240 (in the T52c, the number was only 120).

The movement of the cipher wheels was controlled by their having 47, 53, 59, 61, 64, 65, 67, 69, 71, and 73 teeth; at each step all wheels were moved by one tooth, which gave a kind of a regular wheel movement with a period of $47 \cdot 53 \cdot 59 \cdot 61 \cdot 64 \cdot 65 \cdot 67 \cdot 69 \cdot 71 \cdot 73$, i.e., about 10^{18}.

The T 52d and T 52e (introduced in 1943 and 1944) were variants of T 52a and T 52c, respectively, featuring more "irregular" intermittent wheel movements and supporting an optional *Klartextfunktion*. The T 52b (1934) was different from the T 52a only with respect to improved interference suppression.

Towards the end of the war, the cipher teletype machine T 43 was built in a few copies by Siemens. It used an individual, one-time key and is presumably identical with the machine the British gave the code-name *thrasher*.

9.1.4 SZ. Cryptologically simpler was the cipher teletype machine SZ 40, SZ 42, SZ 42a (*Schlüsselzusatz*, British code-name *tunny*) made by Lorenz. It performed only VERNAM substitutions, correspondingly the encryption was self-reciprocal. In the SZ 42 (Plate N) a first group of five cipher wheels with 41, 31, 29, 26, and 23 teeth (called χ-wheels by the British) operated with VERNAM steps on the 5-bit code groups; at each step all χ-wheels were moved by one tooth. A second group of five cipher wheels with 43, 47, 51, 53, and 59 teeth (called ψ-wheels) operating likewise with VERNAM steps on the 5-bit code groups, followed serially. Two more wheels (called motor-wheels) served for irregular movement only; one, with 61 teeth, moving with the χ-wheels, controlled another one with 37 teeth, which in turn controlled the *simultaneous* (a weakness!) movement of the ψ-wheels. The period was more than 10^{19}. All wheels could be arbitrarily provided with pegs controlling the VERNAM switches and could also be brought to arbitrary initial settings.

9.1.5 Olivetti. Much less is known about the practical use of this cipher teletype machine (Italian Patent 387 482, January 30, 1941), which had only five cipher wheels and two motor wheels, causing a weak irregularity.

9.2 Superencryption

9.2.1 Superencryption. Also called superenciphering (U.S.), reciphering (U.K.), or closing (French *surchiffrement*, German *Überchiffrierung*, jargon also *Überschlüsselung*), it is a common case of a product encryption: a literal or numeral code is encrypted again. VIGENÈRE over $\mathbb{Z}_{10}$, i.e., with $N=10$, is used for numeral codes; the corresponding addition *modulo* 10, i.e., without carry, which in military jargon was occasionally called symbolic addition, can be performed on a mutilated adding machine (Sect. 8.3.3). As early as

1780, Benedict Arnold, a spy for Britain in the New England states, used the overall addition of 7 *modulo* 10 to code groups, i.e., an ordinary CAESAR addition, for superencryption. If instead, for groups of width m, i.e., in the crypt width of the code, a number is added *modulo* 10^m (a polygraphic CAESAR addition), one speaks of an additive. The use of additives in connection with codes became widely known in the 19th century when commercial codebooks began to appear. One particular kind of double superencryption went as follows: Codebooks that had both numerical and literal codegroups (e.g., Fig. 39) were used to retranslate the numeral code obtained by adding the additive into literal code. This is of particular interest if the additive, for ease of numerical computation, is very special, like 02000. In 1876 J. N. H. Patrick was convicted of having used such a system in a corruption affair in the U.S. Congress. The U.S. Navy used the system in the Spanish-American War in 1898, and it was considered to be the most secure and advanced code system of the day—provided the additive was changed at rather short intervals.

The Foreign Office of the *Deutsches Reich* used from 1919 a double superenciphering of a 5-digit numeral code ("Deutsches Satzbuch", DESAB). It used two additives, each one covering six five-digit groups. 5000 of these additives were available. The British tried for a long time without success to break into what they had given the code-name GEC or FLORADORA. However, in May 1940 in the German Consulate in Reykjavik (Iceland) they seized cipher documents including the two codebooks Nos. 22 and 46 and ten additives. At first this gave no more than slight progress achieved by testing on stereotypical expressions. Then, in 1942, the British Consul in Lourenço Marques, the capital of portuguese Moçambique, obtained by lucky circumstances the additives for the next two months. In 1944, A. G. Denniston, P. W. ('Bill') Filby and the reactivated E. C. Fetterlein in Bletchley Park, together with S. Kullback and T. A. Waggoner, Jr. in Washington, succeeded in a complete break.

Transposition may also be used for superencryption: F. J. Sittler, one of the most successful code makers, recommended shuffling the four figures of his code groups. If this transposition is kept fixed, however, the effect of superencryption is just a new code, no more secure than the old one. If a transposition is used, its width should be prime relative to the codelength.

9.2.2 Need for superencryption. In particularly sensitive cases—dates and clock times, coordinates, names and so on—composition of a code with a rather independent superencryption method is indicated. Superencryption of some code with a bipartite bigram substitution is an example—it was used for a 3-figure front-line code (*Schlüsselheft*) in the First World War by the German Army after March 1918. ENIGMA superencryption was used for the map grid (Sect. 2.5.2.1) of the *Kriegsmarine* in the Second World War.

9.2.3 Plugboard. A fixed superencryption of the ENIGMA encryption was accomplished by the plugboard (Sect. 7.3.3). The substitution was self-reciprocal, but this was not necessary, for an arbitrary substitution would have preserved the self-reciprocal character of the ENIGMA, but disallowed Welchman's diagonal board. However, other cryptanalytic methods of the Polish and the British (Zygalski sheets, Turing bombs) were insensitive to 'steckering' and would have worked for an arbitrary plugboard substitution.

9.2.4 ADFGVX. An early case of a thorough amalgamation by a product encryption is the ADFGVX system of the German Army, invented in the First World War by Lieutenant Fritz Nebel (in the Second World War signal and communications officer in the *Luftwaffe*) and introduced under Ludendorff on the Western Front in 1918, with a bipartite 6×6 Polybios substitution (Sect. 3.3.1) in the alphabet {A, D, F, G, V, X} of clearly distinguishable Morse signals (Sect. 2.5.2) and a transposition of width 20. The key was changed every day, and it cost the French cryptologist Georges-Jean Painvin at least a full day to decrypt the signals—if he could solve them at all.

9.2.5 ENIGMA superencryption. In the ENIGMA network of the *Kriegsmarine*, messages of particular importance were superencrypted a second time with the ENIGMA; this was to be marked with the plaintext discriminant (German *Kenngruppe*) *offizier*. The reason was cryptological in nature, but directed against another audience: it screened information from the rank and file. Late in the evening of July 20, 1944, a signal was circulated to all German ships, and was decrypted in Bletchley Park:

```
OKMMM ANANA LLEXX EINSA TZJWA LKUER EJNUR DURCH OFFIZ IERZU ENTZI
FFERN OFFIZ IERJD ORAJD ERFUE HRERJ ADOLF HITLE RJIST TOTXD ERNEU
EFUEH RERIS TFELD MARSC HALLJ VONWI TZLEB ENJ ......
```

Walter Eytan [Ettinghausen], in charge of Z Watch in Hut 4, Bletchley Park, did not know how macabre the wrong news was; anyhow he kept it secret from the ever-present *wrens*, young ladies of the Women's Royal Naval Service.

9.3 Similarity of Encryption Methods

Shannon calls two classes of encryption methods $\mathcal{S}$, $\mathcal{T}$ similar, if there exists (independent of the keys) a one-to-one mapping A of the set of cryptotext words of $\mathcal{T}$ into the set of cryptotext words of $\mathcal{S}$ such that

for all $T \in \mathcal{T}$ there exists $S \in \mathcal{S}$: $S = A\,T$, i.e., $S(x) = A\,T(x)$ for all x.

Encryption methods from similar classes are cryptanalytically equivalent: one can assume that A is known (Kerckhoffs' admonition as stated by Shannon: "The enemy knows the system being used"). A possible way to break $\mathcal{T}$ is then also suitable to break $\mathcal{S}$. Classes of similar encryption methods are:

CAESAR methods and reversed CAESAR methods
(A is the 'inverting' substitution (Sect. 3.2.1) of the cryptotext),

VIGENÈRE methods and BEAUFORT methods
(A is again the 'inverting' substitution of the cryptotext),

simple columnar transposition methods and block transposition methods
(A is the matrix transposition of the cryptotext).

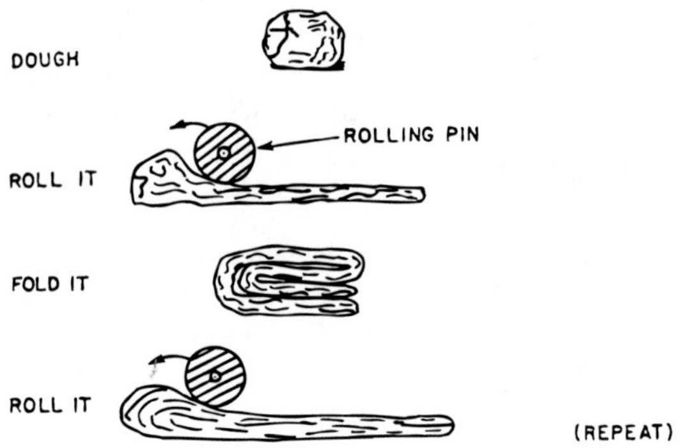

Fig. 70. Production of pastry dough

9.4 Shannon's 'Pastry Dough Mixing'

The composition of a multipartite monographic substitution and a transposition is not commutative. The composition of a proper polygraphic substitution of width k and a transposition of width k is not commutative. The composition of a simple substitution and a VIGENÈRE method is not commutative. Shannon has pointed out in 1945 that the composition of non-commuting encryption methods works like a thorough "pastry dough mixing"[1] (Fig. 70), as studied by Eberhard Hopf in compact spaces[2].

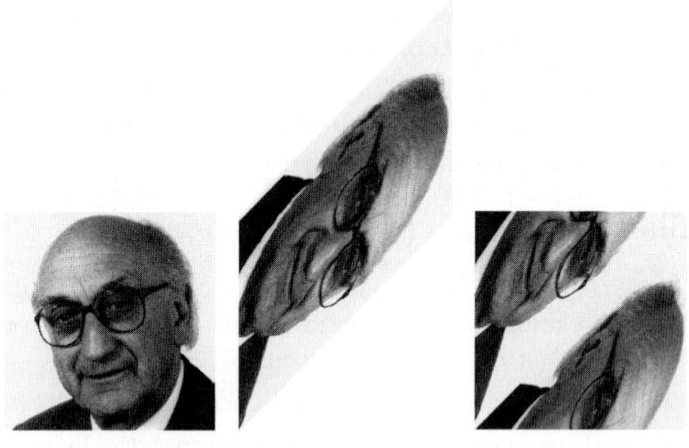

Fig. 71. Modular transformation

9.4.1 Confusion and diffusion. Intuitively, a composition will be efficacious, if the composed methods not only do not commute, but the one is rather independent of the other, like transposition, performing a 'diffusion', and linear polygraphic substitution, performing a 'confusion'. If the product

[1] N. J. A. Sloane, *Encrypting by Random Rotations.* Lecture Notes in Computer Science 434, Springer 1990.

[2] Eberhard Hopf, *On Causality, Statistics and Probability*, Journal of Mathematics and Physics **13**, pp. 51–102 (1934).

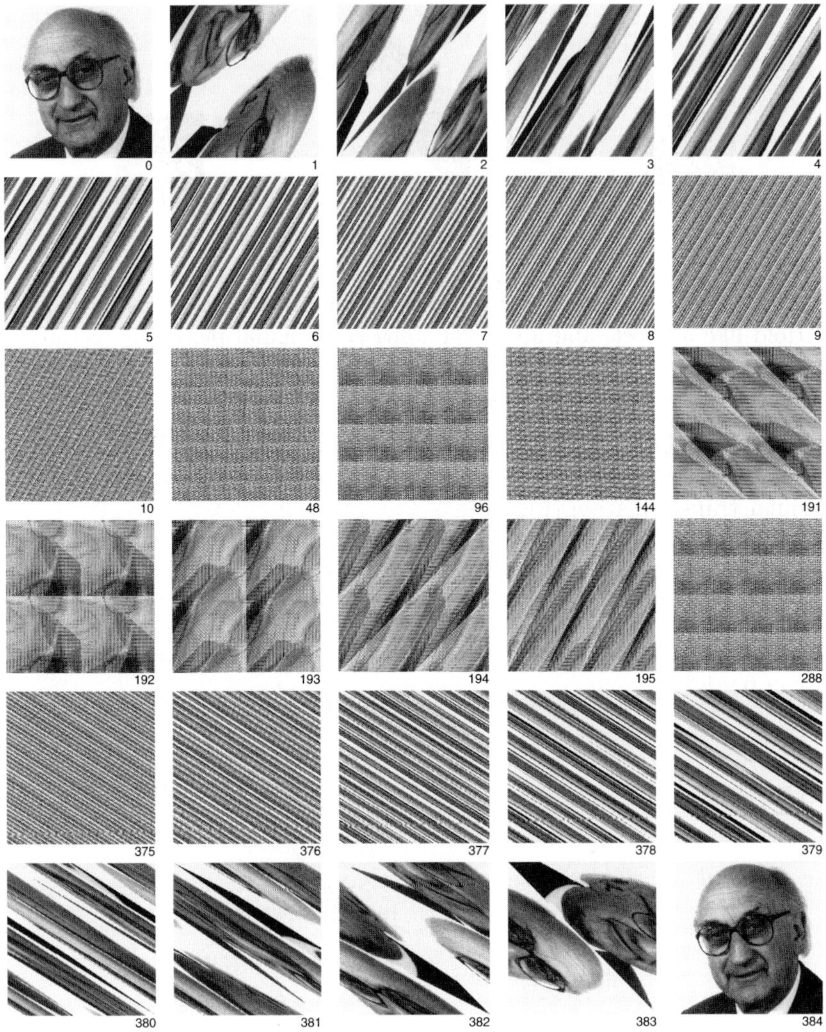

Fig. 72. FLB's resurrection

encryption is not a group, it may be iterated and its combinatorial complexity further increased. In the discrete spaces of encryption, however, any iteration of a fixed transformation is finally periodic and in the end the Hopf mixing is an illusion. This is shown in the following example of an iterated two-dimensional picture transformation which in the first steps displays quite convincingly its amalgamation character.

The transformation step consists of a reflection with affine distortion, followed by a reduction to the basic format by cutting off and pasting back protruding corners (Fig. 71). The result of successive transformation steps is shown in Fig. 72. At first, it looks as if the portrait of FLB is going to be totally mixed

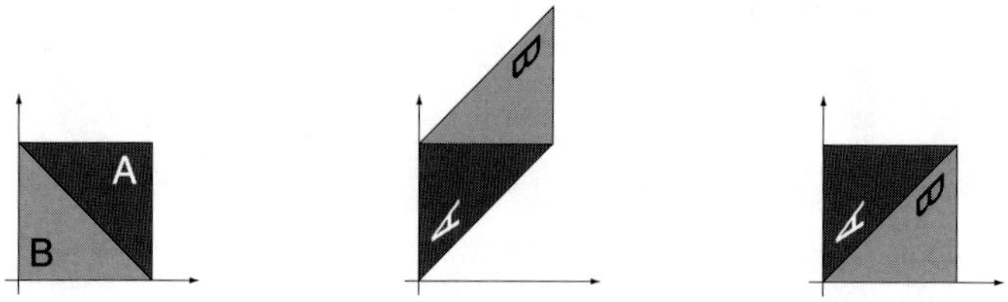

Fig. 73. Modular transformation T

up, but after 48 steps a texture turns up and after 192 steps it reappears fourfold like a ghost, and after 384 steps the original picture is restored. A picture encryption of this kind with a high number of iterations carries the danger of not concealing anything.

9.4.2 Heureka! The phenomenon can be explained: We consider the square $Q: \ 0 \leq x < 1 \ , \ 0 \leq y < 1$ with toroidal connection and on it the modular transformation (Fig. 73):

$$T: \begin{cases} x' = y \\ y' = \begin{cases} x+y-1 & \text{if} \quad x+y \geq 1 \\ x+y & \text{if} \quad 0 \leq x+y < 1 \end{cases} \end{cases} .$$

The local affine distortion, the reflection included, is given by the matrix $T = \begin{pmatrix} 0 & 1 \\ 1 & 1 \end{pmatrix}$, which has shown up already in Sect. 8.6.1. Note that

$\begin{pmatrix} 0 & 1 \\ 1 & 1 \end{pmatrix}^n = \begin{pmatrix} F_{n-1} & F_n \\ F_n & F_{n+1} \end{pmatrix}$, where F_i is the i-th Fibonacci number.[3]

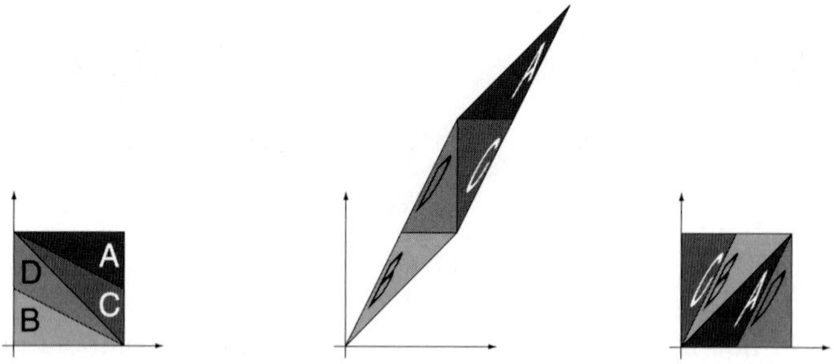

Fig. 74. Modular transformation T^2

Figure 74 shows the effect of T^2 with the local affine distortion due to the matrix

$$\begin{pmatrix} 0 & 1 \\ 1 & 1 \end{pmatrix}^2 = \begin{pmatrix} 1 & 1 \\ 1 & 2 \end{pmatrix} ,$$

[3] See F. L. Bauer, *Efficient Solution of a Non-Monotonic Inverse Problem.* In: W. H. J. Feijen et al. (eds.), *Beauty is our Business.* Springer 1990, pp. 19–26.

Figure 75 finally shows the effect of T^4 with

$$\begin{pmatrix} 0 & 1 \\ 1 & 1 \end{pmatrix}^4 = \begin{pmatrix} 2 & 3 \\ 3 & 5 \end{pmatrix} = \begin{pmatrix} -1 & 0 \\ 0 & -1 \end{pmatrix} + 3 \cdot \begin{pmatrix} 1 & 1 \\ 1 & 2 \end{pmatrix} .$$

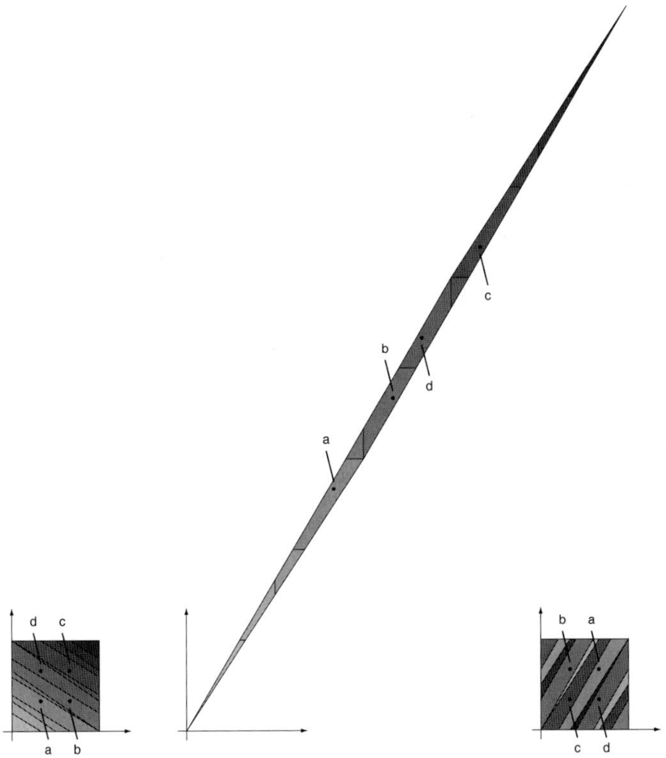

Fig. 75. Modular transformation T^4

Here it can already be seen that the pattern of the four points

$$a = \begin{pmatrix} \frac{1}{3} \\ \frac{1}{3} \end{pmatrix} , \quad b = \begin{pmatrix} \frac{2}{3} \\ \frac{1}{3} \end{pmatrix} , \quad c = \begin{pmatrix} \frac{2}{3} \\ \frac{2}{3} \end{pmatrix} , \quad d = \begin{pmatrix} \frac{1}{3} \\ \frac{2}{3} \end{pmatrix}$$

is rotated by 180° . Correspondingly, these four points are already fixpoints for T^8 with the matrix

$$\begin{pmatrix} 0 & 1 \\ 1 & 1 \end{pmatrix}^8 = \begin{pmatrix} 13 & 21 \\ 21 & 34 \end{pmatrix} = \begin{pmatrix} 1 & 0 \\ 0 & 1 \end{pmatrix} + 3 \cdot \begin{pmatrix} 4 & 7 \\ 7 & 11 \end{pmatrix}$$

For T^{16} with the matrix

$$\begin{pmatrix} 0 & 1 \\ 1 & 1 \end{pmatrix}^{16} = \begin{pmatrix} 610 & 987 \\ 987 & 1597 \end{pmatrix} = \begin{pmatrix} 1 & 0 \\ 0 & 1 \end{pmatrix} + 21 \cdot \begin{pmatrix} 29 & 47 \\ 47 & 76 \end{pmatrix}$$

additional fixpoints appear, in fact

$$\begin{pmatrix} 0 & 1 \\ 1 & 1 \end{pmatrix}^{16} \begin{pmatrix} \frac{i}{21} \\ \frac{k}{21} \end{pmatrix} = \begin{pmatrix} \frac{i}{21} \\ \frac{k}{21} \end{pmatrix} + \begin{pmatrix} 29\,i + 47\,k \\ 47\,i + 76\,k \end{pmatrix} .$$

Thus, all 400 points with the coordinates $(\frac{i}{21}, \frac{k}{21})$, $0 < i < 21, 0 < k < 21$ are fixpoints of T^{16}. T^{48} has the matrix

$$\begin{pmatrix} 0 & 1 \\ 1 & 1 \end{pmatrix}^{48} = \begin{pmatrix} 1 & 0 \\ 0 & 1 \end{pmatrix} + 46\,368 \cdot \begin{pmatrix} 64\,079 & 103\,682 \\ 103\,682 & 167761 \end{pmatrix} ,$$

$$\begin{pmatrix} 0 & 1 \\ 1 & 1 \end{pmatrix}^{48} \begin{pmatrix} \frac{i}{46368} \\ \frac{k}{46368} \end{pmatrix} = \begin{pmatrix} \frac{i}{46368} \\ \frac{k}{46368} \end{pmatrix} + \begin{pmatrix} 64079\,i + 103682\,k \\ 103682\,i + 167761\,k \end{pmatrix} .$$

This results in $46367^2 = 199^2 \cdot 233^2 \approx 2.15 \cdot 10^9$ fixpoints. Outside these points there is a thorough amalgamation. However, if the blackening is restricted by screening to a grid, then the resurrection of the picture is understandable when the set of fixpoints fits the set of mosaic points. Note that $46368 = 2^5 \cdot 3^2 \cdot 7 \cdot 23$. Thus,

$$\begin{pmatrix} 0 & 1 \\ 1 & 1 \end{pmatrix}^{48} \bmod 2^5 = \begin{pmatrix} 1 & 0 \\ 0 & 1 \end{pmatrix} .$$

This means that in the example of Fig. 72 a screening process of 32×32 mosaic points would have led to resurrection after 48 steps. Actually, the screening was done with 256×256 mosaic points. Now,

$$\begin{pmatrix} 0 & 1 \\ 1 & 1 \end{pmatrix}^{48} \bmod 2^8 = \begin{pmatrix} 2\,971\,215\,073 & 4\,807\,526\,976 \\ 4\,807\,526\,976 & 7\,778\,742\,049 \end{pmatrix} \bmod 2^8 = \begin{pmatrix} 225 & 64 \\ 64 & 33 \end{pmatrix}$$

$$\begin{pmatrix} 0 & 1 \\ 1 & 1 \end{pmatrix}^{96} \bmod 2^8 = \begin{pmatrix} 225 & 64 \\ 64 & 33 \end{pmatrix}^2 \bmod 2^8 = \begin{pmatrix} 193 & 128 \\ 128 & 65 \end{pmatrix}$$

$$\begin{pmatrix} 0 & 1 \\ 1 & 1 \end{pmatrix}^{192} \bmod 2^8 = \begin{pmatrix} 193 & 128 \\ 128 & 65 \end{pmatrix}^2 \bmod 2^8 = \begin{pmatrix} 129 & 0 \\ 0 & 129 \end{pmatrix}$$

$$\begin{pmatrix} 0 & 1 \\ 1 & 1 \end{pmatrix}^{384} \bmod 2^8 = \begin{pmatrix} 129 & 0 \\ 0 & 129 \end{pmatrix}^2 \bmod 2^8 = \begin{pmatrix} 1 & 0 \\ 0 & 1 \end{pmatrix}$$

Thus the result that supplements the table in Sect. 8.6.1: $r(T, 256) = 384$.

The chosen example of a two-dimensional picture encryption could be carried over to text encryption by transposition, of course.

9.4.3 Shannon. He recommended quite generally compositions $\mathcal{S}\,\mathrm{F}\,\mathcal{T}$, where $\mathcal{S}$ and $\mathcal{T}$ are classes of relatively simple methods and F is a (fixed) transformation (a barrier) achieving a thorough amalgamation. In the example of tomographic methods (Sect. 4.2) this would be the sandwiched transposition, in the example of mixed-rows columnar transposition (Sect. 6.2.3) the matrix transposition. In modern applications it could be a chip, defining a family of 64 polygraphic substitutions of width 64 bits. A warning seems to be appropriate: blind confidence in the efficacy of such barriers is not justified, for there is always the danger of an illusory complication. Furthermore, the better the amalgamation, the more a local encryption error will propagate over the whole cryptotext. Much worse, a local error in the transmitted cryptotext will spread over the whole decrypted plaintext ("avalanche effect"), making it completely unreadable; with all the bad consequences (Chapter 11) in case of a repetition. In this sense, good amalgamation is dangerously good.

9.4.4 Tomographic methods. They achieve a simple, practical and rather effective amalgamation by using first a multipartite substitution, where the barrier is a special transposition, namely a cutting into pieces and reassembling of the intermediate cryptotext, and finally by applying a multigraphic substitution. It seems that the idea originated with Honoré Gabriel Riqueti Comte de Mirabeau (1749–1791), a French publicist and politician before the time of the French Revolution. After a bipartite, one-to-one Polybios substitution (Sect. 3.3.1) $\mathbb{Z}_{25} \longrightarrow \mathbb{Z}_5 \times \mathbb{Z}_5$ he grouped together all the first digits, then all the second digits. Finally he applied the inverse Polybios substitution $\mathbb{Z}_5 \times \mathbb{Z}_5 \longrightarrow \mathbb{Z}_{25}$. (This was presumably followed by a steganographic method—Bazeries, claiming that he had decrypted some authentic letters of the Marquise Sophie de Monnier to her famous lover Honoré de Mirabeau and mentioning the *teneur pornographique*, does not give further details, but makes the remark in connection with Boetzel and O'Keenan, Sect. 1.2). The to and fro Polybios substitution $\mathbb{Z}_{25} \longleftrightarrow \mathbb{Z}_5 \times \mathbb{Z}_5$ also interested young Lewis Carroll (diary note of February 26, 1858).

9.4.5 Polybios. Instead of a fixed transposition, the barrier can also be a family of linear transformations, e.g., a VIGENÈRE or BEAUFORT over $\mathbb{Z}_5$. This (without the back substitution) is the basic idea of the "Nihilist ciphers" which can be operated with a periodic or with a running key. The to and fro Polybios substitution was moreover used in the early cipher machines B-21, B-211 of Boris Hagelin. For the barrier, Hagelin used two of Damm's half-rotors with ten positions to obtain for each of the two $\mathbb{Z}_5$ ten different alphabets (two quintuplets of shifted ones); altogether 100 different alphabets resulted. The rotor movement was accomplished by two pairs of pin wheels with 17, 19, 21 and 23 teeth. The B-211 also had a plugboard.

9.4.6 Koehl. The tomographic method Alexis Koehl proposed (Sect. 4.2.3) was based, according to Gyldén, on $\mathbb{Z}_{25} \longrightarrow \mathbb{Z}_{10} \times \mathbb{Z}_{10}$. Delastelle[4] also discussed a tomographic method more general than his local one mentioned in Sect. 4.2.3: the regrouping of larger pieces, as Mirabeau had done. For the regrouping he proposed a *longueur de sériation*, e.g., 7 in the following example with a Polybios substitution derived from the password BORDEAUX:

e n v o y e z	u n b a t a i	l l o n i n f	a n t e r i e
1 4 5 1 5 1 5	2 4 1 2 5 2 3	4 4 1 4 3 4 2	2 4 5 1 1 3 1
5 3 3 2 4 5 5	2 3 1 1 2 1 3	1 1 2 3 3 3 5	1 3 2 5 3 3 5
14 51 51 55 33 24 55	24 12 52 32 31 12 13	44 14 34 21 12 33 35	24 51 13 11 32 53 35
D S S Z I C Z	C O T H G O R	P D J A O I K	C S R B H V K

[4] *Félix Marie Delastelle*, 1840–1902. Verfasser des *Traité Élémentaire de Cryptographie*, Gauthier-Villars, Paris 1902.

Delastelle was practical enough to choose a seriation length (*longueur de sériation*) that was not too large: an encryption error would otherwise spread over the whole message (see Sect. 11.3). But the danger of unauthorized decryption is higher.

9.4.7 Other methods. Delastelle also discussed tomographic methods on the basis of a tripartite substitution (Sect. 4.1.3). Other tomographic methods used the ternary Morse code, e.g., a method called POLLUX and a reversed *Kulissenverfahren* by M. E. Ohaver, which in the following example uses encryption width 7:

	s	e	n	d	s	u	p
Morse symbols	···	·	–·	–··	···	··–	·––·
code length	3	1	2	3	3	3	4
reversed	4	3	3	3	2	1	3
regrouped symbols	····	–·–	···	···	·–	·	––·
	H	K	S	S	A	E	G

9.5 Confusion and Diffusion by Arithmetical Operations

A thorough amalgamation is accomplished in particular by arithmetical operations. A method liked by mathematicians, recently rediscovered, is based on an arbitrary monoalphabetic block encryption of a message as a sequence of numbers, followed by an encryption of each one of these numbers by arithmetical operations *modulo* a suitable number q, possibly with re-encryption into literal form ('symbolic addition', 'symbolic multiplication').

Addition *modulo* q, as well as multiplication by a factor h *modulo* q, was discussed in connection with linear substitutions (Sect. 5.7). Likewise the r-th power *modulo* q can be formed. These operations are increasingly amalgamating: multiplication as iterated addition, powering as iterated multiplication. We will see that if the inverse operations exist, they provide authorized decryption for roughly the same effort as needed for encryption.

For given q, a plaintext block can be encrypted whose number equivalent x fulfills the condition $0 \leq x < q$. The encryption as a number can even be performed by customary numeral codes; this brings about a compression that can be considerable for stereotyped texts: standard commercial codes comprise on average 8.5 plaintext letters per five-digit group.

The number representation can be in any number system for the basis B with $B \geq |V|$ that is convenient and allows fast carrying out of the arithmetical operations. With $V = \mathbb{Z}_{26}$ frequently the basis $B = 100$, i.e., essentially decimal arithmetic ($\mathbb{Z}_{10}^2$) with digit pairs was used, or the basis $B = 32$, i.e., essentially binary arithmetic ($\mathbb{Z}_2^5$) with five-bit groups. Today, bytes ($B = 256$), 16-bit groups ($B = 2^{16}$), 32-bit groups ($B = 2^{32}$), and 64-bit groups ($B = 2^{64}$) are commonly used.

9.5.1 Residual arithmetic. For the method of multiplication with a factor h *modulo* q,

$$M_h(x) = x \cdot h \mod q \, ,$$

the necessary preparations have been made in Sect. 5.7.

For prime $q = p$, the multiplication *modulo* p forms a group; for every $h \not\equiv 0 \mod p$ there exists an inverse h' such that $h \cdot h' \mod p = 1$, thus for the multiplication in the Galois field $\mathbb{F}(p)$:

$$M_{h'}(M_h(x)) = x \quad .$$

For non-prime q, h has an inverse (h is regular with respect to q) if and only if it is relatively prime to q (Sect. 5.6).

For technical reasons, q is frequently chosen to be of the form $q = 2^k$ or $q = 2^k - 1$ if computation is done in the binary system; in the decimal system correspondingly $q = 10^k$ or $q = 10^k - 1$. In the first cases the result is directly found in the lower part of the accumulator (Sect. 5.7.1). In the case $q = 2^k$, only the odd numbers have inverses; in the case $q = 2^k - 1$, the non-prime q should be avoided.

For large values of q the determination of the inverse h' of a given h looks non-trivial only at first sight: the division algorithm by successive subtraction functions also for cycle numbers; in fact, an analogue to the fast division algorithm we customarily perform in a positional system for $\mathbb{Z}$ was given in Sect. 5.7.1 for $\mathbb{Z}_q = \{0, 1, 2, 3, \ldots, q-1\} \subset \mathbb{Z}$. It can be brought into the form shown in the following example:

$$17 \cdot h' \equiv 1 \mod 1000$$

$$\begin{array}{r}
-1 \\
\left.\begin{array}{r} 16 \\ 33 \\ 50 \end{array}\right\} 3 \\
\left.\begin{array}{r} 220 \\ 390 \\ 560 \\ 730 \\ 900 \end{array}\right\} 5 \\
\left.\begin{array}{r} 2600 \\ 4300 \\ 6000 \end{array}\right\} 3
\end{array}$$

$$17 \cdot 353 = 6001 \equiv 1 \mod 1000 \, .$$

It is easy to program a microprocessor to do this efficiently. Once h' is determined, decryption needs the same effort as encryption.

If q is the product of two (different) primes, $q = p' \cdot p''$, and if $h \cdot h_1' \equiv 1 \mod p'$ and $h \cdot h_2' \equiv 1 \mod p''$, then $h \cdot h' \equiv 1 \mod q$, where $h' \equiv h_1' \mod p'$ and $h' \equiv h_2' \mod p''$.

The residual arithmetic reduces the effort for the determination of h' considerably.

9.5.2 Powering. The following can be stated for the method of raising a number to a fixed power h *modulo* q,

$$P_h(x) = x^h \mod q \quad (x^0 = 1):$$

For prime $q = p$, if for some h' $h \cdot h' \equiv 1 \mod p{-}1$ (therefore h relatively prime to $p{-}1$), then $P_{h'}(x)$ is inverse to $P_h(x)$. Thus, for the raising to a power in the Galois field $\mathbb{F}(p)$

$$P_{h'}(P_h(x)) = x \quad .$$

Proof: To begin with,

$$\begin{aligned} P_{h'}(P_h(x)) &= x^{h \cdot h'} \mod p = x^{h\cdot(h' \bmod p-1+\alpha\cdot(p-1))} \mod p \quad \text{for suitable } \alpha \\ &= x^{h\cdot h' \bmod p-1} \cdot x^{h\alpha\cdot(p-1)} \mod p \\ &= x^1 \cdot (x^{p-1} \mod p)^{h\alpha} \end{aligned}$$

From Fermat's theorem, $x^{p-1} \mod p = 1$, thus
$P_{h'}(P_h(x)) = x$. ⋈

Examples: Mutually reciprocal pairs (h, h') can be found in Sect. 5.5, Table 1, e.g.,

for $p = 11$: (3,7) and (9,9) $(N{=}10)$;
for $p = 23$: (3,15), (5,9), (7,19), (13,17), and (21,21) $(N{=}22)$;
for $p = 31$: (7,13), (17,23), (11,11), (19,19), and (29,29) $(N{=}30)$.

Case $p = 11$:

x^3 mod 11 has the cycle representation (0) (1) (2 8 6 7) (3 5 4 9) (10) ,
x^9 mod 11 has the cycle representation (0) (1) (2 6) (8 7) (3 4) (5 9) (10) .
x, x^3 mod 11, x^9 mod 11, and x^7 mod 11 form the cyclic group $\mathcal{C}_4$ of order 4.

Case $p = 31$:

x^7 mod 31 has the cycle representation A of order 4
(0) (1) (5) (25) (9 10 20 18) (17 12 24 3) (2 4 16 8)
(6) (26) (14 19 7 28) (22 21 11 13) (15 23 29 27) (30)

x^{11} mod 31 has the cycle representation B of order 2
(0) (1) (5 25) (9 14) (10 19) (17 22) (12 21) (2) (16) (4) (8)
(6 26) (20 7) (18 28) (24 11) (3 13) (15) (29) (23) (27) (30)

x^{17} mod 31 has the cycle representation AB of order 4
(0) (1) (5 25) (14 10 7 18) (22 12 11 3) (2 4 16 8)
(6 26) (9 19 20 28) (17 21 24 13) (15 23 29 27) (30)

x^{19} mod 31 has the cycle representation A^2 of order 2
(0) (1) (5) (25) (9 20) (10 18) (17 24) (12 3) (2 16) (4 8)
(6) (26) (14 7) (19 28) (22 11) (21 13) (15 29) (23 27) (30)

x^{29} mod 31 has the cycle representation A^2B of order 2
(0) (1) (5 25) (9 7) (10 28) (17 11) (12 13) (2 16) (4 8)
(6 26) (14 20) (19 18) (22 24) (21 3) (15 29) (23 27) (30)

x^7 mod 31 and x^{11} mod 31 generate the group $\mathcal{C}_4 \times \mathcal{C}_2$ of order 8.

Case $p = 23$: $x^7 \bmod 23$ has the cycle representation

(0) (1) (2 13 9 4 8 12 16 18 6 3) (5 17 20 21 10 14 19 15 11 7) (22)

and generates the cyclic group $\mathcal{C}_{10}$ of order 10.

Generally, for given odd prime p, the set $\{P_h : h$ regular w. r. t. $p-1\}$ forms an Abelian (commutative) group $\mathcal{M}_{p-1}$, depending on p. Polyalphabetic encryption with this group as key group is possible. The group is of order $\frac{(p-1)}{2} - 1$, if $\frac{(p-1)}{2}$ is prime. A prime p such that $p' = \frac{(p-1)}{2}$ is also a prime, is called a safe (or 'strong') prime (Bob and G. R. Blakely 1978): p' is then called a *Sophie Germain* prime.

Safe primes are 5, 7, 11, 23, 47, 59, 83, 107, 167, 179, 227, 263, 347, 359, 383, 467, 479, 503, 563, 587, 719, 839, 863, 887, 983, 1019, 1187, 1283, 1307, 1319, 1367, 1439, 1487, ... ; but there are also big ones like $45 \cdot 2^{37} - 1$ and $10^{100} - 166517$. Apart from 5 and 7, all safe primes are of the form $12a - 1$.

$P_h(x)$ has the trivial fixpoint $x = 0$ and the two normal fixpoints $x = 1$, $x = p - 1$, besides possibly other ones. The powers $P_h(x)$, $P'_h(x)$ may be obtained as products of repeated squares; a binary representation of h and h' indicates how this is to be done.

For $p=11$, since $3_{10} = 11_2$ and $7_{10} = 111_2$; $9_{10} = 1001_2$:

$$P_3(x) = x \cdot x^2 , \qquad P_7(x) = x \cdot x^2 \cdot (x^2)^2 , \qquad P_9(x) = x \cdot ((x^2)^2)^2 ,$$

where $. \cdot .$ indicates multiplication and $.^2$ squaring, each time *modulo* 11. In fact, for n-bit numbers, with $2^n < p < 2^{n+1}$, raising to a power *modulo* p takes roughly the same effort as n multiplications do. With the present tendency to displace encryption steps into microprocessor chips, arithmetical methods will become more and more important in the future.

Especially for primes of the form $p = 2^{2^k} + 1$ (Fermat primes) one arrives at the problem of reciprocal pairs *modulo* 2^{2^k} ; special solutions exist.

For non-prime q, the situation is more complicated. The special case where q is a product of two (different) primes, $q = p' \cdot p''$, will be treated in Sect. 10.3.

9.5.3 Two-way communication. Since h and h' in Sects. 9.5.1 and 9.5.2 are interchangeable, in the mutual communication of two partners A and B the one can use h both for encryption *and* decryption and the other h' likewise both for encryption *and* decryption (Sect. 2.6.2).

9.5.4 Pliny Earle Chase. A harbinger for these arithmetical methods was Pliny Earle Chase; in 1859 he described in the newly founded *Mathematical Monthly* the following method: After some bipartite injective substitution $V \to W^2$ with $W = Z_{10}$, one forms a number x, as Mirabeau did (Sect. 9.4.4), from the first figures and another one y from the second figures. Then one performs simple arithmetical operations, like multiplying x by seven and y by nine, and finally retranslates the result into V. This simple system offered more security than many customary schemes, although it did not find practical use.

9.6 DES and IDEA®

The *Data Encryption Standard* (DES) algorithm was promulgated in 1977 by the National Bureau of Standards (NBS) in the U.S.A. for use with "unclassified computer data."[5] The DES method is a block encryption for octograms of bytes. A sequence of fixed transpositions and key-dependent, multipartite, non-linear substitutions produces a thorough amalgamation. DES is a tomographic method; this can be seen best from the original proposal LUCIFER by Horst Feistel, an employee of IBM (Fig. 76). Quite obviously, the impression is given that Shannon (Sect. 9.4.3) is the godfather. The key has eight bytes, but in fact this includes eight parity bits, there are only 56 genuine key bits. A short effective key length was desirable for the NSA.

9.6.1 The DES Algorithm. We give only a sketch of the method; for details the official source[5] may be consulted.

9.6.1.1 Encryption. The principal construction of the DES encryption step is shown in Fig. 77: The 8-byte plaintext block is first subjected to a (key-independent) initial transposition T and subsequently split into two 4-byte blocks L_0 and R_0. Next are 16 rounds $(i = 1, 2, 3, \ldots, 16)$ with

$$L_i = R_{i-1} \quad \text{and} \quad R_i = L_{i-1} \oplus f(R_{i-1}, K_i) \ .$$

The symbol $\oplus$ is used for addition *modulo* 2. K_i is a 48-bit key, generated via a selection function by the given key. The final transposition T^{-1}, inverse to T, ends the DES encryption step.

The function f is the central part of the algorithm (Fig. 78). The 32-bit block R_{i-1} is expanded into a 48-bit block $E(R_{i-1})$ by duplication of certain bit positions and added *modulo* 2 to K_i. The resulting 48-bit block is split into eight 6-bit groups, serving as input for each one of the eight substitution modules S_1, S_2 ... S_8 ('S-boxes'). Each of these modules implements four different nonlinear substitutions. The following table shows these substitutions for S_1.

		0	1	2	3	4	5	6	7	8	9	10	11	12	13	14	15
	0	14	4	13	1	2	15	11	8	3	10	6	12	5	9	0	7
S_1:	1	0	15	7	4	14	2	13	1	10	6	12	11	9	5	3	8
	2	4	1	14	8	13	6	2	11	15	12	9	7	3	10	5	0
	3	15	12	8	2	4	9	1	7	5	11	3	14	10	0	6	13

Bit 1 and bit 6 of the 6-bit group, interpreted as a binary number, determine the row (and thus a substitution), bits 2 to 5 the column. For the input

[5] Federal Information Processing Standards Publication 46, National Technical Information Service, Springfield, VA, April 1977. Federal Register, March 17, 1975 and August 1, 1975. For the presentation of background information (from the point of view of N.B.S.) see Smid M. E., Branstad D. K.: *The Data Encryption Standard: Past and Future*, Proceedings of the IEEE, Vol. 76, No. 5, May 1988.

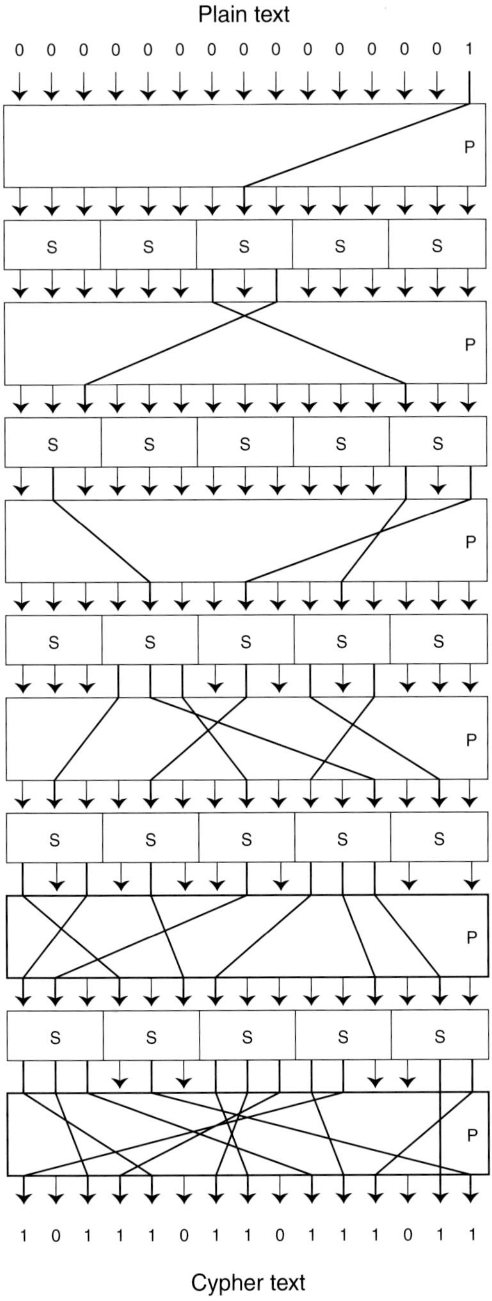

Fig. 76 LUCIFER encryption (Feistel 1973)
A plain text input of a single 1 and fourteen 0's is transformed by the non-linear S-boxes into an avalanche of eleven 1's.

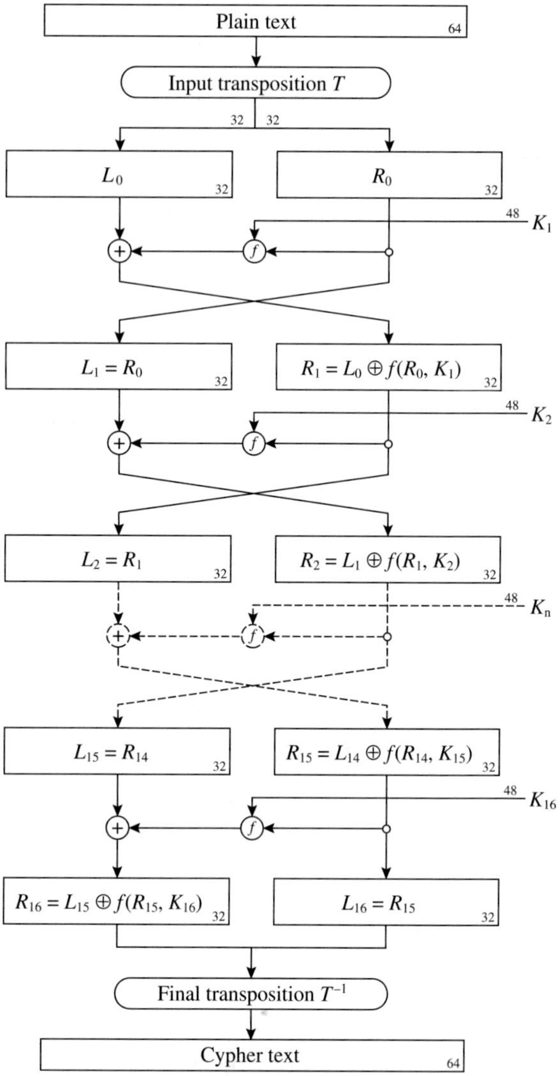

Fig. 77.
DES encryption step

110010 (row 2, column 9 in the table) the module S_1 issues the bit group 1100. The eight 4-bit output blocks of the substitution modules S_1, S_2 ... S_8 are concatenated and subjected to a fixed final transposition P ('P-box').

There remains the question of the derivation of the subkeys. First, the parity bits of the key specified by the user are removed, then the remaining 56 bits are transposed according to a fixed prescription and split into two 28-bit blocks. These blocks are cyclically shifted to the left in each round by one or more positions—depending on the index of the round. From the two of them, according to some specified rule, a 48-bit subkey (K_i) is generated.

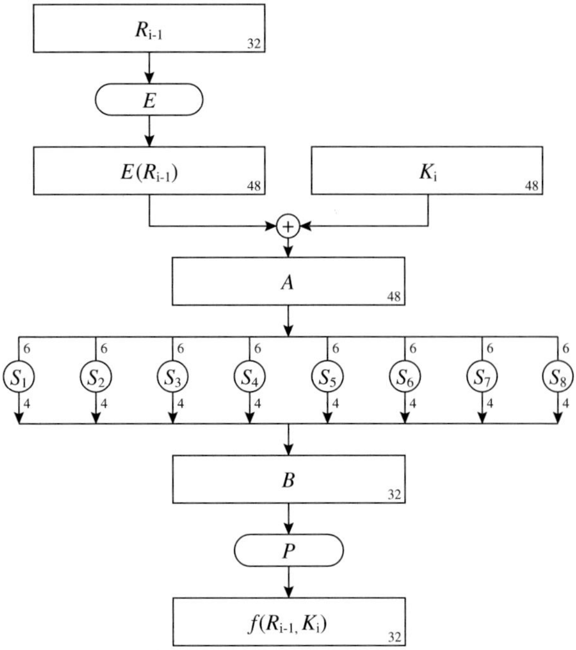

Fig. 78.
The function f

9.6.1.2 Decryption. For decryption the same algorithm is applied, now using the subkeys (K_i) in reverse order. The algorithm is essentially key-symmetric: the same key is used for encryption and decryption. The rounds of the encryption can be described by the self-reciprocal mappings

$$h_i : (R,\ L) \longmapsto (R,\ L \oplus f(R,\ K_i)) \quad \text{(processing)} ,$$

$$g\ : (R,\ L) \longmapsto (L,\ R) \quad \text{(swapping)} .$$

g is obviously an involution, which for h_i follows from the identity

$$L \oplus f(R,\ K_i) \oplus f(R,\ K_i) = L .$$

While the entire encryption is described by

$$DES \equiv T^{-1} \circ h_{16} \circ g \circ h_{15} \circ g \circ \ \dots \ \circ h_2 \circ g \circ h_1 \circ T$$

(in the last round are no swaps), the order of the subkeys is simply reversed for decryption:

$$DES^{-1} \equiv T^{-1} \circ h_1 \circ g \circ h_2 \circ g \circ \ \dots \ \circ h_{15} \circ g \circ h_{16} \circ T \ .$$

Since all the mappings are self-reciprocal, composition of DES and DES^{-1} yields identity.

9.6.2 Avalanche Effect. It turns out that after a few rounds each bit of the intermediate result depends on each bit of the plaintext and of the key. Minimal changes in the plaintext or in the key have the expected effect that about 50 % of the bits change ("avalanche effect").

9.6.3 Modes of Operation for DES. For a plaintext of more than eight bytes, block encryption means dividing the plaintext into 8-byte blocks. In some applications—e.g., if information becomes available step by step and is to be transmitted without delay—plaintext can be shorter than eight bytes. For both cases, a variety of modes of operation is conceivable, with differences with respect to speed of encryption and error propagation. In the end, the intended application will decide which mode is preferable.

The National Institute of Standards and Technology (N.I.S.T., formerly the N.B.S.) has standardized four different modes of operation for use in the U.S.A.—two for each of the principal applications mentioned above.[6]

The ECB mode (Electronic Code Book) treats all 8-byte blocks independently. Identical plaintext blocks result in identical cryptotext blocks. This mode with its strictly monoalphabetic use of the DES algorithm should be avoided as far as possible.

The CBC mode (Cipher Block Chaining) depends on the encryptment's history. The starting point is an initialization block c_0 (*session key*) to be agreed among the partners.

Encryption of the plaintext blocks $m_1, m_2, m_3, \ldots$ results in the following cryptotext blocks $c_1, c_2, c_3, \ldots$ with

$$c_1 = DES(m_1 \oplus c_0) \qquad c_2 = DES(m_2 \oplus c_1) \qquad c_3 = DES(m_3 \oplus c_2) \ .$$

Decryption is performed by

$$m_1 = DES^{-1}(c_1) \oplus c_0 \quad m_2 = DES^{-1}(c_2) \oplus c_1 \quad m_3 = DES^{-1}(c_3) \oplus c_2 \ .$$

The encryption method is now polyalphabetic, but with rather regular construction of the alphabets, in fact it is an autokey method with a priming key, protected only by the non-linearity of the barrier DES.

Apart from these stream-oriented direct encryptions are modes of operation using the DES algorithm for the generation of pseudo-random keytext.

The CFB (Cipher Feedback) mode offers a choice of a 1-bit, 8-bit, 16-bit, 32-bit, or 64-bit output for subsequent use with another encryption method.

The OFB (Output Feedback) mode has internal feedback, the feedback mechanism being independent of both the plaintext and the ciphertext stream. It is normally used to produce an 8-byte (64-bit) output. It has found application in connection with authentication.

---

[6] DES Modes of Operation, National Bureau of Standards (US), Federal Information Processing Standards Publication 81, National Technical Information Service, Springfield, VA, December 1980.

9.6.4 Security of DES. Since the very first publications about the envisaged and later on the agreed-upon standard DES there were discussions and criticism. The number of internal rounds, 16, was considered to be rather low. The main points of attack were and still are:

- The design criteria of the S-boxes have not been disclosed—first not at all, later rather vaguely; however, they were finally published by Don Coppersmith in the 1990s. These barriers, essential for security, could contain 'trapdoors', making unauthorized decryption easy (or at least easier).
- The key length is relatively small. There are only $Z = 2^{56} \approx 72 \cdot 10^{15}$ different keys possible (for the original LUCIFER design—and this comparison is certainly appropriate—the number of keys was much larger, at $2^{128} \approx 340 \cdot 10^{36}$).
- In the ECB mode, the key is kept fixed for quite a while; this monoalphabetic use allows classical attacks ('building a depth', Sect. 19.1).

On the other hand, DES is a rather fast encryption algorithm. A larger key length, many more rounds and other things would have slowed down the chip. The worldwide acceptance of DES as a *de facto* standard justifies to some extent the design.

But in part this discussion was conducted against a background of deep mistrust: The American National Standards Institute was suspected (and even privily accused) of acting as the long arm of the National Security Agency, which was supposed to have an interest in breaking encryptions. Official announcements were not helpful in reducing suspicion.

Even today no trapdoors are (publicly) known. But there is also no proof of their nonexistence. Certain surprising properties of the DES algorithm have been found, like a symmetry under complementation: If both plaintext and key are complemented, the resulting cryptotext is complemented, too. There could be other symmetries undiscovered so far. A residue of distrust has remained. In fact, it is to be hoped that with massive support by faster and faster machines DES can be broken by state authorities, if the national security of the U.S.A. makes it necessary. Private initiative should not be and most likely is not able to solve DES.

Given that DES lacks security—in particular with the ECB mode, which has long been disadvocated but is still used, even commercially, by mediocre vendors—the remedy may be multiple DES encryption with independent keys. But there is the danger of an illusory complication.

An upper limit for the effort to break a method is the brute force attack. It should be kept in mind that DES is available for unlimited tests and thus susceptible to this attack. In 1990, Eli Biham and Adi Shamir found a cryptanalytic countermeasure for attacking amalgamation methods, using some small variations of the plaintext ('differential cryptanalysis'). For DES, it turns out that the brute force attack can indeed be shortened—from 2^{56} tests with full exhaustion to 2^{47}. This is only theoretically interesting—but tempting. In the meantime, Donald Coppersmith has disclosed that as early as 1974 the designers of DES tried their best to prevent such an attack. Several years later, Mitsuri Matsui could shorten this to 2^{43} with his 'linear cryptanalysis'.[7]

Since about 1980, special chips for DES have been on the market. In the first half of the 1990s, they encrypted and decrypted at 10–20 Mbits/sec. Figure 79 shows a chip from the early times (1979).

The DES method has become the worldwide leader on the market (irrespective of American export limitations). It is used by banks for electronic cash transfer, it protects civil satellite communications and UNIX paswords. Whether successors of DES will be equally successful remains to be seen.

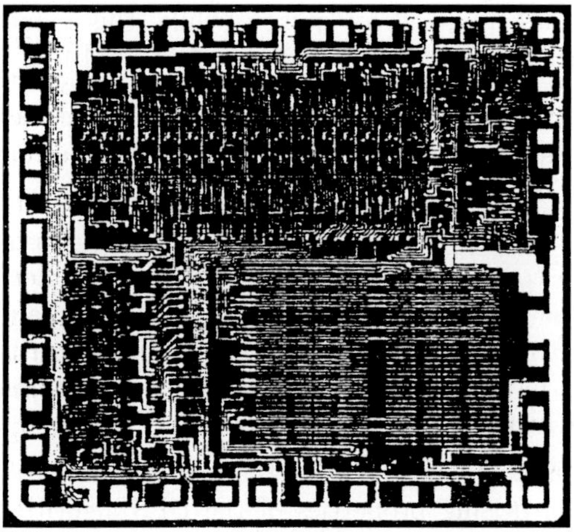

Fig. 79. DES chip from 1979

9.6.5 Successors for DES. Export is free for the 40-bit algorithms RC2 and RC4 of RSA Data Security, Inc., which came on the market in 1993. "If you get permission from the U.S. [for a license to export an encryption algorithm] that probably means it's too easy to decrypt" (Ralph Spencer Poore). The widespread opinion (Otto Horak, in 1996: "DES is nearing the end of its credibility") that the 56-bit DES, intended only for a decade, is to be replaced soon – after all, it can be assumed that in the 20 years from 1977 to 1997 the maximal attainable speed has increased by a factor around 2^{10} – was supported in 1998 by a successful brute force break of exhaustive nature. As a consequence, the N.I.S.T. (National Institute of Standards and Technology) concluded that it could no longer support the use of DES for many applications, and recommended in February 1999 an interim triple-DES (FIPS 46-3) for a few more years, until the new *Advanced Encryption Standard* (AES)[7], with key lengths of 128, 192 or 256 bits, was finalized as

[7] Details are given by Susan Landau, Notices AMS Vol. 47, p. 341 and p. 450.

a Federal Information Processing Standard. (The proposal RIJNDAEL, by Joan Daemen and Vincent Rijmen, was selected October 2, 2000.)

Compared to DES, an essential improvement is offered, too, by the SKIPJACK algorithm with an 80-bit key, working in 32 rounds, which is likewise key-symmetric. Starting from the observation that with a special computer for $ 100 000 in 1993 exhaustion of the key set of DES would have needed[8] 3.5 h; for SKIPJACK, this would be reached within 3.5 h only in 2029, in 1993 it would need $26.2 \cdot 2^{20}$ years, in 2005 still $6.55 \cdot 2^{10}$ years and in 2017 1.64 years—and this for the simple exhaustive ('brute force') attack; every professional decryptor would find ways and means to do it faster.

Alas! Under U.S. law, SKIPJACK, which can be obtained as a tamper-resistant chip (MYK-78 in the Clipper system) programmed by Mykotronx, Torrence, CA, U.S.A., with a throughput up to 20 Mbits/sec and for a price of some $ 10, was until June 1998 classified and not obtainable as software. This meant, of course, that its applicability within computer networks and webs was very restricted, not to speak of political considerations regarding commerce and civil rights. The role of a *de facto* standard that DES has acquired is unlikely to be achieved by SKIPJACK even after declassification.

Europe, commonly more liberal than the U.S.A., does not bind itself to the action of the U.S. government. Among the various unsponsored attempts to create a successor to DES maybe the most promising is IDEA® (*International Data Encryption Algorithm*), developed by J. L. Massey and others since 1990, patented and registered by the Swiss Ascom Tech AG, Solothurn. IDEA has been sold since 1993 as VLSI chip and as software without known commercial restrictions. With a key space of 128 bits, IDEA will measure up for the next century to a brute force attack, although for other cryptanalytic methods, particularly those taking advantage of the enemy's encryption errors, it is just as vulnerable as SKIPJACK and DES. All these encryption algorithms work with 8-byte plaintext blocks.

Sometimes, however, not all existing bits are used cryptologically. Thus, it turned out in early 1998 that in a 64-bits key used worldwide for protection of access to GSM mobile telephones (D1, D2, E-plus) the last ten bits were constantly set to *O*. Consequently, a brute force attack is shortened by a factor 1000 and reaches into the hours-till-days region.

9.6.6 Crypto systems and cryptochips. A lot of excitement was caused by a crypto system which was distributed in mid-1991, a software system called PGP ('Pretty Good Privacy', jokingly also called 'Pretty Good Piracy'). As well as encryption and decryption algorithms (IDEA, more recently also TripleDES and others), PGP contains means for secure key negotiation and for authentication (Sects. 10.5). PGP has grown within a surprisingly short

---

[8] According to M. J. Wiener, *Efficient DES Key Search.* CRYPTO '93, Santa Barbara, CA, Aug. 22-26, 1993.

time into a *de facto* standard for e-mail on the Internet. PGP escaped its originator, Philip R. Zimmermann and slipped through the meshes of U.S. law, very much to the dismay of the N.S.A., but also very much to the malicious glee of an anarchic worldwide cypherpunk movement. In January 1996, the U.S. government dropped its prosecution of Zimmermann. Meanwhile, PGP software has been marketed in the U.S.A. for something like $100 .

Anyway, national laws can not find great attention in an open world, An example was given in March 1998 by Netscape with the release of its *WWW Software Communicator* "mozilla" in source code. In order to obey the U.S. export laws (*International Traffic in Arms Regulations, ITAR*), the *secure sockets layer* (SSL), which allows secure data exchange, was withheld. Within a couple of days an Australian source made "cryptozilla" openly available, a version of the Netscape Communicator equipped with 128-Bit-algorithms for encryption.

Anyhow, a weakness is key negotiation, in general the weakest part of all existing crypto systems. This was a lesson the Polish cryptanalysts taught the Germans. Once key negotiation is broken, the whole encryption algorithm is worthless.

Meanwhile, microprocessor chips are becoming more and more powerful. A recent (1996) general purpose 64-bit processor chip codenamed Alpha-AXP (211 64), made by Digital Equipment Corporation, works with a pulse frequency of 300 MHz, comprises 9.3 million transistors and processes $1.2 \cdot 10^9$ instructions per second. Initially, it has been manufactured in 0.5μ technology, i.e., the electrical connections have had a width of 0.0005 mm. Up to 1999, the pulse frequency was increased; for the successor Alpha 212 64 to about 600 MHz, using a 0.35μ technology. DEC has annonced pulse frequencies higher than 1 GHz for the year 2000 and aims at 0.25μ and 0.18μ technology.

Microprocessors are widely used nowadays in servers, desktop, and laptop computers and are frequently connected into networks, and thus more and more the need to protect their data cryptologically is felt. Crypto AG, Zug (Switzerland) has offered in 1996 a crypto board for stand-alone or networked desktop PCs and notebooks, which provides user identification and access control, encryption of hard disks, floppy disks, directories, and files at a rate of at least 38 Mbits/sec . It has its own tamper-proof key carrier and password storage and works with individually generated pseudo-random keys in a symmetric block cipher algorithm. The key management uses a multi-level key hierarchy. Master keys, data encryption keys, file keys and disk keys have a key variety of $2^{124} = 2 \times 10^{37}$. With dimensions 85 mm by 54 mm and only 3.3 mm thick, the crypto board, as shown in Plate P, is geometrically extremely small. Nevertheless, if it is used properly, it can be expected to withstand the efforts of the largest supercomputer for quite some time.

'Strong' cryptography, which is unbreakable for a while, is possible and is worthwhile; thus, it is going to be winning.

10 Open Encryption Key Systems

The polyalphabetic encryption methods discussed so far use a key for encryption and a key for decryption. Crypto systems with self-reciprocal encryption steps use the same key for both, of course. In general, there are two possibilities:

(1) There is only one key. The same key character has its particular meaning for encryption and for decryption. This is the case for DES (Sect. 9.6.1).

Using crypto machines, this requires a switch that allows a choice between encryption mode and decryption mode.

(2) There are two keys, the encryption key and the decryption key.

The crypto machine needs only one mode, but the derivation of the decryption key from the encryption key needs extra effort.

This can be illustrated with a classical VIGENÈRE method: Case (1) uses for encryption E and decryption D

$$E = \{\chi_{k_j}\} : \quad \chi_{k_j}(x) \quad = x + k_j \bmod N^n$$
$$D = \{\chi_{k_j}^{-1}\} : \quad \chi_{k_j}^{-1}(x) = x - k_j \bmod N^n$$

while case (2) uses for decryption

$$D = \{\chi_{k_j}^{-1}\} : \quad \chi_{k_j}^{-1}(x) = x + (-k_j) \bmod N^n$$

i.e., $\quad \chi_{k_j}^{-1} = \chi_{(k_j)^{-1}} \quad$ where $\quad (k_j)^{-1} = -k_j \, .$

The derivation of $(k_j)^{-1}$ from (k_j) is simple enough in this example, and since $(k_j)^{-1}$, the decryption key, is to be kept secret, (k_j), the encryption key, is to be kept secret as well. But if it were as difficult to derive $(k_j)^{-1}$ from (k_j) as to break the cryptotext with any other means, the encryption key (k_j) could be made public. If so, we would have an open encryption key system. Surprisingly, such cryptosystems exist.

The question is: does such an open encryption key system (public key system for short) offer advantages? And if so, why was such a simple idea presented so late—the mid-1970s—in the long history of cryptology? The answer is that the use of cryptographic methods in commercial networks has characteristics that were missing in the classical two-partner situation. In fact, there are indeed advantages in the case of a many-partner network, as well as advantages if authentication is given equal or even more weight than secrecy, a situation that exists in modern global financial transactions.

10.1 Symmetric and Asymmetric Encryption Methods

10.1.1 Symmetric Methods (Private Key Methods). The key agreed upon by two partners determines in classical cryptosystems both the encryption step and the decryption step in a simple way, which is symmetric in the sense that both times essentially the same effort is needed. Moreover, usually the two partners are at different times both sender and receiver, and in cases where encryption and decryption commute (Sects. 2.6.2, 9.5.3), each needs just one, his or her private key.

These symmetric, private key methods did not cease to exist with the advent of the electronic age, say around 1950. The DES method, discussed in Sect. 9.6, the best known example of modern block encryption, is in this class; there is (case (1) above) only one key, and encryption and decryption differ only with respect to the order in which the rounds generated from the key are applied. Encryption and decryption is fast: in 1995, about 20 Mbits/s.

Thus, cryptanalytic security depends on the secrecy of this key. Furthermore, if the user hopes that the would-be unauthorized decryptor even with knowledge of the method class will never find this key, it goes almost without saying that under real conditions nobody would be able to encrypt a fake message such that the recipient could decrypt it without becoming suspicious. Authentication is not a problem, and it is guaranteed as long as secrecy is guaranteed, but also only if secrecy is guaranteed (but see Sect. 10.5).

However, there are certain disadvantages:

(1) It is impossible for the sender of a message to prove to his partner or a third person that he has sent a particular message. This lack of judicial protection is a handicap for the transmission of orders and for financial transactions.

(2) The keys have to be communicated or negotiated on a channel whose cryptanalytic security is much higher than the security of the channel used for normal transmissions. Spontaneous secure communication may not be possible.

(3) With a large number of partners wanting secure communication, the number of two-way channels and therefore the number of keys becomes quite large. For a network with n partners, each one wanting to exchange messages safely with everyone, $\binom{n}{2} = n \cdot (n-1)/2$ self-reciprocal keys or $n \cdot (n-1)$ symmetric keys are necessary. For $n = 1000$, the numbers are 499 500 and 999 000.

10.1.2 Asymmetric Methods (Public Key Methods). Dispensing with the symmetry of the cryptosystem, the decryption key, of course, is to be protected; but asymmetry may go so far that the encryption key is not only unprotected but even published (public key). On a two-way channel with partners $\mathcal{A}$ and $\mathcal{B}$, there are now four keys: an open encryption key for $\mathcal{B}$ and a corresponding private decryption key for $\mathcal{A}$, an open encryption key for $\mathcal{A}$ and a corresponding private decryption key for $\mathcal{B}$. This is twice the number before. But $\mathcal{A}$ may now receive messages from many other partners, all knowing for their encryption the public key of $\mathcal{A}$ and trusting that only the authorized receiver $\mathcal{A}$ can decrypt it. Thus, each partner has an open (public) key and a private key, and the total number of keys for a network of 1000 partners drops from nearly a million to two thousand.

This eliminates the disadvantage (3) above. As for (1), the solution will be taken up below and in Sect. 10.5. And the problem under (2) disappears, for a published key does not have to be negotiated; whenever a partner $\mathcal{A}$ decides to open a communication channel with $\mathcal{B}$, he may do so after consulting the directory with the keys of all participants.

The concept of an open encryption key system was published in 1976 by Whitfield Diffie and Martin E. Hellman.[1]

10.1.3 Encryption and Signature Methods. Let KP_i denote the public key of the i-th partner and KC_i his private key. KP_i determines an encryption E_i and KC_i determines a decryption D_i. Both E_i and D_i have effective implementations, but $\{KP_i\}$ is a public directory and KC_i is only known to the i-th partner. For all partners, it is impossible (or, practically speaking, intractable) to derive KC_i from KP_i.

If E_i and D_i fulfill the property

$(*)\quad D_i\,(E_i\,(x)) = x\,,$

we speak of an (asymmetric) encryption method serving secrecy.

If moreover E_i and D_i fulfill the property

$(**)\quad E_i\,(D_i\,(x)) = x\,,$

we speak of an (asymmetric) signature method serving authentication.

The asymmetric encryption method works as follows: If partner $\mathcal{A}$ wants to send an encrypted message m to $\mathcal{B}$, he takes from the directory under the heading $\mathcal{B}$ the key $KP_{\mathcal{B}}$ (this determines $E_{\mathcal{B}}$):

$(A)\qquad c = E_{\mathcal{B}}(m)$

and sends the cryptotext c over the public channel to $\mathcal{B}$.

$\mathcal{B}$ uses his private key $KC_{\mathcal{B}}$ (which determines $D_{\mathcal{B}}$) to recover the message m:

$(B)\qquad D_{\mathcal{B}}(c) = D_{\mathcal{B}}(E_{\mathcal{B}}(m)) = m \qquad$ (because of $(*)$)

[1] *New Directions in Cryptography*, IEEE Transactions on Information Theory, IT-22, Vol. 6, pp. 644–654 (1976).

The asymmetric encryption and signature method works as follows: If partner $\mathcal{A}$ wants to send an encrypted message m signed with his signature "A" to $\mathcal{B}$, he first encrypts m with *his private key* $KC_{\mathcal{A}}$ (this determines $D_{\mathcal{A}}$):

$$(A1) \qquad d = D_{\mathcal{A}}(m) ,$$

and joins to d his signature "A" . Next he takes from the directory under the heading $\mathcal{B}$ the key $KP_{\mathcal{B}}$ (this determines $E_{\mathcal{B}}$) and encrypts the pair ("A", d):

$$(A2) \qquad e = E_{\mathcal{B}}(\text{“}A\text{”}, d) = E_{\mathcal{B}}(\text{“}A\text{”}, D_{\mathcal{A}}(m)) .$$

$\mathcal{A}$ sends the cryptotext e over the public channel to $\mathcal{B}$.

$\mathcal{B}$ uses his private key $KC_{\mathcal{B}}$ (which determines $D_{\mathcal{B}}$) to recover the pair:

$$(B1) \qquad D_{\mathcal{B}}(e) = D_{\mathcal{B}}(E_{\mathcal{B}}(\text{“}A\text{”}, d)) = (\text{“}A\text{”}, d) \qquad \text{(because of } (*)\text{)} .$$

$\mathcal{B}$ recognizes from the part "A" , that $\mathcal{A}$ is the sender. $\mathcal{B}$ now uses the public key $KP_{\mathcal{A}}$ of $\mathcal{A}$ (which determines $E_{\mathcal{A}}$) to recover from d the message m:

$$(B2) \qquad E_{\mathcal{A}}(d) = E_{\mathcal{A}}(D_{\mathcal{A}}(m)) = m \qquad \text{(because of } (**)\text{)} .$$

By obtaining meaningful text, $\mathcal{B}$ is assured that he did get the message from $\mathcal{A}$ since no other partner could have encrypted it with $D_{\mathcal{A}}$.

10.2 One-Way Functions

The whole success of asymmetric open encryption key systems rests on the question: How can it be accomplished that D_i, i.e., E_i^{-1}, cannot be easily obtained from E_i, that breaking the encryption is practically intractable?

10.2.1 Strict One-Way Functions. An injective function $f : X \to Y$ is called a strict one-way function if the following holds:

> There is an efficient[2] method to compute $f(x)$ for all $x \in X$, but there is no efficient method to compute x from the relation $y = f(x)$ for all $y \in f[X]$.

Arto Salomaa has given a striking example of a one-way function. An encryption with homophonic encryption steps $Z_{26} \dashrightarrow Z_{10}^7$ is defined as follows: For a letter X, *some* name commencing with this letter X is looked up in the telephone directory of a large city $\mathcal{Z}$, and a 7-digit telephone number listed under this name is the cryptotext. To be concrete: for encrypting /kindergarten/, the steps are

k	$\mapsto$	Koch	$\mapsto$	8202310	i	$\mapsto$	Ivanisevic	$\mapsto$	8119896
n	$\mapsto$	Nadler	$\mapsto$	6926286	d	$\mapsto$	Dicklberger	$\mapsto$	5702035
e	$\mapsto$	Esau	$\mapsto$	8348578	r	$\mapsto$	Remy	$\mapsto$	7256575
g	$\mapsto$	Geith	$\mapsto$	2730661	a	$\mapsto$	Aranyi-Gabor	$\mapsto$	2603760
r	$\mapsto$	Rexroth	$\mapsto$	5328563	t	$\mapsto$	Tecins	$\mapsto$	6703008
e	$\mapsto$	Eisenhauer	$\mapsto$	7913174	n	$\mapsto$	Neunzig	$\mapsto$	3002123

[2] Efficient: With a computational effort which is polynomial in $\log |X|$. If P = NP would hold, no strict one-way function could exist at all.

The following encryption of /kindergarten/ , a sequence of 12 7-digit code-groups:

8202310 8119896 6926286 5702035 8348578 7256575 2730661 2603760 5328563 6703008 7913174 3002123 ,

is obtained by a human operator in less than one minute of time. The decryption is unique, but needs hours and hours if done with the help of the telephone directory of about 2000 pages. A one-way function poses as an encryption for the authorized decryptor already unsurmountable difficulties.

Thus, strict one-way functions cannot be used in a reasonable way for encryption of messages followed up by decryption. However, strict one-way functions without homophones can be used for identification and authentication: A password is encrypted by a strict one-way function and is stored in this form. Any time access is required the password presented is encrypted, too, and the cryptotexts are compared. This scheme is applied in the widely used operating system UNIX[3], but based only on a variant of the DES algorithm which does not qualify as a strict one-way function.

However, it is possible that the legal user of this system is in the possession of an inverse telephone directory—either obtained illegally from the post office or purchased. Such a directory makes the decryption process as simple as encryption is. It is a hidden suspension of the one-way direction like a *trapdoor*: an unsuspecting person cannot go back once he or she has passed through, but the initiated person knows where to find the hidden knob.

10.2.2 Trapdoor One-Way Functions. For data security, which is essentially cryptologic, trapdoor one-way functions are needed, allowing data access by decryption for the *authorized* user.

An injective function $f : X \to Y$ is called a trapdoor one-way function if the following holds:

There is an efficient method to compute $f(x)$ for all $x \in X$.

There is an efficient method to compute $f^{-1}(y)$ for all $y \in f[X]$, but it cannot be derived efficiently from the relation $y = f(x)$ for all $y \in f[X]$: secret additional information, the trapdoor, is necessary.

The trapdoor in Salomaa's example is the inverse directory. It can be established legally by the user who can afford the time to do so, if he needs it frequently (or if he can sell it), and if storage complexity does not preclude it. This sort of preprocessing is one of the best strategies to break asymmetric encryption systems, since they are normally used for some time.

10.2.3 The Efficiency Boundary, Function inversion breaks down for one-way functions only because of lack of time and storage space, i.e., because of time and storage complexity. Technological progress shifts the border line between 'intractable' and 'efficient'; at present roughly every two years the

[3] UNIX is a registered trade mark.

speed of the fastest single computer is doubled and roughly every 15 months computer costs are halved. This latter trend allows increasing parallelization.

The cryptologist counteracts this technological progress by suitably increasing some of the encryption parameters, those that influence the encryption in a sensitive way, and thus he prevents cryptanalysts surmounting the barrier. To give an example: for some methods, the inversion of a one-way function amounts to the decomposition of a number n into its prime factors, a huge task compared with the multiplication of the factors to form the number n. One of the fastest known algorithms, the 'Quadratic Sieve',[4] has 'subexponential complexity', i.e., needs asymptotically of the order of magnitude

$$e^{\sqrt{\ln n \cdot \ln(\ln n)}} = n^{\sqrt{\ln(\ln n)/\ln n}}$$

operations. For $n = 10^{70}$, the exponent $\sqrt{\ln(\ln n)/\ln n}$ has the value ≈ 0.178, correspondingly $n^{\sqrt{\ln(\ln n)/\ln n}} = 2.69 \cdot 10^{12}$. The observation that in 1984 the factorization of $(10^{71}-1)/9$ needed 9.5 h on a CRAY X-MP may serve for calibration; this amounts to roughly $80 \cdot 10^6$ 'macro-computing steps' per second, and gives by extrapolation the following picture (assuming every two years a doubling of the maximal attainable speed of a single computer):

n	bits	$e^{\sqrt{\ln n \cdot \ln(\ln n)}}$	Time 1984	Time 1994	Time 2004
10^{50}	166	$1.42 \cdot 10^{10}$	181 s	5.66 s	181 ms
10^{70}	232	$2.69 \cdot 10^{12}$	9.5 h	0.297 h	34.2 s
10^{100}	332	$2.34 \cdot 10^{15}$	344 d	10.75 d	8.26 h
10^{120}	399	$1.31 \cdot 10^{17}$	52.57 a	600 d	19.3 d
10^{140}	465	$5.49 \cdot 10^{18}$	$2.2 \cdot 10^3$ a	68.75 a	803 d
10^{154}	512	$6.69 \cdot 10^{19}$	$2.7 \cdot 10^4$ a	837 a	26.77 a
10^{200}	664	$1.20 \cdot 10^{23}$	$4.8 \cdot 10^7$ a	$1.5 \cdot 10^6$ a	$4.8 \cdot 10^4$ a

The efficiency boundary is indicated by the 'yearlong' work of several hundred days. It has moved from (in 1984) $n = 10^{100} \approx 2^{332}$ to (in 1994) $n = 10^{120} \approx 2^{399}$ and is expected in 2004 to be $n = 10^{140} \approx 2^{465}$, still below 2^{512}.

A sensation was stirred in 1994 when a 129 decimal digits number (429 bits) was factored into its two prime factors of 65 decimal digits each. According to the extrapolation above a single supercomputer would have needed 3330 days. In fact, the total work was distributed on 1600 (less powerful) computers connected by the Internet and was finished in 8 months time. In 1999, a number with 465 bits was factorized. Starting around the year 2004, numbers with 512 binary bits will no longer give sufficient security against factorization.

More and more special computers with a high degree of parallelization are coming into use. Whatever efforts in this way are made, there are limits

[4] Based on early work by M. Kraitchik, improved by C. Pomerance (1985), P. Montgomery (1987), R. D. Silverman (1987). The fastest version of this algorithm is called 'Double Large Prime Variation of the Multiple Polynomial Quadratic Sieve' (ppmpqs). Neither it nor the even better 'Number Field Sieve' method by John Pollard (1988), asymptotically needing a number of steps of order $e^{((\ln n)^{1/3} \cdot (\ln(\ln n))^{2/3})} = n^{(\ln(\ln n)/\ln n)^{2/3}}$, are 'efficient' in the sense of 10.2.1.

to the mastery of storage and time complexity which cannot be surpassed for reasons of physics. For example, according to our present knowledge a 10^{60}-bit store would need the mass of our entire solar system, likewise 10^{70} operations would take more time than has elapsed since the birth of the known universe in the Big Bang, some 10^{18} sec ago, even if each operation took no longer than 10^{-43} sec, the Planck time, the shortest time interval that is meaningful in terms of known physics.

But these large numbers can be deceptive. There is no proof of the nonexistence of an algorithm much faster than the Quadratic Sieve or other roughly comparable ones. It could happen that prime factorization of n can be done within a time increasing with a low polynomial in n. But it is not likely. As is frequently said, people have been investigating factorization for more than two thousand years. Generally, the non-existence of other trapdoors than the ones already known is hard to prove. And complexity theory in its present state is little help, for it usually gives only upper bounds for the effort needed. "There are no provable lower bounds for the amount of work of a cryptanalyst analyzing a public-key cryptosystem" (Salomaa 1990). A newly found trapdoor could endanger the security of a cryptosystem as much as a direct decryption attack, bypassing the function inversion altogether. This is a *principal* risk of asymmetric methods and such a big disadvantage that their use in highly sensitive areas is rather questionable.

10.2.4 Known Examples of One-Way Functions. A proof for the existence of strict one-way functions is hampered by the lack of sufficiently good lower bounds for the known methods. But there are good candidates, based upon the operations multiplication and exponentiation over the Galois field $\mathbb{F}(p)$, where p is prime.

10.2.4.1 A One-Way Function without trapdoor: Multiplication of Primes

As Turing remarked in 1937 (Sect. 5.7), it is relatively simple to multiply two numbers of ten thousand decimal digits and therefore also two primes of this size; on a home computer it takes only seconds. But today there is (see Sect. 10.2.3) no efficient method (publicly) known to decompose a 200-digit decimal number into its prime factors (apart from special cases).

Let $X = \{(x_1, x_2) \mid x_1, x_2 \text{ prime}, K \leq x_1 \leq x_2\}$ for a sufficiently large K. The injective function

$$f : X \to \mathbb{N} \quad \text{defined by} \quad f(x_1, x_2) = x_1 \cdot x_2$$

is therefore a one-way function. No trapdoors are known.

10.2.4.2 A One-Way Function without trapdoor: Exponentiation in $\mathbb{F}(p)$

Let p be prime. For a fixed a the a-exponential function in $\mathbb{F}(p)$

$$F_a : \mathbb{Z}_{p-1} \to \mathbb{Z}_p \backslash \{0\} \quad \text{defined by} \quad F_a(n) = a^n \bmod p$$

is for sufficiently large p and a a one-way function (for $\mathbb{Z}_p$ see Chap. 5).

Example: $p = 7$, $\mathbb{Z}_p\backslash\{0\} = \{1,2,3,4,5,6\}$, $a = 2$:

n	0	1	2	3	4	5	6≡0
2^n	1	2	4	8	16	32	64
$2^n \bmod 7$	1	2	4	1	2	4	1

The computational effort for F_a is within endurable bounds even for values of p and a surpassing 10^{200}. The basic idea of repeated squaring and multiplying is as in Sect. 9.5.2; it is demonstrated by the following example where $.\cdot.$ indicates multiplication and $.^2$ squaring, each time *modulo* p:

$$a^{25} = \left(\left(\left(a^2 \cdot a\right)^2\right)^2\right)^2 \cdot a \ , \quad \text{since} \quad 25_{10} = 11001_2 \ .$$

The example $a = 2$, $p = 7$ shows that the a-exponential function is not necessarily injective. If F_a is injective over $\mathbb{Z}_p\backslash\{0\}$ and thus is a group isomorphism of $\mathbb{Z}_{p-1}$ and $\mathbb{Z}_p\backslash\{0\}$, then a is called a primitive root of $\mathbb{Z}_p$; like $a = 3$, $a = 11$, $a = 12$, $a = 22$ for $p = 31$:

$3^n \bmod 31$ gives a permutation with a (13+9+8) cycle decomposition
(1 3 27 23 11 13 24 2 9 29 21 15 30) (6 16 28 7 17 22 14 10 25) (4 19 12 8 20 5 26 18)

$11^n \bmod 31$ gives a permutation with a (26+3+1) cycle decomposition
(1 11 24 8 19 22 18 2 28 10 5 6 4 9 23 12 16 20 25 26 7 13 21 27 15 30) (3 29 17) (14)

$12^n \bmod 31$ gives a cyclic permutation with the cycle of length 30
(1 12 20 23 28 26 2 24 9 15 25 21 4 17 18 30 19 11 8 3 5 29 7 22 16 6 10 27 14 13)

$22^n \bmod 31$ gives a permutation with a (24+5+1) cycle decomposition
(1 22 10 5 6 8 28 18 16 9 27 29 24 4 20 25 26 14 7 21 23 3 15 30) (2 19 11 17 12) (13)

It can be shown that for each prime p there is always at least one primitive root. In fact their number is $\varphi(p-1)$, others for $p = 31$ are 21, 17, 13, 24. For special p there may be peculiarities. For example, for $p = 17, 257, 65537$ and all larger primes p (if any) of the form $p = 2^{2^k} + 1$ (Fermat primes), 3 and 7 are always primitive roots (Albert H. Beiler, Armin Leutbecher).

If a is indeed a primitive root, F_a has an inverse F_a^{-1}, named the (discrete) a-logarithm function or index in $\mathbb{Z}_p\backslash\{0\}$. While the exponentiation in $\mathbb{Z}_p$ is quite efficient, it is hard to get efficient algorithms for the computation of the discrete logarithm.

Among the known algorithms for the discrete logarithm over a multiplicative group such as $\mathbb{Z}_p\backslash\{0\}$, even good ones like the 'Giant-Step-Baby-Step' algorithm[5] (Daniel Shanks 1971) need an effort proportional
$\sqrt{|\mathbb{Z}_p|} = \sqrt{p} = e^{\frac{1}{2}\ln p}$ and thus are not efficient.

The better 'Index Calculus' method—which requires finding a suitable basis of the multiplicative group, usually the first t primes and thus a huge data

[5] For a running program, see Otto Forster, *Algorithmische Zahlentheorie*, Vieweg, Braunschweig 1996.

base to be precomputed—works only in special cases, but has still subexponential complexity, i.e., needs an effort of the same order of magnitude $e^{\sqrt{\ln p \cdot \ln(\ln p)}}$ as prime decomposition by the Quadratic Sieve does.

$(\mathbb{Z}_p, +, \times)$ is a field, the Galois field $\mathbb{F}(p)$ of characteristic p. More generally, consider the Galois field $\mathbb{F}(p^k)$, an extension of $\mathbb{F}(p)$, and its multiplicative group $\mathbb{F}(p^k)\backslash\{0\}$ which is still a cyclic group. It is generated by some element x, which is a nontrivial root of the equation $x^{p^k} - x = 0$. The elements of $\mathbb{F}(p^k)$ are the p^k polynomials of degree at most $k-1$ over the field $\mathbb{F}(p)$ and can be implemented as a k-dimensional vector space.

Example: $p = 2$, $k = 3$, $\mathbb{F}(2^3) = \{0, 1, x, x+1, x^2, x^2+1, x^2+x, x^2+x+1\}$.
$x^8 - x$ has an irreducible factor $x^3 + x + 1$, powers reduced by $x^3 \mapsto x + 1$.

It is only the multiplicative group of $\mathbb{F}(p^k)\backslash\{0\}$ that matters. For $k>1$, this multiplication is different from that of the modular arithmetic. Thus, we finally have the group isomorphism

$$F_a : \mathbb{Z}_{p^k-1} \rightarrow \mathbb{F}(p^k)\backslash\{0\} \quad \text{defined by} \quad F_a(n) = a^n \text{ in } \mathbb{F}(p^k) \ .$$

In the example $\mathbb{F}(2^3)$ with $a = x$ and with $a = x+1$:

n	0	1	2	3	4	5	6	$7{\equiv}0$
x^n	1	x	x^2	x^3	x^4	x^5	x^6	x^7
x^n red.	1	x	x^2	$x+1$	x^2+x	x^2+x+1	x^2+1	1

n	0	1	2	3	4	5	6	$7{\equiv}0$
$(x+1)^n$	1	$x+1$	$(x+1)^2$	$(x+1)^3$	$(x+1)^4$	$(x+1)^5$	$(x+1)^6$	$(x+1)^7$
$(x+1)^n$ red.	1	$x+1$	x^2+1	x^2	x^2+x+1	x	x^2+x	1

The case $p = 2$ and k large has particular interest. In 1988 John Pollard gave a variant *Number Field Sieve* (NFS) of the Index Calculus method, where primes are to be replaced by irreducible polynomials—the complexity being determined by $e^{((\ln p)^{1/3} \cdot (\ln \ln p)^{2/3})}$. With massive parallel computation, preparatory work has been done for such ambitious problems as $\mathbb{F}(2^{503})$ (Don Coppersmith 1986, Kevin McCurley 1990, Dan Gordon and McCurley 1993).

A natural next step in the use of group isomorphisms is made with the elliptic curve method (ECM), developed by Neil Koblitz (1985), Victor S. Miller (1985), Hendrik W. Lenstra, jr. (1986). It uses the known theory of certain algebraic curves of third order ('elliptic curves') in the projective plane over a finite field K, in particular over $\mathbb{F}(p^k)$. Some algorithms, like the Giant-Step-Baby-Step algorithm, can be extended to the point group of an elliptic curve. It seems that for the rather good Index Calculus method, finding a basis is hopeless in the case of elliptic curves. Thus, the elliptic curve method possibly offers greater security for identification and authentication, which makes it for the time being an interesting topic of research. Elliptic curves over $\mathbb{F}(2^k)$ (case $p = 2$) are particularly advantageous, since the arithmetic processors for the underlying field are easy to construct and relatively simple to implement for large n.

No trapdoors are known for the methods considered so far, not even for the case $\mathbb{F}(2^k)$, $k > 1$. For composite q, $F_a(n) = a^n \bmod q$ has a trapdoor, if q is a product of two different primes: the prime factorization of q allows the preparation of a table, which with the help of the Chinese Remainder Theorem (Sect. 10.4.3) makes calculating the discrete logarithm easier.

10.2.4.3 A Trapdoor One-Way Function: Raising to a Power *modulo* q

In Sect. 9.5.2, raising to a fixed power restricted to the Galois field $\mathbb{F}(p)$ was discussed. Now q may be composite,

$$P_h(x) = x^h \mod q .$$

There still exist suitable pairs (h, h') of fixed numbers from $\mathbb{Z}_q \setminus \{0\}$ such that

(1) there exists an efficient method to compute $P_h(x)$ for all $x \in \mathbb{Z}_q$,

(2) there exists an efficient method to compute $P'_h(x)$ for all $x \in \mathbb{Z}_q$,

such that $P_{h'}(P_h(x)) = x$ and $P_h(P_{h'}(x)) = x$.

But if only h and q are known and q is large enough, say $q > 10^{200}$, then there is no efficient method (publicly) known to compute h' efficiently.

There is a trapdoor. The derivation of h' is much easier if q is composite and a factorization of q into two factors, both rather large, is known. This will be discussed in more detail in Sect. 10.3.

10.2.4.4 A Trapdoor One-Way Function: Squaring *modulo* $q = p' \cdot p''$

This is the important special case $h = 2$ of Sect. 10.2.4.3, using 'quadratic residues', square roots *modulo* q, for which there is a theory that goes back to Legendre and Gauss. An application for open encryption key systems was studied in 1985 by H. C. Williams.

We consider first the case $q = p$, p prime. Table 1 in Sect. 5.5 shows for odd p under $N = p-1$ no entries for $h = 2$. P_2 in Sect. 10.2.4.3 is neither injective nor surjective. For the equivocal inversion of P_2 we may write $\sqrt{\ }$. To give an example, say for $p = 17$, by inversion of the function table we obtain:

$$\sqrt{1}=\pm 1 \quad \sqrt{2}=\pm 6 \quad \sqrt{4}=\pm 2 \quad \sqrt{8}=\pm 5$$
$$\sqrt{9}=\pm 3 \quad \sqrt{13}=\pm 8 \quad \sqrt{15}=\pm 7 \quad \sqrt{16}=\pm 4 .$$

For prime p, there are efficient methods for the calculation of the square root *modulo* p, based on Gauss's Golden Theorem, the Law of Quadratic Reciprocity.

The situation is different for composite q, say $q = p' \cdot p''$. If the prime decomposition of q is known, then square roots $\pm u$ of a *modulo* p' and square roots $\pm v$ of a *modulo* p'' can be obtained efficiently and $\sqrt{a}$ *modulo* q can be calculated easily. But for anyone who does not know the prime decomposition of q, the computation of $\sqrt{a}$ *modulo* q has been proven by M. O. Rabin (1979) as hard as the factorization of q.

10.3 RSA Method

The RSA method is the best known among the open encryption key methods[6] and is named after Ronald L. Rivest, Adi Shamir, and Leonard M. Adleman (1978), the U.S. patent holded until Sept. 20, 2000. The RSA method is based on the widely accepted conjecture that under certain conditions, raising to a fixed power *modulo* q is a one-way function with trapdoor (Sect. 10.2.4.3).

10.3.1 For the i-th partner in an asymmetric encryption network, let

(1) $q_i = p'_i \cdot p''_i$, where p'_i , p''_i odd primes, $p'_i \neq p''_i$.

(2) e_i , $d_i \in \{1, 2, \ldots, \psi(q_i) - 1\} \subset \mathbb{Z}_{q_i} \setminus \{0\}$, with

(2a) $\mathrm{ggt}\,(e_i, \psi(q_i)) = 1$, (2b) $\mathrm{ggt}\,(d_i, \psi(q_i)) = 1$,

(2c) $e_i \cdot d_i \bmod \psi(q_i) = 1$,

where ψ denotes the Carmichael function,[7]

$$\psi(p'_i \cdot p''_i) = \mathrm{lcm}\,(p'_i - 1\,, p''_i - 1) = 2 \cdot \mathrm{lcm}\,(\tfrac{p'_i - 1}{2}\,, \tfrac{p''_i - 1}{2})\,.$$ [8]

The RSA method is a highly polygraphic, monoalphabetic block encryption, with plaintext characters $p_j \in \mathbb{Z}_{q_i}$ and cryptotext characters $c_j \in \mathbb{Z}_{q_i}$.

The following keys are used with the i-th partner:

public e_i (for encryption)[9] , q_i

private d_i (for decryption) .

The encryption step is defined through the one-way function $E_i : \mathbb{Z}_{q_i} \to \mathbb{Z}_{q_i}$,

$$E_i(m_j) \;=\; m_j^{e_i} \bmod q_i \;=\; c_j\,.$$

The decryption step is defined through the one-way function $D_i : \mathbb{Z}_{q_i} \to \mathbb{Z}_{q_i}$,

$$D_i(c_j) \;=\; c_j^{d_i} \bmod q_i \;=\; m_j\,.$$

This gives an asymmetric signature system, since

$$D_i(E_i(x)) = E_i(D_i(x)) = x \quad \text{for all } x \in \mathbb{Z}_{q_i}\,.$$

The proof follows the one in Sect. 9.5.2. It can be based on the following corollary of Carmichael's theorem for relatively prime a , n:

If $b \equiv b' \; mod \; \psi(n)$, then $a^b \equiv a^{b'} \; mod \; n$.

---

[6] U.S. Patent No. 4 405 829, September 20, 1983.

[7] In the original publication, instead of Carmichael's function Euler's function φ is used, $\varphi(p'_i \cdot p''_i) = (p'_i - 1) \cdot (p''_i - 1)$. The conditions stated guarantee the conditions of the original method.

[8] $\psi(2 \cdot p'_i \cdot p''_i) = \mathrm{lcm}(1\,, p'_i - 1\,, p''_i - 1) = \mathrm{lcm}(p'_i - 1\,, p''_i - 1)$.
For a general definition of $\psi(n)$ see Arnold Scholz and Bruno Schoeneberg, *Einführung in die Zahlentheorie*, 5th ed., de Gruyter, Berlin 1973. $\psi(n)$ is divisor of $\varphi(n)$. Carmichael's theorem states: For relatively prime a, n , $a^{\psi(n)} \; mod \; n = 1$ holds; $\psi(n)$ is the least exponent x such that $a^x \; mod \; n = 1$ for all a relatively prime to n. The theorem of Carmichael is "... a very useful, but often forgotten, generalization of Euler's theorem" (H. Riesel, *Prime Numbers and Computer Methods for Factorization*. Birkhäuser, Basel 1985). Indeed, Rivest, Shamir, and Adleman do not use the theorem and Salomaa also does not point to it in his 1990 book. Scholz and Schoeneberg do not give Robert D. Carmichael (1879–1967) as source.

[9] Since $2 \mid \psi(p'_i \cdot p''_i)$, $e = 2$ is excluded.

10.3.2 Example (Salomaa, revised):

$$(e_i\,,\,d_i) = (1031\,,\,31\,963\,885\,304\,131\,991)$$
$$q_i = 32\,954\,765\,761\,773\,295\,963 = 3\,336\,670\,033 \cdot 9\,876\,543\,211\,;$$
$$\psi(q_i) = \;5\,492\,460\,958\,093\,347\,120 = 3\,336\,670\,032 \cdot 9\,876\,543\,210\,/\,6\;.$$

Even this example with large numbers is unrealistic for practical security. For demonstration we use small numbers, suitable for a hand-held calculator.

In designing a RSA cryptosystem, the trapdoor information is used: we start with two prime numbers

$$p'_i = 47\;,\;\; p''_i = 59\;.$$

This results in

$$q_i = p'_i \cdot p''_i = 47 \cdot 59 = 2773\,,$$
$$\psi(q_i) = \psi(2773) = \mathrm{lcm}\,(46\,,\,58) = 2 \cdot 23 \cdot 29 = 1334\;.$$

Now e_i is to be found, such that $gcd\,(e_i\,,\,1334) = 1$. There are many possible choices of numbers, like $e_i = 3\,,5\,,7\,,\ldots\,,19\,,21\,,25\,,27\,,33\,,35\,,37\,,39\,,\ldots$. If a set $\{e_i^{(j)}\}$ is chosen, even polyalphabetic encryption is possible.
Take $e_i = 17$. d_i is obtained from $e_i \cdot d_i \equiv 1\; mod\; 1334$ with the fast division algorithm, which gives $d_i = 157$. (d_i should not become too small, otherwise simple trial and error may help to find it (Sect. 10.4.3). It may therefore be preferable to choose d_i and to determine e_i .)

Thus, the encryption step is

$$E_i(m_j) \;=\; m_j^{17} \bmod 2773\;.$$

Because of $17_{10} = 10001_2$, this step can be performed efficiently by

$$\left(\left(\left((m^2 \bmod 2773)^2 \bmod 2773\right)^2 \bmod 2773\right)^2 \bmod 2773\right) \cdot m \bmod 2773\,.$$

The decryption step is

$$D_i(c_j) \;=\; c_j^{157} \bmod 2773\;.$$

Encoding the literal characters ␣ (space)[10] , a , b , ... , z with the bigrams 00 , 01 , 02 , ... , 26 , allows plaintext bigrams to be encoded with numbers from $\mathbb{Z}_{2773}$, since $2626 < 2773$.

The message errare␣humanum␣est

is first encoded in blocks of length two:

0518 1801 1805 0008 2113 0114 2113 0005 1920

and then encrypted:

1787 2003 2423 0596 0340 1684 0340 0508 2109 .

Identical plaintext blocks lead to identical cryptotext blocks—the encryption is blockwise monoalphabetic. This ECB mode (in DES jargon, Sect. 9.6.3) should be protected at least by an autokey as in the CBC mode. Even periodic truly polyalphabetic encryption would be far better.

---

[10] Contrary to classical custom, Rivest, Shamir and Adleman did not suppress the word spacing. The literature on the RSA method follows them on this.

10.4 Cryptanalytic Attack upon RSA

Nothing should prevent the cryptanalyst from trying all classical methods (see Part II) against RSA encryption. There are also some specific weaknesses:

10.4.1 Attack by Factorization of q_i. The cryptanalyst who finds the factorization of q_i: $q_i = p'_i \cdot p''_i$, can calculate $\psi(q_i) = 2 \cdot \mathrm{lcm}\,(\frac{p'_i - 1}{2}, \frac{p''_i - 1}{2})$ and from knowing e_i also d_i. To protect the RSA method against this attack, i.e., to make the factorization of q_i intractable (the actual factorization is frequently more difficult than the mere compositeness proof), the following conditions should be fulfilled:

(1) $q_i = p'_i \cdot p''_i > 10^{200}$.

(2) p'_i and p''_i differ in length as dual or decimal numbers by a few digits.

(3) Neither p'_i nor p''_i is small, or is taken from some table of primes, or is of some special form.

Condition (1) prevents (Sect. 10.2.3) a brute force attack.
Condition (2) thwarts exhausive search for a representation of q_i as a difference of two squares (a technique going back to Fermat 1643):

$$q_i = p'_i \cdot p''_i = (\frac{p'_i + p''_i}{2})^2 - (\frac{p'_i - p''_i}{2})^2$$

with values for $\frac{p'_i + p''_i}{2}$ going upwards from $\sqrt{q_i}$.

Condition (3) thwarts exhausive search in a rather small set of primes that are possibly factors. None of these attacks have been reported so far as having been successful, probably because it is easy enough to obey these safeguard measures.

10.4.2 Attack by Iteration (Sect. 9.4.2). Let

$c^{(0)} = m_j$ (plaintext block), $c^{(1)} = c_j = E_i(m_j)$ (cryptotext block) .

Form the sequence

$$c^{(\kappa+1)} = E_i(c^{(\kappa)}) .$$

The least $k \geq 1$ with $c^{(k+1)} = c^{(1)}$ is called the iteration exponent s_{m_j} of m_j; s_{m_j} indicates the length of the cycle to which m_j belongs. $s_{m_j} - 1$ is called the recovery exponent of m_j.

Example 1: As above in Sect. 10.3.2, with $m_j = 0518$.

$(e_i, d_i) = (17, 157)$ $\quad q_i = 2773 = 47 \cdot 59$; $\quad \psi(q_i) = 1334 = 2 \cdot 23 \cdot 29$

$$\begin{array}{lllll} c^{(0)} = & m_j = & & 0518 & (= 11 \cdot 47 + 1) \\ c^{(1)} = & c_j = & 0518^{17} \bmod 2773 = & 1787 & (= 38 \cdot 47 + 1) \\ c^{(2)} = & & 1787^{17} \bmod 2773 = & 0894 & (= 19 \cdot 47 + 1) \\ c^{(3)} = & & 0894^{17} \bmod 2773 = & 1364 & (= 29 \cdot 47 + 1) \\ c^{(4)} = & & 1364^{17} \bmod 2773 = & 0518 & = m_j \end{array}$$

Here we have arrived at plaintext, with a recovery exponent $s_{m_j} = 3$. The unauthorized decryptor cannot yet know this; he finds at the next iteration step:

$$c^{(5)} = \quad 0518^{17} \bmod 2773 = 1787 = c_j \ .$$

From the corollary of Carmichael's theorem (see Sect. 10.3.1):

$$c^{(k)} = m_j^{17^k} \bmod 2773 = m_j^{17^k \bmod 1334} \bmod 2773 \ .$$

But $17^{44} \bmod 1334 = 1$. Therefore $c^{(44)} = m_j^1 \bmod 2773 = m_j$, and 44 is an upper bound for the longest period that can occur with $e_i = 17$. Note that 44 is a divisor of $\psi(\psi(47 \cdot 59)) = \psi(2 \cdot 23 \cdot 29) = 2 \cdot 11 \cdot 14 = 308$. In fact, the total of 2773 elements of $\mathbb{Z}_{2773}$ is partitioned into

9 cycles of length 1 (fixpoints)
42 cycles of length 4—including the cycle starting with 0518
6 cycles of length 22
56 cycles of length 44 .

Example 2:

$(e_i, d_i) = (7, 23) \qquad q_i = 55 = 5 \cdot 11 \ ; \quad \psi(q_i) = 20 = 2 \cdot 2 \cdot 5$

The example is small enough that all cycles can be listed:

9 fixpoints:
(0) (1) (10) (11) (21) (34) (44) (45) (54)

3 cycles of length 2:
(12, 23) (22, 33) (32, 43)

10 cycles of length 4:
(2, 18, 17, 8) (3, 42, 48, 27) (4, 49, 14, 9) (5, 25, 20, 15) (6, 41, 46, 51)
(7, 28, 52, 13) (16, 36, 31, 26) (19, 24, 29, 39) (30, 35, 40, 50) (37, 38, 47, 53)

Note that $7^4 \bmod 20 = 1$. 4 is an upper bound for the longest period that can occur with $e_i = 7$. Note also that 4 coincides with $\psi(\psi(55)) = \psi(20) = 4$.

Example 3:

$(e_i, d_i) = (3, 675) \qquad q_i = 1081 = 23 \cdot 47 \ ; \quad \psi(q_i) = 506 = 2 \cdot 11 \cdot 23 \ .$

Now $3^{55} \bmod 506 = 1$. 55 is an upper bound for the longest period that can occur with $e_i = 3$. Note that 55 is $\frac{1}{2} \cdot \psi(\psi(1081)) = \frac{1}{2} \cdot \psi(506) = \frac{1}{2} \cdot 110$.

Such a cycle of length 55 is

(512, 768, 430, 531, 629, 98, 722, 683, 209, 284, 995, 653, 16, 853,
813, 535, 239, 1051, 25, 491, 190, 55, 982, 439, 54, 719, 676, 568,
393, 307, 397, 331, 384, 324, 721, 1041, 860, 1005, 991, 675, 213, 538,
660, 807, 606, 627, 101, 108, 347, 192, 581, 354, 867, 2, 8) .

There are typically 16 cycles of length 55, 12 cycles of length 11, 12 cycles of length 5 and again 9 cycles of length 1 (fixpoints)
(0) (1) (46) (47) (93) (988) (1034) (1035) (1080) .

Protecting the RSA method against attack by iteration means achieving a large recovery exponent for a large majority[11] of elements $m_j \in \mathbb{Z}_{q_i}$. To allow this, $\psi(\psi(q_i))$ should be as large as possible. In fact, there is the

Main theorem on the iteration attack. For all e_i that are relatively prime to $\psi(q_i)$, the iteration exponent is a divisor of $\psi(\psi(q))$; this bound for the iteration exponent can be attained for suitable e_i.

The proof is based on the corollary of Carmichael's theorem:

$$c^{(k)} = m_j^{e_i^k} \bmod q_i = (m_j^{e_i^k \bmod \psi(q_i)}) \bmod q_i$$
$$= (m_j^{(e_i^k \bmod \psi(\psi(q_i))) \bmod \psi(q_i)}) \bmod q_i \ .$$

With $k = \psi(\psi(q_i))$, the result is

$$c^{(\psi(\psi(q_i)))} = m_j^{e_i^{\psi(\psi(q_i))}} \bmod q_i = (m_j^{e_i^0 \bmod \psi(q_i)}) \bmod q_i = m_j^1 \bmod q_i = c^{(0)} .$$

Thus, $\psi(\psi(q_i))$ is a period and a multiple of the iteration exponent.

To prevent at least $\psi(q_i) = \psi(p_i' \cdot p_i'') = 2 \cdot \mathrm{lcm}\,(\frac{p_i'-1}{2}, \frac{p_i''-1}{2})$ from becoming small, the following conditions for p_i' and p_i'' should also hold:

(4) both $\frac{p_i'-1}{2}$ and $\frac{p_i''-1}{2}$ contain large prime factors.

(5) $\gcd\,(\frac{p_i'-1}{2}, \frac{p_i''-1}{2})$ is small.

Conditions (4) and (5) are optimally fulfilled, if p_i' and p_i'' are safe primes (Sect. 9.5.2): then $\frac{p_i'-1}{2}$ and $\frac{p_i''-1}{2}$ are prime, $\psi(q_i) = 2 \cdot \frac{p_i'-1}{2} \cdot \frac{p_i''-1}{2} \approx q_i/2$. The effort to find safe primes may be worthwhile, but it is an open problem whether or not there are infinitely many safe primes.

Furthermore, preventing also $\psi(\psi(q_i))$ from becoming small, in view of $\psi(2 \cdot \frac{p_i'-1}{2} \cdot \frac{p_i''-1}{2}) = 2 \cdot \mathrm{lcm}\,((\frac{p_i'-1}{2} - 1)/2, (\frac{p_i''-1}{2} - 1)/2) = 2 \cdot \mathrm{lcm}\,(\frac{p_i'-3}{4}, \frac{p_i''-3}{4})$ means that the following conditions for p_i' and p_i'' should hold, too:

(6) both $\frac{p_i'-3}{4}$ and $\frac{p_i''-3}{4}$ contain large prime factors.

(7) $\gcd\,(\frac{p_i'-3}{4}, \frac{p_i''-3}{4})$ is small.

Conditions (6) and (7) are optimally fulfilled, if in addition $\frac{p_i'-1}{2}$ and $\frac{p_i''-1}{2}$ are safe primes, i.e., p_i' and p_i'' are *doubly safe primes*; then $\frac{p_i'-3}{4}$ and $\frac{p_i''-3}{4}$ are prime, $\psi(\psi(q_i)) = 2 \cdot \frac{p_i'-3}{4} \cdot \frac{p_i''-3}{4}$.

Doubly safe primes are 11, 23, 47, 167, 359, 719, 1439, 2039, 2879, 4079, 4127, 4919, 5639, 5807, 5927, 6047, 7247, 7559, 7607, 7727, 9839, 10799, 11279, 13799, 13967, 14159, 15287, 15647, 20327, 21599, 21767, ... ; also 2 684 999, 5 369 999, and 10 739 999. Apart from 11, all doubly safe primes are of the form $24a - 1$. For doubly safe primes p_i', p_i'', $\psi(\psi(q_i)) \approx q_i/8$.

---

[11] More cannot be expected, since there are always even fixpoints of the iteration; in fact Salomaa has shown that there always exist nine of them.

10.4.3 Attack in Case of Small e_i **.** The effort involved in RSA encryption is small if e_i is small—in the extreme $e_i = 3$. This may be advantageous if the sender has limited computing power, e.g., in case it is a smart card, and if the receiver does not suffer from a rather big d_i , e.g., in case it is a central computer.

Using small d_i invites an exhaustive decryption attack and should therefore be avoided. But using small e_i is dangerous, too, in the case that one and the same plaintext message block m_j using the same power $e_1 = e_2 = \ldots = e_s = e$ is sent to many different receivers with (presumably pairwise relatively prime) q_1 , q_2 , ... q_s , the cryptotexts being $m_j^e \bmod q_1$, $m_j^e \bmod q_2$, ... $m_j^e \bmod q_s$. From these intercepted cryptotexts, with the help of the Chinese Remainder Theorem, the value of $m' = m_j^e \bmod q_1{\cdot}q_2{\cdot}\ldots{\cdot}q_s$ can be computed. But since m' is less than each of the individual moduli, the equation $m_j^e = m'$ holds. This equation with known m' and known small e (although involving rather big numbers) can be solved for m_j .[12] The break is not complete: d_i is still left open.

10.4.4 Risks. There are not only certain plaintext blocks that should be avoided since they lead to very short recovery exponents. Example 2 shows that there are also choices of e_i that lower the maximal cycle length. Certain choices of e_i are to be avoided totally: $e_i = \psi(q_i)+1$ means that $d_i = \psi(q_i)+1$ and $E_i(m_j) = D_i(m_j)$ is the identity, so all m_j become fixpoints.

There are also surprising findings: If for a given product $q_i = p_i' \cdot p_i''$ of two primes p_i', p_i'' $\psi(q_i)$ can be computed, then the factorization of q_i can be computed. In fact, if $\frac{p_i'-1}{2}$ and $\frac{p_i''-1}{2}$ are relatively prime, the equations $q_i - 2 \cdot \psi(q_i) + 1 = p_i' + p_i''$ and $q_i = p_i' \cdot p_i''$ determine the two factors.

10.4.5 Shortcomings. The RSA method is widely considered as practically secure, provided the conditions stated above are observed; at least no serious successful attacks have been published.

But the RSA method has disadvantages:

RSA needs relatively long keys q_i, in near future of 1024 or more bits.

RSA is slower than DES by a factor of about a thousand.

10.5 Secrecy Versus Authentication

Because it is a public key system, an open encryption key system is confronted with a problem that classical, symmetric encryption methods have neglected for a long time: their proponents were only concerned about the passive enemy's reactions, reading or eavesdropping encrypted messages transmitted by rather unprotected channels like wire, radio signals, or optical and acoustic

---

[12] M. J. Wiener, *Cryptanalysis of short RSA secret exponents.* EUROCRYPT '89 Proceedings. Lecture Notes in Computer Science 434, Springer 1990.
Also: IEEE Transactions on Information Theory, Vol. 36 No. 3, May 1990, pp. 553–558.

means. The goal of encryption was to make such cryptanalysis as difficult as possible. The possibility of active influence on the message channel was not taken very seriously, and the impossibility of penetration was simply assumed. This was careless.

Wireless contact with spies, however, showed the problem at its human end: a spy could be captured and somebody else could operate his radio set. Such a *Funkspiel* occurred in 1942 and 1943 between the Germans and the British, involving a Dutch underground agent. It is always necessary to insist on a signature from the operator, although even that would not help if the operator was 'turned around', too. If however the operator were working under pressure, he would often omit the 'security checks' he was supposed to intersperse regularly as an authenticator. All this was quite similar to civil use of signatures. In sensitive matters, authentication is as important as classical secrecy.

But there exists a deep conflict as the following example shows: A message with the character of an alarm can be recorded and infiltrated later, which allows the release of false alarms. This can be suspended by a time indication within the message—but this causes a plaintext-cryptotext compromise (Sect. 11.2.5) with the danger of a break. Secrecy and authentication are two different things, and one does not imply the other.

The conflict is further illustrated by the role redundancy plays. An encrypted message is better protected against cryptanalysis, the less redundancy it contains; against counterfeit, the more redundancy it contains. This can be learned from banknotes and handwritten signatures. Secrecy is antagonistic to authentication, and to achieve both requires two measures that are independent of each other. This can be seen in the definition of a signature method (Sect. 10.1.3) as opposed to a secrecy-only method.

Encryption methods that are also signature methods offer additional identifying information according to a prearranged etiquette ('protocol') and prior error-detecting and error-correcting (Sect. 4.4.6) coding.

Asymmetric methods show their strength particularly in authentication and are useful for key negotiation, the dangerous part of key management. This was pointed out from the beginning by Diffie and Hellman. The large amount of time asymmetric methods require (it can be larger than symmetric methods by several powers of ten) is not only justified, it can also be afforded for signatures as well as for keys, since they are usually short compared with messages. Asymmetric, open encryption key systems and classical symmetric systems are not antagonistic, but supplement each other. In international banking, the data are usually only weakly encrypted, but authentication is given high priority (and is highly profitable).

The *Digital Signature Standard* (DSS) of the National Institute of Standards and Technology (N.I.S.T.) of the U.S.A. is based on the *Digital Signature Algorithm* (DSA), which, infringing on patents of Schnorr and ElGamal, uses as one-way function the discrete logarithm function (Sect. 10.2.4.2). There was criticism, in particular that it was not the RSA method that was standardized. Quite generally, cryptanalysis by preprocessing is advocated.

All one-way functions mentioned so far come from residue classes in arithmetic. Another one-way function, which is mathematically highly interesting, comes from the 'knapsack problem', a problem in integer programming.

The standardised *Secure Hash Algorithm* of the N.I.S.T. allows on both sides of the transmission line the formation of 160-bit check groups.

10.6 Security of Public Key Systems

Shannon certainly did not want his admonitory maxim "The enemy knows the system being used" to be interpreted in the sense that the enemy should be given the complete machinery. 'Open encryption key systems' says better than 'public key systems' that decryption keys and the rest are kept secret. This openness has technical reasons, not political ones, and a necessity is made into a virtue (nowadays also with symmetric methods, which do not need to be open). Among the public, the expression 'public key' may have given the impression that cryptanalysis is more than ever in the public domain.

This is not so, of course. Cryptanalysis is still wrapped in a mystery inside an enigma. Nevertheless, I cannot help observing that the commercially used open key systems must be a great joy for the professional cryptanalyst. Apart from the system-oriented kinds of attack, all classical attack routes are wide open. In particular, these systems will be used too long and will be used under heavy traffic with one and the same running key started over and over again. Clever ideas may lead to illusory complications, for example the use of doubly safe primes could open an avenue of cryptanalytic attack.[13]

The pretence of security that is given to the user is often wrongly based on nothing but combinatorial complexity. The situation in complexity theory—a rather difficult part of mathematics—is characterized by the fact that it gives almost exclusively upper bounds; lower bounds, say for the effort of factorization into primes, are not obtainable.

The elimination of the classical crypto clerk, the 'cipher clerk' or 'code clerk', his replacement by a computer plus a typist, makes security even harder: the elimination of encryption errors, which were once the privilege of the crypto clerks, is by far counterbalanced by the lack of experience and shrewd intellect that are the only remedy for dangerous cryptological mistakes.

Thus, it cannot be expected that the proposed public encryption systems are out of the reach of expert cryptanalysts, especially in the executive authorities. The professionals, however, are very reserved and do not brag about their competence, which they are more inclined to understate.

[13] Anton Gerold has shown that conclusions can be drawn from the module on the structure of the doubly safe prime factors.

11 Encryption Security

Even in cryptology, silence is golden.
Laurence D. Smith

Passwords serve to select a method from a class of methods, and keys especially to select encryption steps from an encryption system. It is wise to assume pessimistically that the enemy knows what method has been chosen—there are not too many of them, and most cryptographers are familiar with only a few. The 'basic law of cryptology', which Kerckhoffs[1] had formulated as "*il faut qu'il puisse sans inconvénient tomber entre les mains de l'ennemi*" was expressed more succinctly by Shannon in 1949: "the enemy knows the system being used." It follows that one must be particularly careful in the choice of a key. It is a serious mistake to use obvious words. Porta gave the express warning: "the further removed the key words are from common knowledge, the greater the security they provide." The use of keys had hardly become common practice before unauthorized persons succeeded in decrypting messages by guessing the key word.

Porta reported having solved a message within a few minutes, by guessing at the key OMNIA VINCIT AMOR. Giovanni Batista Argenti also made the lucky guess IN PRINCIPIO ERAT VERBUM. Words such as TORCH and LIBERTY, GLOIRE and PATRIE, KAISER and VATERLAND, expressing noble patriotic sentiments, may be very good for boosting morale, but are most unsuitable as cryptographic keys. (It is astonishing how many people choose their name or date of birth as a computer password. Perhaps they are incapable of remembering anything else.)

11.1 Cryptographic Faults

By faults we mean infringements of security; not just the use of an obvious key, but anything which makes life easier for an unauthorized decryptor.

11.1.1 That includes, of course, encryption errors. These make the work of the authorized decrypter difficult or even impossible. In the latter case, disaster is just round the corner: he must ask for the message to be repeated. If the original wording is encrypted with the same key (correctly, this time), then it is an easy matter to compare the two messages, which will generally be identical up to the point where the error occurred, and some 'differential cryptanalysis' of this 'plaintext-plaintext compromise' is indicated. If a different key is used on the same message ('cryptotext-cryptotext compromise'), suitable procedures can occasionally yield the key—even if the key was a progressive

[1] Auguste Kerckhoffs (1835–1903), Flemish professor (*La cryptographie militaire*, 1883).

one in which the alphabets were not yet repeated. Incredible as it may seem, in the Second World War the Germans frequently radioed the same orders to several units, belonging to different key nets, using different encryption methods or keys—the identical length inevitably aroused suspicion. The only solution is to rewrite the message using different words and phrases. Not even the Russian method (Sect. 3.4.2) of cutting the message somewhere in the middle and joining the parts in reverse order can help in such cases.

11.1.2 Another classic technical mistake is to repeat an encrypted message in plain, for example because the recipient has not yet received the new key. Not only can anyone read the message: the method and the key can now be reconstructed. That may compromise not just the key for the day, but also the basic method used for constructing or selecting the daily key. For that reason, "woe betide anyone who transmits plain text" was a cast-iron rule of Lieutenant Jäger (Sect. 4.4), who was a favorite of the Allied cryptanalysis groups. It is obviously a climax in the life of a professional cryptanalyst to experience a compromise, and equally understandable that the secret services employ all their cunning to try and provoke such an occurrence. In 1941 a senior Japanese civil servant managed to slip the American ambassador Joseph C. Grew a note, with the remark that a member of the Japanese government wanted to communicate the message to the U.S. government but was afraid that the military leaders might get wind of it, and would he therefore transmit it in the most secret diplomatic code. That was M-138-A, and so the encrypted text of a known message flashed across the ether. Nevertheless, it was said that the Japanese failed to break M-138-A.

There was a similar story at the time of the Dreyfus affair. When Alfred Dreyfus was arrested in 1894, on the flimsiest of accusations, and *La Libre Parole* joyfully trumpeted the news, General Panizzardi, the Italian military attaché at the Quai d'Orsay, sent a telegram back to Rome. The French cryptanalysts, who were passed a copy, had reason to believe that Panizzardi had used the commercial Baravelli code (Sect. 4.4.3) which operated with groups of one, two, three, and four, and that the code was then superencrypted. Searching for the sequence /dreyfus/, which would have to be encoded as 227 1 98 306, they found the pattern 527 3 88 706 and so knew that the recrypting affected only the first digit of each group (it was achieved by renumbering the pages of the code book). They were able to decrypt the message with the exception of the last four groups. It was suspected that these signified *uffiziale rimane prevenuto emissaria*, which was taken (by Sandherr, the chief of intelligence) as evidence of Dreyfus' guilt. The next day they worked out the system of page renumbering, which yielded *uffizialmente evidare commenti stampa.* This exonerated Dreyfus, but Sandherr was unconvinced. "These things are always somewhat imprecise," he commented. So Matton, one of Sandherr's subordinates, had the idea of palming a message on Panizzardi. A double agent leaked him a text made to look like an important message, and Panizzardi passed it on almost word for word. The cryptanalysts were not aware of what was going on, and decrypted the message almost immediately; Matton was now convinced he had been right. All the same, a falsified version was presented in court, and it took until 1906 before Dreyfus was acquitted. France has still not got over its Dreyfus scandal: in February 1994 the French Defense Minister François Léotard dismissed the head of the Armed Forces historical archive, Colonel Paul Gaujac, for publishing an 'unacceptably tendentious analysis' of the Dreyfus case.

The Austro-Hungarian empire also had its triumph. After Figl's team had analysed 150 words of an Italian diplomatic code used between Rome and Constantinople, they increased their knowledge step by step by a process of smuggling fragments of information of military relevance into an Italian newspaper published in Constantinople. Within a month they were able to extend their vocabulary to 2000 words.

An even simpler method is one the Russians are famous for: stealing the plaintext from the ambassador. Italy, too, had its *penetrazione squadra.* After such a theft the diplomat is quick to assure his government that the code in use, which is now compromised, was not a very important one.

An example of an obsolete code was the U.S. State Department's GRAY code (meaning the color gray and not Gray's method of binary encoding). When it came into use at the end of the First World War to replace the outdated and already compromised RED, BLUE, and GREEN, nobody thought it would remain in use for two decades. The Foreign Service officers were so familiar with it that they could deliver extempore speeches in GRAY. On December 6, 1941 Franklin Roosevelt sent a memo to Cordell Hull: "Dear Cordell—shoot this to Grew [the American ambassador in Tokyo]—I think it can go in gray code—saves time—I don't mind if it gets picked up. FDR." It was too late to achieve the desired effect; it took time to decrypt the text, and the personal peace overture which Roosevelt wanted to communicate to Tenno would in any case not have prevented the attack on Pearl Harbor.

11.1.3 These episodes show up a general method of cryptanalysis, that of the probable word. Such words are often based on current events; then the message must be rephrased. In the First World War, French troops carried out attacks on German positions simply to trigger certain 'probable words' in German radio transmissions—it is a good thing that soldiers seldom know what they are risking their lives for. In the Second World War the British sank a lighted buoy which marked a channel through the otherwise mined entrance to Calais, merely in order to trigger a German message containing the sequence /leuchttonne/ (Sect. 14.1).

Besides words such as attack, bombardment, etc., military communications contain a treasure of conspicuous words and stereotyped phrases such as headquarters and general staff, division, and radio station. The same message repeated daily—even if it only reports "nothing to report"—can have a devastating effect. It provides a chink for applying the method of the probable word, just like the words love, heart, fire, flame, burning, life, death, which

Porta listed as being the immutable building blocks of love letters. Stereotyped phrases can nullify the advantages of a change of key: the new key can rapidly be deduced from the repeated sequences. Not everyone, of course, will be as lucky as Lieutenant Hugo A. Berthold of office G.2A.6 of the American Expeditionary Forces, who intercepted a radio transmission at 07:40 on March 11, 1918, which consisted of a string of digits and was evidently in a new key; a few hours later he heard a message of the same length but in letters—the recipient had not received the new ciphering instructions and had requested a repeat transmission in the old key. And with the PLAYFAIR manual key used by the German *Afrika-Korps*, a similar compromise took place when the key was changed on January 1, 1942.

It had far-reaching consequences when the existence of the forthcoming 4-rotor ENIGMA was revealed late in 1941 by several practice transmissions in parallel with a message encrypted using the 3-rotor ENIGMA. As a result, Bletchley Park was able to work out the wiring of the new ('Greek') rotor β before the new ENIGMA M4 was officially introduced on February 1, 1942.

Even in peacetime it is important to master the craft: because nobody knows any better, standard texts and obvious phrases are transmitted on manœuvres. If the wording is sufficiently unimaginative, the entire crypto system can be revealed before a single shot is fired. As Hüttenhain wrote: "It is a mistake to take as the main encrypting method one that has already been used by a small circle over an extended period." As far as the—sometimes unavoidable—stereotyped beginnings and endings are concerned ('For Murphy': Sect. 4.4), even the method of 'Russian copulation' is of little help; nevertheless, it can put inexperienced codebreakers off the scent.

11.1.4 The very fact that a message is being transmitted can be significant. Knowing that the communications channels are likely to be heavily loaded during a major military operation, staff officers tend to send their personal messages a few days beforehand. This is called the underwear effect. If conditions allow, the communications channels should be kept open all the time, and 'dummy filling' sent during quiet periods—not test phrases or excerpts from newspaper articles, but irrelevant and nonperiodic sequences, if possible random text or better synthetic language with a multigram letter frequency similar to some natural language ('traffic padding'). Long nonperiodic sequences can be generated by starting at random or irregular points in a text of, say, 10 000 words. Still better, using a method proposed by Küpfmüller in the 1950s, an n-gram approximation is obtained from a master text by using a shift register process: Taking the last $n-1$ characters, the next word in the master text is sought that contains these consecutive characters; then the following character is adjoined and the process is repeated. With a tetragram approximation, from the first chapter of a famous novel by Thomas Mann the following synthetic nonsense text was derived:

thomas ist daher mit mein hand zeigen augen von geschaeftig im kreissigen mauemdisellschaeftwar zur seligen durchterlich hier familie hierheben herz igkeit mit eindrinnen tonyzu plaudertfuenf uhr erzaehlung ich regeshaehm die konnte neigte sie dern ich was stuetzte heissgetuebrige wahrend tause

Traffic flow security by 'padding' with nonsense text greatly increases the load on the unauthorized decryptor and delays his decryption of a genuine message. On the other hand, the intended recipient must be on his guard not to overlook an occasional genuine message in the flood of garbage.

11.1.5 Filling with /x/s, repeating a word or the use of letter doublets can present a risk. The solution is to rephrase the sentence or use synonyms or homophones (chosen at random!)—this includes the use of nulls to conceal partial repetition in the vicinity of homophones. And it is one of the basic rules of professional cryptography to suppress not only punctuation but also word spacing. Thus it is horrifying to think that it was common practice (if not explicit orders) in the German *Wehrmacht* to insert /x/ for stop, /y/ for comma, /j/ for quotation marks, and even /xx/ for colon, /yy/ for hyphen. Sometimes, numbers encoded by letters were bracketed: /y/.../y/. Important words were doubled, i.e. /anan/ and /vonvon/ for 'to' and 'from', /krkr/ for *Kriegstelegramm* ('very urgent') and even tripling of letters was used: /bduuu/ for *Befehlshaber der U-Boote*, /okmmm/ for *Oberkommando der Marine* (on the other hand, German /ch/ was frequently replaced by /q/). Together with the inevitable 'by order of the Führer' and 'Heil Hitler', which nobody dared suppress, that greatly aided the British in breaking ENIGMA. They were so used to these foolish German habits that they became quite indignant when decrypted ENIGMA transmissions produced meaningless sequences (in British jargon *quatsch*) at the beginning and end of a message.

More attention was paid to these things at the time of the Argentis than in the 19th and 20th centuries, when people had become overconfident of the uncrackability of superencrypted codes and other combined methods. Suppression of letter doublets is only one of the 'deliberate spelling mistakes' which the Argentis recommended. As Porta so wisely wrote in 1563, "It is better for a scribe to be thought ignorant than to pay the price for the revelation of one's plans." Unfortunately, the more senior the officer, the less he can be expected to show the insight needed to put up with disfigured texts. The ideal crypto clerk would possess cold-blooded intelligence combined with poetic imagination and a total disregard of conventional spelling. Of course it is tempting to encipher 'radio' and 'station' separately, or even spell them out in letters, like the Austrian clerk who was too lazy to look up the right combination; that provided Luigi Sacco with a break in 1918 (Sect. 13.4.1). The same applies to misusing nulls as word spacings; some members of the French *résistance* used 'tabac' as a dummy, which may have brought the required reinforcements but also led to the cracking of a double transposition. One careless mistake can have disastrous consequences. "The sending of this one message must certainly have cost the lives of thousands of Germans" wrote Moorman, the chief of G.2A.6, about Berthold's episode (Sect. 11.1.3), which revealed the plans for the German offensive of March 21, 1918.

11.1.6 It is the mark of a good signal officer that he explains to his subordinates how the slightest encrypting mistake plays into the hands of the enemy, and he also monitors their efforts. Givierge[2] writes "encode well or do not encode at all. In transmitting cleartext, you give only a piece of information to the enemy, and you know what it is; in encoding badly, you permit him to read all your correspondence and that of your friends." However, this well-meant advice should not be interpreted so literally as to transmit radio messages without encrypting them. That happened at the end of August 1914 with Rennenkampf's Russian Narev army in East Prussia, because the troops had not yet received the code books and the telephone lines were overloaded or non-existent. Hindenburg and Ludendorff won the battle of Tannenberg and became popular heroes as a result. At the other extreme, the Germans encrypted weather reports in the Second World War; as the prevailing wind is westerly in Europe, that often provided the 'probable word'. It would have been better to transmit such low-priority messages in clear.

Rohrbach recommended including vulnerability to errors when assessing the security of a method, on the principle that humans err. Rules of intelligence security and counter-intelligence also play an important role, of course.

11.1.7 The use of easily memorized passwords and keys provides the unauthorized decryptor with added confirmation if he succeeds in reconstructing a key. That is particularly true if the key has some special significance for the originator of the message.

11.1.8 The organizational inconvenience of maintaining crypto security must not be underestimated. Regular changes of key make work for all concerned. Even so, it is hard to understand why the U.S. State Department was still using such short key words as PEKIN and POKES as late as 1917, though Porta used CASTUM FODERAT LUCRETIA PECTUS ALGAZEL; the Argentis had used keys such as FUNDAMENTA EIUS IN MONTIBUS SANCTIS or GLORIA DICENTUR DE TE QUIA POTENTER AGIS. As Vigenère wrote: "the longer the key, the harder is the cipher to break."

A necessary condition of polyalphabetic encryption is that it be quasi-nonperiodic; that is to say, if the key is periodic it must not be significantly shorter than the message. If necessary, a long message must be chopped up and the parts encrypted with different keys. Hitt's warning that "no message is safe unless the key phrase is comparable in length with the message itself" does not mean that the message may be as long as the period of the key—whether or not an encrypting machine is used. Messages of over 1000 characters are in any case at risk, since automatic decryption techniques for the M-209, for example, work well with a message of about 800 characters or more (pure cryptanalysis, Sect. 22.2.3.1). Messages of 200–300 characters are normally safe from such an approach; a maximum length of 500 charac-

[2] Marcel Givierge, French general, successful cryptanalyst in the Second World War, author of *Cours de Cryptographie*, Paris 1925.

ters was allowed with the M-209. A limit of 180 characters applied to the ENIGMA increased to 250 when rotors VI–VIII were introduced (Sect. 8.5.3).

The use of an individual key represents additional organizational effort and requires extra security and counter-espionage measures. There are many situations where it becomes impossible, for example in isolated positions where the (safe) supply of new keys cannot be guaranteed, or where a stock of individual keys might fall into the hands of the enemy and be used for deception.

11.1.9 This raises the question of how a receiving station can tell whether a radio message originates from a legitimate partner or an impostor. The cryptographic measures mentioned in Sect. 10.5 can be supplemented by steganographic measures (security checks), such as inserting particular null characters at specific points in the cryptotext, or making deliberate spelling mistakes at agreed points in the plain text—quite apart from the 'fist', the individual transmitting style (a 'radio fingerprint') of the operator who does the *Funkspiel* 'playback'.

11.1.10 It is worth mentioning the most banal and most brutal cryptanalytic method: the capture of crypt documents by espionage, theft, robbery, or as spoils of battle. The best way of protecting oneself against that is obvious, yet frequently ignored: *What no longer exists cannot fall into unauthorized hands!* (Karl Weierstraß took that to heart with Sofia Kovalevskaya's letters.) The maxim applies particularly to individual keys: the output from the cipher machine should go straight into the shredder (Sect. 8.8.2). Planning for multiple use of individual keys (Sect. 8.8.7) is absurd.

11.1.11 It is a truism that war can bring rich booty. That applies particularly to cryptological material. For example the German submarine *U-33* was captured by the Royal Navy in the Firth of Clyde on February 12, 1940. The otherwise reliable radio operator Kumpf forgot to throw the ENIGMA rotors overboard. The Poles had already worked out the wiring of the first five, but rotors VI and VII were new to the British. In August 1940 rotor VIII was captured, too. On April 26, 1940 the German trawler *Polares (VP 26)* was seized off Ålesund. The British found matching plaintext and cryptotext for the previous four days, although this was not enough to allow the encryption of the naval ENIGMA to be fully broken. The operating instructions captured from the submarine *U-13* in June were also of little help. The breakthrough came the next year: on March 4, 1941 the capture of the trawler *Krebs* in the Norwegian Vestfjord produced not only two familiar rotors but also the complete keys for the previous month. This allowed BP to read in March 1941 all of the February *Kriegsmarine* signals. As a consequence, the reconstruction of the bigram tables used was possible. On May 7, a special attack on the weather ship *München* provided complete keys for June (those for the current month, and the machine itself, had been thrown overboard) and also the 'brief weather key'. The *U-110* was forced to surface in a depth-charge attack off the west coast of Ireland on May 9, 1941 and was boarded by a crew of the destroyer HMS Bulldog; the booty included besides another ENIGMA machine a golden treasury of rules for its use, including the BACH bigram table (Sect. 4.1.2) for encoding the indicator, and also the *Kurzsignalheft*. Finally, there was another planned attack on a weather ship, the *Lauenburg*, on June 28, 1941. The Germans managed to ditch the ENIGMA, but the British captured the complete keys for July. That gave Bletchley Park a breakthrough for the 3-rotor naval ENIGMA; from then on, they could eavesdrop regularly on radio communications to and from submarines, with only a few hours' delay. British shipping losses were correspondingly less.

The introduction of the 4-rotor ENIGMA on February 1, 1942 was a setback, but by December of that year the signals could again be decrypted on a regular basis, and the Allies gained the upper hand in the U-boat war. This was again achieved by capture, which brought to light an incredible stupidity on the part of Eberhard Maertens and Ludwig Stummel, chiefs of the *B-Dienst* of the *Kriegsmarine*. The seizing of the *U-559* off Port Said on October 30, 1942 by HMS Petard provided a new edition of the *Kurzsignalheft* and a second impression of the *Wetterkurzschlüssel*, which would have been a fair prize in itself. In addition, Philip E. Archer managed on December 13 to decrypt a message, which showed that when the 4-rotor ENIGMA was communicating with coastal stations that had only a 3-rotor machine, the fourth rotor (the *Griechenwalze*) was simply placed in the neutral position. That was a convention which made communication possible. The stupidity was that the three-letter ring setting of the 3-rotor ENIGMA was always the same as the first three letters of the ring setting for the 4-rotor machine. That was not necessary, but was done purely for convenience in producing the monthly orders. It meant that if the enemy knew the ring setting for the 3-rotor machine, then only 26 attempts were needed to find the setting for the Greek rotor. Thus, starting with December 13, 1942 the British finally cracked the 4-rotor ENIGMA for the entire TRITON key net of the submarines (introduced in 1941); even the introduction of a second Greek rotor on July 1, 1943 did little to alter their complete mastery of ENIGMA traffic until the end of the war.

However, the British had losses to contend with, too. In 1940, the German auxiliary cruiser *Komet* captured bigram ciphers and code books from the Merchant Navy (Sects. 4.1.2, 4.4.5). The Allies did not find out about it until they studied the German archives after the war.

11.1.12 It is not only the major things which help the enemy; even the smallest details can betray information of great significance. In August 1941 the German submarine *U-570* fell into British hands off the coast of Iceland, almost without a scratch. The wooden box for the ENIGMA was empty, but there was a slot for a fourth rotor. That was confirmation of what they already suspected from references to the 4-rotor ENIGMA in manuals that had been captured, that the introduction of this version was imminent. It is such a wealth of minor details which weave the tapestry that keeps cryptanalysis going. Every break in the thread is a setback to decryption, for a shorter or longer time, possibly for ever.

11.2 Maxims of Cryptology

The [ENIGMA] machine, as it was,
would have been impregnable,
if it had been used properly.

Gordon Welchman 1982

No cipher machine alone can do its job properly,
if used carelessly. During World War II,
carelessness abounded, particularly on the Axis side.

Cipher A. Deavours, Louis Kruh 1985

Over the centuries cryptology has collected a treasury of experiences—even the open literature shows this. These experiences, normally scattered, can be concentrated into a few maxims for cryptographic work, in particular for defense against unauthorized decryption. Especially now, in the era of computers, these maxims have importance for a wider circle than ever.

11.2.1 The native abilities of man include confidence, fortitude, and withstanding danger. These positive qualities have the side effect that man is inclined to overrate his abilities. But

Maxim No. 1: One should not underrate the adversary.

As we have seen, the German authorities did not suspect that the Allies could have penetrated their crypto systems. There were isolated cases[3] of apprehension, but the official opinion was held stubbornly. The *Kriegsmarine* was the only arm of the service that improved its crypto machines decisively by a transition on February 1, 1942 from the 3-rotor ENIGMA to the 4-rotor ENIGMA, and by providing since 1938/1939 altogether eight rotors, compared to five for the Army and Air Force ENIGMAs. Thus it accepted that it was worthwhile to do more for its security. The German *Generalstab* was confident of victory and was intellectually not prepared to take warnings serious. But even in the Navy there was a deep-rooted belief in the unbreakability of the ENIGMA. For example, still in 1970, *Kapitän zur See* [Commodore] Heinz Bonatz, once Staff Officer in the *B-Dienst* of the *Kriegsmarine*, published in a book his naïve belief that the Allies, although they had seized some ENIGMAs, had not broken the German crypto system—at worst, the Allies would have been able to read German signals for a limited time.

It was not only the Germans who were unsuspecting. The U.S. cryptologists, too, could not imagine that Rohrbach had broken their M-138-A. And the Signal Security Agency of the U.S. Army had tried to break their new M-134-C (SIGABA), a rotor machine, without success. What would that

---

[3] Already in 1930, Lieutenant Henno Lucan, Second Signals Officer of the battleship *Elsaß*, pointed out in a study a weakness of the ENIGMA. With the introduction of the plugboard, the worries seemed to be banished.

mean? Why couldn't the Germans do as well as the English, who had conquered the ENIGMA? Typically, it was Roosevelt, the intellectual among the Allied leaders, who always slightly distrusted the assertions of the cryptologists. Did he know better the deep-rooted human habit of ignoring the undesirable?

It took the Royal Navy three years to find out that the *B-Dienst* of the *Kriegsmarine* did read some of their encryptions. ENIGMA decrypts gave finally the proof that at least Naval Cypher No. 3, since June 1941 the main crypto system for convoy formations in the North Atlantic (German code name *Frankfurt*), was broken. It was replaced by Naval Cypher No. 5, and thus from the middle of June 1943 as was made known after the war, the Germans were rather cut off from the well. What would have happened if the Germans had found out in the same way about the insecurity of their ENIGMAs?

Perhaps nothing, as a particularly crass case of the permanent underrating of the British by Rear Admiral Eberhard Maertens and his Chief of Staff Ludwig Stummel shows. It happened in mid-1943: Decrypts of signals from Allied convoys showed that the Americans supposed there were twenty German submarines in a narrow map square. Indeed, the wolfpack *Meise* with its 18 boats was in the square. The *Befehlshaber der Untersee-Boote*, Großadmiral Dönitz (1891–1980), ordered Maertens to investigate, as he had done in 1941 when *U-570* was seized. Again Maertens exculpated ENIGMA. The British U-boat situation reports themselves, he said, stated that the Allies' information on submarine locations was coming from direction-finding. Maertens saved his head by explaining falsely that they had been located by the H2S (German code name *Rotterdam-Gerät*), a radar bombing aid working on a wavelength of 9.7 centimeters, found February 2, 1943 in a British bomber shot down over Rotterdam. Dönitz had to comply, but remained suspicious and finally fired Maertens after an accident around the convoy SC 127 on March 12, 1944 was again explained by either treason or lack of cipher security. It is known today that poor Maertens was the victim of tricky British disinformation.

The Russians also managed to penetrate the ENIGMA encryption. They raised *U-250* after she was sunk in the Gulf of Finland on July 30, 1944 and recovered her ENIGMA. Opinion is divided on how far the Russians succeeded. While in a German document from January 1943 it is stated "It is certainly true that in individual cases the Russians succeeded in decrypting ENIGMA messages," E. E. Thomas said in 1978 that after ten years of detailed study he found nowhere any evidence to show that the Russians at any time could decrypt the German radio traffic.

Whether the Soviets penetrated U.S. crypto systems was often debated, particularly after Isaac Don Levine, the Russian-born journalist who specialized in Soviet affairs, became "convinced by mid-1939 from numerous conversations he had with General Walter Krivitsky, the defected head of Soviet military intelligence for Western Europe, that the Communist cryptanalysts were reading American codes" (Kahn).

11.2.2 More harmless, but also more imperiled, are the inventors of crypto systems. "Nearly every inventor of a cipher system has been convinced of the unsolvability of his brainchild," writes Kahn. A rather tragicomic example was offered by Bazeries himself. Working for the French government and army, he had ruined a number of inventions by breaking test samples he had asked for. Finally, he invented his own system and promptly dubbed it absolutely secure. The Marquis de Viaris, whose invention Bazeries had smashed a short while before, took revenge. He even invented a method of cryptanalysis (Sect. 14.3) applicable for a wide class of instruments, from Jefferson and Bazeries to M-94 and M-138-A, all using families of unrelated alphabets. Here we are led to Kerckhoffs' maxim,

Maxim No. 2: Only a cryptanalyst, if anybody, can judge the security of a crypto system.

This knowledge can be found with Porta, and was formulated by Kerckhoffs in 1883. He criticized judging the encryption security of a method by counting how many centuries it would take to exhaust all possible combinations. Indeed, such combinatorial counts can only give a bound for the effort necessary in the worst case, for the crudest of all cryptanalytic methods, exhaustive search, also called 'brute force attack'.

Everywhere in the civilized world therefore the governmental services (and some non-governmental ones) have the double duty to design secure crypto systems and to break allegedly secure ones. "With code breakers and code makers all in the same agency, N.S.A. has more expertise in cryptography than any other entity in the country, public or private," wrote Stewart A. Baker, not without pride. He is a famous lawyer, who was for a few years the top lawyer at the National Security Agency. His praise would sound even better from a neutral source.

11.2.3 Kerckhoffs was one of the first to deal with cryptography from a practical point of view. In discussing questions of ease of handling (to be treated later) he wrote: "It is well to distinguish between a crypto system intended for a brief exchange of letters between a few isolated people and a method of cryptography designed to regulate for an unlimited time the correspondence between different army commanders". He distinguished between the crypto system as a class of methods (French *système*) and the key in the narrower sense and postulated, as mentioned above, "*Il faut qu'il puisse* [*le système*] *sans inconvénient tomber entre les mains de l'ennemi.*" [No inconvenience should occur if the system falls into the hands of the enemy.] This brought Shannon to formulate more precisely

Maxim No. 3: In judging the encryption security of a class of methods, one has to take into account that the adversary knows the class of methods ("The enemy knows the system being used", *Shannon*).

Otto J. Horak expressed it this way: "Security of a weak cipher method is not increased by trying to keep it secret".

For practical reasons, in certain situations certain methods are used preferentially, others not at all. In particular, the ingrained conservatism of the established apparatus creates certain preferences, which cannot be hidden from the adversary ('encryption philosophy'). Moreover, a rough differentiation, like one between transposition, monoalphabetic and polyalphabetic encryption, is possible on the basis of simple tests. There are also rules of thumb, like Sacco's criterium that a short cryptotext of not more than, say, 200 characters embracing all characters of the alphabet is most likely polyalphabetically encrypted.

Machines and other devices, including encryption documents, can fall in combat into the hands of the enemy or can be stolen. This includes machines like the ENIGMA. Following Kerckhoffs' doctrine strictly, the ENIGMA should have been extended at the beginning of the Second World War in one big step to a 5-rotor machine; the rotor position (*Walzenlage*) should have been changed from the first day every 6 hours (not merely three times a day from 1942 on); and every three months the rotor set should have been exchanged completely. Admittedly, this would not have been easy, in view of the tens of thousands of ENIGMAs, but it would have been appropriate in retrospect.

But, as Kahn wrote, "the Germans had no monopoly on cryptographic failure. In this respect the British were just as illogical as the Germans." And the Americans were illogical, too. Their cipher machine M-209, constructed by Hagelin and built under license, was considerably less secure than the ENIGMA and was also used by the Italian navy (C-38m), an Axis partner. No wonder the Germans in North Africa in 1942 and 1943 often knew the goals and times of American attacks. And the British, who could solve the C-38m too, knew all they needed about the supply situation of Field Marshal Erwin Rommel.

11.2.4 The desire of the cryptographer not to make it too easy for the adversary leads to the introduction of complications of known methods. The composition of methods (Chapter 9) has long been used for this purpose, mainly the combination of essentially different methods, like transposition of a monoalphabetic substitution or superencryption of code by polyalphabetic encryption. Specific cryptanalytic methods, however, are frequently insensitive to such complications. At best nothing is gained, at worst the combination offers an unforeseen entry. According to Givierge (1924),

Maxim No. 4: Superficial complications can be illusory, for they can provide the cryptographer with a false sense of security.

In a typical case, someone excludes with the very best intention, but quite unnecessarily, the identity as encryption step in a VIGENÈRE, under the impression that no letter should be left untouched, and thus no letter may represent itself. But this property allows one to determine for a sufficiently long probable word a few positions where it could be found (Chapter 14). The same property is shared by crypto systems with monocyclic alphabets, all cylinder devices from Jefferson and Bazeries to the M-94, and all strip devices from Hitt to the M-138-A. Moreover, all crypto systems with genuine self-reciprocal alphabets have the property too, including the ENIGMA, thanks to the invention of the reflecting rotor, which was a masterpiece of *complication illusoire.* To this, Welchman remarked "It would also have been possible, though more difficult, to have designed an Enigma-like machine with the self-encipherment feature, which would have knocked out much of our methodology, including 'females' [Sect. 19.6.2.1]."

11.2.5 Finally, the last and perhaps the most important point is human weakness. The encryption security is no better than the crypto clerk. The unauthorized decryptor feeds on the faults mentioned at the beginning of this chapter.

First of all, there are the compromises:

plaintext-cryptotext compromise: repetition of the transmission in clear,
plaintext-plaintext compromise of the key: transmission of two different plaintexts, encrypted with the same key text,
cryptotext-cryptotext compromise of the keys: transmission of two 'isologs', i.e., the same plaintext, encrypted with a different key text (in particular, public keys invite this compromise).

Next, there are the faults enabling a 'probable word' attack:
the frequent use of stereotype words and phrases (which flourish not only in diplomatic and military language),
the use of a common word for a sudden or unforeseen event,
the use of short passwords and keys that can easily be guessed.

Moreover, there are the elementary rules a of good cryptographic language:

not to use double letters and frequent letter combinations like /ch/ and /qu/ , to suppress punctuation marks and in particular to suppress the word spacing, to use homophones and nulls prophylactically against probable word attacks.

Plaintext prepared optimally for encryption is orthographically wrong, linguistically meager, and stylistically horrible. Which commanding general would like to phrase an order in this way, which ambassador would send such a report to the head of his government? The answer is simple: they should not do it themselves, but their crypto officers should have to do it for them. Both Roosevelt and Churchill complied in the Second World War with the needs of crypto security. Only Murphy did not.

In addition, ambassadors and generals are normally disinclined to take the time to supervise their crypto clerks; indeed most of them do not understand their needs and are cryptologically ignorant. When Wheatstone invented a special bigram substitution that was later called PLAYFAIR (Sect. 4.2.1), he could not overcome the Foreign Office's dislike of complicated encryption. Napoleon's generals encrypted their messages only partly, and so did still in 1916 the Italians at the Isonzo battle.

An important principle for communication services is therefore that monitoring and surveillance of their own units is at least as important as listening to the adversary's. To this, Erich Hüttenhain remarked: "*Ein Verbündeter, der keine sicheren Chiffrierungen verwendet, stellt ein potentielles Risiko dar.*" [An ally who does not use secure cryptography represents a potential risk.]

It is frequently said that *a cryptographer's error is the cryptanalyst's only hope*. This hope is justified: there is always nervous stress for the crypto clerk in diplomatic and military service and encryption errors are likely to happen. The more complicated the method, the more mutilated will be the plaintext that is eventually decrypted. Under pressure of time, the dangerous repetition of a message without careful paraphrasing may then seem unavoidable. Givierge's advice was: *Chiffrez bien, ou ne chiffrez pas.* Rohrbach's was:

Maxim No. 5: In judging the encryption security of a class of methods, cryptographic faults and other infringements of security discipline are to be taken into account.

The good cryptologist knows that he cannot rely on anything, not even on the adversary continuing his mistakes. He is particularly critical about his own possible mistakes. Surveillance of one's own encryption habits by an *advocatus diaboli* is absolutely necessary, as the experiences of the Germans in the Second World War showed only too clearly. Sir Stuart Milner-Barry wrote "Had it not been for human error, compounded by a single design quirk, the Enigma was intrinsically a perfectly secure machine."

The design quirk, the seemingly clever idea (7.3.2), was the properly self-reciprocal character of the enciphering caused by the introduction of the reflector. For a facilitation of the operation a high price was paid by allowing a dangerous encroachment.

11.3 Shannon's Yardsticks

If somebody is willing to follow the advice given so far, there remains the question of which method to take. The answer depends on the one side on the degree of security desired, on the other side on the effort invested. Claude E. Shannon (1916–2001) listed[4] five yardsticks for measuring a class of cryptographic methods:

(1) Degree of required encryption security	How much does the adversary gain from receiving a certain amount of material?
(2) Key length	How short is the key, how simple is its manipulation?
(3) Practical execution of encryption and decryption	How much work is necessary?

[4] Claude E. Shannon, *A Mathematical Theory of Cryptography*. Internal Report, September 1, 1945. Published in: Communication Theory of Secrecy Systems. Bell System Technical Journal **28**, 656-715 (October 1949).

(4) Inflation of cryptotext	How much longer than plaintext is the cryptotext ?
(5) Spreading of encryption errors	How far do encryption errors spread?

These yardsticks are contradictory to the extent that no crypto system is known (and presumably none can exist logically) that fullfills the maximal requirements in all points. On the other hand, no point can be absolutely ignored.

If point (1) is dropped completely, then even plaintext is acceptable. If point (2) is dropped completely, then an individual key is acceptable. If points (3) and (4) are dropped completely, then there exist exotic crypto systems that fulfill all other points optimally. If point (5) is dropped completely, then methods performing a thorough amalgamation can optimally approximate all other points.

Modern cryptography tends, depending on the situation, to use individual keys (which require uninterrupted and secure key distribution) or amalgamation methods (which require noise-free, i.e., error-correcting, communication channels).

In situations of utmost secrecy, say between heads of states in emergency situations, the use of individual keys is quite normal, since there are usually not very many messages. It may be appropriate even in case of heavy traffic. Frederick W. Winterbotham, responsible for the security of the material that came from the Bletchley Park decrypts of the German ENIGMA and SZ42 radio traffic and was distributed under the cover name ULTRA to the field units, insisted that it was encrypted for this purpose with an individual key. This shows on the one hand how priceless the secret material was, in view of the trouble connected with the use of individual keys, and on the other hand, how safe the British rated individual keys, certainly rightfully so. Did any German officer have a chance to impose such strong regulations?

In the commercial field, with DES an amalgamation method has been in use for two decades now. The continuing criticism of the security of the present *de facto* standard would already be mollified if the key length, which many consider to be too short, were increased.

11.4 Cryptology and Human Rights

Since cryptographic methods are in use, even amateurs try to break them. Today, an amateur with access even to a middle-sized computer will find it difficult to penetrate an encryption that satisfies professional standards. The National Security Agency (N.S.A.) of the U.S.A., however, wishes to retain a surveillance capability over any commercial message channel that comes under suspicion. It can be expected that the U.S. Government will not allow the intelligence service of a potential adversary—there are still some—to build up a communication network under the cover of a private commercial undertaking. The times are over when Henry L. Stimson, Secretary of State of President Herbert Hoover, could send the Black Chamber of the State Department to the desert (1929!) and then justify this in his autobiography (1948) with the reason "Gentlemen do not read each other's mail". Not even President Carter showed such moral scruples. Or does this show that the Americans under Carter did not succeed in reading Russian traffic? The end of the Cold War means only a reduction and not a cessation of the latent danger of being spied out.

11.4.1 But cryptology is not only an issue concerning the diplomatic and military authorities of different states. One should not forget that the permanent conflict of interests between the citizen or the individual and the state representing society at large is affected by cryptology. On the one side there is the undeniable right of the citizen (or of a corporation) to protect his or her private sphere (or its commercial interests) by efficient crypto systems, on the other side there is the constitutional duty of the state to protect its internal and external security, which may require penetration of encrypted messages for intelligence purposes.

The position of the state was expressed by Charles A. Hawkins, Acting Assistant Secretary of Defense, U.S.A. on May 3, 1993, as follows: "The law enforcement and national security communications argue that if the public's right to privacy prevails and free use of cryptography is allowed, criminals and spies will avoid wire taps and other intercepts." The privacy of letters is not absolute even in civilized countries, and in cases that are regulated by laws it can be suspended for the benefit of the state—but not for the benefit of private persons. Encrypted messages are no exception—just the use of cryptography creates a certain initial suspicion.

On the other side, precisely in the U.S.A. where most citizens see possession of firearms as their constitutional right, the possession of the crypto weapon is also not seen to be a state monopoly. Europe, with its somewhat different history, does not go so far in this respect.

Whitfield Diffie distilled it to a short formula: '... an individual's privacy as opposed to Government secrecy'. In Europe there is reason enough to insist on freedom from the authoritarian state. Thus, there is a need to find within the framework of each political constitution a means to regulate governmental cryptanalysis; a borderline is to be defined. This is already required by the existing legal framework. Strangely, the larger countries have more difficulties here in achieving results than the smaller ones; Austria, for example, being more advanced than Germany.

A solution is also necessary in the interest of world trade. In the U.S.A., the rule for trading cryptological equipment with foreign partners is: "Encryption for the purpose of message authentication is widely allowed, whereas encryption for the purpose of keeping information private raises eyebrows" (David

S. Bernstein). As a case in point, Philip R. Zimmermann, according to his lawyer, was in 1994 facing a charge for violating the *International Traffic in Arms Regulations*, because he fed into the Internet and thus made freely available the crypto system PGP (*Pretty Good Privacy*, Sect. 9.6.6), which counts as war material ('cryptographic devices, as well as classified and unclassified data related to cryptographic devices', Category XIII). The charge has been dropped in 1996, but the situation is unsatisfactory.

11.4.2 To regulate the conflict between the protection of the private sphere of the law-abiding citizen, guaranteeing the confidentiality of his or her messages on the one side, and on the other the fulfillment of the functions of the state, several schemes have been drawn up:

(1) A limitation of the use of crypto systems in the civilian domain by a requirement to seek official approval, either in individual cases or for types of usage, imposed upon commercial vendors (the inhibition of certain methods alone, except the use of individual keys, is not sufficient, since it invites circumvention).

(2a) A restriction of encryption security by regulating the availability of suitable crypto systems in the civilian domain. The agency that makes the crypto systems available can at the same time give the citizen a cryptanalytic guarantee and thus can increase the incentive for voluntary conformity (a commercial vendor may serve as market leader with state support).

(2b) Like (2a), but in conjunction with inhibition of the use of other crypto systems in the civilian domain.

(3) An escrow system, requiring the deposition of the complete data for each crypto system used in the civilian domain, the escrow agency being independent and required to maintain confidentiality.

Further proposals may come up, as well as mixtures between the ones listed above. It is to be expected that different democratic states will come to different solutions within the scope of their sovereignty. In France, for example, a solution along the lines of (1) which could be considered undemocratic is already established, and the Netherlands flirted for a while with such a regulation. In Germany, there is for quite some time a tendency for a solution like (2a), with a recently created *Bundesamt für Sicherheit in der Informationstechnik* (BSI), subordinate to the Ministry of the Interior; one might guess that it could develop into a solution like (2b). This liberal fundamental position was in June 1999 confirmed by the new Federal Cabinett. It is not yet apparent what solution the United Kingdom with its 'Official Secrets Act' will adopt. In the U.S.A., in 1993, a solution along the lines of (3) was advocated by the Clinton administration (a *key escrow system*[5], see below). It caused loud protest and a modification in the direction of (2a) with a kind

---

[5] *Escrowed Encryption Standard* (EES), Federal Information Processing Standards Publication (FIPS PUB) 185, Feb. 9, 1994.

of voluntary submission was still under discussion in 1996. For the European Union, insofar as it legislates on this question at all, neither (1) nor (3) are feasible options.

Furthermore, it can be imagined what such a disorder of different regulations means for international commercial vendors. Trade in cryptographic devices crossing international borders is already difficult enough: "International use of encryption plunges the user headfirst into legal morass of import, export and privacy regulations that are often obscure and sometimes contradictory" (David S. Bernstein). International travel with a laptop computer was potentially punishable. *Martha Harris*, Deputy Assistant Secretary of State for Political-Military Affairs, stated on February 4, 1994: "We will no longer require that U.S. citizens obtain an export license prior to taking encryption products out of the U.S. temporarily for their own personal use. In the past, this requirement caused delays and inconvenience for business travellers."

11.4.3 The *Escrowed Encryption Standard* concerns the encryption algorithm SKIPJACK (Sect. 9.6.5) within the CLIPPER chip. To be stored in escrow by two separate escrow agents are two 'chip unique key components'. These components are released to an authorized government official only in conjunction with authorized electronic surveillance and only in accordance with procedures issued and approved by the Attorney General. The key components are needed to construct by addition *modulo* 2 the 'chip unique key'. An 80-bit message setting ('session key') KS, negotiated between partners or distributed according to a security device, serves as in DES to form an initialization block c_0 for the encryption process which can be used monoalphabetically in a mode corresponding to *Electronic Code Book* or chained in an autokey way in a mode corresponding to *Cipher Block Chaining* (Sect. 9.6.3). Most important, the chip contains an emergency trapdoor, the 'Law Enforcement Access Field' LEAF, where the current session key is stored in encrypted form; by using the chip unique key the session key is obtained. Thus, following a court order, a government-controlled decrypt device can survey the channel. Every time a new conversation starts with a new session key, the decrypt device will be able to extract and decrypt the session key from the LEAF. Except for an initial delay getting the keys, intercepted communications can be encrypted in real time for the duration of the surveillance. Thus, even voice communication in digitized form can be surveyed.

Unlike DES, the SKIPJACK algorithm itself was kept secret by the authorities "to protect the LEAF" even though security against a cryptanalytic attack as such does not require the algorithm to be kept secret. Moreover, the SKIPJACK algorithm was classified as SECRET — NOT RELEASABLE TO FOREIGN NATIONALS. It therefore was not suitable as an international *de facto* standard. In 1998, these restrictions finally have been dropped.

In fact, the 'Law Enforcement Access Field' trapdoor is rather primitive. It is openly accessible along the transmission line so that unauthorized decryptors can give it their best efforts. Dorothy E. Denning has studied some of the practical questions that arise. Silvio Micali ('Fair Cryptosystems', U.S. Patent 5 276 737, January 4, 1994) has proposed improved cryptosystems that cannot be misused either by criminals or by state officials.

For real-time, interactive communications, Thomas Beth and others proposed in 1994 to make the investigative law enforcement agency an active participant in the protocol used by the sender and receiver to establish the session key, in such a way that the two parties cannot detect the participation of the agency. The novelty of this approach, however, lies in the possibility that in case of noninterception the network provider can prove this fact.

11.4.4 Mistrust felt by some citizens (or legal corporations) against state power is not diminished by some recent experiences, for example in the U.S.A. with actual or alleged encroachments of the National Security Agency into the development of encryption algorithms like DES. It was said that giving N.S.A. responsibility for approving and recommending encryption algorithms is "like putting the fox in charge of guarding the hen house."

In 1957, there were also reports of close contacts between William F. Friedman and Boris Hagelin, good friends in the wartime, that aroused suspicions.

Boris Hagelin
(1892–1983)

Thus, a third party is mentioned, who stands outside the philosophy of balancing constitutional rights on privacy and state rights on law enforcement but cannot be overlooked in view of his economic importance: the commercial vendor. It is in the interest of this party to have good relations with both the citizen as potential client and the state as supervisor (and sometimes client, too). At best, the vendor is an honest broker between the other two parties.

However, this role is impeded by a certain dishonesty the state authorities force upon commercial vendors by making injunctions upon their trading with foreign partners that do not hold for their inland trade with the state itself. This hardly accords with the rules of global free trade.

11.4.5 One cannot help feeling that cryptology at the beginning of the third millennium is still kept within a *Black Chamber*. The state authorities are to that extent impenetrable and can cling to their last shreds of omnipotence. But there is a firm foundation of rights the state authorities cannot give up, for there has to be a balance of power. Not only the sole remaining superpower, the U.S.A., but also the smaller powers in Europe will find it necessary that civilian and commercial cryptography and cryptanalysis come to an agreement with the state. The United Kingdom, with its long tradition of democracy yet its hitherto very tight security in matters of cryptanalysis, adheres to the motto that he who does not protect his own security endangers the security of his friends. But the claims of the civilian and commercial world are to be taken seriously. It would hardly be politically acceptable if in the U.S.A. patent applications for cryptosystems were blocked under the authority of the Invention Secrecy Act of 1940 or the National Security Act of 1947. Likewise, sensitivity about the protection of private or personal data is a political factor that cannot be overlooked. In the U.S.A., policy on domestic controls is still inconclusive, as was shown by the furor over FIDNET (Federal Intrusion Detection Network) and CESA (Cyberspace Electronic Security Act) in 1999.

11.4.6 Liberalization was pushed in 1997 by international organizations like OECD and EU. In December 1998, in the scope of the *Wassenaar Arrangement on Export Control for Conventional Arms and Dual-Use Goods and Technologies*, comprising 28 nations, some guide-lines for a rather liberal export control of cryptographic products were achieved. In particular, exports of 64-bits encryption algorithms were decontrolled by the member countries of the Wassenaar Arrangement.

Then on September 16, 1999, the Clinton administration announced its intention of further liberalization, allowing "the export and reexport of any encryption commodity or software to individuals, commercial firms, and other non-government end-users in all destinations". The new policy will simplify U.S. encryption export rules and rests on the following three principles: a technical review of encryption products in advance of sale, a streamlined post-export reporting system, and a process that permits the government to review exports of strong encryption to foreign governments. This relieves the feelings of the U.S. commerce. "Restrictions on terrorist-supporting states, their nationals and other sanctioned entities are not changed by this rule." This may console the U.S. Department of Justice. How necessary it is was demonstrated by the September 11, 2001 terrorist attack on the United States of America.

Altogether, the U.S. government expects that "the full range of national interests continue to be served by this new policy: supporting law enforcement and national security, protecting privacy, and promoting electronic commerce". And on January 12, 2000 the *Bureau of Export Administration* (BXA) published the new liberal regulations for software export.

11.4.7 Still, it is to be hoped, that in the long run there will be a victory of common sense. Above all, the aim of scientific work on cryptosystems for civilian and commercial channels is more than ever *to find lower bounds for the complexity of unauthorized decryption using a precisely defined type of computer, under realistic assumptions on the lack of discipline among non-professional users.*

It is a worthwhile task to give the user of a cryptosystem a guaranteed amount of security. This includes the need for open source code of the cryptosystem, since every crypto system with an unpublished algorithm may contain unpleasant surprises.

The disk of Phaistos, a Cretan-Minoan clay disk of about 160 mm diameter from the 17th century B.C. (the plate shows the side $\mathcal{A}$), is covered with graphemes with clear word spacing. A decryption that is generally accepted does not seem to exist. "The torn and short text does not reveal its meaning without further clues" (J. Friedrichs).

Plate A

Plate B Two cipher disks, presumably from the 18th/19th century. The upper disk is in the style of a nomenclator. On the plaintext side it contains both alphabet letters and several syllables and frequently occurring words; on the cryptotext side it uses two-place decimal numbers.

The 'Cryptograph' of Wheatstone, a device in the form of a clock, was shown for the first time at the Paris World Exhibition in 1867. It is a polyalphabetic cipher device: the hand is moved clockwise each time to the next cleartext letter, which slowly moves the disk with the mixed cryptotext alphabet.

Plate C

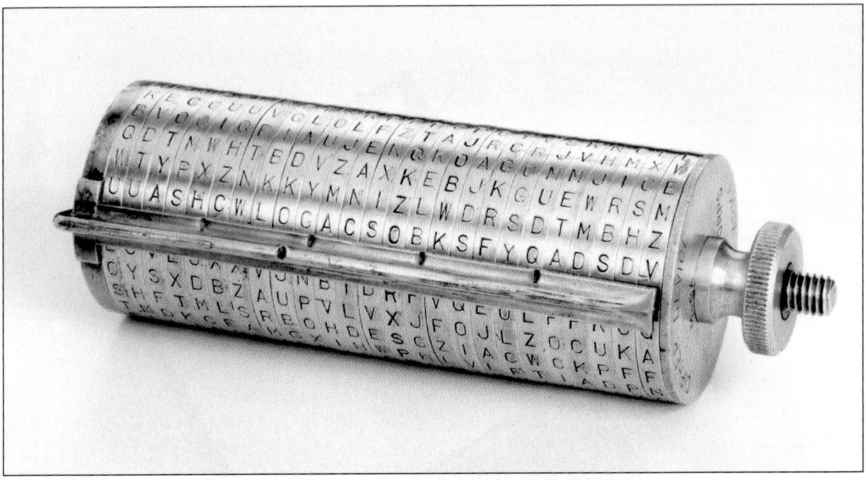

Plate D The U.S. Army cipher device M-94 in cylindrical form with 25 aluminum disks of 35 mm diameter, with alphabet letters engraved on the rim, goes back to the models of Jefferson and Bazeries. Introduced in 1922 under the influence of W. F. Friedman for lower-level military communications, it was in wide use until 1942.

Strip cipher M-138-T4 used by the U.S. Army and U.S. Navy in the Second World War, based on a proposal by Parker Hitt in 1914. The 25 removable paper strips were numbered and used in prearranged order. The encryption was cryptologically equivalent to the M-94 .

Plate E

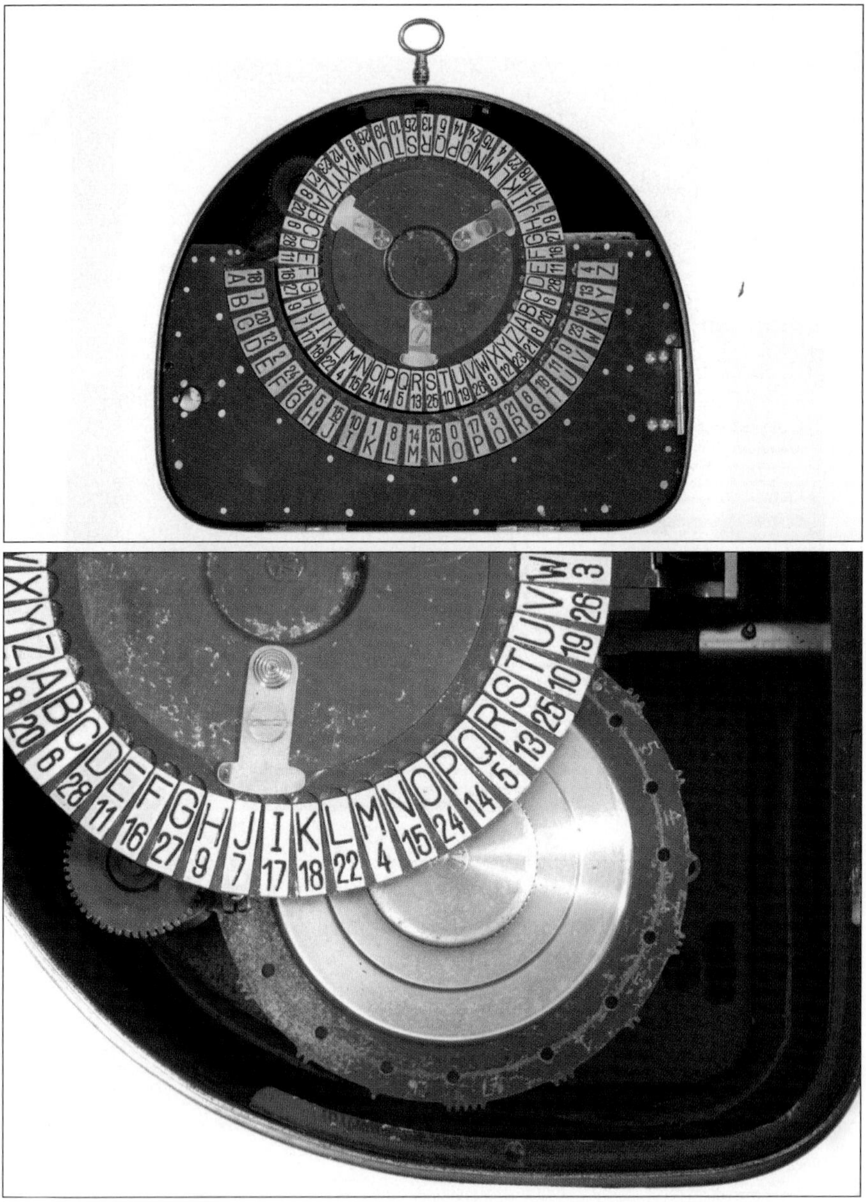

Plate F The cipher machine 'Kryha' was invented by Alexander von Kryha, Berlin-Charlottenburg, in about 1926. It is a polyalphabetic cipher device with a fixed periodic key of length 442. Irregular movement of the cryptotext disks is achieved by a wheel with a varying number of teeth. Despite its cryptological weakness, this neat machine sold well in many countries.

The Hagelin cipher machine 'Cryptographer' C-36, made by the Aktiebolaget Cryptoteknik, Stockholm, in 1936, has self-reciprocal encryption by BEAUFORT encryption steps performed by the 'lug cage', an invention of Boris Hagelin. The irregular movement is based on the use of keying wheels with different graduation, namely with 17, 19, 21, 23, and 25 teeth, which gives a key period of length 3 900 225 . For such a purely mechanical machine, this was a pioneering achievement.

Plate G

Plate H The M-209 was an improved Hagelin C-36. Under Hagelin license, it was manufactured by Smith-Corona for the U.S. Army; it had an additional keying wheel with 26 teeth which increased the period to 101 405 850. When the crank was turned, the lettered wheels moved pins and lugs that shifted bars in the cylindrical cage; the bars acted like cogs that turned a wheel to print the cipher letter on the roll of tape behind the knob.

Rotor cipher machine ENIGMA, as invented by Arthur Scherbius in 1919, with light bulb display and plugboard (in front); 4-rotor version M4 for the *Kriegsmarine*, 1944. It enciphered with 3 (out of 8) normal rotors and 1 (out of 2) reflecting rotors (*Griechenwalzen* β, γ), the introduction of which stopped the British reading German U-boat signals from February to December 1942.

Plate I

Plate K ENIGMA rotors: The internal wiring has 26 electrical connections between the contacts on the one side and those on the other side.
Above: Rotor I with visible setting ring.
Below: Rotor VIII with two notches.

The British TYPEX was an improved copy of the German 3-rotor ENIGMA; it had two extra rotors (not movable during operation) that made penetration much more difficult. It was actively used in British communications, and also to help decrypt German signals after their key was broken. The plate shows a TYPEX Mark III Serial No. 376.

Plate L

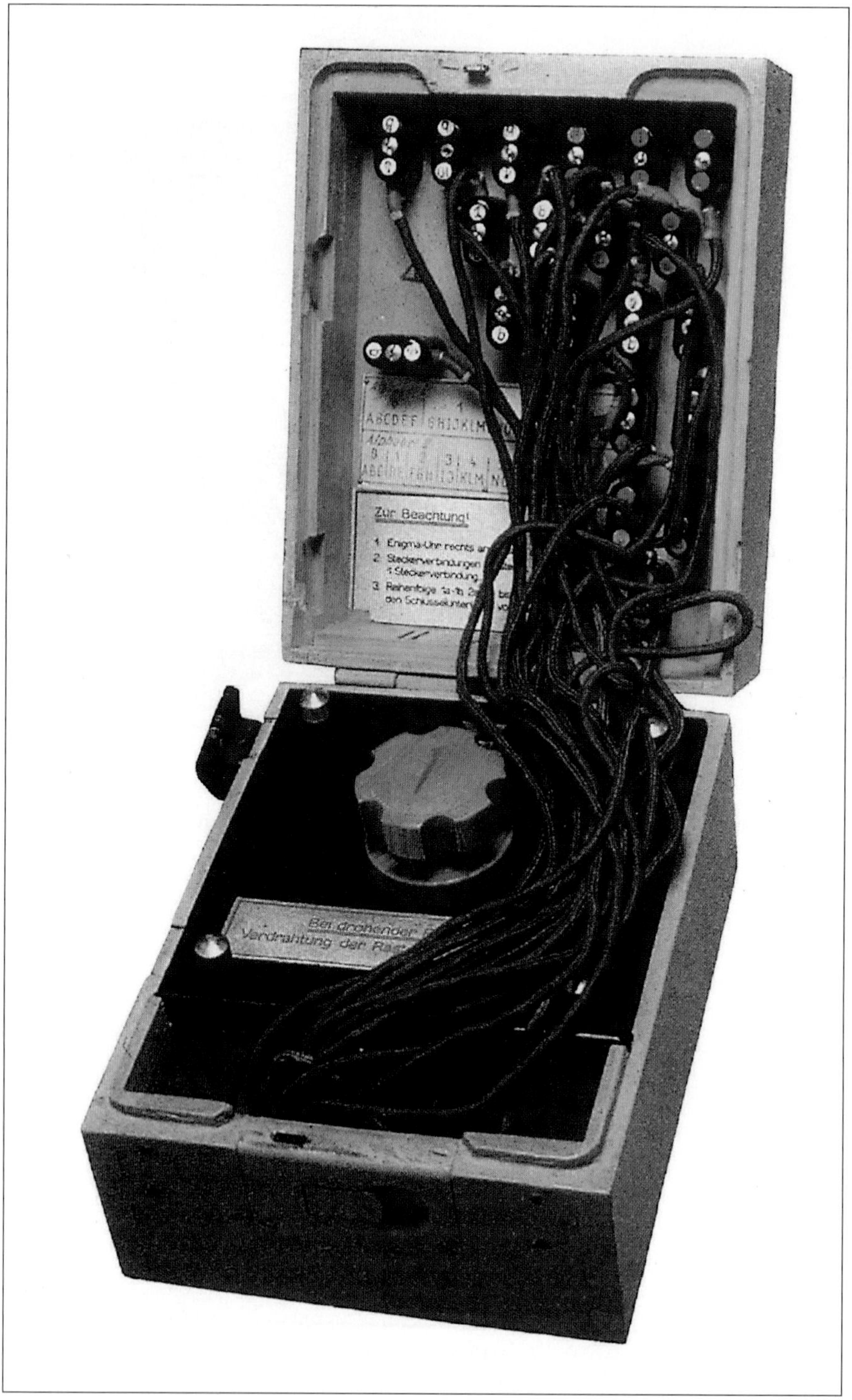
Zur Beachtung!
Bei drohender

On-line cipher teletype machine Lorenz SZ 42 *Schlüsselzusatz*, made by C. Lorenz A.G., Berlin, about 1943. A cipher machine for teletype Baudot signals, British cover name 'tunny', it was used at the strategic level down to Army headquarters. Twelve keying wheels with different graduation, using (from left to right) 43, 47, 51, 53, 59, 37, 61, 41, 31, 29, 26, 23 teeth, and irregularly spaced pegs, produce a key of very high period. Five pairs of wheels each control five VERNAM substitutions of the 5-bit code; two wheels ('motor wheels') serve for irregular movement only. The SZ 40/SZ 42 encryption was penetrated by the British due to an encryption fault on the German side and was then read regularly using the electronic COLOSSUS machines.

Plate N

Plate M

The *Uhr* box was used to replace the steckering of the *Wehrmacht* ENIGMA plugboard by a non-reciprocal substitution, which also could be changed easily by turning the knob (presumably every hour) selecting one out of 40 positions. First use by the German Air Force in July 1944, by the German Army in September 1944. Despite the extra security it added, the *Uhr* box was not widely used.

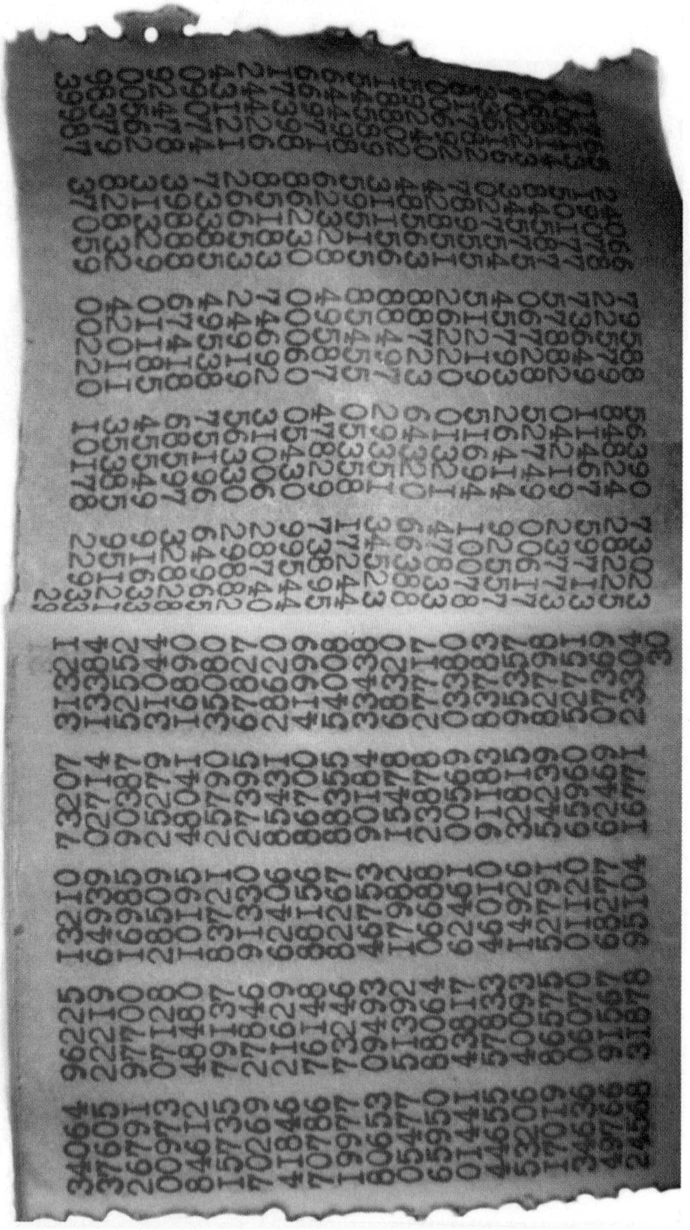

Plate O One-time pad of Russian origin, small enough to fit in the palm of a hand. The typewritten numbers have figures in Russian style.

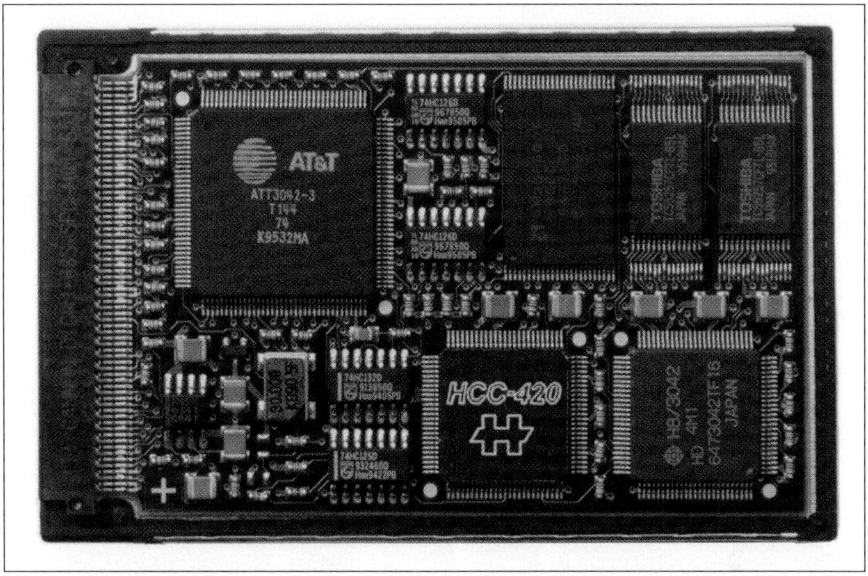

Crypto board, manufactured 1996 by Crypto AG, Zug (Switzerland), to be used for stand-alone or networked computers to provide access protection, secrecy of information, integrity of information, and virus protection. This highly reliable hardware with very long mean time between failure can be stored without batteries.

Plate P

Plate Q CRAY-1 S (1979). CRAY Supercomputers originated from the famous CRAY-1, designed by Seymour Cray (1928–1996) and in use since 1976, when it had a market price of $ 8 million. Supercomputers contain a very large number of integrated circuits allowing highly parallel work, but requiring very compact technology. They work at extremely high speed and need extensive cooling. First used for cryptanalytic tasks; civil versions have been available under certain limitations since 1979. The series continued with CRAY-2, CRAY X-MP, CRAY Y-MP, CRAY C 90, CRAY J90 leading to CRAY T 90, whose configuration T932 comprises 32 processors. A massively parallel line was opened by the model CRAY T3D, the more recent model CRAY T3E (July 1996) is liquid-cooled and has up to 2048 processors, using the DEC Alpha EV-5 (211 64) chip, each one running at 600 megaflops, 1.2 teraflops peak performance (1998: T3E-1200E 2.4 teraflops).

Part II: Cryptanalysis

Il ne faut alors ni se buter,
ni se rebuter,
et faire comme en politique:
changer son fusil d'épaule.

[One should neither run
one's head against a wall
nor get scared away
and should act like in politics:
change one's opinion.]

Étienne Bazeries, 1901

Deciphering is, in my opinion,
one of the most fascinating of arts,
and I fear I have wasted upon it
more time than it deserves.

Charles Babbage, 1864

The Machinery

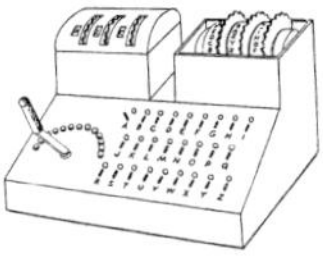

Cyclometer
(Poland)

Bombe
(Great Britain)

Colossus
(Great Britain)

Bazeries' quotation warns with Gallic charm against overestimating the systematics of the cryptanalytic methods we discuss in this second part.

Apart from the simple but generally intractable method of exhaustion, these methods have their roots in the inherent properties of language, which are hard to eradicate even by the most refined encryption. For a systematic treatment, the invariants of cryptological methods are established and used. Both pattern finding and frequency analysis are based on invariants of monoalphabetic (monographic or polygraphic) encryptions. But even polyalphabetic encryption leaves invariant a certain linguistic-statistical parameter, called *Kappa*. This allows the reduction of repeatedly used polyalphabetic encryptions, especially of periodic ones, to monoalphabetic encryptions. For transposition, which is a polygraphic encryption of very special kind, exploiting this specialty leads to the use of contact frequencies.

According to William F. Friedman (who coined the term in 1920), cryptanalysis involves the determination of the language employed, the general crypto system, the specific key, and the plaintext; usually in this order.

Cryptanalysis requires applying the right means at the right place. Givierge expressed it drastically: *Certains rasoirs excellents sont pourtant tout à fait dangereux dans les mains d'un singe.* [Some excellent razors are after all still dangerous in the hands of an ape.]

It should be noted that as a rule active cryptanalytic attacks both against governmental and against commercial communication channels are punishable, and against private communication may provoke a lawsuit for damages. However, since we need knowledge of cryptanalytic methods to be able to reflect on the secure use of cryptographic methods, in particular on avoiding illusory complications, we hope to be excused on scientific grounds for discussing cryptanalysis prophylactically.

Frequently, cryptanalysis is not only a question of material effort, but also of the time available. Much news is, as the saying goes, futile as soon as it is out of date, and in some areas messages become old-fashioned quickly. Patrick Beesly (*Very special intelligence*, London 1977) remarked in this respect:

> It should, however, be emphasized that cryptanalysis must be swift to be of real operational use.

And Friedman stated:

> The best that can be expected is that the degree of security be great enough to delay solution by the enemy for such a length of time that

when the solution is finally reached, the information thus obtained has lost all its ... value.

Sometimes, solutions of cryptograms have been given which were little more than guesswork. The requirements for a successful reliable unauthorized decryption fluctuate, depending on the situation, between a rational reconstruction of 90% of the plaintext (Meyer-Matyas) and a complete disclosure not only of the message, but of the key, too, and of the whole crypto system (Rohrbach).

Encryption devices and machines are no longer of immediate value once they are out of use. Planned obsolescence (Sect. 2.1.1) is a good stratagem—it minimizes the risk in case crypto tools are stolen or seized. The time scale to be applied depends very much on the genre of the messages: ten hours for artillery fire commands may correspond to ten years in diplomatic telegrams. But even outdated codebooks or cipher tables are revealing. Practical cryptanalysis makes use of many tiny details about the adversary's pecularities, habits, and preferences and records them carefully. In this sense, cryptanalysis feeds on the results it has already attained.

Cryptanalysis is nurtured to a good part by encryption errors and even by lesser mistakes of the adversary. Luigi Sacco sarcastically said:

Les chiffreurs se chargent suffisamment d'aider l'ennemi.
[The cipherers are sufficiently occupied to help the enemy.]

Asymmetric crypto systems allow a publication of the encryption keys ('public key') and indicate a liberalization of cryptography ('public cryptography'), which also entails an increased public awareness of cryptanalysis generally—although the cryptologic services of the government agencies see this side of public cryptography rather reluctantly. These agencies do not see themselves as instruments for the education of the masses. To give an example with respect to the modes of operation of standardized DES, Philip Zimmermann reports that "... the authors of a number of [these encryption packages] say they've never heard of CBC or CFB mode. The very fact that they haven't even learned enough cryptography to know these elementary concepts is not reassuring." Cryptanalysis is not going to die out.

Cryptanalysis is also already on the free market. The company Access Data Recovery (87 East 600 South, Orem, Utah 84058, USA)—and this is not the only example—sells for some $100 a program developed by Eric Thompson, that penetrates into the built-in encryptions of WordPerfect, Lotus 1-2-3, MS Excel, Symphony, Quattro Pro, Paradox and MS Word—and this not by exhaustive trial and error, but by genuine cryptanalytic methods. Some people buy it when they have forgotten their password, and police officers use it, too, if they want to read confiscated data.

12 Exhausting Combinatorial Complexity

Gewöhnlich glaubt der Mensch, wenn er nur Worte hört,
es müsse sich dabei doch auch was denken lassen.
Goethe

The cardinal number of a class of methods—corresponding to the number of available keys—is a criterion for the combinatorial complexity of the encryption. As a measure of security against unauthorized decryption, it gives an upper bound on the work required for an exhaustive search under the assumption that the class of methods is known (Shannon's maxim: "The enemy knows the system being used.")

We shall frequently make use of an improved Stirling formula[1] for $n!$

$$n! = (n/e)^n\sqrt{2\pi n}\cdot(1+\frac{1}{12n-\frac{1}{2}}+O(\frac{1}{n^3})) = \sqrt{2\pi e}(n/e)^{n+\frac{1}{2}}\cdot(1+\frac{1}{12n-\frac{1}{2}}+O(\frac{1}{n^3}))$$

with the numerical values

$$\sqrt{2\pi} = 2.506\,628\,275\ldots\ ,$$
$$e = 2.718\,281\,828\ldots\ ,$$
$$\sqrt{2\pi e} = 4.132\,731\,353\ldots$$

and of the asymptotic formula for the base 2 logarithm of the factorial[2]

$$\mathrm{ld}\,n! = (n+\tfrac{1}{2})\,(\mathrm{ld}\,n-\mathrm{ld}\,e)+\tfrac{1}{2}(\mathrm{ld}\,\pi+\mathrm{ld}\,e+1)+\mathrm{ld}\,e\,(\tfrac{1}{12\,n}-\tfrac{1}{360\,n^3}+O(\tfrac{1}{n^5}))$$

with the numerical values

$$\mathrm{ld}\,e = 1.442\,695\,041\ldots\ ,$$
$$\tfrac{1}{2}(\mathrm{ld}\,\pi+\mathrm{ld}\,e+1) = 2.047\,095\,586\ldots\ .$$

$|V|$, the cardinal number of the alphabet V, is abbreviated by N.

$Z = |S|$ denotes the cardinal number of the class of methods S.

In the following the combinatorial complexities Z are compiled for some classes of methods S. $\mathrm{ld}\,Z$, the information of the class of methods S, is measured in [bit]. ${}^{10}\log Z$ is measured in [ban], a unit introduced by Turing, with the practical unit $1\,[\text{deciban}] = 0.1/{}^{10}\log 2\ [\text{bit}] \approx 0.332\ [\text{bit}]$.

---

[1] $26! = 403\,291\,461\,126\,605\,635\,584\,000\,000 = 2^{23}\cdot 3^{10}\cdot 5^6\cdot 7^3\cdot 11^2\cdot 13^2\cdot 17\cdot 19\cdot 23$

[2] $\mathrm{ld}\,x$ denotes the logarithm with the base 2: $\mathrm{ld}\,x = \ln x/\ln 2 = {}^{10}\log x/{}^{10}\log 2$.

12.1 Monoalphabetic Simple Encryptions

Simple substitutions are monographic. Leaving homophones and nulls out of consideration, we can restrict our interest essentially to permutations.

12.1.1 Simple Substitution in General (special case $n=1$ of Sect. 12.2.1)

12.1.1.1 (Simple substitutions, Sect. 3.2)

Permutations $V \longleftrightarrow V$ show the same cardinal number as one-to-one mappings (without homophones) of V *into* $W^{(m)}$, independent of W and m :

$$Z = N! \asymp \sqrt{2\pi e}\left(\frac{N}{e}\right)^{N+\frac{1}{2}} = 4.13 \cdot \left(\frac{N}{e}\right)^{N+\frac{1}{2}}$$

$$\operatorname{ld} Z \asymp \left(N+\frac{1}{2}\right) \cdot (\operatorname{ld} N - 1.44) + 2.05$$

For $N = 26$: $Z \approx 4.03 \cdot 10^{26}$, $\operatorname{ld} Z \approx 88.382$ [bit], $\log Z \approx 266.06$ [deciban] .

12.1.1.2 (monocyclic simple substitutions, Sect. 3.2.3)

Permutations $V \overset{N}{\longleftrightarrow} V$ with exactly one cycle, of the maximal order N:

$$Z = (N-1)! \asymp \sqrt{2\pi e}\left(\frac{N-1}{e}\right)^{N-\frac{1}{2}} = 4.13 \cdot \left(\frac{N-1}{e}\right)^{N-\frac{1}{2}}$$

$$\operatorname{ld} Z \asymp \left(N-\frac{1}{2}\right) \cdot (\operatorname{ld}(N-1) - 1.44) + 2.05$$

For $N = 26$: $Z \approx 1.55 \cdot 10^{25}$, $\operatorname{ld} Z \approx 83.682$ [bit], $\log Z \approx 251.91$ [deciban] .

12.1.1.3 (properly self-reciprocal simple substitutions, Sect. 3.2.1)

N is even, $N = 2\nu$ for a properly self-reciprocal permutation $V \overset{2}{\longleftrightarrow} V$.

$$Z = (N-1)!! \overset{def}{=} (N-1)(N-3)(N-5)\dots \cdot 5 \cdot 3 \cdot 1 \asymp \sqrt{2} \cdot \left(\frac{N}{e}\right)^{\frac{N}{2}}$$

$$\operatorname{ld} Z \asymp \frac{N}{2} \cdot (\operatorname{ld} N - 1.44) + \frac{1}{2}$$

For $N = 26$: $Z \approx 7.91 \cdot 10^{12}$, $\operatorname{ld} Z \approx 42.846$ [bit], $\log Z \approx 128.98$ [deciban] .

12.1.2 Decimated Alphabets (special case $n = 1$ of Sect. 12.2.2)

Assuming a linear cyclic quasiordering of the alphabet, Sinkov's 'decimation by q' (Sect. 5.6) .

$$Z = \varphi(N) \quad \text{, where } \varphi \text{ is the Euler totient function (Sect. 5.6)}$$

$$\operatorname{ld} Z = \operatorname{ld} N + \textstyle\sum_{\mu=1}^{k} \operatorname{ld} \rho(p_\mu, 1) \quad \text{(see Sect. 12.2.2)}$$

For $N = 26$: $Z = 12$ (Sect. 5.5, Table 1b),
$\operatorname{ld} Z \approx 3.58$ [bit], $\log Z \approx 10.79$ [deciban] .

	Z	$\mathrm{ld}\, Z$
12.2.1 Substitution in general	$(26^n)\,!$	$(26^n + \frac{1}{2})(4.70\,n - 1.44) + 2.05$
12.2.2 HILL transformation	$0.265 \cdot 26^{n^2}$	$4.70\,n^2 - 1.916$
12.2.3 CAESAR addition	26^n	$4.70\,n$
12.2.4 Transposition	$n!$	$(n + \frac{1}{2})(\mathrm{ld}\, n - 1.44) + 2.05$

Table 3. Complexity of monoalphabetic (polygraphic) encryption ste

12.1.3 CAESAR Addition (special case $n = 1$ of Sect. 12.2.3)

CAESAR addition $V \overset{+}{\longleftrightarrow} V$, a shift, is the monoalphabetic special case of a VIGENÈRE substitution (Sect. 7.4.1).

$$Z = N$$
$$\mathrm{ld}\, Z = \mathrm{ld}\, N$$

For $N = 26$: $Z = 26$, $\mathrm{ld}\, Z \approx 4.70$ [bit], $\log Z \approx 14.15$ [deciban] .

12.2 Monoalphabetic Polygraphic Encryptions

The combinatorial complexity of polygraphic substitutions depends on the encryption width n .

12.2.1 Polygraphic Substitution in General

Permutations $V^n \longleftrightarrow V^n$ show the same cardinal number as one-to-one mappings of V^n *into* $W^{(m)}$, independent of W and m :

$$Z = (N^n)!$$
$$\mathrm{ld}\, Z \asymp \left(N^n + \frac{1}{2}\right)(n \cdot \mathrm{ld}\, N - 1.44) + 2.05$$

For $N = 26$: $Z = (26^n)!$, $\mathrm{ld}\, Z \approx (26^n + \frac{1}{2})(4.70\, n - 1.44) + 2.05$.

Digraphic substitutions:	$Z \approx 1.88 \cdot 10^{1\,621}$,	$\mathrm{ld}\, Z \approx$ 5387 [bit] ;
Trigraphic substitutions:	$Z \approx 1.19 \cdot 10^{66\,978}$,	$\mathrm{ld}\, Z \approx$ 222500 [bit] ;
Tetragraphic substitutions:	$Z \approx 4.82 \cdot 10^{2\,388\,104}$,	$\mathrm{ld}\, Z \approx$ 7933000 [bit] .

PLAYFAIR substitutions show the same cardinal number as monocyclic simple substitutions with $N = 25$:

$$Z = 25!/(5{\cdot}5) \approx 6.20{\cdot}10^{23}\ ,\ \mathrm{ld}\, Z \approx 79.038\ \text{[bit]},\ \log Z \approx 237.93\ \text{[deciban]}\ .$$

ld Z				
$n=1$	$n=4$	$n=16$	$n=64$	$n=256$
$8.84 \cdot 10^{1}$	$7.93 \cdot 10^{6}$	$3.22 \cdot 10^{24}$	$1.08 \cdot 10^{93}$	$2.07 \cdot 10^{365}$
3.58	73.29	$1.20 \cdot 10^{3}$	$1.93 \cdot 10^{4}$	$3.08 \cdot 10^{5}$
4.70	18.80	75.21	300.83	1 203.31
	4.58	44.25	296.00	1 684.00

: $N=26$ depending on the encryption width n

12.2.2 Polygraphic Homogeneous Linear Substitution (HILL Transformation)

Assuming a linear cyclic quasiordering of the alphabet, from Sect. 5.2.3

$$Z = N^{n^2} \cdot \rho(N,n) \quad , \quad \text{where for } N = p_1^{s_1} p_2^{s_2} \ldots p_k^{s_k}$$
$$\rho(N,n) = \rho(p_1,n)\rho(p_2,n)\ldots\rho(p_k,n).$$
$$\operatorname{ld} Z = n^2 \operatorname{ld} N + \textstyle\sum_{\mu=1}^{k} \operatorname{ld} \rho(p_\mu,n) \quad .$$

For large n approximative values for $\rho(p,n)$ in Sect. 5.2.3;
for large n and not too small p with $\operatorname{ld} e \approx 1.44$,

$$\operatorname{ld} \rho(p,n) \approx 1.44 \,/\, (\tfrac{3}{2} - p) \ .$$

For $N=26$ and large n: $\rho(2,n) \approx 0.289$ and $\rho(13,n) \approx 0.917$, thus $\rho(26,n) \approx 0.289 \cdot 0.917 = 0.265$; altogether:

For $N=26$ and large n:
$Z \approx 0.265 \cdot 26^{n^2}$, $\operatorname{ld} Z \approx 4.70\, n^2 - 1.92$ [bit] , $\log Z \approx 14.15\, n^2 - 5.78$ [deciban] .

12.2.3 Polygraphic Translation (Polygraphic CAESAR Addition)

Polygraphic CAESAR addition $V^n \overset{+}{\leftrightarrow} V^n$ with encryption width n, a shift, is a special case of inhomogeneous linear substitution, where T is the identity matrix.

$$Z = N^n \ , \quad \operatorname{ld} Z = n \operatorname{ld} N$$

For $N=26$: $Z = 26^n$, $\operatorname{ld} Z \approx 4.70 \cdot n$ [bit], $\log Z \approx 14.15 \cdot n$ [deciban] .

12.2.4 Transposition

Transpositions of width n are subsumed (somewhat surprisingly) under linear substitutions, since they are linear substitutions whose matrix is a permutation matrix. The complexity is therefore independent of N.

$$Z = n\,!$$
$$\operatorname{ld} Z = (n + \tfrac{1}{2})(\operatorname{ld} n - 1.44) + 2.05$$

	Z	ld Z
12.3.1 PERMUTE substitution	$(26!)^d$	$88.38 \cdot d$
12.3.2 MULTIPLEX substitution	$(25!)^d$	$83.68 \cdot d$
12.3.3 ALBERTI substitution	$26!\,26^{d-1}$	$4.70 \cdot d + 83.68$
12.3.4 VIGENÈRE substitution	26^d	$4.70 \cdot d$

Table 4. Complexity of polyalphabetic (monographic) crypto syster

12.2.5 Summary on Monoalphabetic Substitutions

The combinatorial complexities of monoalphabetic substitutions are tabulated in Table 3 and shown graphically in Fig. 80.

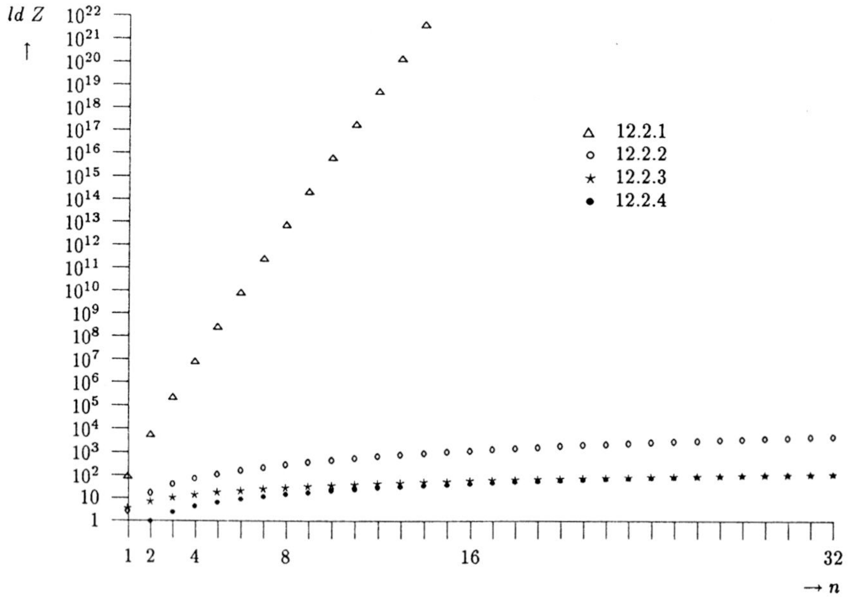

Fig. 80. Combinatorial complexity of polygraphic substitutions of width n, for $N = 26$

Note that the complexity of transposition surpasses the complexity of polygraphic CAESAR addition at about $n = N \cdot e$ (for $N = 26$ just at $n = 68$ with ld $Z \approx 320$).

For $N = 26$, transposition reaches at $n = 26$ simple (monogram) substitution; homogeneous linear substitution with ld $Z \approx 4.70\, n^2 - 1.92$ surpasses

ld Z				
$d=1$	$d=10$	$d=100$	$d=1000$	$d=10000$
88.38	883.82	8 838.20	88 381.95	883 819.53
83.68	836.82	8368.15	83681.51	836 815.36
88.38	130.69	553.73	4 784.12	47 088.08
4.70	47.00	470.04	4 700.44	47 004.40

$N=26$ depending on the number d of alphabets used

at $n=\ 5$ simple (monographic) substitution with $\text{ld}\, Z \approx 88.38$,
at $n=34$ bigraphic substitution with $\text{ld}\, Z \approx 5.387 \cdot 10^3$.

Polygraphic CAESAR addition is surpassed at $n=2$ by homogeneous linear substitution.

Note that a block transposition of width n is polygraphic, but also monoalphabetic—it is obfuscating to call n a 'period'.

12.3 Polyalphabetic Encryptions

The combinatorial complexity of the most general polyalphabetic (periodic) encryption with d unrelated alphabets is the product of the complexities of the different alphabets. For the case of d related alphabets the complexity is correspondingly smaller.

12.3.1 PERMUTE Encryption with d Alphabets

$$Z = (N\,!)^d$$

$$\text{ld}\, Z = d \cdot ((N + \tfrac{1}{2})\,(\text{ld}\, N - 1.44) + 2.05)$$

For $N = 26$: $Z = (4.03 \cdot 10^{26})^d$, $\text{ld}\, Z = 88.382 \cdot d$ [bit] .

12.3.2 MULTIPLEX Encryption with d Alphabets

$$Z = ((N-1)\,!)^d$$

$$\text{ld}\, Z = d \cdot ((N - \frac{1}{2})\,(\text{ld}\,(N-1) - 1.44) + 2.05)$$

For $N=26$: $Z = (1.55 \cdot 10^{25})^d$, $\text{ld}\, Z = 83.682 \cdot d$ [bit] .

12.3.3 ALBERTI Encryption with d Alphabets

$$Z = N!\, N^{d-1}$$

$$\text{ld}\, Z = d \cdot \text{ld}\, N + (N - \tfrac{1}{2})(\text{ld}\,(N-1) - 1.44) + 2.05$$

For $N = 26$: $Z = 1.55 \cdot 10^{25} \cdot 26^d$, $\text{ld}\, Z \approx 4.70 \cdot d + 83.682$ [bit] .

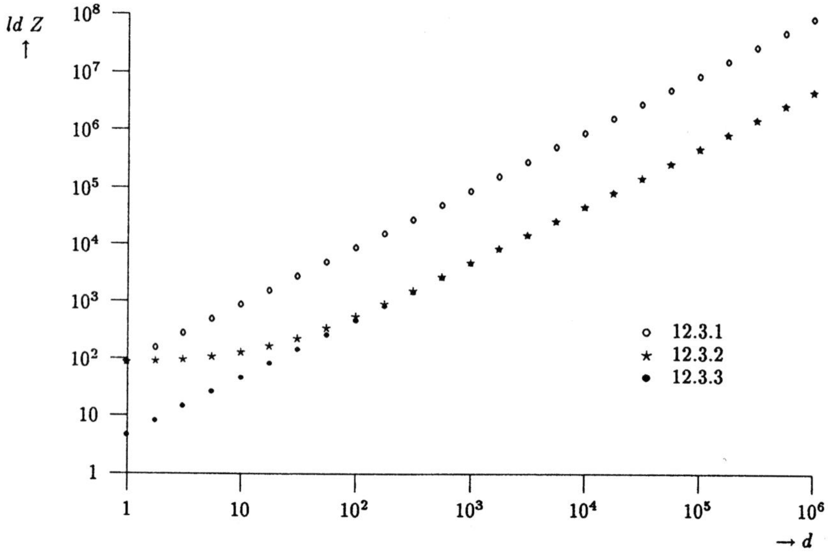

Fig. 81. Combinatorial complexity of polyalphabetic encryption with a number d of alphabets used, for $N = 26$

12.3.4 VIGENÈRE or BEAUFORT Encryption with d Alphabets

$$Z = N^d$$

$$\text{ld}\, Z = d \cdot \text{ld}\, N$$

For $N = 26$: $Z = 26^d$, $\text{ld}\, Z \approx 4.70 \cdot d$ [bit] .

12.3.5 Summary of Polyalphabetic Encryption

The combinatorial complexities of polyalphabetic substitutions are tabulated for the monographic case $n = 1$ in Table 4 and shown graphically in Fig. 81 .

Note that the VIGENÈRE or BEAUFORT encryption and (monoalphabetic) simple substitution have the same complexity for $d \approx N + \frac{1}{2} - \frac{N - \ln 2\pi}{\ln N}$ (i.e., in the case $N = 26$, for $d = 19$; in the case $N = 26^2$, for $d = 573$).

MULTIPLEX encryption with d alphabets and (monoalphabetic) polygraphic substitution with width n have approximately the same complexity for $d \approx n \cdot N^{n-1}$ (more precisely, with bigrams, i.e., $n=2$, in the case $N=26$ for $d = 55$).

The complexity of the VIGENÈRE or BEAUFORT encryption with period $d = h$ and polygraphic CAESAR addition with width $n = h$ coincide. For $N = 10$, an adding machine with h positions can be used, the mechanical carry device of which has been dismantled in the first case, not in the second case (Sect. 5.7.1 and Sect. 8.3.3).

12.4 General Remarks on Combinatorial Complexity

In studying combinatorial complexity theoretically, the whole class of methods is envisaged. Practically, the unauthorized decryptor will often be able to find restrictions caused by the encryptor's habits or stupidity. The circumstances matter, too.

12.4.1 For example, the cylinder of Jefferson and Bazeries has without knowledge of the disks the complexity of a MULTIPLEX encryption; with knowledge of the disks[3] however only of a transposition. For $d = 25$ (M-94) this means a reduction from $Z = (26!)^{25} \approx 1.38 \cdot 10^{665}$ to $Z = 25! \approx 1.55 \cdot 10^{25}$, or from $\mathrm{ld}\, Z \approx 2209$ to $\mathrm{ld}\, Z \approx 83.68$. A similar situation exists with an Alberti disk: If it falls into the hands of the enemy, the ALBERTI encryption collapses to a VIGENÈRE encryption; Z correspondingly drops from $(N!) \cdot N^{d-1}$ to N^d, or for $N = 26$, $\mathrm{ld}\, Z$ from $4.70 \cdot d + 83.68$ to $4.70 \cdot d$.

12.4.2 Note, too, that the complexity of double transposition is $Z = (n!)^2$ and thus is somewhat smaller than that of transposition with doubled width, $Z = (2n)! = (n!)^2 \cdot \binom{2n}{n}$, where asymptotically $\binom{2n}{n} \asymp 4^n / \sqrt{\pi \cdot (n + \frac{1}{4} + \frac{1}{32 \cdot n})}$. But this means only that exhaustion, given a fixed time limit, carries further for double transposition, which does not contradict the empirical fact that in a region of complexity where exhaustion is not tractable, cryptanalysis of double transposition is much more difficult than cryptanalysis of columnar transposition with doubled width.

12.4.3 Finally, it is remarkable that for a VIGENÈRE encryption, Z and $\mathrm{ld}\, Z$ depend only on N^d and thus (for $N = 2^k$) are invariant under transition to binary encoding: $(2^k)^d = 2^{(k \cdot d)}$. In contrast to this, transition to binary encoding reduces the complexity of PERMUTE encryption drastically: $((2^k)!)^d > (2!)^{k \cdot d}$ for $k \geq 2$.

12.5 Cryptanalysis by Exhaustion

It should be clear that combinatorial complexity is a measure of security only in the sense that it is a measure of the effort needed for a particular kind of unauthorized decryption, albeit a very simple and very general one, which we shall call exhaustion attack. After guessing a class of encryption methods, we construct all plaintexts that lead under some encryption process of one of these methods to a given cryptotext (all 'variants'), and then read the 'right' message, or 'gather' it in the true meaning of the word. This attack can lead to more than one gathered message, which shows that the decryption is not unique—more precisely, that encryption is not injective for the encryption method one has guessed at. This can mean that one has to

---

[3] 'A crypto device can fall into the hands of the enemy': Maxim No. 3 (Sect.11.2.3). Bazeries invented his device in 1891, eight years after Kerckhoffs had published his advice.

H V Z D U V F K R Q G X Q N H O D O V L F K L Q E R Q Q D Q N D P L F K C
I W A E V W G L S R H Y R O I P E P W M G L M R F S R R E R O E Q M G L D
J X B F W X H M T S I Z S P J Q F Q X N H M N S G T S S F S P F R N H M E
K Y C G X Y I N U T J A T Q K R G R Y O I N O T H U T T G T Q G S O I N F
L Z D H Y Z J O V U K B U R L S H S Z P J O P U I V U U H U R H T P J O G
M A E I Z A K P W V L C V S M T I T A Q K P Q V J W V V I V S I U Q K P H
N B F J A B L Q X W M D W T N U J U B R L Q R W K X W W J W T J V R L Q I
O C G K B C M R Y X N E X U O V K V C S M R S X L Y X X K X U K W S M R J
P D H L C D N S Z Y O F Y V P W L W D T N S T Y M Z Y Y L Y V L X T N S K
Q E I M D E O T A Z P G Z W Q X M X E U O T U Z N A Z Z M Z W M Y U O T L
R F J N E F P U B A Q H A X R Y N Y F V P U V A O B A A N A X N Z V P U M
S G K O F G Q V C B R I B Y S Z O Z G W Q V W B P C B B O B Y O A W Q V N
T H L P G H R W D C S J C Z T A P A H X R W X C Q D C C P C Z P B X R W O
U I M Q H I S X E D T K D A U B Q B I Y S X Y D R E D D Q D A Q C Y S X P
V J N R I J T Y F E U L E B V C R C J Z T Y Z E S F E E R E B R D Z T Y Q
W K O S J K U Z G F V M F C W D S D K A U Z A F T G F F S F C S E A U Z R
X L P T K L V A H G W N G D X E T E L B V A B G U H G G T G D T F B V A S
Y M Q U L M W B I H X O H E Y F U F M C W B C H V I H H U H E U G C W B T
Z N R V M N X C J I Y P I F Z G V G N D X C D I W J I I V I F V H D X C U
A O S W N O Y D K J Z Q J G A H W H O E Y D E J X K J J W J G W I E Y D V
B P T X O P Z E L K A R K H B I X I P F Z E F K Y L K K X K H X J F Z E W
C Q U Y P Q A F M L B S L I C J Y J Q G A F G L Z M L L Y L I Y K G A F X
D R V Z Q R B G N M C T M J D K Z K R H B G H M A N M M Z M J Z L H B G Y
E S W A R S C H O N D U N K E L A L S I C H I N B O N N A N K A M I C H Z
F T X B S T D I P O E V O L F M B M T J D I J O C P O O B O L B N J D I A
G U Y C T U E J Q P F W P M G N C N U K E J K P D Q P P C P M C O K E J B

Table 5. 26 variants of a CAESAR encryption: H V Z D U V F K R Q ...

look for a narrower class of encryption methods. We shall come back to this phenomenon under the catchword 'unicity distance'. If no message can be gathered, the guess as to the class of encryption methods was erronous—or a mistake was made in the encryption process.

Proceeding by exhaustion is only tractable, of course, if the number of variants to be scrutinized is not gigantic. However, it is not necessary to gather for each scrutiny the full alleged plaintext; an escape should be possible as soon as a part of the plaintext is found to be absurd.

We illustrate exhaustion with two small examples, where the cardinality of the alleged plaintexts which are to be scrutinized is about two dozen variants:

a) CAESAR addition with $\mathbb{Z}_{26}$: 26 variants (Table 5),
b) transposition with width 4: 24 variants (Table 6).

The method of exhaustion is also indicated if the number of 'probable word' keys that are given or guessed is not too large. In the Renaissance the repertoire of familiar quotations was not too big—proverbs like OMNIA VINCIT AMOR, VIRTUTI OMNIA PARENT, SIC ERGO ELEMENTIS, IN PRINCIPIO ERAT VERBUM, to mention a few that are present in the cryptologic literature. Indeed even today one finds from amateurs up to statesmen a predilection for programmatic keywords.

S A E W S H R C N U O D K L N E L I A S H N C I O N B N N A A K I H M C W
A S E W H S R C U N O D L K N E I L A S N H C I N O B N A N A K H I M C N
A E S W H R S C U O N D L N K E I A L S N C H I N B O N A A N K H M I C N
E A S W R H S C O U N D N L K E A I L S C N H I B N O N A A N K M H I C Z
S E A W S R H C N O U D K L N E L A I S H C N I O B N N N A A K I M H C W
E S A W R S H C O N U D N K L E A L I S C H N I B O N N A N A K M I H C Z
S W E A S C R H N D O U K E N L L S A I H I C N O N B N N K A A I C M H W
W S E A C S R H D N O U E K N L S L A I I H C N N O B N K N A A C I M H A
W E S A C R S H D O N U E N K L S A L I I C H N N B O N K A N A C M I H A
E W S A R C S H O D N U N E K L A S L I C I H N B N O N A K N A M C I H Z
E S W A R S C H O N D U N K E L A L S I C H I N B O N N A N K A M I C H Z
S W A E S C H R N D U O K E L N L S I A H I N C O N N B N K A A I C H N W
W S A E C S H R D N U O E K L N S L I A I H N C N O N B K N A A C I H M A
W A S E C H S R U U N D E L K N S I L A I N H C N N D B K A N A C H I M A
A W S E H C S R U D N D L F K N I S L A N I H C N N O B A K M A H C I M N
S A W E S H C R N U D O K L E N L I S A H N I C O N N B N A K A I H C M W
A S W E H S C R U N D D L K E N I L S A N H I C N O N B A N K A H I C M N
A W E S H C R S U D O N L E N K I S A L N I C H N N B O A K A N H C M I N
W A E S C H R S D U O N E L N K S I A L T N C H N N B O K A A N C H M I A
W E A S C R H S D O U N E N L K S A I L I C N H N B N O K A A N C M H I A
E W A S R C H S D D U N N E L K A S I L O I N H A N N D A K A N M C H I Z
A E W S H R C S U D O N L N E K I A S L N C I H N B N O A A K N H M C I N
E A W S R H C S O U D N N L E K A I S L O N I H B N N O A A K N M H C I Z
S E W A S R C H N D D U K N E L L A S I H C I N O B N N N A K A I M C H W

Table 6. 24 variants of a transposition of width 4: S A E W S H R C N U ...

12.6 Unicity Distance

Pursuing stepwise, letter by letter, the buildup of the feasible plaintext fragments leads to the observation that after a certain rather clearly defined length the decision for just one plaintext can be made confidently. The number of characters up to this length is called the empirical unicity distance U of the class of methods in question. Remarkably, in the two examples of Table 5 and Table 6 with almost equal complexity ($Z \approx 25$ and $\mathrm{ld}\, Z \approx 4.64$) the unicity distance is roughly equal, i.e., about four characters. There are for example only very few 4-letter words allowing an ambiguous CAESAR decryption, in English (Z_{26}): mpqy: ADEN, KNOW; aliip: DOLLS, WHEEL; afccq: JOLLY, CHEER; in German (Z_{26}): zydd: BAFF, POTT; qfzg: LAUB, TICK;

qunq: EIBE, OSLO; himy: ABER, NORD, KLOA(KE), (ST)OPSE(L) ($Z_{25}!$). Only words with different letters are here essential.

The unicity distance can be estimated by experienced cryptanalysts for encryptions with much larger complexity Z, like monoalphabetic simple substitution ($Z = 26!$, $\mathrm{ld}\, Z = 88.38$): for clearly shorter cryptotexts, there is ambiguity, for clearly longer cryptotexts, there is a unique solution. In the case of monoalphabetic simple substitution, the empirical unicity distance has been reported to be between 25 and 30: "... the unicity point , at about 27 letters. ... With 30 letters there is always a unique solution to a cryptogram of this type and with 20 it is usually easy to find a number of solutions" (Shannon 1945); "Practically, every example of 25 or more characters representing monoalphabetic encipherment of a 'sensible message' in English can be readily solved" (Friedman 1973). Experimental checks with encryption methods of very large complexity Z support the empirical law:

> *The unicity distance depends (for one and the same natural language) only on the combinatorial complexity Z of the class of methods. Moreover, it is (for not too small Z) proportional to* $\mathrm{ld}\, Z$.

This quantitative result was still unpublished around 1935. Only qualitative insights, like "The key should be comparable in length with the message itself" (Parker Hitt 1914, Sect. 8.8.2) have been known since Kasiski (Sect. 17.4), although presumably Friedman had an inkling. It means that the whole influence of the redundant language underlying the text can only be expressed in the proportionality constant. This was a starting point for the foundation of Claude E. Shannon's information theory, which he wrote as a classified report in 1945. It was released to the public in 1949.

Assuming Friedman's value $U = 25$ for monoalphabetic simple substitution, $\mathrm{ld}\, Z \approx 88.38$ results in an empirical calibration for the proportionality:

$$(*) \qquad U \approx \frac{1}{3.54}\, \mathrm{ld}\, Z \approx \frac{1}{1{,}06}\, {}^{10}\!\log Z .$$

Table 7 has been computed according to Sect. 12.2 for different width n of monoalphabetic polygraphic substitution.

	$n=1$	$n=4$	$n=16$	$n=64$	$n=256$
Substitution in general	**25**	2 240 000	10^{24}	10^{93}	10^{365}
homogeneous linear substitution	(1.02)	22	340	5 500	88 000
CAESAR addition	(1.34)	6	22	86	340
Transposition		(1.30)	13	85	480

Table 7. Empirical unicity distance U, extrapolated according to $(*)$, rounded up (N=26)

The values in parentheses turn out to be too small to be meaningful.[4]

[4] The rule has the following background in information theory:
The value 4.7 = ld 26 [bit/char] is split into 3.5 bit per character (74.5%) redundancy and 1.2 bit per character (24.5%) information (for Z_{26} and the English language).
For the theoretical foundation, see the appendix *Axiomatic Information Theory*.

In particular, there results

for digraphic substitution	$U \approx 1\,530$,
for trigraphic substitution	$U \approx 63\,000$,
for tetragraphic substitution	$U \approx 2\,240\,000$.

For periodic polyalphabetic encryption, the empirical unicity distance of a basic monoalphabetic encryption is to be multiplied with the length d of the period. Thus, for VIGENÈRE encryption, based on CAESAR addition steps

with $d = 10^2$	$U \approx 134$
with $d = 10^4$	$U \approx 13\,400$
with $d = 10^6$	$U \approx 1\,340\,000$.

If for an encryption method an empirical unicity distance exists, it may be expected that by suitable attacks other than exhaustion the breaking of an encryption becomes easier and less uncertain with increasing length of the cryptotext, whereas after some length near the unicity distance the solution becomes unproblematic, provided sufficient effort can be made. For holocryptic ('unbreakable') encryptions (Sect. 8.8.4), no unicity distance exists.

12.7 Practical Execution of Exhaustion

The practical execution of exhaustion proceeds by stepwise increasing the length of the fragments of the texts, cutting out each time the 'impossible' variants and leaving the 'possibly right' ones. The tables of bigrams and trigrams printed in the literature show that in English, French, or German among 676 bigrams about half are 'possible', among the 17 576 trigrams only about a thousand. The execution can easily be carried out interactively with computer help if the number of variants to begin with is not much larger than ten thousand. On the monitor screen groups of five to eight characters are easily picked out at a glance, and at least 100 of those selections can be made in one minute, which means 6 000 initial variants can be scanned in an hour. Later the number of variants remaining is reduced drastically, so in less than two hours the 'right' solution should be found or its nonexistence shown. For the examples in Tables 5 and 6 this can be seen in Figs. 82 and 83. Even a reader who is only vaguely familiar with the language will find it not difficult to weed out the senseless instantiations. In order to eliminate marginal influence, we have started with the 6th column.

Note that according to Sects. 12.3.3 and 12.2.4 ,

for VIGENÈRE (a polyalphabetic CAESAR addition)
$Z = 17\,576$ for period 3 , $Z = 456\,976$ for period 4 ;

for transposition (a special polygraphic encryption)
$Z = 40\,320$ for width 8 , $Z = 362\,880$ for width 9 .

This shows the (restricted) range of the exhaustion attack. For general monoalphabetic simple substitution and PLAYFAIR substitution, with Z

V	VF	VFK?					
W	WG	WGL	WGL S ?				
X	XH	XHM?					
Y	YI	YI N?					
Z	Z J	Z J O	Z J OV	Z J OVU?			
A	AK	AKP?					
B	BL	BL Q?					
C	CM	CMR?					
D	DN	DNS ?					
E	EO	EOT	EOTA	EOTAZ?			
F	FP	FPU	FPUB	FPUBA?			
G	GQ	GQV?					
H	HR	HRW?					
I	I S	I SX	I SXE	I SXED?			
J	J T?						
K	KU	KUZ	KUZ G?				
L	LV	LVA	LVAH	LVAHG?			
M	MW	MWB?					
N	NX	NXC?					
O	OY	OYD	OYDK?				
P	PZ	PZE	PZEL	PZELK	PZELKA	PZELKAR?	
Q	QA?						
R	RB	RBG	RBGN	RBGNM?			
S	S C	S CH	S CHO	S CHON	S CHOND	S CHONDU	S CHONDUN●
T	TD	TDI	TDI P	TDI PO	TDI POE	TDI POEV?	
U	UE	UEJ ?					

Fig. 82. Exhaustion for 26 variants of a CAESAR encryption

H	HR	HRC	HRCN?				
S	S R	S RC?					
R	RS	RS C	RS CU	RS CUO?			
H	HS	HS C	HS CO	HS COU	HS COUN	HS COUND	HS COUNDN?
R	RH	RHC?					
S	S H	S HC?					
C	CR	CRH?					
S	S R	S RH	S RHD?				
R	RS	RS H	RS HD	RS HDO	RS HDON	RS HDONU?	
C	CS ?						
S	S C	S CH	S CHO	S CHON	S CHOND	S CHONDU	S CHONDUN●
C	CH	CHR	CHRN?				
S	S H	S HR	S HRD?				
H	HS	HS R?					
C	CS ?						
H	HC	HCR?					
S	S C	S CR?					
C	CR	CRS ?					
H	HR	HRS	HRS D?				
R	RH	RHS ?					
C	CH	CHS	CHS D	CHS DD?			
R	RC	RCS ?					
H	HC	HCS ?					
R	RC	RCH	RCHN	RCHND?			

Fig. 83. Exhaustion for 24 variants of a transposition of width 4

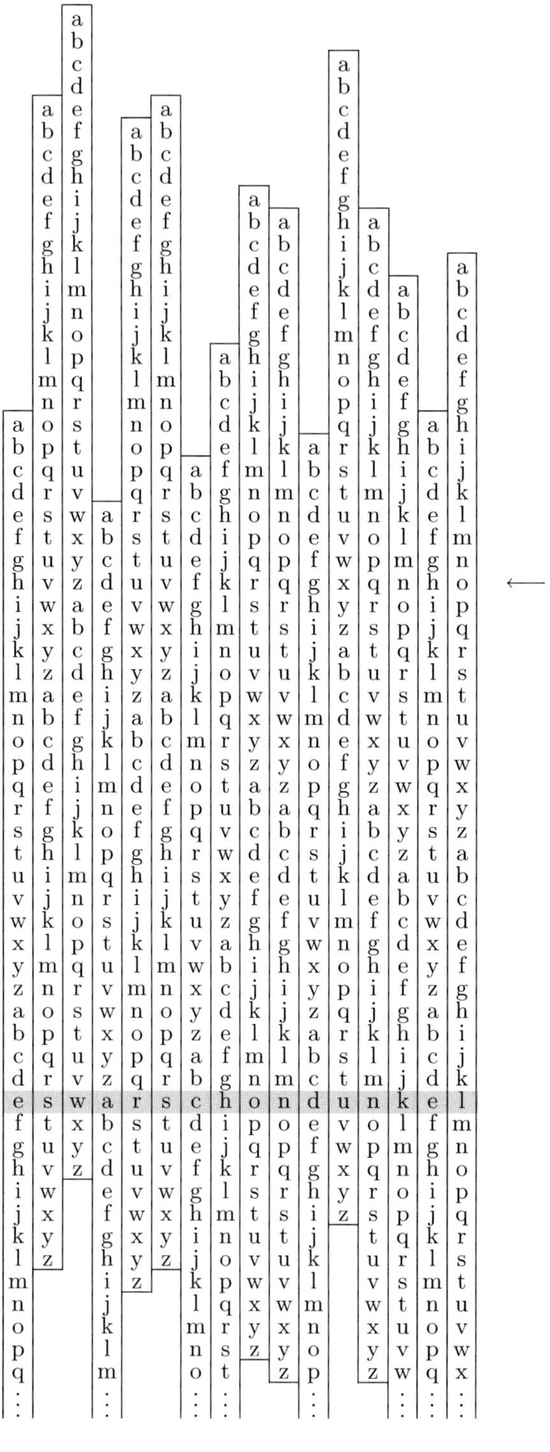

Fig. 84.
Strip method for the solution of a CAESAR encryption

of the order of magnitude 10^{25}, it is useless, at least in its pure form with human interaction. Computers can help to weed out the impossible variants much faster, but this may still not be sufficient. However, if high combinatorial complexity can be drastically lowered by other, suitable means, exhaustion may come within reach. In other words:

> *The exhaustion attack, although by itself alone rather insignificant, is in combination with other, likewise automatic attacks the fundamental method of intelligent cryptanalysis.*

Exhaustion is also used by the authorized decryptor in case of polyphone encryptions. The best known example is polyalphabetic encryption with unrelated monocyclic substitution alphabets, used with the cylinder of Jefferson and Bazeries, where the plaintext is to be found among two dozen variants.

12.8 Mechanizing the Exhaustion

12.8.1 Exhausting substitution. For the exhaustion of a simple CAESAR addition there exists a mechanization by the strip method. Ready-made strips containing the duplicated standard alphabet are used to demonstrate the cryptotext and all plaintext variants as well (Fig. 84). Cylinders containing the standard alphabet on their rim can be used as well. Here, the mechanical decryption aid is nothing but an encryption device, applied backwards. This can be applied to other mechanical devices, too. For example, an ENIGMA imitation can be used to find the one among $26^3 = 17\,576$ rotor positions which gives a cryptotext fragment for a probable word—provided the rotors have fallen into the cryptanalyst's hands.

12.8.2 Exhausting transposition. For the exhaustion of a transposition of known width n, the cryptotext is written horizontally as an array of n columns, then the sheet is cut into n vertical strips, which can be permuted (Fig. 85).

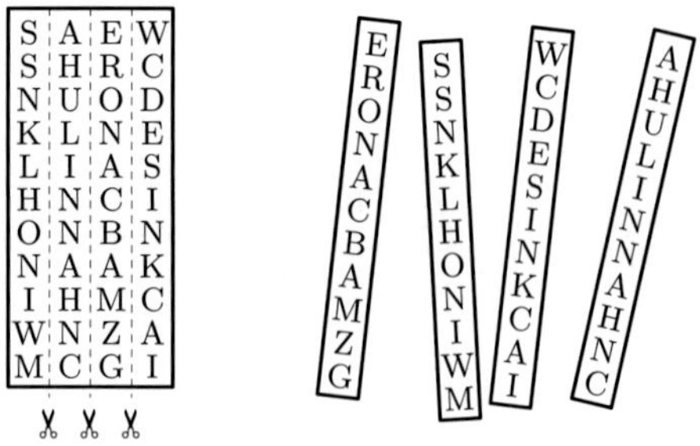

Fig. 85.
Scissors-and-paste method for the solution of a transposition

12.8.3 Brute force contra Invariance. Cryptanalysis by exhaustion is a brute force method and as such is subject to limits of power. In the next chapters, methods of cryptanalysis are discussed, which are more cleverly based on "the 'invariant' characteristics of the cryptographic system employed" (Solomon Kullback, in: Statistical Methods in Cryptanalysis, 1935).

13 Anatomy of Language: Patterns

No matter how resistant the cryptogram, all that is really needed is an *entry*, the identification of one word, or of three or four letters.

Helen Fouché Gaines 1939

Language contains an internal frame of regularities that are hard to extirpate. Particularly resistant are repeated patterns.

13.1 Invariance of Repetition Patterns

Invariance Theorem 1: For all monoalphabetic, functional simple substitutions, especially for all monoalphabetic linear simple substitutions (including CAESAR additions and reversions),
repetition patterns of the individual characters in the text are invariant.

The plaintext	w i n t e r s e m e s t e r
encrypted by a CAESAR addition	Z L Q W H U V H P H V W H U ,
or with a reversed alphabet	D R M G V I H V N V H G V I ,
or with a permuted alphabet	V A H O R M N R G R N O R M ,

has an invariant pattern of character repetition. According to Shannon, patterns are just the 'residue classes' of simple substitutions. Text particles with the same repetition pattern are called 'idiomorphs'.

Monoalphabetic and functional *polygraphic* substitutions $V^{(n)} \longrightarrow W^{(m)}$ leave the patterns of polygram repetitions (observing the hiatuses) invariant.

By contrast, transpositions do not preserve repetition patterns. Homophonic and particularly polyalphabetic substitutions destroy repetition patterns.

Patterns are usually denoted by finite sequences of numbers in normal form, i.e., each number has at its first appearance (from left to right) only smaller numbers to its left. *123324156* is in normal form.

In the cryptotext V A H O R M N R G R N O R M from above,
NRGRN has the pattern *12321*, NRGRNOR has the pattern *1232142*, ORMNRGRNORM has the pattern *12342524123*. The pattern *12321* of NRGRN is particularly conspicuous. Moreover, the word ORM occurs twice. Thus, the pattern *12345675857456* describes some text with groups of 6 and 8 letters that rhyme. In fact, we know that wintersemester is a solution, but there is hardly an idiomorph solution in German, and most likely none in English.

abbacy cabbage cabbala sabbath scabbard baccalaureate maccabee staccato affable affair baggage braggart haggard laggard allah allay ballad ballast fallacy gallant installation mallard palladium parallax wallaby diagrammatic flammable gamma grammar mamma programmatic annalist annals bandanna cannabis hosanna manna savannah appal apparatus apparel apparent kappa arrack arraign arrange arrant arras array barrack barracuda barrage carragheen embarrass narrate tarragon warrant ambassador assail assassin assault assay cassandra massacre massage passage vassal wassail attach attack attain rattan attar battalion coattail rattan regatta wattage piazza beebread boob booby deed deedless indeed doodle ebbed eccentric bedded reddear redden shredder wedded effect effeminate effendi efferent effervesce effete begged bootlegger egged legged pegged trekked aquarelle bagatelle belle chancellery chanterelle driveller dweller excellent feller fontanelle gazelle groveller hellebor hellenic impellent intellect jeweller libeller mademoiselle nacelle pellet propeller repellent seller teller traveller emmet barrenness comedienne fennec fennel jennet kennel rennet tenner pepper stepper zeppelin deterrent ferret interregnum interrelation overreact parterre terrestrial adressee dessert dresser essence essential finesse largesse lessen messenger noblesse quintessence tesselate vessel begetter better burette corvette curette fetter gazette getter letter marionette pirouette rosette roulette setter silhouette geegee googol heehaw capriccio pasticcio forbidding yiddish difficile difficult griffin tiffin biggish bacilli billiard billion brilliant chilli cyrillic fillip illicit illinois illiquid illiberal illiterate illimitable lilliput milliard millibar milligram milliliter millimeter milliner millionaire millivolt penicillin postillion shilling silliness tranquillize trillion trillium vanillin gimmick immigrant imminent immiscible immitigable finnish innings pinniped zinnia pippin irrigate irritate admission commission dissident dissimilar dissipate emission fissile fission fortissimo missile mission missive omission permission permissive acquitting fitting kittiwake civvies noon broccoli sirocco apollo collocate colloid colloquial colloquium follow hollow rollout common accommodate commode commodore commotion connote opponent opportune oppose opposite borrow corroborate corrode horror morrow sorrow blossom crossover blotto bottom cotton grotto lotto motto ottoman risotto glowworm powwow peep poop career seesaw teeter teethe teetotal teetotum toot toothache tootle hubbub succulent succumb succuss pullup nummulite unnumbered chaussure guttural

Fig. 86. Instantiations of the pattern *1221* in English (after Hugh Casement)

Short patterns allow many instantiations by meaningful words or fragments of those. The pattern *1221* allows in English the idiomorphs compiled in Fig. 86. The list (excluding proper names) is intended to contain all the words or fragments (without grammatical variations) listed in *Cassell's English Dictionary*.

Note that this list contains only few words from the military genre like assa(ult), atta(ck), (b)atta(lion), (b)arra(ck), (z)eppe(lin), (sh)ippi(ng), (m)issi(le), (c)ommo(dore); furthermore not very many words from the diplomatic genre like affa(ir), (amb)assa(dor), assa(ssin), (chanc)elle(ry), (sh)illi(ng), immi(grant), (comm)issi(on). The search space for instantiations of a pattern is considerably narrowed down by knowledge of the circumstances.

Apart from *1221*, other interesting patterns of four figures are *1211*, *1212*, *1231*, *1122*, *1112*, *1111*. While for the first three patterns instantiations exist like lull(aby), (r)emem(ber), (b)eave(r), it is hard to find natural ones for the remaining patterns. Note that a pattern like *123245678* means that the eight characters involved are different, otherwise the pattern should better be written $*232*****$ and would not express more than the pattern *121* does. A rather large pattern with more than one repeated figure like *12134253* normally has very few or zero instantiations, in this case the words pipeline, piperine, pipelike.

The conclusion is clear: Words and phrases that form a conspicuous pattern should be eliminated by the encryption clerk, usually by paraphrasing, as was regularly done for the British Admiralty's traffic. A notorious example is *1234135426*, which allows in German nothing but the ominous instantiation *heilhitler*. Who would have dared in the *Reich* of *Hitler* to eliminate this stereotyped ending? Kerckhoffs even pointed out that repetitions like the French *pouvez-vous vous défendre* should be avoided. But in clear contrast to this, it was common practice in military signal units to put emphasis on a group by repeating it, like `OKMMMANAN` (Sect. 9.2.5) in German signals. The Allies did the same: `SC48SC48` in a signal to the Allied Convoy SC.48 (Beesley) or `CHICKEN-WIRE£CHICKEN-WIRE` and `HUDDLE-TIME£HUDDLE-TIME` in a message from BP to operational units, transmitting the decryption of a German signal concerning American passwords and replies (Lewin).

The number of patterns with n elements equals the number of partitions of n into a sum of natural numbers, the Bell number $B(n)$, which grows rather fast with n, as the following table shows:

n	0	1	2	3	4	5	6	7	8	9	10	11	12
$B(n)$	1	1	2	5	15	52	203	877	4 140	21 147	115 975	678 570	4 213 597

13.2 Exclusion of Encryption Methods

Theorem 1 can be used negatively to exclude monoalphabetic, functional simple substitutions—namely if the cryptotext contains no more patterns than a random text. But caution is advised. For example, the lack of doubled characters does not mean much. Since the work of G. B. and M. Argenti, professional cryptographers have known the rule of impeding pattern finding by suppressing doubling of characters even in the plaintext, e.g., writing *sigilo* instead of *sigillo*. Also the classical suppression of the word spacing is meant to diminish the formation of patterns (Sect. 13.6.1). Suppression is a polyphonic step; in rare cases, suppression of the word spacing leads to a violation of injectivity: *the messages that were translated* – *the messages that we retranslated*, or *we came together* – *we came to get her.*

13.3 Pattern Finding

Theorem 1 can be used positively, if there are reasons to assume that a monoalphabetic, functional simple substitution is present.

13.3.1 An example. In the following example by Helen Fouché Gaines (spaces, denoted by ␣ , are not suppressed)

F D R J N U ␣ H V X X U ␣ R D ␣ M D ␣ S K V S O ␣ P J R K ␣ Z D
Y F Z J X ␣ G S R R V T ␣ Q Y R ␣ W D A R W D F V ␣ R K V ␣ D R
K V T ␣ D F ␣ S Z Z D Y F R ␣ D N ␣ N V O V T S X ␣ S A W V Z R

the guess is at simple substitution. Helen Fouché Gaines starts by noticing words with the pattern *1231* , i.e., the two-letter words ␣RD␣ , ␣MD␣ , ␣DF␣ , and ␣DN␣; the occurrence of D in each one suggests trying the instantiations /of/, /on/, /or/, /do/, /go/, /no/, /to/; i.e., the entry D $\widehat{=}$ o. And there are two occurrences of the pattern *12341* , i.e., the three-letter words ␣RKV␣ and ␣QYR␣ have R in common, which also occurs in ␣RD␣ . Among the 3-letter plaintext words that start with /d/, /g/, /n/, or /t/, /the/ is reasonable. Assuming RKV $\widehat{=}$ the , ␣DRKVT␣ becomes ␣otheT␣ and T $\widehat{=}$ r would be almost certain. Thus, five letters are tentatively known and the partial decryption reads

F o t J N U ␣ H e X X U ␣ t o ␣ M o ␣ S h e S O ␣ P J t h ␣ Z o
Y F Z J X ␣ G S t t e r ␣ Q Y t ␣ W o A t W o F e ␣ t h e ␣ o t
h e r ␣ o F ␣ S Z Z o Y F t ␣ o N ␣ N e O e r S X ␣ S A W e Z t

Confirmation comes from GSRRVT turning into GStter . For the other three-letter word QYt /not/ , /got/ , /out/ , /yet/ are disqualified, since /e/ and /o/ are already determined, /but/ would do it. Furthermore there are only the possibilities DF $\widehat{=}$ on and DN $\widehat{=}$ of (or swapped) left, since /r/ is already determined. In the first (happy) case there is now the following fragment

n o t J f U ␣ H e X X U ␣ t o ␣ M o ␣ S h e S O ␣ P J t h ␣ Z o
u n Z J X ␣ G S t t e r ␣ b u t ␣ W o A t W o n e ␣ t h e ␣ o t
h e r ␣ o n ␣ S Z Z o u n t ␣ o f ␣ f e O e r S X ␣ S A W e Z t

Now, SZZount is read as /account/ and this leads also to a solution for ZounZJX , namely councJX as /council/ . The following fragment is obtained:

n o t i f U ␣ H e l l U ␣ t o ␣ M o ␣ a h e a O ␣ P i t h ␣ c o
u n c i l ␣ G a t t e r ␣ b u t ␣ W o A t W o n e ␣ t h e ␣ o t
h e r ␣ o n ␣ a c c o u n t ␣ o f ␣ f e O e r a l ␣ S A W e c t

This can be read in plaintext at first sight, maybe apart from /Helly/, which could be a proper name.

The phases of decryption shown here could be called 'pace' (until after the entry three to five characters are tentatively found), 'trot' (until about eight to ten characters are found and there is no doubt any more) and 'gallop' (the remaining work). This is reflected in the build-up of the decryption table:

A	B	C	D	E	F	G	H	I	J	K	L	M	N	O	P	Q	R	S	T	U	V	W	X	Y	Z
			o							h							t		r		e				
					n								f			b								u	
									i									a					l		c
s						m						g		d	w					y		p			

Only H is still unclear, as B, C, E, I, L do not occur in the cryptotext. At this last phase one should try to reconstruct the full decryption table. The reader may have noticed that N and F were standing for each other; oF, oN becoming /on/, /of/. The same is seemingly true for A and S, D and O, P and W, R and T, U and Y. If the encryption were self-reciprocal, then K $\hat{=}$ h would imply H $\hat{=}$ k, and Helly would be plaintext /kelly/. The whole encryption table would be

A	B	C	D	E	F	G	H	I	J	K	L	M	N	O	P	Q	R	S	T	U	V	W	X	Y	Z
$\underline{s}$	$\underline{q}$	z	$\underline{o}$	v	$\underline{n}$	$\underline{m}$	$\underline{k}$	$\underline{j}$	$\underline{i}$	$\underline{h}$	x	$\underline{g}$	$\underline{f}$	$\underline{d}$	w	$\underline{b}$	t	$\underline{a}$	r	y	e	p	l	u	c

Since the underlined letters of the lower row run backwards, there is presumably a construction of the substitution alphabet from a memorable password. In fact, reordering produces the involutory substitution[1]

$$\updownarrow \begin{array}{cccccccccccc} c & u & l & p & e & r & a & b & d & f & g & h & i \\ z & y & x & w & v & t & s & q & o & n & m & k & j \end{array} .$$

A district attorney could base an indictment on this absolutely plausible decryption revealing the system completely. A "systematic and exact reconstruction of the encryption method and of the passwords and keys used" (Hans Rohrbach 1946) is required if cryptanalysts are witnesses for the prosecution, like Bazeries in 1898 in the lawsuit against the Duke of Orléans, or Elizebeth Friedman, the wife of W. F. Friedman, in a trial against the Consolidated Exporters Company, a smuggling organization at the time of prohibition.

13.3.2 Aristocrats. It should be clear that the decryption of the example above was this easy because word spacings have not been suppressed, contrary to professional tradition. Not to suppress spaces is among the rules of the game 'Cryptos' found in American newspapers (Fig. 87), at least for the

---

[1] Samuel Woodhull and Robert Townsend provided in 1779 General Washington with valuable informations from New York that was occupied by English troops; they used as cover names CULPER SR. und CULPER JR., (Sect. 4.4.1). Was this shortened from *CULPEPER*, which is sometimes used in the cryptographic literature for the construction of keys? Edmund Culpeper, 1660–1738, was a famous English instrument maker.

Cryptoquip

K I O S P X F I E V B O S F E F P M H

Y I M K K J X F I E V B J F K Y F I -

K O M H

Yesterday's Cryptoquip— GLUM GOLFER TODAY STUDIES SNOWMEN ON FAIRWAY.

Today's Cryptoquip clue: S equals C

The Cryptoquip is a simple substitution cypher in which each letter used stands for another. If you think that X equals O, it will equal O throughout the puzzle. Single letters, short words, and words using an apostrophe can give you clues to locating vowels. Solution is accomplished by trial and error.

Cryptoquip

K R K K R L H P L R U I O Z G K A Y M -

M G O R A U L Y P Q , Q R U A H U L Z I U

Yesterday's Cryptoquip— TRICK HARMONICA MAKES PRETTY HARMONY AT PARTIES.

Today's Cryptoquip clue: I equals M

The Cryptoquip is a simple substitution cypher in which each letter used stands for another. If you think that X equals O, it will equal O throughout the puzzle. Single letters, short words, and words using an apostrophe can give you clues to locating vowels. Solution is accomplished by trial and error.

Fig. 87. Cryptoquips from Los Angeles Times, 1977

sort that goes under the name 'aristocrats': Spaces and punctuation marks remain strictly untouched, only letters are allowed in the cryptotext character vocabulary, and no letter may represent itself. The length of the cryptotext in genuine aristocrats (without 'clues') is 75–100 characters, i.e., rather long in view of a unicity distance of ≈ 25 for a simple substitution with permuted alphabet; in return for this the cryptotext may contain the most extraordinary and queer American words (but no foreign words) and apart from being formally grammatically correct, does not need to make sense; to understand it may be as difficult as to understand the cryptogram itself. Words from biology like *pterodactyl*, *ichtyomancy*, and *syzygy* may occur, but also *yclept*, *crwth*, and *cwm* may be found. The text can be chosen such that the normal frequencies of letters and phrases are completely faked, which means that the methods based on frequency analysis, to be discussed in following chapters, are useless ("the encipherer's full attention has been given to manipulation of letter characteristics", H. F. Gaines).

Kahn gives the solution of a cryptogram of the sort 'aristocrat' that describes itself: *Tough cryptos contain traps snaring unwary solvers: abnormal frequencies, consonantal combinations unthinkable, terminals freakish, quaint twisters like 'myrrh'.*

13.3.3 Lipograms. There are texts (lipograms) written totally without /e/ ; most famous is the (artistically unpretentious) novel *Gadsby* (Fig. 88) by Ernest Vincent Wright (Wetzel Publishing Co., Los Angeles 1939, 287 pp.). Wright wrote in the preface that he had fixed the /e/ key on his typewriter, because now and then an /e/ wanted to slip into the manuscript.

Along this line, but with higher pretension, was also Georges Perec (1936–1982) with his 1969 novel *La disparation* (English translation *A Void* by Gilbert Adair, HarperCollins 1995, 285 pp.). Perec, who also played with acronyms, acrostics, anagrams, and palindromes and indulged in linguistic

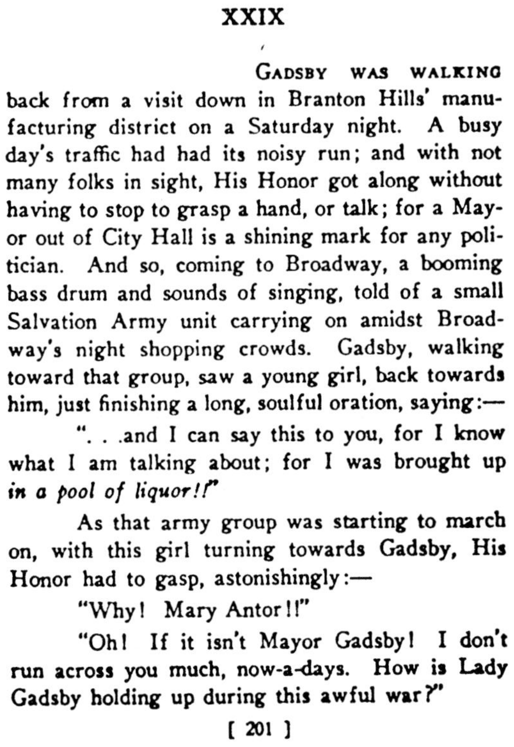

XXIX

GADSBY WAS WALKING back from a visit down in Branton Hills' manufacturing district on a Saturday night. A busy day's traffic had had its noisy run; and with not many folks in sight, His Honor got along without having to stop to grasp a hand, or talk; for a Mayor out of City Hall is a shining mark for any politician. And so, coming to Broadway, a booming bass drum and sounds of singing, told of a small Salvation Army unit carrying on amidst Broadway's night shopping crowds. Gadsby, walking toward that group, saw a young girl, back towards him, just finishing a long, soulful oration, saying:—

". . .and I can say this to you, for I know what I am talking about; for I was brought up *in a pool of liquor!!*"

As that army group was starting to march on, with this girl turning towards Gadsby, His Honor had to gasp, astonishingly:—

"Why! Mary Antor!!"

"Oh! If it isn't Mayor Gadsby! I don't run across you much, now-a-days. How is Lady Gadsby holding up during this awful war?"

[201]

Fig. 88. Page from *Gadsby* by Ernest Vincent Wright

activism—he used computer programs and presented in 1969 a palindrome of 5 000 letters— published a history of lipograms in 1973. Earlier, in 1820, a Dr. Franz Rittler in Vienna published the novel *Die Zwillinge*, written totally without /r/ . Even earlier, in 1800, Gavrila Romanovich Dershavin, an important Russian poet (1743–1816), wrote the novel *A Waggish Wish* completely without /r/ and with only very few /o/ .

James Joyce, too, wrote cryptic prose. The last words in *Finnegans Wake*:

End here. Us then. Finn, again!
Take. Bussoftlee, mememormee!
Till thousendsthee. The keys to. Given!
Lps. A way a lone a last a loved a long the.

if given to a decryptor, would cause him great trouble. Joyce's earlier novel *Ulysses* already contained plenty of cryptological puzzles.

Cryptologically, these curiosities have little importance, of course, any more than the cryptological decorations Vladimir Nabokov included in his works. Vaclav Havel took the right point of view when he made fun of the Marxist-Leninist party (secret) language, the *Ptydepe* and its bureaucratic successor, the *Chorukor*.

13.4 Finding of Polygraphic Patterns

One of the reasons for the use of codes is the suppression of conspicuous patterns. But wrongly designed codes that do not pay regard to frequently used longer phrases, and particularly wrong use of codes, will ruin this.

13.4.1 Luigi Sacco, who became in 1916, at age 32, chief of the *Reparto crittografico* of the Italian headquarter at the Isonzo and Piave front in northern Italy, received on June 30, 1918 two radio signals with the same ending

....4 92073 06583 47295 89255 07325 58347 29264 .

This was a grave mistake of the Austrians; but even worse, the fragment 073**5834729 was repeated in the short distance of 18 digits. This gave a clear indication for a 3-figure code ending with

492 073 065 834 729 589 255 073 255 834 729 264 .

Sacco had some experience with Austrian habits and reason to conjecture that carelessly a longer word had been encoded letter by letter. The code group pattern was $1\underline{2}3\underline{45}6\overline{7}\underline{2}\overline{7}\underline{45}8$ and Sacco, an engineer, had the splendid idea to read it `r a d i o s t a t i o n` . The lazy Austrian code clerk had not seen a need to look up the code groups for `r a d i o` and `s t a t i o n` . And if in exceptional cases—e.g., for proper names—letter by letter encoding was unavoidable, then such a disclosure of the code for single letters should not happen at the beginning or the end of the text.

Anyhow, this gave Sacco an entry to decrypt other words encrypted letter by letter and thus to break the whole code. To be sure, the Austrians did cryptanalytically at least as well. In the *Kriegschiffrengruppe* under the command of Colonel Ronge was an Italian section in which Major (later Colonel) Andreas Figl did excellent work, like his colleague Hermann Pokorny in the Russian section—aided by the adversary's stupidity.

13.4.2 But even without Sacco's imagination there would have been an immediate entry using a prefabricated list of patterns and their instantiations. For the pattern *123456727458* of length 12 with 4 repetitions, the instantiation `r a d i o s t a t i o n` is very likely unique, and if not, trial and error with only a few instantiations would give immediate results.

13.5 The Method of the Probable Word

So far, only formal considerations have been used, assuming no more than a guess at the natural language underlying the crypt. We now take other information into account. Much beloved for an entry into a monoalphabetic simple substitution is the method of the probable word (French *mot probable*, German *wahrscheinliches Wort*). It is not a conspicuous pattern in the text that we seek (and look later for its instantiations); instead, a search for the pattern of the probable word is made in the cryptotext—whether, and if so, where it may occur. Each such cryptotext fragment together with the probable word forms a 'crib'.

13.5.1 Cribs. This method was already described by Giovanni Battista Porta (1535–1615) in *De furtivis*, 1563. If according to the circumstances *division* is a probable word, a search for the pattern *12131* of (d)ivisi(on) is indicated. From Fig. 95 it can be seen that in the military genre the danger of finding a wrong word with this pattern is rather small, although the word is short. Instead of a word, whole phrases can be used like 'Oberkommando der Wehrmacht' or 'Combined Chiefs of Staff'. Particularly suited are stereotyped expressions frequently used at the beginning and ending of plaintexts in both the commercial and the military world. Examples like

reference to your letter

Hochachtungsvoll Ihr

An SS-Gruppenführer Generalleutnant der Waffen-SS Berger, Berlin W.35, SS-Hauptamt, mit der Bitte um absprachegemässe Weitergabe

From Algeria to Washington, 21. 7. To the State Department in Washington. Strictly confidential. Most urgent and personal for Deputy Under State Secretary. From Murphy

show that there are usually enough cribs. Even Russian copulation, the arbitrary cutting of the text and recombining it in the wrong order, is of no help, for it does not remove patterns at all. Moreover, insight into the situation of the adversary and empathy can initiate a chain reaction that Jack Good has described well as "success leading to more success." More in Sect. 19.7. And if no cribs turn up, they can be provoked: In the sequel of certain war actions, words like *attack* or *bombardment* are to be expected.

13.5.2 Murphy and Jäger. Immortal credit for a success on the German side in the Second World War was gained by the American diplomat and later Deputy Secretary of State Robert Daniel Murphy (1894–1978), who insisted on underlining his importance in his telegrams by always using the expressions 'From Murphy' or 'For Murphy'. Nevertheless, Lieutenant Jäger, also mentioned in Sect. 4.4, stole the show. Obedience is no substitute for discipline, which requires brains and is therefore rare. To report regularly 'Nothing to report' is contradictory.

13.5.3 Führerbefehl. The following fictitious example (Fig. 89) by U. Kratzer, based on an infamous *Führerbefehl* in the year 1939, shows how far a single probable word can carry. According to the circumstances it could be guessed that the year '1939' occurs in the plaintext and in view of the bombastic style Hitler's generals had adopted, it could not even be excluded that despite all precautions '**neun**zehnhundert**neun**unddreissig' would occur literally. This would suggest a search for the pattern *1231* of '**neun**', although it is very short and many mishits ('blind hits') were to be expected.

Indeed, this pattern occurs a few times (Fig. 90), in particular as **HQGH** four times, as **QHXQ** twice in the third line from below and once in the second line from below. For '**neun**zehnhundert**neun**', the second occurrence in the

J H K H L P H N R P P D Q G R V D F K H Z H L V X Q J Q U H
L Q V I X H U G L H N U L H J V I X H K U X Q J Q D F K G H
P D O O H S R O L W L V F K H Q P R H J O L F K N H L W H Q
H U V F K R H S I W V L Q G X P D X I I U L H G O L F K H P
Z H J H H L Q H I X H U G H X W V F K O D Q G X Q H U W U D
H J O L F K H O D J H D Q V H L Q H U R V W J U H Q C H C X
E H V H L W L J H Q K D E H L F K P L F K C X U J H Z D O W
V D P H Q O R H V X Q J H Q W V F K O R V V H Q G H U D Q J
U L I I D X I S R O H Q L V W Q D F K G H Q I X H U G H Q I
D O O Z H L V V J H W U R I I H Q H Q Y R U E H U H L W X Q
J H Q C X I X H K U H Q P L W G H Q D E D H Q G H U X Q J H
Q G L H V L F K E H L P K H H U G X U F K G H Q L Q C Z L V
F K H Q I D V W Y R O O H Q G H W H Q D X I P D U V F K H U
J H E H Q D X I J D E H Q Y H U W H L O X Q J X Q G R S H U
D W L R Q V C L H O E O H L E H Q X Q Y H U D H Q G H U W D
Q J U L I I V W D J H U V W H U Q H X Q W H U Q H X Q C H K
Q K X Q G H U W Q H X Q X Q G G U H L C L J D Q J U L I I V
C H L W Y L H U X K U I X H Q I X Q G Y L H U C L J

Fig. 89. Fictitious encryption of a *Führerbefehl* in the year 1939

J H K **H L P H** N R P P D Q G R V D F K H Z H L V X Q J Q U H
L Q V I X H U G L H N U L H J V I X H K U X Q J Q D F K G H
P D O O H S R O L W L V F K H Q P R H J O L F K N **H L W H** Q
H U V F K R H S I W V L Q G **X P D X** I I U L H G O L F K **H P**
Z H J H **H L Q H I X H U G H** X W V F K O D **Q G X Q** H U W U D
H J O L F K H O **D J H D** Q V **H L Q H** U R V W J U **H Q C H** C X
E H V H L W L J H Q K D E H L F K P L F K C X U J H Z D O W
V D P H Q O R H V X **Q J H Q** W V F K O R V V **H Q G H** U D Q J
U L I **I D X I** S R O H Q L V W Q D F K G H Q I X **H U G H** Q I
D O O Z H L V V J H W U R I I H Q H Q Y R **U E H U** H L W X **Q**
J H Q C X I X **H K U H** Q P L W G H Q D E D **H Q G H** U X Q J H
Q G **L H V L** F K E H L P K H H **U G X U** F K G H Q L Q C Z L V
F K H Q I D V W Y R O O **H Q G H** W H Q D X I P D U V F K **H U**
J H E H Q D X I J D E **H Q Y H U W H** L O X **Q J X Q** G R S H U
D W L R Q V C L H O E O **H L E H** Q X Q Y **H U D H Q G H** U W D
Q J U L I I V W D J H U V W H U **Q H X Q** W H U **Q H X Q** C H K
Q K X Q G H U W **Q H X Q** X Q G G U H L C L J D Q J U L I I V
C H L W Y L H U X K U I X H Q I X Q G Y L H U C L J

Fig. 90. Occurrences of the pattern *1231*

third line from below has just the right distance from the occurrence in the second line from below. This entry, provoked by the repetition of **QHXQ**, provides a first tentative decryption in Fig. 91. Obviously, there are more dates at the end of the text. It does not need much imagination to read the

```
J e h e L P e N R P P D n d R V D F h e Z e L V u n J n r e
L n V I u e r d L e N r L e J V I u e h r u n J n D F h d e
P D O O e S R O L t L V F h e n P R e J O L F h N e L t e n
e r V F h R e S I t V L Q d u P D u I I r L e d O L F h e P
Z e J e e L n e I u e r d e u t V F h O D n d u n e r t r D
e J O L F h e O D J e D n V e L n e r R V t J r e n z e z u
E e V e L t L J e n h D E e L F h P L F h z u r J e Z D O t
V D P e n O R e V u n J e n t V F h O R V V e n d e r D n J
r L I I D u I S R O e n L V t n D F h d e n I u e r d e n I
D O O Z e L V V J e t r R I I e n e n Y R r E e r e L t u n
J e n z u I u e h r e n P L t d e n D E D e n d e r u n J e
n d L e V L F h E e L P h e e r d u r F h d e n L n z Z L V
F h e n I D V t Y R O O e n d e t e n D u I P D r V F h e r
J e E e n D u I J D E e n Y e r t e L O u n J u n d R S e r
D t L R n V z L e O E O e L E e n u n Y e r D e n d e r t D
n J r L I I V t D J e r V t e r n e u n t e r n e u n z e h
n h u n d e r t n e u n u n d d r e L z L J D n J r L I I V
z e L t Y L e r u h r I u e n I u n d Y L e r z L J
```

Fig. 91. Fragmentary decryption with the help of ‘neunzehnhundertneun’

```
g e h e i P e N R P P D n d R V D F h e Z e i V u n g n r e
i n V f u e r d i e N r i e g V f u e h r u n g n D F h d e
P D O O e S R O i t i V F h e n P R e g O i F h N e i t e n
e r V F h R e S f t V i Q d u P D u f f r i e d O i F h e P
Z e g e e i n e f u e r d e u t V F h O D n d u n e r t r D
e g O i F h e O D g e D n V e i n e r R V t g r e n z e z u
E e V e i t i g e n h D E e i F h P i F h z u r g e Z D O t
V D P e n O R e V u n g e n t V F h O R V V e n d e r D n g
r i f f D u f S R O e n i V t n D F h d e n f u e r d e n f
D O O Z e i V V g e t r R f f e n e n v R r E e r e i t u n
g e n z u f u e h r e n P i t d e n D E D e n d e r u n g e
n d i e V i F h E e i P h e e r d u r F h d e n i n z Z i s
F h e n f D V t v R O O e n d e t e n D u f P D r V F h e r
g e E e n D u f g D E e n v e r t e i O u n g u n d R S e r
D t i R n V z i e O E O e i E e n u n v e r D e n d e r t D
n g r i f f V t D g e r V t e r n e u n t e r n e u n z e h
n h u n d e r t n e u n u n d d r e i z i g D n g r i f f V
z e i t v i e r u h r f u e n f u n d v i e r z i g
```

Fig. 92. Further fragmentary decryption with the help of ‘vieruhrfuenfundvierzig’

very end as ‘**vi**eruhr**f**uen**f**und**vi**erz**ig**’. This gives the fragmentary decryption of Fig. 92, which means there are already a dozen characters decrypted:

```
. . . d e f g h i . . . . n . . . r . t u v . . . z
. . . G H I J K L . . . . Q . . . U . W X Y . . . C
```

The text in Fig. 92 can be read quite fluently. This results in /m/ for P, /a/ for D, /s/ for V, /o/ for R and /c/ for F and confirms that we are on the right path. Now 17 characters are reconstructed:

a	.	c	d	e	f	g	h	i	.	.	.	m	n	o	.	.	r	s	t	u	v	.	.	.	z
D	.	F	G	H	I	J	K	L	.	.	.	P	Q	R	.	.	U	V	W	X	Y	.	.	.	C

Fig. 93 shows this last intermediate result and Fig. 94 presents the final result, the *Weisung Nr. 1 für die Kriegsführung*.

Here at the latest it turns out that the encryption is a CAESAR addition. If we had assumed this and used an exhaustive search, we would have been sure after a few steps. But who could know it? The example is fictitious, and an order of this significance would not be encrypted by a CAESAR addition—and it would not go by radio, but by courier. However, with enough imagination one could perhaps see Admiral Wilhelm Canaris (1887–1945), the head of the *Abwehr*, the counterespionage organisation of the O.K.W. and conspirator against Hitler, passing on the text so that a simple agent could radio it to Sweden.

g e h e i m e N o m m a n d o s a c h e Z e i s u n g n o e
i n s f u e r d i e N r i e g s f u e h r u n g n a c h d e
m a O O e S o O i t i s c h e n m o e g O i c h N e i t e n
e r s c h o e S f t s i n d u m a u f f r i e d O i c h e m
Z e g e e i n e f u e r d e u t s c h O a n d u n e r t r a
e g O i c h e O a g e a n s e i n e r o s t g r e n z e z u
E e s e i t i g e n h a E e i c h m i c h z u r g e Z a O t
s a m e n O o e s u n g e n t s c h O o s s e n d e r a n g
r i f f a u f S o O e n i s t n a c h d e n f u e r d e n f
a O O Z e i s s g e t r o f f e n e n v o r E e r e i t u n
g e n z u f u e h r e n m i t d e n a E a e n d e r u n g e
n d i e s i c h E e i m h e e r d u r c h d e n i n z Z i s
c h e n f a s t v o O O e n d e t e n a u f m a r s c h e r
g e E e n a u f g a E e n v e r t e i O u n g u n d o S e r
a t i o n s z i e O E O e i E e n u n v e r a e n d e r t a
n g r i f f s t a g e r s t e r n e u n t e r n e u n z e h
n h u n d e r t n e u n u n d d r e i z i g a n g r i f f s
z e i t v i e r u h r f u e n f u n d v i e r z i g

Fig. 93. Last intermediate decryption

13.5.4 Invariance against choice of substitution. The example would have been treated in exactly the same way if any other monoalphabetic substitution were present. This shows that the pattern finding method is totally independent of the kind of (simple) substitution it is up against.

```
g e h e i me k o mma n d o s a c h e we i s u n g n o e
i ns f u e r d i e k r i e g s f u e h r u n g n a c h d e
ma l l e p o l i t i s c h e n mo e g l i c h k e i t e n
e r s c h o e p f t s i n d u ma u f f r i e d l i c h e m
we g e e i n e f u e r d e u t s c h l a n d u n e r t r a
e g l i c h e l a g e a n s e i n e r o s t g r e n z e z u
b e s e i t i g e n h a b e i c h mi c h z u r g e wa l t
s a me n l o e s u n g e n t s c h l o s s e n d e r a n g
r i f f a u f p o l e n i s t n a c h d e n f u e r d e n f
a l l we i s s g e t r o f f e n e n v o r b e r e i t u n
g e n z u f u e h r e n mi t d e n a b a e n d e r u n g e
n d i e s i c h b e i mh e e r d u r c h d e n i n z wi s
c h e n f a s t v o l l e n d e t e n a u f ma r s c h e r
g e b e n a u f g a b e n v e r t e i l u n g u n d o p e r
a t i o n s z i e l b l e i b e n u n v e r a e n d e r t a
n g r i f f s t a g e r s t e r n e u n t e r n e u n z e h
n h u n d e r t n e u n u n d d r e i z i g a n g r i f f s
z e i t v i e r u h r f u e n f u n d v i e r z i g
```

Fig. 94. Final decryption: *Weisung Nr. 1 für die Kriegsführung*

13.6 Automatic Exhaustion of the Instantiations of a Pattern

Helen Fouché Gaines points out that prefabricated lists of words with the same pattern can help to solve the most confounded monoalphabetic substitutions.

13.6.1 Listings. It can be safely assumed that the professional cryptanalytic bureaus know this and that they have made practical use of it, at least since computers with large magnetic tape storage became available in about 1955. More recently, by private initiative, tables specifying English instantiations for patterns of up to 12 letters were published 1971, 1972 by Jack Levine, and for patterns of up to 15 letters 1977, 1982, 1983 by Richard V. Andree. The listing is appropriately done in the KWIC ('Key Word in Context') way, printing the left and right context in parenthesis. Figure 95 shows a somewhat multilingual example for the pattern *12131* of (d)ivisi(on). Note that anana(s) and (r)ococo do not belong to the pattern *12131*, but to the pattern *12121*.

Such collections of patterns can be produced mechanically on the basis of a dictionary of the language or languages in question, today even by optical scanning. In this way, however, the word spacing prevents contacts between words; patterns originating from the suppression of the space are not taken into account. Also grammatical endings may be neglected.

It is therefore better to start from a large text base of the genre in question, comprising up to a billion characters—say a newspaper year on a CD.

(m) acada (m)	ebene	(fr)igidi(ty)	
(m) ahara (ni)	(l) edere (inband)	(r)igidi(ty)	(l) oboto (my)
alaba (ma)	(h)egeme(ister)	(n) ihili (sm)	(s)olomo(n)
(m) alaga	(v)eheme(nt)	(b)ikini	(d) oloro (sa)
(c) alama (ry)	(b) elebe (n)	(m) iliti (a)	(g) onoko (ccus)
(p) alata (l)	(b) elege (n)	imiti(eren)	(m) onolo (gue)
(m) alaya	(g) elege (n)	(l)imiti(eren)	(m)onopo(ly)
(t)amara	eleme (nt)	(d) irigi (eren)	(m) onoto (ny)
(p)anama	(t) eleme (try)	(v) isiti (eren)	(t) opolo (gy)
(s) araba (nd)	(h) elene	(c) ivili (an)	(d) oxolo (gy)
(f) arada (y)	(s) elene	(d)ividi(eren)	
(k) araja (n)	(g) elese (n)	(d) ivisi (on)	
(c) arapa (ce)	eleve (n)		(c)umulu(s)
(c) arava (n)	eleve		
(c)atama(ran)	(h) exere (i)		
(c) atara (ct)			
(c) atafa (lque)			(s) tatut

Fig. 95. Two-language KWIC list of words with the pattern *12131* of (d)ivisi(on)

13.6.2 Search for patterns. Computer-aided, interactive work is useful when for a given probable word instantiations of the pattern of this word—to be displayed on the screen—are looked for. Computer help is particularly necessary if no probable word is available and no pattern is given, but rare patterns or repeated patterns[2] in the cryptotext are to be extracted. If some exist (i.e., if the text is long enough), this almost certainly leads to an entry. It goes without saying that subsequent computer-aided fragmentary decryption can be done semi-automatically, with little interactive intervention.

If it is done systematically, this intuition-free method of pure pattern finding can be fully automated; working without semantic assumptions, it is a first example of a ciphertext-only attack ('pure cryptanalysis'). The problem is to keep the search space small and the number of permissible variants low, and thus to reduce the exhaustive element in the method. To this end, several refinements of pattern finding can be applied. One of them uses coupled pattern finding in the following sense.

13.6.3 Coupling of patterns. If two or more patterns are investigated, it frequently happens that some instantiations are mutually exclusive. This reduces the search space. To give an example, in the cryptotext

S E N Z E I S E J P A N O A I A O P A N C A H A O A J
1 2 3 2 1 4 2 (under O A I A O P A) *1 2 1 3 1* (under A H A O A)

the patterns *1232142* and *12131* occur; the instantiation s e m e s t e(r) for *1232142*, mentioned in Sect. 13.1, allows only a few of the instantiations for *12131*, namely those that are compatible with e H e s e ; from the list of Fig. 95 this is only (g)e l e s e(n) . But there is another instantiation for

[2] The search for repeated patterns will be taken up again in Sect. 17.4.

1232142 , namely g e r e g n e(t) ; with this instantiation are compatible only those instantiations for *12131* that are compatible with e H e g e ; which from the list of Fig. 95 are (g) e l e g e (n) and (b) e l e g e (n) . P$\hat{=}$n from the instantiation g e r e g n e(t) collides with J$\hat{=}$ n both from (g) e l e g e (n) and from (b) e l e g e (n) . The attempt aborts.

This shows how two short patterns can be coupled. The background of this consideration is the finding of one unified pattern

S E N Z E I S E J P A N O A I A O P A N C A H A O A J
1 2 3 2 1 4 2 5 6 2 7 2 1 2 8

and its instantiations.

Coming back to the decryption, with s e m e s t e r for O A I A O P A and g e l e s e n for C A H A O A J we have now tentatively found seven letters and the following fragment:

S E r Z E m S E n t e r s e m e s t e r g e l e s e n

For S , E and Z, the choice of plaintext characters has $19 \cdot 18 \cdot 17 = 5814$ possibilities. The search space could be reduced further by investigating the dozen or so instantiations for S E n t e r, each time trying 17 cases of instantiations of Z . These $12{\cdot}17 \approx 200$ computer-aided tests take only a few seconds. One decryption obtained this way is

w i r d i m w i n t e r s e m e s t e r g e l e s e n

The reader who has doubts about this decryption (after all, not everybody is so very familiar with a text in an obscure foreign language) or who thinks that a text much shorter than the unicity distance for monoalphabetic simple encryption may allow more than one decryption will become more confident when following Rohrbach's advice and finding out that the encryption is a CAESAR addition with a key $22 \overset{26}{\simeq} -4$. That should do it. It is to be noted that again in the decryption method no use was made of the peculiarities of a CAESAR addition.

13.6.4 Reduction of the search space. It should be expected that in a monoalphabetically encrypted cryptotext the number of patterns of length, say, up to 15 is proportional to the length of the text. The number of couplings between the patterns, however, grows at least quadratically with the length, such that the restrictions arising from couplings increase rapidly and reduce the search space correspondingly, which means there is a length of text for which the pure pattern finding method regularly succeeds.

13.7 Pangrams

A special case of patterns consists of those containing no repeated characters, especially long patterns of the form *123456789*... $\mathcal{N}$. Necessarily, $\mathcal{N} \leq N$, the cardinality of the plaintext vocabulary. Instantiations of these patterns are called non-pattern words or pangrams.[3]

---

[3] Richard V. Andree, *Nonpattern Words of 3 to 14 Letters*, Raja Press, Norman, Oklahoma 1982.

Andree lists about 6000 non-pattern words each of length 6 and length 7, 4200 of length 8, 2400 of length 9, and 1050 of length 10. There are still several hundred non-pattern words of length 11 such as

'abolishment' 'atmospheric' 'comradeship' 'exculpation' 'filamentous'
'hypogastric' 'nightwalker' 'questionary' 'slotmachine' 'spaceflight'

and some dozen of length 12 such as

'ambidextrous' 'bakingpowder' 'bodysnatcher' 'disreputably'
'housewarming' 'hydrosulfite' 'springbeauty' 'talcumpowder' .

Even some non-pattern words of length 13 are listed: 'bowstringhemp' 'doubleparking' 'doublespacing' 'groupdynamics' 'publicservant'

and one non-pattern word of length 14: 'ambidextrously'.

Note that in these examples word spacing is suppressed. There are, of course, also longer non-pattern sentences. Non-pattern words or sentences of some rather large length $\mathcal{N}$ should be avoided or suppressed in the plaintext, since they at once expose a decryption of $\mathcal{N}$ letters to an exhaustive search in a rather small search space.

Genuine pangrams are sentences containing every letter just once ($\mathcal{N} = N$). In English, genuine pangrams in very free language are possible, for example

cwm, fjord-bank glyphs vext quiz (Dmitri Borgmann),
squdgy fez, blank jimp, crwth vox (Claude E. Shannon),
Zing! Vext cwm fly jabs Kurd qoph (author unknown).

Good approximations are

waltz, nymph, for quick jigs vex bud (28 characters),
jackdaws love my big sphinx of quartz (31 characters),
pack my box with five dozen liquor jugs (32 characters).

In German or French, no genuine pangram is known. Approximations are

sylvia wagt quick den jux bei pforzheim (33 characters),
bayerische jagdwitze von maxl querkopf (34 characters),
zwei boxkaempfer jagen eva quer durch sylt (36 characters).
Qui, flamboyant, guida Zéphire sur ses eaux (35 characters; Guyot, 1772).

Internationally known for many years are the test texts for teletype lines

kaufen sie jede woche vier gute bequeme pelze
the quick brown fox jumps over the lazy dog
voyez le brick geant que j'examine pres du wharf.

The French language is particularly rich in vowel contacts, like in *ouïe*, and therefore suitable for vowel-pangrams, containing every vowel just once. Good examples are *ultraviolet, trouvaille, autrefois, ossuaire, oripeau, ouaille,* and with only six letters *oiseau*.

14 Polyalphabetic Case: Probable Words

14.1 Non-Coincidence Exhaustion of Probable Word Position

Pattern finding, using the positive coincidence of two text patterns, is necessarily restricted to monoalphabetic encryptions. But for a wide class of polyalphabetic encryptions, namely for those with fixpoint-free encryption steps, whose alphabets fulfill the condition "no letter may represent itself," there is no coincidence between plaintext and cryptotext. This allows us to exclude certain positions of a probable word and thus establishes the remaining ones as possible positions. It is a probable word attack by exhausting positions. Exhaustion runs only over the length of the text and is feasible.

The precondition that no letter may represent itself holds more often than one might think at first. It may happen that an encryptor avoids fixpoints with the very best intentions. Monoalphabetic simple substitutions do this regularly, and for 'aristocrats' (Sect. 13.3.2) it is even prescribed. Furthermore, polyalphabetic substitutions using a collection of such alphabets fulfill the condition—irrespective of any *complication illusoire.*

Moreover, all polyalphabetic substitutions with genuinely self-reciprocal alphabets fulfill the condition. This includes in particular methods with PORTA encryption steps (Sect. 7.4.5) (requiring $N = |V|$ even), but not those with BEAUFORT encryption steps (Sect. 7.4.3).

The non-coincidence exhaustion attack usually allows several possible positions of a tentative probable word, which need to be investigated exhaustively. If for a Shannon crypto system (Sect. 9.1.1) the alphabets are known and if the probable word really does occur, this gives an entry, leading to the reconstruction of a part of the key. In case of a key with known construction principle, that's it; in case of a periodic key, large parts of the plaintext are disclosed. In the monoalphabetic case, of course, non-coincidence exhaustion works, too.

For the phrase "*Erloschen ist Leuchttonne*" (Sect. 11.1.3) there are in the following cryptotext under fixpoint-free encryptions, e.g., ENIGMA steps, only two positions possible (all others lead to a crash, marked by boldface):

YOAQUTHNCHWS YTI WHTOJ QMTCF KUS L Z VS MF NGTDUQNYAVH

erlosc**h**enistleuc**ht**tonne
erloscheni**s**tleuchttonne
erloschenis**t**leuchttonne
erlos**ch**enistleuchttonne
erloschenistleucht**t**onne
erloschen**i**stleuch**t**tonne
erloschenis**t**leuchttonne
erlo**s**chenistleuchttonne
erloschenistleu**c**httonne
→ erloschenistleuchttonne
erlosc**h**enistleuchttonne
erloschenis**t**leuchttonne
erloschenistle**u**chttonne
erloschenistleuchtton**n**e
erloschenistleuchtto**n**ne
erl**o**schenistleuchttonne
erloschenist**l**euchttonne
erloscheni**s**tleuchttonne
erlos**c**henistleucht**t**onne
erloschenistleuch**t**ton**n**e
erloschenistleuchtto**n**ne
erloscheni**s**tleuchttonne
→ erloschenistleuchttonne

Quite a number of mechanical crypto systems are susceptible to this exhaustion attack. First of all, there are the strip and cylinder multiplex encryptions (Sect. 7.5.3), whose non-identical alphabets are monocyclic substitutions (the only substitution violating the condition is identity, which is avoided anyhow).

Second, there are machines working 'for facilitation' self-reciprocally. Then every alphabet is self-reciprocal. If 1-cycles are excluded for technical reasons, then the precondition "no letter may represent itself" holds.

While the crypto machines of Boris Hagelin, working with mechanical BEAUFORT steps, did allow 1-cycles, the electrical ENIGMA did not, suffering from a weakness which was a consequence of introducing the reflecting rotor—meant to be a particular refinement. It would be hard to believe that Group IV (Erich Hüttenhain) of the Cipher Branch OKW, watching over the security of their own systems, did not know of *this* possibility for a break, but presumably they underestimated it. In any case, it was an essential piece of good luck for the work of the Polish Biuro Szyfrów and of the British decryptors in Bletchley Park. But the strip cipher CSP-642, used equally carelessly in the U.S. Navy, was also polyalphabetic with multiple mixed alphabets and it was also endangered; the Japanese made use of this after they had seized the strip devices on the Wake and Kiska islands.

Short probable words allow many hits, of course, and thus also many mishits. The total probability of hits in looking for a word of length n is $P_n = (1-1/N)^n \asymp e^{-n/N}$; n should be large enough to make P_n somewhat smaller than the frequency of the probable word. Figure 96 shows some values of P_n for the usual case $N = 26$.

n	P_n [%]	n	P_n [%]
1	96.15	12	62.45
2	92.45	16	53.39
3	88.90	24	39.01
4	85.48	32	28.51
5	82.19	48	15.22
6	79.03	64	8.13
8	73.07	100	1.98
10	67.56	128	0.66

Fig. 96. Total probability of hits for non-coincidence exhaustion ($N=26$)

The following plaintext of length 60 (Konheim) is encrypted by ENIGMA:

```
m a n y o   r g a n i   z a t i o   n s r e l   y o n c o   m p u t e   r s a
G R S U Z   T L D S Z   N K W N E   R D P F B   O V V Q N   O B K Y I   Q N J
```

Non-coincidence exhaustion for the 8-character word /computer/ should show for the first 26 positions about 19 ($26 \cdot 0.7307$) possible positions. In fact, 21 of the positions cannot be excluded and thus there are actually 20 mishits:

```
    G R S U Z T L D S Z N K W N E R D P F B O V V Q N O B K Y I Q N J
    c o m p u t e r
→     c o m p u t e r
→       c o m p u t e r
→         c o m p u t e r
→           c o m p u t e r
→             c o m p u t e r
→               c o m p u t e r
→                 c o m p u t e r
                    c o m p u t e r
→                     c o m p u t e r
→                       c o m p u t e r
→                         c o m p u t e r
→                           c o m p u t e r
→                             c o m p u t e r
                                c o m p u t e r
→                                 c o m p u t e r
→                                   c o m p u t e r
→                                     c o m p u t e r
→                                       c o m p u t e r
                                          c o m p u t e r
→                                           c o m p u t e r
→                                             c o m p u t e r
→                                               c o m p u t e r
→                                                 c o m p u t e r
                                                    c o m p u t e r
→                                                     c o m p u t e r
```

/computer/ is indeed a very short word, /oberkommandoderwehrmacht/ (24 characters) would do much better (Sect. 19.7.1). A count gives for the first 36 positions of this word 14 possible positions (all mishits), compared with the expected $36 \cdot 0.3901 = 14.04$. A probable word of length well over 100 has a good chance to avoid mishits in a text fragment of length 300.

14.2 Binary Non-Coincidence Exhaustion of Probable Word Position

More information about the encryptions of a polyalphabetic cryptosystem may improve the situation with respect to mishits. To give an example, assume polyalphabetic PORTA encryption steps mapping one half of the alphabet into the other half and vice versa (Sect. 7.4.5), say with $V = Z_{26}$

$$\{\text{a b c d} \dots \text{l m}\} \overset{2}{\longleftrightarrow} \{\text{n o p q} \dots \text{y z}\} \ .$$

The binary pattern of a text is obtained if every character is replaced by 0 or 1, depending on whether it is from the first or the second half of the alphabet.

Now assume that the same text as above is somehow PORTA encrypted:

```
m a n y o   r g a n i   z a t i o   n s r e l   y o n c o   m p u t e   r s a
P R G B F   I O Z G P   L Y C N E   E D G W S   A I K Q D   O B K J Q   C M P
```

/computer/ has the binary pattern 01011101 , thus in the cryptotext complete non-coincidences with this pattern are to be looked at:

```
P R G B F I O Z G P L Y C N E E D G W S A I K Q D O B K J Q C M P
1 1 0 0 0 0 1 1 0 1 0 1 0 1 0 0 0 0 1 1 0 0 0 1 0 1 0 0 0 1 0 0 1

0 1 0 1 1 1 0 1
  0 1 0 1 1 1 0 1
    0 1 0 1 1 1 0 1
      0 1 0 1 1 1 0 1
        0 1 0 1 1 1 0 1
          0 1 0 1 1 1 0 1
            0 1 0 1 1 1 0 1
              0 1 0 1 1 1 0 1
                0 1 0 1 1 1 0 1
                  0 1 0 1 1 1 0 1
                    0 1 0 1 1 1 0 1
                      0 1 0 1 1 1 0 1
                        0 1 0 1 1 1 0 1
                          0 1 0 1 1 1 0 1
                            0 1 0 1 1 1 0 1
                              0 1 0 1 1 1 0 1
                                0 1 0 1 1 1 0 1
                                  0 1 0 1 1 1 0 1
                                    0 1 0 1 1 1 0 1
                                      0 1 0 1 1 1 0 1
                                        0 1 0 1 1 1 0 1
                                          0 1 0 1 1 1 0 1
                                            0 1 0 1 1 1 0 1
                                        →     0 1 0 1 1 1 0 1
                                                0 1 0 1 1 1 0 1
                                                  0 1 0 1 1 1 0 1
```

In this example, even with a rather short probable word, the binary coincidence exhaustion has no mishits, while the non-coincidence exhaustion does not exclude 21 of the positions and thus has 20 mishits. Even the shorter word /comp/ with the binary pattern 0101 has only three mishits—compared with the expected $26 \cdot (\frac{1}{2})^4 = 1.62$ possible positions. For PORTA encryptions, probable words of a few characters are in general already sufficient to distinguish between a genuine hit or a total miss.

The word /oberkommandoderwehrmacht/ contains the 11-character word /kommandoder/ with the pattern

01000101001

It cannot be present in the plain text of this example, since every position leads to a crash:

```
P R G B F I O Z G P L Y C N E E D G W S A I K Q D O B K J Q C M P
1 1 0 0 0 0 1 1 0 1 0 1 0 1 0 0 0 0 1 1 0 0 0 1 0 1 0 0 0 1 0 0 1

0 1 0 0 0 1 0 1 0 0 1
  0 1 0 0 0 1 0 1 0 0 1
    0 1 0 0 0 1 0 1 0 0 1
      0 1 0 0 0 1 0 1 0 0 1
        0 1 0 0 0 1 0 1 0 0 1
          0 1 0 0 0 1 0 1 0 0 1
            0 1 0 0 0 1 0 1 0 0 1
              0 1 0 0 0 1 0 1 0 0 1
                0 1 0 0 0 1 0 1 0 0 1
                  0 1 0 0 0 1 0 1 0 0 1
                    0 1 0 0 0 1 0 1 0 0 1
                      0 1 0 0 0 1 0 1 0 0 1
                        0 1 0 0 0 1 0 1 0 0 1
                          0 1 0 0 0 1 0 1 0 0 1
                            0 1 0 0 0 1 0 1 0 0 1
                              0 1 0 0 0 1 0 1 0 0 1
                                0 1 0 0 0 1 0 1 0 0 1
                                  0 1 0 0 0 1 0 1 0 0 1
                                    0 1 0 0 0 1 0 1 0 0 1
                                      0 1 0 0 0 1 0 1 0 0 1
                                        0 1 0 0 0 1 0 1 0 0 1
                                          0 1 0 0 0 1 0 1 0 0 1
                                            0 1 0 0 0 1 0 1 0 0 1
```

14.3 The De Viaris Attack

Even against the strip and cylinder multiplex encryptions, working with monocyclic unrelated alphabets allowing non-coincidence exhaustion according to Sect. 14.1, the unauthorized decryptor can do more. This discovery was made unhappily by Étienne Bazeries, the great practitioner of cryptanalysis, when his allegedly unbreakable device was ridiculed by his opponent De Viaris. Incidentally, the attack of De Viaris in 1893 and Friedman in 1918 does not presuppose that the alphabets are monocyclic.

14.3.1 But even the general method requires that the device is in the hands of the unauthorized decryptor. Since nothing prevents the number of disks or strips from being two dozen or more, there is always a number of permutations big enough to exclude trivial exhaustion. The maximal period is determined by the system and the known alphabets could be tested against columns of monographically encrypted characters (Sect. 18.2.5); but the depth of the material as a rule is not sufficient to succeed with a frequency analysis.

For the special case of the strip and cylinder devices there is the seeming complication by homophony. It will turn out that homophony does not hinder the unauthorized decryptor much more than the authorized one. For the moment, let us assume that the (homophonic) cryptotext fragment is read from the k-th row after the plaintext row (in the k-th generatrix, Sect. 7.5.3). Beginning with small values of k, there would be in the worst case two dozen trial and error cases. We also assume, for simplicity, that a probable text will be short enough not to be cut by the period hiatus of the encryption.

Now, for fixed k and a given probable plaintext word, we determine the set of all characters occurring on the disks or strips in the k-th generatrix. With this basic information we investigate all positions of the probable word to find out for which ones the cryptotext could have been obtained at all. For a short probable word we expect there to be several possibilities to follow up. If the probable word is long enough, it may happen that no possibility is found, in which case transition to another generatrix is indicated. If this is unsuccesful for every generatrix, then the probable word is not present in the plaintext despite our assumption—which may also mean that it was interrupted.

For Bazeries' cylinder with 20 disks and an example of a cryptotext that goes back to the military genre of Givierge,

```
FSAMC RDNFE YHLOE RTXVZ
LRMQU UXRGZ NBOML NDNPV
RTMUK HRDOX LAXOD CREEH
VREXZ GUGLA BSEST VFNGH
```

the De Viaris decryption attack goes as follows:

Let the probable word be /division/. For the 20 cycles of Bazeries (Fig. 56), Fig. 97 (a) displays the encryptions of /division/ for the first generatrix. Thus, the sets of crypto characters that occur are to be read vertically under the plaintext characters /d/, /i/, /v/, ... of /division/.

Sliding a paper strip with the cryptotext along these sets, we can decide for every position of the probable word whether all the corresponding letters in the cryptotext are found among the available ones. For example, this is not the case for the following position, with the fragment FSAMCRDN,

d i v i s i o n
F S A **M C** **R** D N F E Y H L O E R T X V Z

where only the four letters in boldface type (instead of eight) are found. The same is true for the next position, with the fragment SAMCRDNF,

  d i v i s i o n
F S A M **C** **R** D **N F** E Y H L O E R T X V Z

(a)

	d	i	v	i	s	i	o	n
1	E	J	X	J	T	J	P	O
2	**F**	O	X	O	T	O	U	P
3	**F**	O	X	O	T	O	J	P
4	C	H	U	H	R	H	N	M
5	C	Q	T	Q	R	Q	I	M
6	C	E	T	E	R	E	I	M
7	P	J	B	J	E	J	N	S
8	T	E	D	E	P	E	Y	H
9	E	J	X	J	L	J	P	O
10	I	E	X	E	V	E	T	C
11	B	T	I	T	**C**	T	U	D
12	G	C	Y	C	A	C	R	P
13	N	R	Y	R	A	**R**	I	S
14	**F**	B	X	B	V	B	N	E
15	**F**	N	X	N	T	N	P	S
16	K	M	X	**M**	V	M	G	F
17	**F**	E	X	E	O	E	N	A
18	I	U	X	U	T	U	M	Q
19	**F**	J	L	J	U	J	N	T
20	G	J	X	J	T	J	U	O

(b)

	d	i	v	i	s	i	o	n
1	**H**	M	A	M	X	A	S	R
2	J	B	E	B	Z	B	E	S
3	I	**L**	E	L	Z	L	M	Q
4	Z	E	R	**E**	O	E	K	J
5	U	M	**O**	M	Q	M	N	E
6	U	X	Q	X	N	X	Z	J
7	J	V	K	V	D	V	F	T
8	M	U	J	U	D	U	F	X
9	L	N	H	N	I	N	V	U
10	P	R	D	R	Z	R	A	J
11	**H**	S	L	S	**R**	S	N	G
12	K	B	R	B	U	B	Z	**V**
13	U	T	L	T	B	**T**	M	X
14	K	F	H	F	Z	F	R	T
15	K	R	N	R	E	R	**X**	U
16	S	O	J	O	Z	O	R	H
17	J	O	Y	O	B	O	C	D
18	B	C	L	C	U	C	R	Y
19	J	Q	F	Q	M	Q	Z	A
20	J	P	N	P	A	P	E	T

Fig. 97. Encryptions of /division/, (a) 1st generatrix (with **F**SA**MCR**DN), (b) 4th generatrix (with **HLOERTXV**)

where again only the four letters in boldface type (instead of eight) are found. For the next but one position, with the fragment AMCRDNFE,

```
  d i v  i s i o n
F S A M C  R D N F E  Y H L O E  R T X V Z
```

(bold in the second line: M, R, N, E)

there is also no hit. Continuing in this way, the first generatrix can be excluded for all positions of /division/.

Next we turn to another generatrix. In Fig. 97 (b) the encryptions of the word /division/ for the fourth generatrix are displayed. Again the sets of crypto characters that occur are to be read vertically under the plaintext characters /d/, /i/, /v/, ... of /division/. Beginning again from the left, we get a hit for the twelfth position with the fragment HLOERTXV for the first time,

```
                         d i v i  s i o n
F S A M C  R D N F E  Y H L O E  R T X V Z
```

(bold in the second line: H L O E R T X V)

All eight letters (in boldface type) are found among the available ones and seven of them just once, and they determine the corresponding alphabet. However, for **H** there is a choice between the first and the eleventh alphabet, as Fig. 97 shows.

14.3.2 At this moment some additional knowledge about the system can be made use of. In principle, any alphabet, unrelated or accompanying, could be used several times. For VIGENÈRE encryption steps this would be quite normal. Progressive encryption (Sect. 8.4.3) limits this, in order to prevent accumulation of material encrypted with the same alphabet. For cylinder and strip devices progressive encryption is systemic; it seems to increase security to have each cylinder or strip available only once and thus to use it within the period exactly once. But as soon as the alphabets fall into the hands of the foe, this is actually a *complication illusoire*.

Under the assumption of progressive encryption, which holds for Bazeries' cylinder, H $\hat{=}$ d excludes the eleventh alphabet, since this is already needed uniquely for R $\hat{=}$ s. Thus, so far the order of the cylinders is partly determined as follows:

∗ ∗ ∗ ∗ ∗   ∗ ∗ ∗ ∗ ∗   ∗ 1 3 5 4   11 13 15 12 ∗

Furthermore, to exclude the possibility of a mishit, we investigate whether in a distance of 20 characters a meaningful decryption results. For the fragment BOMLNDNV the result

L R M Q U	U X R G Z	N B O M L	N D N P V		
		1 3 5 4	11 13 15 12		
		z h p n	r m y k		24.
		a i n m	a t i n	→	0.
		B O M L	N D N P		**1.**
		c j l k	d n s q		2.
		d k k j	b s t t		3.

shows a convincing decryption /ainmatin/: for this round the first generatrix was used. Furthermore, in a distance of 40 characters for the fragment AXODCREE the result

R T M U K	H R D O X	L A X O D	C R E E H		
		1 3 5 4	11 13 15 12		
		A X O D	C R E E		**22.**
		b z i c	o e z z		23.
		c a q b	u m l l		24.
		d e p a	r t a s	→	0.
		e b n z	a d j a		1.

produces the convincing decryption /departas/: for this round the 22nd generatrix was used.

The plaintext fragment /departas/ can be supplemented in two ways: to /departasixheures/ or to /departaseptheures/. Considering the fact that 6 o'clock would be rather early, we try /departaseptheures/ as next probable word, which is cut into /departase/ and /ptheures/. This last one is treated in Fig. 98; for the third generatrix the encryptions of /ptheures/ are displayed. In the position immediately following, with the fragment VREXZGUG (chances for a hit are 1:206) there is indeed a genuine hit

p t h e u   r e s
V R E X Z   G U G L A   B S E S T   V F N G H

which determines the positions of four more, not yet treated cylinders: R $\hat{=}$ t requires the 7th, X $\hat{=}$ e the 6th, G $\hat{=}$ r the 10th, G $\hat{=}$ s the 9th cylinder. In Fig. 98, the cylinders so far determined are marked. From the remaining ones, Z $\hat{=}$ u requires uniquely the 17th cylinder, while three cases are left open: V $\hat{=}$ p requires the 16th or 20th, E $\hat{=}$ h the 14th or 18th, U $\hat{=}$ e the 2nd or 8th cylinder.

		p	t	h	e	u	r	e	s
•	1	S	X	K	H	Y	U	H	V
	2	S	Z	L	U	C	V	**U**	X
•	3	Q	Z	J	D	R	V	D	X
•	4	M	Q	**E**	B	R	O	B	P
•	5	L	O	D	H	V	Q	H	I
•	6	L	Q	D	**X**	E	N	X	P
•	7	J	**R**	Q	D	B	U	D	T
	8	D	M	X	U	L	S	**U**	V
•	9	**V**	F	Y	L	**Z**	D	L	**G**
•	10	T	A	M	R	O	**G**	R	Y
•	11	Y	S	M	T	N	D	T	U
•	12	**V**	F	N	S	D	Z	S	C
•	13	Y	S	P	D	C	T	D	X
	14	B	I	**E**	T	P	A	T	Y
•	15	X	E	O	A	L	Z	A	U
	16	**V**	B	C	G	D	U	G	Y
	17	U	X	P	O	**Z**	L	O	A
	18	T	U	**E**	S	C	D	S	I
	19	S	A	B	C	M	X	C	Y
	20	**V**	A	K	C	E	Y	C	L

Total probability of hit

$$\frac{12}{25} \times \frac{13}{25} \times \frac{14}{25} \times \frac{13}{25} \times \frac{13}{25} \times \frac{14}{25} \times \frac{13}{25} \times \frac{11}{25}$$

$$\approx 1 : 206$$

$e \mapsto \{A, B, C, D, G, H, L, O, R, S, T, U, X\}$

$h \mapsto \{B, C, D, E, J, K, L, M, N, O, P, Q, X, Y\}$

$p \mapsto \{B, D, J, L, M, Q, S, T, U, V, X, Y\}$

$r \mapsto \{A, D, G, L, N, O, Q, S, T, U, V, X, Y, Z\}$

$s \mapsto \{A, C, G, I, L, P, T, U, V, X, Y\}$

$t \mapsto \{A, B, E, F, I, M, O, Q, R, S, U, X, Z\}$

$u \mapsto \{A, B, E, F, I, M, O, Q, R, S, U, X, Z\}$

Fig. 98. Encryptions of /ptheures/ , 3rd generatrix (with **VREXZGUG**)

The result is the following distribution of 19 of the total of 20 cylinders:

$\substack{16\\20}\ 7\ \substack{14\\18}\ 6\ 17 \quad 10\ \substack{2\\8}\ 9 * * \quad * \ 1\ 3\ 5\ 4 \quad 11\ 13\ 15\ 12 *$

The remaining decryption is a trifling matter: with the 13 cylinders whose situation is determined so far, there is a fragmentary decryption

```
F S A M C   R D N F E   Y H L O E   R T X V Z
* a * r o   i * i * *   * d i v i   s i o n *

L R M Q U   U X R G Z   N B O M L   N D N P V
* p * r t   e * a * *   * a i n m   a t i n *

R T M U K   H R D O X   L A X O D   C R E E H
* r * e i   m * s * *   * d e p a   r t a s e

V R E X Z   G U G L A   B S E S T   V F N G H
p t h e u   r e s * *   * p x x x   x x x x *
```

which immediately suggests two further fragments: /la troisieme/, /demain/, allowing us to fill all but the 20th position, and after that the 20th position, too. The complete order of the cylinders, the password, can be reconstructed:

16 7 18 6 17   10 8 9 20 19   2 1 3 5 4   11 13 15 12 14

The complete decryption is (note the patching nulls at the end):

```
F S A M C   R D N F E   Y H L O E   R T X V Z
l a t r o   i s i e m   e d i v i   s i o n s

L R M Q U   U X R G Z   N B O M L   N D N P V
e p o r t   e r a d e   m a i n m   a t i n s

R T M U K   H R D O X   L A X O D   C R E E H
u r r e i   m s s t o   p d e p a   r t a s e

V R E X Z   G U G L A   B S E S T   V F N G H
p t h e u   r e s s t   o p x x x   x x x x x
```

14.3.3 Even if a probable word is missing, the De Viaris attack may work. Following Givierge (1925), frequent bigrams, trigrams, and tetragrams are used. We shall show this for the French and English standard ending /ation/. For each generatrix, for each plaintext letter, the sets of possible cryptotext letters are prefabricated. Figure 99 shows this for the first generatrix. Since the associated sets comprise only roughly half of the letters, the danger of mishits is again not too great.

	a	t	i	o	n
1	B	U	J	P	O
2	E	V	O	U	P
3	E	V	O	J	P
4	Z	S	H	N	M
5	J	S	Q	I	M
6	Z	S	E	I	M
7	L	D	J	N	S
8	V	O	E	Y	H
9	R	S	J	P	O
10	F	G	E	T	C
11	N	Z	T	U	D
12	I	V	C	R	P
13	U	D	R	I	S
14	I	P	B	N	E
15	J	R	N	P	S
16	I	H	M	G	F
17	B	U	E	N	A
18	B	D	U	M	Q
19	C	E	J	N	T
20	C	E	J	U	O

Total probability of hit

$$\frac{12}{25} \times \frac{11}{25} \times \frac{12}{25} \times \frac{10}{25} \times \frac{12}{25} \approx 1 : 51$$

a $\mapsto$ {B, C, E, F, I, J, L, N, R, U, V, Z}
t $\mapsto$ {D, E, G, H, O, P, R, S, U, V, Z}
i $\mapsto$ {B, C, E, H, J, M, N, O, Q, R, T, U}
o $\mapsto$ {G, I, J, M, N, P, R, T, U, Y}
n $\mapsto$ {A, C, D, E, F, H, M, O, P, Q, S, T}

Fig. 99. Encryptions of /ation/, 1st generatrix

Methodically, the attack of De Viaris and Friedman and in particular the variant of Givierge try to find many small islets which can be enlarged into archipelagos, which in turn can be merged into continents, and so on.

14.3.4 As noted above, the general De Viaris attack does not presuppose the alphabets to be monocyclic. We can now see clearly that both binary coincidence exhaustion (Sect. 14.2) and non-coincidence exhaustion (Sect. 14.1) are special cases where the cryptotext character sets associated with a plaintext character are formed systematically. In order to avoid mishits, the smaller the associated crypto character sets are, the better for the decryptor. On the other hand, for this reason the general De Viaris attack breaks down if each of the associated sets is the full cryptotext vocabulary. A crypto system with this defensive property we shall call transitive. Necessarily then the number of alphabets is greater than or equal to N. The Bazeries cylinder of 20 disks violated this condition. The M-138-A from the U.S.A. used 30 strips, and it can be expected that the alphabets were always selected (out of 50 or 100) to give a transitive cryptosystem.

If the number of alphabets equals N, then for a transitive MULTIPLEX cryptosystem the alphabets of N characters each form a Latin square (Sect. 7.5.4). The M-94, with 25 disks, had alphabets (almost) constructed this way

(Table 2). Equivalently, for each pair of plaintext and cryptotext characters the corresponding alphabet is even uniquely determined. But this is just the condition characterizing a Shannon crypto system (Sect. 2.6.4).

14.3.5 The French Marquis Gaëtan Henri Léon de Viaris (gallicized di Lesegno) was born on February 13, 1847 at Cherbourg, son of a captain of artillery. At age 19, De Viaris entered the famous *École Polytechnique*, at age 21 he went to sea; later he became prefect of police and finally infantry officer. His interest in cryptology arose around 1885; he first made his reputation inventing a printing cipher machine. He was, after Babbage, the first to use mathematical relations in cryptology, namely when characterizing linear substitutions in a series of articles in 1888. In 1893 he wrote the cryptanalytic essay *L'art de chiffrer et déchiffrer les dépêches secrètes* which made him famous. In 1898, he also published a commercial code. He died on February 18, 1901.

Marcel Givierge was Major and assistant to Colonel Cartier when he started in 1914, after the outbreak of the First World War, to build up the decryption bureau of the French General Staff. In 1925, when his book *Cours de Cryptographie* was published, he was Colonel; later he was promoted General. Not without pride he remarked that one *mot probable* was worth quintillions of trials.

Under Givierge worked Major Georges-Jean Painvin, a genius of a decryptor who had studied paleontology and after the war became an important tycoon.

14.3.6 One of the few cases in the 20th century of an enlightening and open report on successful professional cryptanalysis is due to the peculiar situation after the end of the Second World War, when the *FIAT Review of German Science* was written. In the series on applied mathematics, Hans Rohrbach reported on cryptology, and this included details of breaking the 'American Strip Cipher O-2', as Rohrbach calls it, a variant of the M-138-A for the diplomatic service of the U.S. State Department, which was accomplished in the German *Sonderdienst Dahlem* of the *Auswärtiges Amt*. In 1979, a quite detailed report, written in the second half of 1945, was published for the first time. It comprised the work of the mathematicians Werner Kunze, Hans Rohrbach, Annelise Hünke, Erika Pannwitz, Hansgeorg Krug, Helmut Grunsky, and Klaus Schultz—not to forget the linguists Hans-Kurt Müller, Asta Friedrichs, Annemarie Schimmel, Joachim Ziegenrücker, and Ottfried Deubner.

The work started in November 1943 with collecting and sorting a rich legacy of cryptotexts, mainly addressed to or sent from the U.S. Embassy in Berne, Switzerland (where Allen W. Dulles, Office of Strategic Services (O.S.S.), Chief of the U.S. espionage network O.S.S.(S.I.) in Europe, was stationed), which showed

(1) frequent parallels, including longer ones, but never longer than 30 characters and frequently of length 15,

(2) frequent parallels between messages of the same day, but never in two messages of different days in the same month,
(3) no parallels between two messages, if one was before and the other after August 1, 1942.

The conclusion was that after 15, sometimes after 30 characters a change in the encryption was made, that the password was changed daily, and that on August 1, 1942 a more fundamental change in the encryption system was made. This allowed the working hypothesis of a polyalphabetic monographic encryption of period 15. The encryption system used was unknown to Rohrbach, but it was known that the U.S. cryptologists had a liking for cylinder and strip ciphers. However, even so, Rohrbach did not have the alphabets. It was therefore not possible to start with a plain De Viaris attack.

Further studies using Hollerith punch card machines showed that
(4) if the messages were broken into blocks ('*Zeilen*') of 15 characters, all repetitions of at least 8 characters appeared vertically in the same columns.
This confirmed the assumption of a polyalphabetic monographic encryption of period 15; moreover from stereotyped repetitions ('From Murphy', 'Strictly Confidential') at the beginning of the messages it could be deduced that no letter could represent itself, and that the same plaintext in the same position would give cryptotexts without coincidences. This all focused the suspicion onto a polyphonic encryption with monocyclic alphabets, as done by a cylinder or strip cipher, and not by a machine. Thus, 15 or 30 alphabets had to be determined for each day.

But the legacy was rich, with a daily average of 15 messages, each with 40 blocks of 15 characters. These blocks had to be grouped in 'families' encrypted with the same set of alphabets, presumably selected from a supply of more than 15, in the same order. Moreover, the blocks had to be grouped in classes according to the generatrix to which they belonged. Once the blocks were coordinated in this way, the cryptanalysts dealt with monoalphabetic encryptions. With massive use of Hollerith punch card machines and of special equipment built by Krug, the coordination proceeded in small steps of forming 'nuclei'; for this task they used the *Chi* test (Chapter 16, Sect. 18.2.5). The most voluminous class ('Class III') finally comprised 3000 blocks, grouped into 25 families of between 60 and 150 blocks. For the reconstruction of the alphabets finally they used the probable word fragments /tion/ and /ation/, supported by the bigram triplets /in/, /an/, /on/ and /in/, /an/, /un/. Rohrbach describes vividly the process of crystallization in this task. After about one year, the undertaking ran under its own power. Class III was first worked out fully and its 2×15 strips determined. Some of these strips occurred in other classes, too, for example 18 in Class I. In the end, Rohrbach's group found out that 50 strips altogether had been used—we know today that this corresponds to the facts. The classical De Viaris method then gave the selection and order of the strips belonging to the daily passwords, of which 40 were identified.

Thus, all messages encrypted with O-2 could be read. Unfortunately for the German side, fortunately for the Anglo-American one, shortly after full use could be made of the results of the break, the State Department changed in mid-1944 to the more modern and secure SIGTOT machines with individual keys, provided by the U.S. Army. By September 1944 the well was dry. Moreover, the efficiency of *Sonderdienst Dahlem* was suffering under Allied bombing. Then, the Russian army came closer and closer to its evacuation site in Silesia. Towards the end, it moved to a castle in Thuringia, and was transported to Marburg when the Western Allies left Thuringia. Meanwhile, the *Forschungsamt des Reichsluftfahrtministeriums*, Göring's eavesdropping agency, was not much better off.

The Japanese tried also to break the CSP-642, but not very successfully. Friedman had armored it against a De Viaris and Givierge attack, and only a non-coincidence exhaustion of probable word positions was feasible. How well the Russians managed is unknown.

14.4 Zig-Zag Exhaustion of Probable Word Position

Some methods use a probable word to reconstruct the key, which is only possible, of course, if plaintext and cryptotext determine the key uniquely. This is trivially so with monoalphabetic encryption, even if polygraphic with a large width. It is also the case with polyalphabetic encryptions obeying the Shannon condition (Sect. 9.1.1). Most prominent representatives are all linear substitutions, where the key—be it periodic or not—can be simply calculated by subtraction provided the alphabetic order is known. This possibility was studied in 1846 by Babbage for VIGENÈRE and BEAUFORT encryptions. Among the non-linear substitutions that obey the Shannon condition, ALBERTI encryptions and PORTA encryptions bring no complications provided the reference alphabets are known.

It is doubly dangerous to use a meaningful keytext in a common language. If for a polyalphabetic Shannon crypto system—periodic or not—the encryption steps are known, then the possible positions of a probable word in the plaintext are those that give reasonable keytext fragments; they can be exhaustively determined.

In this case, however, the role of plaintext and keytext can also be exchanged. In a meaningful keytext there is most likely also a probable word that gives a reasonable plaintext fragment.

As an example, we assume the cryptotext

B A W I S M E W O O P G V R S F I B B T J T W L H W W A H T M J V B

has been encrypted over $\mathbb{Z}_{26}$ with VIGENÈRE steps. As a probable word in the keytext we assume the frequent word *THAT* which holds rank 7 in the frequency list for English. Exhausting the positions of this word in the keytext gives the following fragments of plaintext,

itwp hpiz dbst plml zfed txwv lpov <u>dhow</u> vhpn ... <u>dtha</u> <u>hatt</u> , among which the eighth, the last but one, and the last one look promising.

Guessing now that dhow can be continued dhowever results in the prolonged key fragment *THATCANB*. The last two fragments dtha and hatt overlap and are mutually exclusive, but it will turn out that dtha is the right one.

On the plaintext side, words can be guessed, too; e.g., should leads to

JTIOHJ , *IPUYBB* , *EBESTT*, *QLYKLL* , ... ,

where the third position fits. Thus should and dhowever overlap to form together shouldhowever, corresponding to *EBESTTHATCANB*. Extended to *THEBESTTHATCANBE* , this gives a plaintext extension that fits and reads itshouldhoweverb, which in turn suggests an extension to itshouldhoweverbe and so on.

Such a zig-zag interplay will frequently result in a complete decryption (Friedman 1918); in the given example plaintext and keytext read as follows (compare the quotations in the introduction to Part II):

i t s h o u l d h o w e v e r b e e m p h a s i z e d t h a t c r y
T H E B E S T T H A T C A N B E E X P E C T E D I S T H A T T H E D

Nonperiodic keytext does not prevent zig-zag exhaustion; what matters is that both the keytext and the plaintext have clearly more than 50% redundancy.

14.5 The Method of Isomorphs

Encryptions based on rotated alphabets also suffer from the defect that the alphabets may not form a Latin square. Thus, an avenue of attack is opened.

14.5.1 Knox and Candela. This can be demonstrated by the method of isomorphs for breaking ROTOR encryptions, used already in 1935, if not earlier, by Dillwyn Knox in a break of the Italian commercial ENIGMA and later against Franco in Spain, but described in the open literature only in 1946 by Rosario Candela. For the case of the commercial ENIGMA without plugboard, the method of isomorphs was called *méthode des bâtons* or rodding ('cliques on the rods', 'rodding'); its existence was the main reason for the introduction (as early as 1930!) of the plugboard in the *Wehrmacht* ENIGMA. In the 1938–1939 Spanish civil war, all sides used commercial ENIGMAs—the Italian Navy even in 1941—and the method of isomorphs served British, German, Italian, and Spanish republican cryptanalytic efforts.

Assume the polyalphabetic substitution is of the form (p_i plaintext character, c_i cryptotext character)

$$c_i = p_i \; S_i \; U \; S_i^{-1}$$

with known alphabets S_i, whose order is known, too—the unknown key is the starting index of the sequence and possibly U . With the isomorphic (Sect. 2.6.3) sequences $c_i S_i$ and $p_i S_i$,

$$c_i S_i = p_i S_i \cdot U \; ;$$

i.e., the sequence $(c_i S_i)$ is the monoalphabetic image of $(p_i S_i)$ under U.

In general, two arbitrarily chosen sequences are not isomorphic: the sequences (a l l e ...) and (g a n g ...) are not isomorphs, because, e.g., the pairs (l,a) and (l,n), as well as the pairs (a,g) and (e,g), are contradictory (they 'scritch').

Cryptanalysis needs for a given probable word $(p_i, p_{i+1}, \ldots p_{i+k})$ a suitable initial index i such that the sequences $(p_i S_i, p_{i+1} S_{i+1}, \ldots p_{i+k} S_{i+k},)$ and $(c_i S_i, c_{i+1} S_{i+1}, \ldots c_{i+k} S_{i+k},)$ are isomorphs. Contradictions lead to an exclusion of the index. Among the suitable indexes is certainly the right one, if the probable word occurs; the longer the probable word is, the fewer mishits are to be expected.

The precondition above is fulfilled in the case (b°) of Sect. 7.2.2 with $S_i = \rho^i$. This situation occurs specifically with the commercial machines ENIGMA C and ENIGMA D without plugboard, with 3 rotors and fixed or movable reflectors (Sect. 7.3.2, where the plugboard T is the identity), with

$$S_{(i_1,i_2,i_3)} = \rho^{-i_1} R_N \, \rho^{i_1-i_2} R_M \, \rho^{i_2-i_3} R_L \, \rho^{i_3} \quad , \text{ or}$$
$$S_{(i_1,i_2,i_3,i_4)} = \rho^{-i_1} R_N \, \rho^{i_1-i_2} R_M \, \rho^{i_2-i_3} R_L \, \rho^{i_3-i_4} \quad ,$$

as soon as all rotors used are explored—this is so with a commercial machine anyhow—and their order is known (in the worst case, for a 3-rotor ENIGMA there are six orders of the rotors to be tested). Moreover, U is self-reciprocal in the ENIGMA case; this leads to some further possibilities of 'scritching' and the exclusion of an index, i.e., a rotor position, as well as to a positive confirmation of a suitable index by the appearance of a self-reciprocal substitution. The possibility of mishits is reduced; the involutory character of the ENIGMA helps the decryptor.

Moreover, thanks to the regularity (Sect. 8.4.4) of the ENIGMA rotor movement it is normally not necessary to test all $26^3 = 17\,576$ or $26^4 = 456\,976$ rotor alphabets. It usually suffices to consider only the 26 positions of the fast rotor R_N lying between two steps of the medium rotor R_M. The two other rotors remain fixed for the while and form together with U a pseudo-reflector

$$U'_{(i_2,i_3)} = \rho^{-i_2} R_M \, \rho^{i_2-i_3} R_L \, \rho^{i_3} \cdot U \cdot \rho^{-i_3} R_M \, \rho^{i_3-i_2} R_L \, \rho^{i_2} \quad , \text{ or}$$
$$U'_{(i_2,i_3,i_4)} = \rho^{-i_2} R_M \, \rho^{i_2-i_3} R_L \, \rho^{i_3-i_4} \cdot U \cdot \rho^{i_4-i_3} R_M \, \rho^{i_3-i_2} R_L \, \rho^{i_2} \; ;$$

with $S_i = \rho^{-i} R_N \, \rho^i \; : \; c_i S_i = p_i S_i \cdot U'_{(i_2,i_3)}$ or $c_i S_i = p_i S_i \cdot U'_{(i_2,i_3,i_4)}$.

14.5.2 A strip method. For the practical performance of the method of isomorphs there exists again a strip method, the strips carrying the columns of the rotated alphabets. Following an example by Deavours and Kruh, the following probable word is to be compared with a fragment of the cryptotext:

```
r e c o n n a i s s a n c e
U P Y T E J O J Z E G B O T
```

The test is to be made with the rotor I of the *Wehrmacht* ENIGMA, the columns are given in Sect. 7.3.5. Plaintext word and cryptotext fragment are formed with the strips. The confrontation is character by character, as

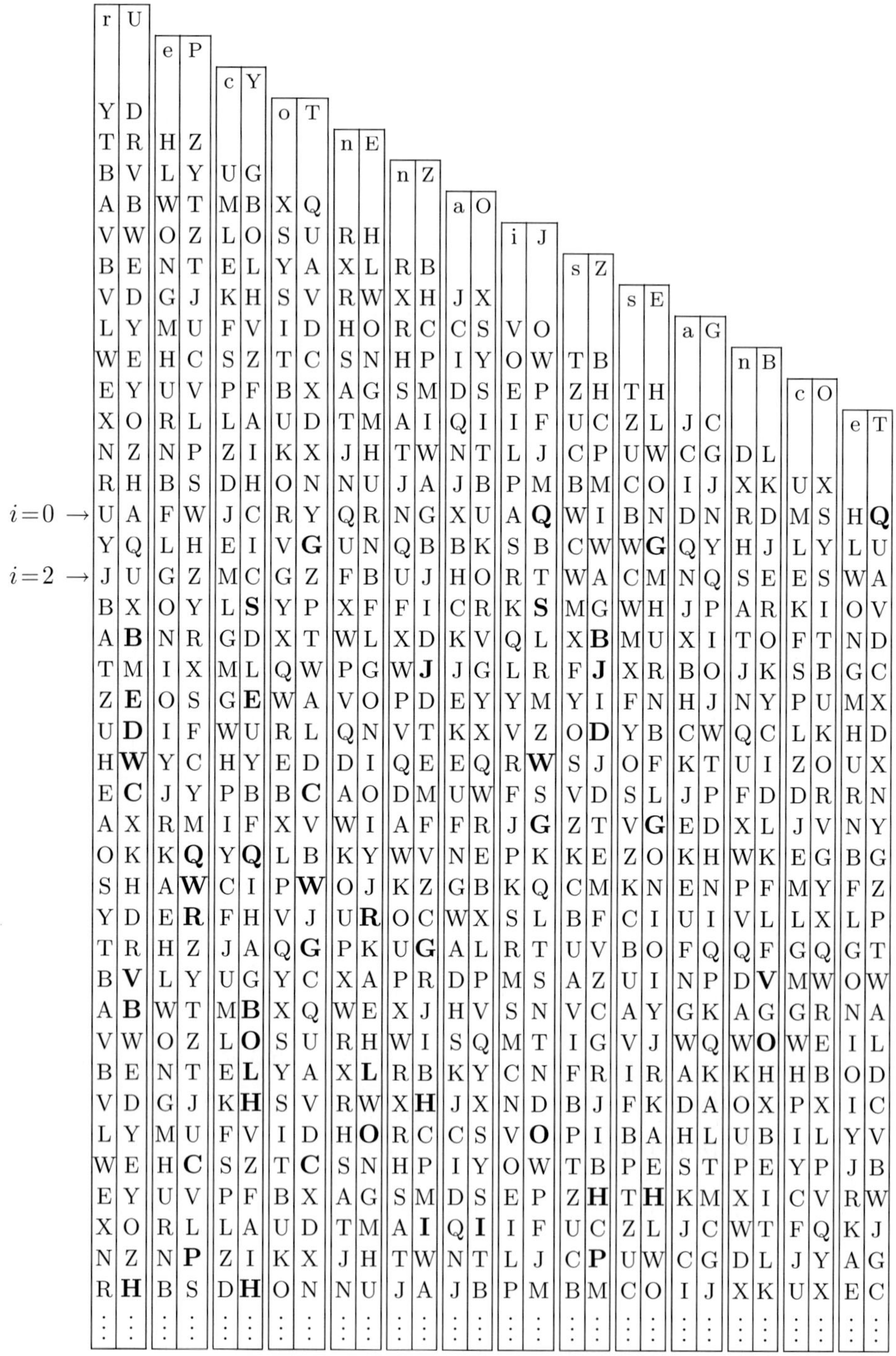

Fig. 100. Method of isomorphs with strips (*méthode des bâtons*)

shown in Fig. 100. In each line, except the one denoted by the rotor position $i = 2$, there are contradictions (one of them is always marked by bold type). For example, in the line $i = 0$, the pairs A Q and H Q as well as B N and D N violate injectivity; the pairs R Y and R D violate uniqueness of the encryption step; and the pairs U A and A Q, F W and W I, Q R and R D, X U and U A, A Q and Q R, B N and N G, D N and N G, H Q and Q R violate the involutory property. On the other hand, in the line $i = 2$, there are the 2-cycles (J U), (M C), (S E) and the involutory property is not violated; this single hit gives the following pair of isomorphs:

j	g	m	g	f	u	h	r	w	c	n	s	e	w
U	Z	C	Z	B	J	O	T	A	M	Q	E	S	A

Thus, rotor I is confirmed as 'fast' rotor R_N. Moreover, the 14 pairs of entry and exit characters define already nine 2-cycles of the pseudo-reflector $U'_{(i_2,i_3)}$ or $U'_{(i_2,i_3,i_4)}$, namely (A W), (B F), (C M), (E S), (G Z), (H O), (J U), (N Q), (R T). The method obviously does not require very long probable words.

From a prefabricated catalogue with $2 \times 26^2 = 1352$ entries of all $U'_{(i_2,i_3)}$ or $2 \times 26^3 = 35152$ entries of all $U'_{(i_2,i_3,i_4)}$ the position and order of the two rotors II and III serving for R_M and R_L can be determined. With such an indicator setting the decryption can be carried through on an ENIGMA replica. The method is characterized as 'meet in the middle'. Switzerland, like other small nations, used ENIGMAs without a plugboard (with changed rotor wiring) during (and partly after) the Second World War (U.S. codename INDIGO). With the help of catalogues, the Germans thus read all their news.

14.5.3 Investigation in parts. The analysis above is based on the assumption that the medium rotor R_M, given the shortness of the probable word, does not move. If it does, however, then on the hiatus the pseudo-reflector is changed, and the investigation decomposes into two parts, without becoming essentially more difficult. In the *Wehrmacht* ENIGMA, there is even the advantage that the position of the notch and thus the ring-setting is disclosed (in the commercial ENIGMA, the notch was fixed to the rotor, and for each rotor the position of the notch was known). If there are two isomorphic texts (c', p') and (c'', p'') before and after the hiatus, then some 2-cycles of the pseudo-reflector $U^{(1)}$ before and some 2-cycles of the pseudo-reflector $U^{(2)}$ after the hiatus are known; this helps to find the position and order of the medium rotor R_M. For an ENIGMA D this reduces the volume of the catalogue to $2 \times 26^2 = 1352$ entries. All this is within easy reach.

An example (Deavours) may illustrate this: We seek to investigate the following pair of (rather long) probable word and a cryptotext fragment:

g	e	n	e	r	a	l	f	e	l	d	m	a	r	s	c	h	a	l	l	k	e	s	s	e	l	r	i	n	g

L	S	H	X	B	T	F	W	U	I	O	V	B	C	A	R	X	S	N	C	V	Z	Y	X	N	J	B	F	W	B

We assume that the investigation of the fragment g e n e r a l at the positions $i=17$ to $i=23$ has given a hit and two isomorphs. Continuation until $i=26$

(but not any further) confirms this and gives

e x o v l y l x r u
M R H F D T D R X G

i.e., the 2-cycles (E M), (R X), (H O), (F V), (D L), (T Y), (G U) as parts of $U^{(1)}$. The remaining text is to be linked up with $i = 1$ to $i = 10$, which yields indeed the two isomorphs

b d a q r w r l j b s p f q c h m o o z
N R W X D A D J L N M Y H X I F S E E T

and the 2-cycles (B N), (D R), (A W), (Q X), (J L), (S M), (P Y), (F H), (C I), (E O), (T Z) as parts of $U^{(2)}$. The two sets have in common the eleven characters D, E, F, H, L, M, O, R, T, X, Y .

For the rotor that is supposed to be the medium rotor the following table of rotated P alphabets is assumed:

i	a	b	c	d	e	f	g	h	i	j	k	l	m	n	o	p	q	r	s	t	u	v	w	x	y	z
0	L	W	F	T	B	A	X	J	D	S	C	K	P	R	Z	Q	Y	O	E	H	U	G	M	I	V	N
1	O	M	X	G	U	C	B	Y	K	E	T	D	L	Q	S	A	R	Z	P	F	I	V	H	N	J	W
2	X	P	N	Y	H	V	D	C	Z	L	F	U	E	M	R	T	B	S	A	Q	G	J	W	I	O	K
3	L	Y	Q	O	Z	I	W	E	D	A	M	G	V	F	N	S	U	C	T	B	R	H	K	X	J	P
:	:	:	:	:	:	:	:	:	:	:	:	:	:	:	:	:	:	:	:	:	:	:	:	:	:	:

For the eleven common characters D, E, F, H, L, M, O, R, T, X, Y there result the following tables of $U^{(1)}$ and $U^{(2)}$ images:

$U^{(1)}$:	a	b	c	d	e	f	g	h	i	j	k	l	m	n	o	p	q	r	s	t	u	v	w	x	y	z
i				l	m	v		o				d	e		h			x		y				r	t	
0				K	P	G		Z				T	B		J			I		V				O	H	
1				D	L	V		S				G	U		Y			N		J				Z	F	
2				U	E	J		R				Y	H		C			I		O				S	Q	
3				G	V	H		N				O	Z		E			X		J				C	B	
:				:	:	:		:				:	:		:			:		:				:	:	

$U^{(2)}$:	a	b	c	d	e	f	g	h	i	j	k	l	m	n	o	p	q	r	s	t	u	v	w	x	y	z
i				r	o	h		f				j	s		e			d		z				q	p	
0				O	Z	J		A				S	E		B			T		N				Y	Q	
1				Z	S	Y		C				E	P		U			G		W				R	A	
2				S	R	C		V				L	A		H			Y		K				B	T	
3				C	N	E		I				A	T		Z			O		P				U	S	
:				:	:	:		:				:	:		:			:		:				:	:	

Comparing now $i = 0$ of $U^{(1)}$ and $i = 1$ of $U^{(2)}$, we find the common letter Z in the line $i = 0$ under h, in the line $i = 1$ under d:

i	a	b	c	d	e	f	g	h	i	j	k	l	m	n	o	p	q	r	s	t	u	v	w	x	y	z
0				K	P	G		**Z**				T	B		J			I		V				O	H	
1				**Z**	S	Y		C				E	P		U			G		W				R	A	

However, in the corresponding cutting from the table of rotated P alphabets

i	a	b	c	d	e	f	g	h	i	j	k	l	m	n	o	p	q	r	s	t	u	v	w	x	y	z
0	L	W	F	T	B	A	X	**J**	D	S	C	K	P	R	Z	Q	Y	O	E	H	U	G	M	I	V	N
1	O	M	X	**G**	U	C	B	Y	K	E	T	D	L	Q	S	A	R	Z	P	F	I	V	H	N	J	W

there is no coincidence. Thus, this rotor position 'scritches'.

Comparing on the other hand $i = 1$ of $U^{(1)}$ and $i = 2$ of $U^{(2)}$, one finds
the letter L in the line $i = 1$ under e, in the line $i = 2$ under l,
the letter V in the line $i = 1$ under f, in the line $i = 2$ under h,
the letter S in the line $i = 1$ under h, in the line $i = 2$ under d,
the letter Y in the line $i = 1$ under o, in the line $i = 2$ under r :

i	a	b	c	d	e	f	g	h	i	j	k	l	m	n	o	p	q	r	s	t	u	v	w	x	y	z
1				D	**L**	**V**		**S**				G	U		**Y**			N		J				Z	F	
2				**S**	R	C		**V**				**L**	A		H			**Y**		K				B	T	

In the corresponding cutting from the table of rotated P alphabets

i	a	b	c	d	e	f	g	h	i	j	k	l	m	n	o	p	q	r	s	t	u	v	w	x	y	z
1	O	M	X	G	**U**	**C**	B	**Y**	K	E	T	D	L	Q	**S**	A	R	Z	P	F	I	V	H	N	J	W
2	X	P	N	**Y**	H	V	D	**C**	Z	L	F	**U**	E	M	R	T	B	**S**	A	Q	G	J	W	I	O	K

there is coincidence in all cases (with **Y** for **S**, **U** for **L**, and so on). Thus, this rotor position is a hit and the right position of the medium rotor is found.

Practically, this determination of the rotor position can be carried through with strips carrying the columns of the rotated alphabets, too.

14.5.4 Pluggable reflector. A variant of the method allows one to determine all 2-cycles of the true reflector U . This became necessary for the Allies, when in early 1944 the Germans from time to time used a 'pluggable' reflector in the *Wehrmacht* ENIGMA (see Sect. 7.3.3). Now, the method of isomorphs is carried through for all $26^3 = 17\,576$ initial indexes, with suitable probable words. This was not feasible manually and required special machines. The relay machine AUTOSCRITCHER (workable 1944) and the electronic machine SUPERSCRITCHER (workable 1946) were built in the U.S.A. by the F Branch of the Army Signal Security Agency under the command of Colonel Leo Rosen.[1]

14.5.5 Opposing the steckering. The plugboard ruins the method of isomorphs, because the unknown plugboard connection ('steckering') veils the probable plaintext word. Now, it is necessary to find repeated pairs of plaintext and corresponding cryptotext characters. Each group of such pairs is mapped under the plugboard substitution into a group of corresponding characters from two isomorphs. Isomorphism requires that the groups are not split when all rotor positions are tested. The machines AUTOSCRITCHER and SUPERSCRITCHER were designed to carry out this task, too. The U.S. Navy's DUENNA and the British GIANT were related.

---

[1] In the open literature, the words *scritch*, *scritchmus* were used without detailed explanation, e.g., by Derek Taunt (1993) when describing the atmosphere of the work at Bletchley Park with reference to the duties of Dennis Babbage, mentioning also the pluggable reflector. The origin of the term is therefore to be sought in Great Britain. David J. Crawford and Philip E. Fox reported in 1992 that they built the AUTOSCRITCHER and SUPERSCRITCHER, but were not informed about the cryptanalytic background. Recent work by Cipher A. Deavours (1995) has established the connection with the method of isomorphs.—*Scritch* is a dialect variant of *screech*.

A manual procedure ('Hand-Duenna') was described in 1944 by C. H. O'D. Alexander in an internal Bletchley Park report ('Fried Reports').

14.6 Covert Plaintext-Cryptotext Compromise

The probable word methods have the aim of recovering the plaintext. A plaintext-cryptotext compromise[2] does not need this, although in fortunate cases it gives the chance to recover the key and thus far more than just one plaintext. Technically, all methods that work for probable words are applicable, and comfortably long words can be chosen.

It can be suspected (or hoped—depending on what side one takes) that direct plaintext-cryptotext compromises are not too frequent. But there are indirect ones, where the plaintext has been obtained by decryption and one is now confronted with a cryptotext obtained by encryption of the same plaintext with another cryptosystem. There are many kinds of negligence and stupidity that can lead to such a situation, which starts out as a harmless-looking cryptotext-cryptotext compromise.

There is an immense number of possible ways that such a compromise can occur. One can be found in the organizational problems of key supply. A radical change in the crypto system cannot always be carried through smoothly and it may happen that a message, still sent in the old key, is repeated in the new key. How serious this danger is can be judged from the proverb: "The risk that a cryptosystem is broken is never greater than at the end of its lifetime." Erich Hüttenhain reported that between 1942 and September 1944 a number of so-called CQ signals ('call to quarters', signals of general interest), sent from the State Department in Washington to its diplomatic outposts, were read by the Germans. The CQ strip sets for the M-138 were identical for all embassies. Thus, once the cipher was broken, a compromise was almost bound to happen when a transition to new strip sets was made.

Moreover, for many practically used methods, as soon as the system is known a plaintext-cryptotext compromise is particularly dangerous because even the key is exposed and thus a deep break into the cryptosystem is possible. Hüttenhain concluded in retrospect (1978) that no encryption method should be used that is susceptible to plaintext-cryptotext compromise („*Es dürfen also keine Chiffrierverfahren verwendet werden, die gegen Klar-Geheim-Kompromisse anfällig sind*"). In combination with Kerckhoffs' maxim, Hüttenhain's maxim excludes many beloved classical cryptosystems; it excludes all those having the Shannon property (see Sect. 2.6.4).

Errors are bound to happen everywhere. Kahn remarked to this "the Germans had no monopoly on cryptographic failure. In this respect the British were just as illogical as the Germans".

---

[2] The verb 'to compromise' means, according to Merriam-Webster, among other things 'to put in jeopardy, to endanger by some act that cannot be recalled, to expose to some mischief'. In cryptology, the use of the word 'compromise' has this particular flavor.

15 Anatomy of Language: Frequencies

We can only say that the decryptment
of any cipher even the simplest will at times
include a number of wonderings.
Helen Fouché Gaines 1939

Decryption as discussed so far, based on patterns, uses the skeleton of the language underlying the plaintext. The decryption strategy to be discussed now uses the internal organs. It aims at the stochastic laws of the language, particularly at character and multigram frequencies. This aspect of cryptography goes back to the arab philosopher al-Kịndī (about 800–870) and was already published by Leone Battista Alberti (*Trattati in cifra*, 1470).

First of all, there is the

> **Invariance Theorem 2**: For all simple transpositions,
> *frequencies of the individual characters in the text are invariant.*

15.1 Exclusion of Encryption Methods

Theorem 2 can be used negatively to exclude transpositions—namely if the cryptotext shows individual character frequencies which are definitely not those of the presumed language of the plaintext. But caution is also advised. For example, data on technical measurements may well have frequencies different from those of usual natural languages.

15.1.1 The cryptotext

```
FDRJN  UHVXX  URDMD  SKVSO  PJRKZ  DYFZJ
XGSRR  VTQYR  WDARW  DFVRK  VDRKV  TDFSZ
ZDYFR  DNNVO  VTSXS  AWVZR
```

shows R, D, V, S as the most frequent characters, while B, C, E, I, L are very rare, indeed they are missing. It cannot be obtained from an English, German, French, or Italian plaintext by transposition. In fact, it is a simple substitution, see Sect. 13.3.1 .

15.1.2 On the other hand, it cannot be excluded that the cryptotext (Sect. 12.5, Table 6)

```
SAEWS  HRCNU  ODKLN  ELIAS  HNCIO  NBNNA
AKIHM  CWNZA  MCGIM  IHEEN  NAUFK  NNCTI
TIHMD  RTEWO  ATAIM  TALKB  UEAFZ  LNUSE
ASDEN  .....
```

with E and T among the most frequent individual characters and with V, P, J, Q, X and Y missing originated from transposition of a German text.

15.1.3 Theorem 2 is also used (logically inadmissibly) in the sense of plausible reasoning: If the frequency distribution of the individual characters is that of a suitable natural language, then *presumably* transposition has happened. The naive argument is: What else—which other procedure would leave invariant the frequency distribution of the individual characters? This may be plausible if one is sure that nobody would have taken the trouble to devise an encryption completely different from transposition, which leaves the frequency distribution invariant, but this is no proof and a judgement could not be based on it. In fact, a homophonic polygraphic substitution, say a code, can easily be made to imitate prescribed character frequencies. W. B. Homan described in 1948 a coding method that gives all characters equal frequencies ('equifrequency cipher'). The same is achieved with straddling by Shannon's and Huffman's redundancy-eliminating 'optimal' coding.

Such tricks, however, will not dupe the professional unauthorized decryptor for long. Nevertheless, in 1892 the great Bazeries, attempting to break a message seized from a group of French anarchists, was delayed for a fortnight because he was misled by six nulls adjoined to the beginning and to the end, and by several rare letters sprinkled into the message. In fact, it was a VIGENÈRE with period 6, otherwise a triviality for Étienne Bazeries. It may have been a mistake to adjoin just as many letters as the period was, but maybe he could not imagine such a stupidity. Part of the deadly message, by the way, read: *La femme et lui sont des mouchards, s'il m'arrive quelque chose, songe à les supprimer* [He and the woman are spies; if anything happens to me, take care of their suppression].

15.2 Invariance of Partitions

Partitions are to this chapter what patterns were to Chapter 13: the abstract vehicle for the invariance of frequencies. A partition is a decomposition of a natural number M into a sum of natural numbers m_i,

$$M = m_1 + m_2 + m_3 + \ldots + m_N .$$

To every text of length M there belongs a partition of M, namely the number of occurrences of the N individual characters in the text with the vocabulary Z_N. Zeros are usually suppressed, thus the text

w i n t e r s e m e s t e r

has the partition 14 = 4+2+2+2+1+1+1+1 .

Therefore, we speak of a partition of the number of characters in the text.

There is a fundamental theorem parallel to Theorem 1 (Sect. 13.1):

> **Invariance Theorem 3**: For all monoalphabetic, functional simple substitutions, especially for all monoalphabetic linear simple substitutions (including CAESAR additions and reversals),
> *partitions of the individual characters in the text are invariant.*

The monoalphabetic encryption of w i n t e r s e m e s t e r by functional simple substitutions, whatever they may be, consists of 4 specimens of some character, 2 specimens of some other character, 2 specimens of some third character and so on. The partition 4+2+2+2+1+1+1+1 is invariant.

Given the encryption

Z L Q W H U V H P H V W H U

and assuming that the unauthorized decryptor knows the frequencies of certain plaintext letters, namely /e/ four times, /t/ twice, /r/ twice, /s/ twice, /n/ once, /i/ once, /w/ once, /m/ once, then he knows that

$$\mathrm{H} \mathrel{\hat{=}} \mathrm{e}\,,\ \{\mathrm{UVW}\} \mathrel{\hat{=}} \{\mathrm{r\ s\ t}\}\,,\ \{\mathrm{LPQZ}\} \mathrel{\hat{=}} \{\mathrm{i\ m\ n\ w}\}\,.$$

and has a polyphonic decryption

$$\begin{array}{cccccccccccccc}
\mathrm{i}&\mathrm{i}&\mathrm{i}&\mathrm{r}&&\mathrm{r}&\mathrm{r}&&\mathrm{i}&&\mathrm{r}&\mathrm{r}&&\mathrm{r}\\
\mathrm{m}&\mathrm{m}&\mathrm{m}&\mathrm{s}&\mathrm{e}&\mathrm{s}&\mathrm{s}&\mathrm{e}&\mathrm{m}&\mathrm{e}&\mathrm{s}&\mathrm{s}&\mathrm{e}&\mathrm{s}\\
\mathrm{n}&\mathrm{n}&\mathrm{n}&\mathrm{t}&&\mathrm{t}&\mathrm{t}&&\mathrm{n}&&\mathrm{t}&\mathrm{t}&&\mathrm{t}\\
\mathrm{w}&\mathrm{w}&\mathrm{w}&&&&&&\mathrm{w}&&&&&
\end{array}$$

In fact, the unauthorized decryptor knows a little bit less, for he knows the frequencies of all the plaintext letters only approximately. According to the inherent rules of a language, each character χ_i appears with a certain probability p_i (a 'stochastic source' Q), such that the frequency $m_i = Q[\chi_i]$ of its occurrence is close to $M \cdot p_i$.

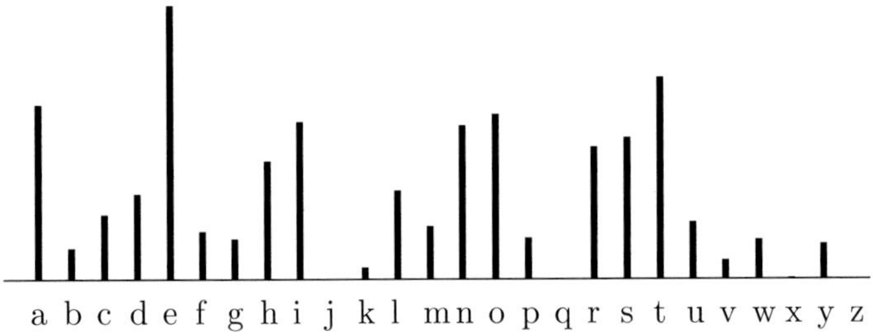

Fig. 101. Frequency profile, English language

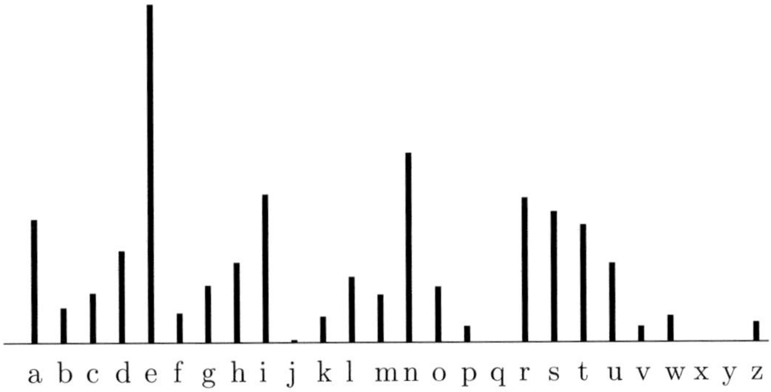

Fig. 102. Frequency profile, German language

15.3 Intuitive Method: Frequency Profile

To arrive at an intuitive method of decrypting monoalphabetic substitutions, it is recommended to visualize the 'frequency profile' of the language under consideration.

In the English language (Fig. 101), the frequency profile shows a marked e-peak and a somewhat smaller a-peak. There is also the marked elevation, of the r-s-t ridge, and two smaller ones, the l-m-n-o ridge and the h-i ridge.

In the German language (Fig. 102), the frequency profile is rather similar, but the e-peak is more marked, there is a wider r-s-t-u ridge, and a wider f-g-h-i ridge. Both languages show a j-k depression and a p-q depression and a very marked v-w-x-y-z lowland.

The discrepancies between any of the major European languages, like French, Italian, or Spanish, are no greater than those between English and German. In the Romance languages, the a-peak is more marked and there is an isolated i-peak. At a glance, they are rather alike.

15.3.1 For a transposition, a frequency count gives a profile close to that of the language under consideration. But also a monoalphabetic linear simple substitution with $q = 1$, a CAESAR addition, is detected at first glance:

> **Invariance Theorem 4:** For all CAESAR additions,
> *the frequency profile of the text is simply cyclically shifted.*

The cryptotext of 349 characters

```
HVZDU  VFKRQ  GXQNH  ODOVL  FKLQE  RQQDQ
NDPLF  KCZDQ  JPLFK  PHLQH  DQNXQ  IWQLF
KWPLW  GHUDX  WRPDW  LNDEO  DXIHQ  CXODV
VHQGL  HVLFK  LQIXH  QIMDH  KULJH  PXQWH
UZHJV  VHLQK  HUDXV  JHELO  GHWKD  WEDKQ
VWHLJ  WUHSS  HUXQW  HUEDK  QVWHL  JWUHS
SHUDX  IUHLV  HWDVF  KHDEV  WHOOH  QIDKU
NDUWH  DXVGH  UPDQW  HOWDV  FKHQH  KPHQU
HLVHW  DVFKH  DXIQH  KPHQI  DKUND  UWHDE
JHEHQ  CXPCH  LWXQJ  VVWDQ  GDEHQ  GCHLW
XQJHQ  NDXIH  QQDFK  GUDXV  VHQJH  KHQXQ
GHLQW  DALKH  UDQZL  QNHQ
```

has the frequency profile in Fig. 103. Obviously, the encryption is a CAESAR addition with a shift by 3. (The decrypted text is from a novel by Heinrich Böll.)

15.3.2 But note that Theorem 4 cannot be reversed: A cyclically shifted frequency profile is compatible with a composition of a transposition and a CAESAR addition.

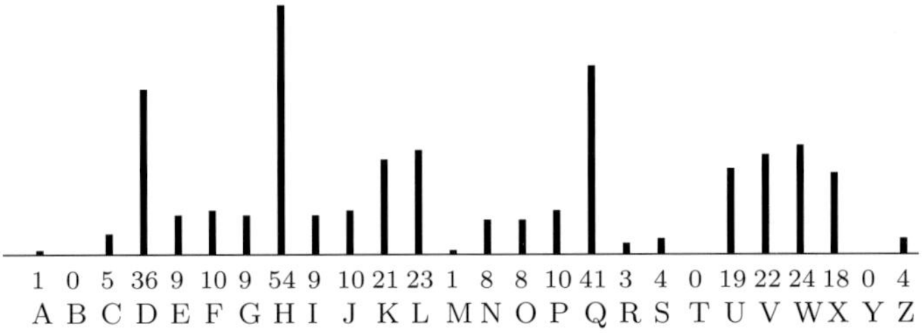

Fig. 103. Frequency profile for the cryptotext *Böll* of Sect. 15.3.1

15.3.3 The intuitive method may be misleading under exceptional circumstances, e.g., with the following cryptotext of 175 characters:

V Q P O U	T K T K B	I K T C B	N H P K O	H U P T I	P X Z P V
I P X B C	V O D I P	G C S K H	I U Z P V	O H G P M	L T E K E
G K O E B	D I B N Q	K P O B N	B O X K U	I C P Z T	B O E H K
S M T P G	I K T P X	O B N B O	P G T P E	P N K O U	K O H B O
E I B Q Q	Z K O E K	W K E V B	M K U Z U	I B U Z P	V U I K T
E S B X O	U P I K N	B T K T B	G M Z U P	B T V H B	S C P X M

The frequency profile (Fig. 104) shows a considerable deviation from that in Figs. 101 or 102. One could think of an exotic language. The suspicion that the encryption is also a CAESAR addition (over Z_{25}) with a shift by 1: "upon this basis i am going to show you ....."—is raised by the strip method of Sect. 12.7, which already gives 'upon' for the first four characters without reasonable doubt. The breakdown of the frequency-oriented intuitive method comes from the fact that the frequencies of the characters are distorted: the text is a lipogram taken from *Gadsby* (Sect. 13.3.2, Fig. 88).

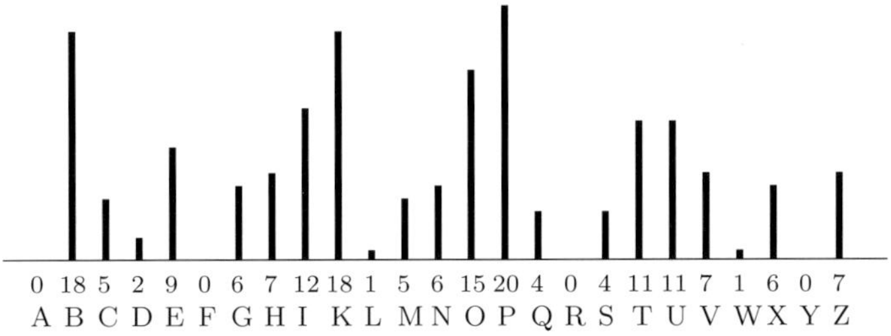

Fig. 104. Frequency profile for the cryptotext of Sect. 15.3.3

15.4 Frequency Ordering

For a monoalphabetic linear simple substitution with $q = -1$ (especially for a reversal) the frequency profile is simply right-left reflected. For values of q different from 1 and -1 and for non-linear simple substitutions, the frequency profile is useless: the letter neighborhoods are torn. Naive intuition uses the frequency ordering in this case: The most frequent character in the cryptotext should correspond to the most frequent letter of the language under consideration. After removing this cryptotext and plaintext character pair, the procedure is repeated until all characters are exhausted and all the encryption steps are established.

15.4.1 Drawbacks of frequency ordering. Theoretically, the method should work, at least for sufficiently long texts—sufficiently long would mean that the few lipograms that may exist would also be submerged in the mass of 'normal' texts. But the example given in Sect. 15.2, leading to a polyphonic situation even if the true frequencies of the plaintext letters are known, shows a fundamental limitation of this procedure: There may be cryptotext letters of the same frequency, and the choice is then non-deterministic.

Moreover, even long texts normally show considerable fluctuations of character frequencies. 'The' frequency distribution of English is a fiction, and at best the military, diplomatic, commercial, or literary sublanguages show some homogeneity; indeed even the same person may speak a different language depending on the circumstances. Correspondingly, statistics on letter frequencies in different languages are quite variable. Moreover, most of the older counts were based on texts of only 10 000 or fewer letters. For the frequency ordering there are already great differences in the literature:

For the English language:

eaoidhnrstuycfglmwbkpqxz	(E. A. Poe 1843)
etaoinshrdlucmfwypvbgkqjxz	(O. Mergenthaler 1884)
etoanirshdlcfumpywgbvkxjqz	(P. Valério 1893)
etaonisrhldcupfmwybgvkqxjz	(H. F. Gaines, O. P. Meaker 1939)
etoanirshdlcwumfygpbvkxqjz	(L. D. Smith 1943)
etoanirshdlufcmpywgbvkxzjq	(L. Sacco 1951)
etaonirshdlucmpfywgbvjkqxz	(D. Kahn 1967)
etaonrishdlfcmugpywbvkxjqz	(A. G. Konheim 1981)
etaoinsrhldcumfpgwybvkxjqz	(C. H. Meyer, S. M. Matyas 1982)

For the French language:

eusranilotdpmcbvghxqfjyzkw	(Ch. Vesin de Romanini 1840)
ensautorilcdvpmqfgbhxyjzkx	(F. W. Kasiski 1863)
esriantouldmcpvfqgxjbhzykw	(A. Kerckhoffs 1883)
easintrulodcpmvqfgbhjxyzkw	(G. de Viaris 1893)
enairstuoldcmpvfbgqhxjyzkw	(P. Valério 1893, M. Givierge 1925)
eaistnrulodmpcvqgbfjhzxykw	(H. F. Gaines 1939)
etainroshdlcfumgpwbyvkqxjz	(Ch. Eyraud 1953)

For the German language:

enrisdutaghlobmfzkcwvjpqxy	(Ch. Vesin de Romanini 1840)
enirsahtudlcgmwfbozkpjvqxy	(F. W. Kasiski 1863)
enirstudahgolbmfzcwkvpjqxy	(E. B. Fleissner von Wostrowitz 1881)
enritsduahlcgozmbwfkvpjqxy	(P. Valério 1893)

enrisatdhulcgmobzwfkvpjyqx	(F. W. Kaeding 1898)
enritsduahlcgozmbwfkvpjyqx	(M. Givierge 1925)
enirstudahgolbmfczwkvpjqxy	(A. Figl 1926)
enirsadtugholbmcfwzkvpjyqx	(H. F. Gaines, J. Arthold 1939)
enristudahglocmbzfwkvpjqxy	(L. D. Smith 1943)
enritsudahlcgozmbwfkvpjqxy	(L. Sacco 1951)
enisrtahduglcofmbwkzvpjyqx	(Ch. Eyraud 1953)
enristdhaulcgmobzwfkvüpäöjyqx	(K. Küpfmüller, H. Zemanek 1954)
enisratduhglcmwobfzkvpjqxy	(W. Jensen 1955)
enirsatdhulgocmbfwkzpvjyxq	(F. L. Bauer 1993)

The count by K. Küpfmüller and H. Zemanek, comprising the modified vowels, is of course cryptologically rather irrelevant, but is given for comparison. Figures for Italian, Spanish, Dutch, and Latin can be found in the book by André Lange and E.-A. Soudart, 1935.

For the first dozen or so letters there exist pretty mnemonic strophes, like

English:	etaoinshrdlu	(LINOTYPE)
French:	esarintulo	(Bazeries, Givierge)
German:	enirstaduhl	(Hüttenhain)
Italian:	eiaorlnts	(Sacco)

The frequency distribution in English was reflected already in the length of the Morse code symbols of telegraphy—Morse counted the letters in the type-case of a printer's shop in Philadelphia and found: 12000 /e/, 9000 /t/, 8000 /a/, /i/, /n/, /o/, /s/, 6400 /h/. For technical reasons, the frequency distribution of letters in English also influenced the arrangement on the keyboard of the type-setting machine LINOTYPE (Fig. 105) of Ottmar Mergenthaler (1854–1899).

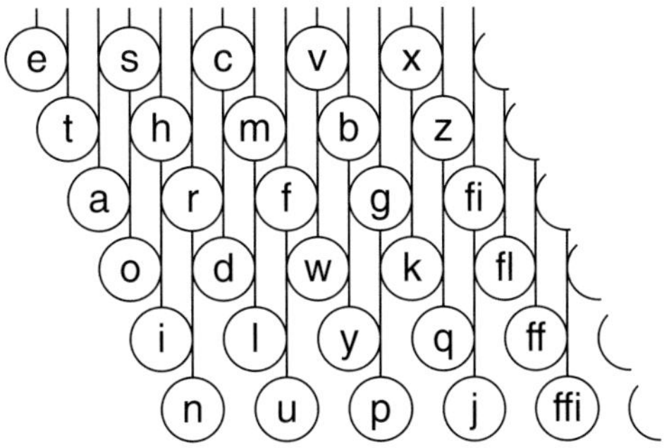

Fig. 105.
Original keyboard
of the LINOTYPE
(Ottmar Mergenthaler 1884)

15.4.2 Frequency counts. For the German language, in 1898 F. W. Kaeding made an extensive frequency count. For the purpose of stenography he studied texts comprising altogether 20 million syllables and thus had 62 069 452 letters (with ä, ö, ü replaced by ae, oe, ue). We can presume that this count is large enough to avoid bias.

The frequency ordering mentioned above is based on this count. If we confronted it with the frequency ordering for the cryptotext *Böll* in Sect. 15.3.1 (Fig. 103), putting letters with the same frequency in alphabetic order, we obtain the decryption table

54	41	36	24	23	22	21	19	18	10	10	10	9	9	9	8	8	5	4	4	3	1	1	0	0	0
H	Q	D	W	L	V	K	U	X	F	J	P	E	G	I	N	O	C	S	Z	R	A	M	B	T	Y
e	n	r	i	s	a	t	d	h	u	l	c	g	m	o	b	z	w	f	k	v	p	j	y	q	x .

Decrypting the beginning of the cryptotext *Böll*,

HVZDU VFKRQ GXQNH ODOVL FKLQE RQQDQ ,

with this table produces a totally unacceptable plaintext:

eakrd autvn mhnbe zrzai uting vnnrn .

Taking the cryptotext letters of equal frequency in a different order does not improve the situation. In fact, the decryption table is obtained by counting backwards three letters and reads

H	Q	D	W	L	V	K	U	X	F	J	P	E	G	I	N	O	C	S	Z	R	A	M	B	T	Y
e	n	a	t	i	s	h	r	u	c	g	m	b	d	f	k	l	z	p	w	o	x	j	y	q	v ,

so the true decryption is

eswar schon dunke lalsi chinb onnan .

15.5 Cliques and Matching of Partitions

Fig. 106 shows that in the example above the true frequency ordering differs considerably from the one based theoretically on probabilities: there are local permutations, where /r/ and /v/ jump by 5 positions, and /d/, /l/, and /o/ by 6. Others jump only by one or two—but whether large or small, every crossover ruins the right association of plaintext and cryptotext characters. Only a few letters—among others /e/ and /n/—are paired correctly. Certainly, the shortness of the cryptotext *Böll* is responsible for fluctuations, but not completely, as we shall see in a moment.

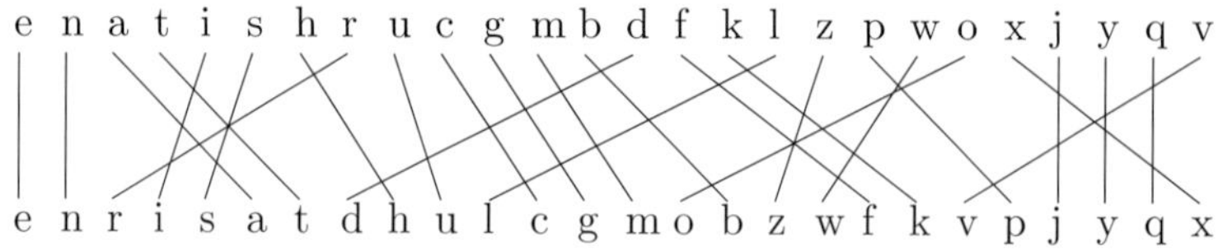

Fig. 106. Confrontation between observed frequencies and probability based frequencies

There is no fully automatic decryption on the basis of frequency ordering. The reason is that even longer texts show fluctuations, and the empirically determined probabilities fluctuate as well. This leads to crossovers in the frequency order.

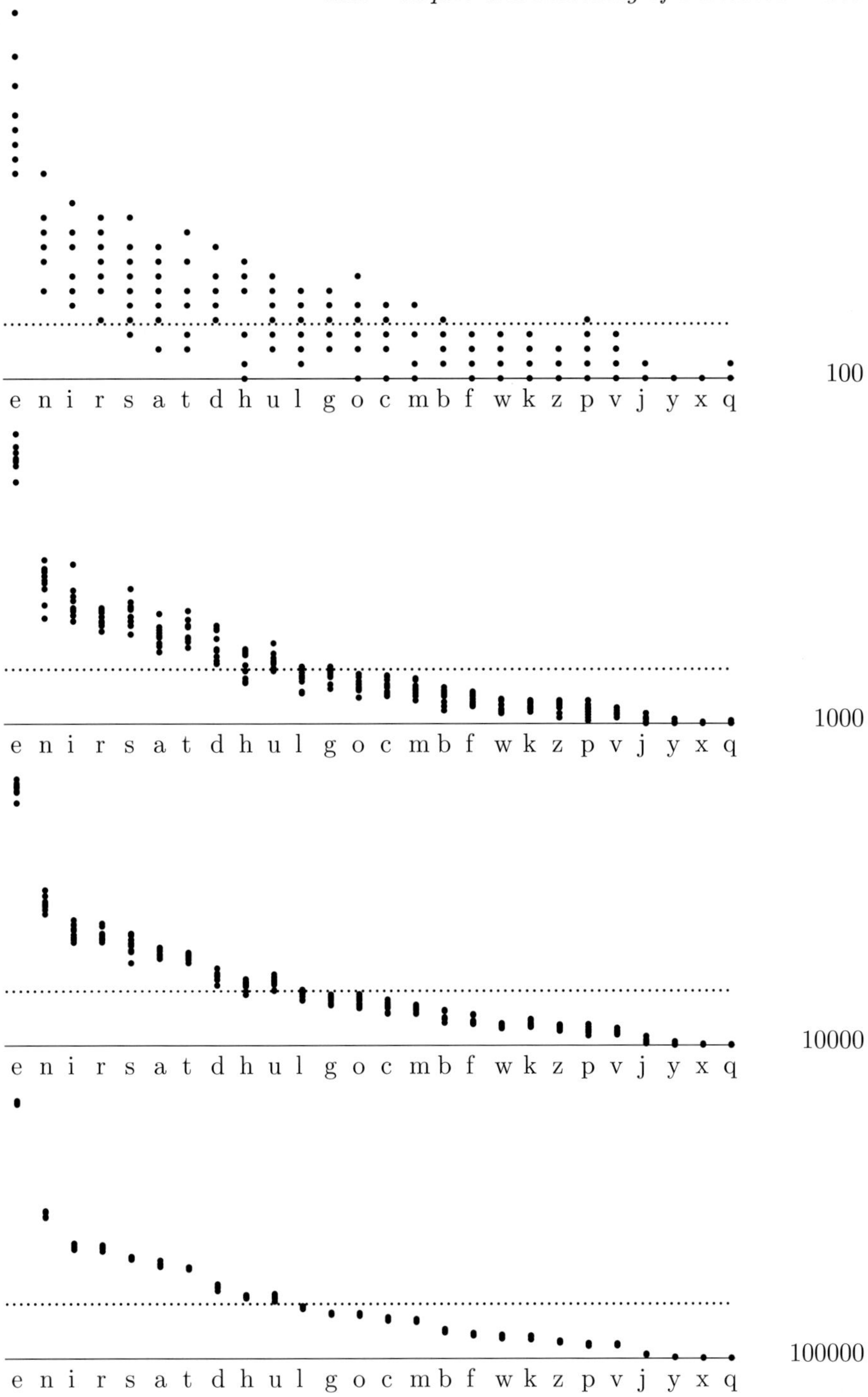

Fig. 107. Fluctuations of the frequency of the individual letters in newspaper German

15.5.1 Fluctuations. In fact, not only the frequency order, but also the individual frequencies given in the literature show deviations. We therefore investigated the fluctuations that are to be expected in German texts of 100, 1 000, 10 000, and 100 000 characters. A typical result is given in Fig. 107 (the dotted line gives the mean frequency). The text basis of 681 972 characters was a collection of all political commentaries taken from a daily newspaper in March 1992 (henceforth called SZ3-92). It clearly shows the overlapping of the fluctuation regions and how it decreases, the longer the text is. The fluctuation itself decreases roughly with the square root of the length of the text, as Fig. 108 shows for the letter /e/.

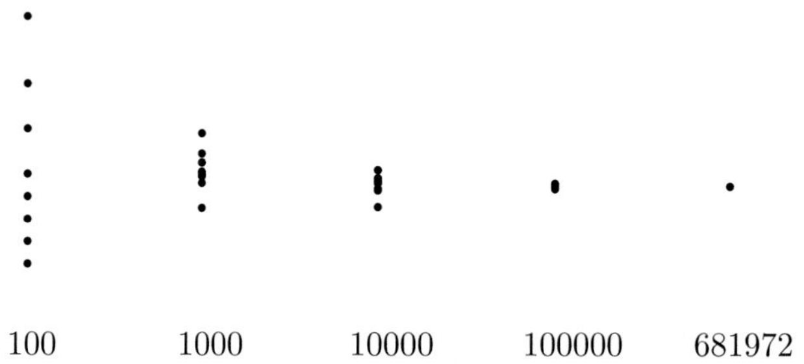

Fig. 108. Fluctuations of the frequency of the letter /e/ in German depending on the length of the text

Specifically, we made the experiment of confronting the frequency order of Meyer-Matyas (Sect. 15.4.1) with a rather long English text of 29 272 characters (taken from this book), whose letter frequencies are given in Fig. 109 .

3879	2697	2240	2151	2133	2082	1910	1907	1415	1095	1035	995	780
e	t	a	n	o	i	r	s	h	d	l	c	m
765	719	687	620	551	469	404	277	230	101	55	45	30
u	f	p	y	g	w	b	v	k	x	z	q	j

Fig. 109. Frequency distribution in an English text of 29 272 characters

Figure 110 shows the resulting confrontation. There are fewer crossovers, but there are still some.

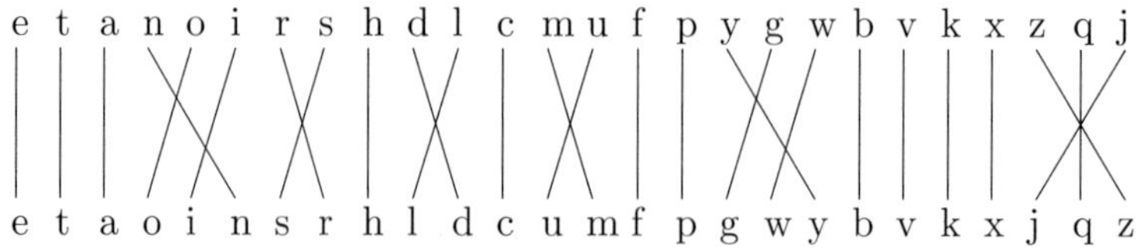

Fig. 110. Confrontation between observed frequencies in an English text of 29 272 characters and probability-based frequencies (Meyer-Matyas)

If the long English text is subjected to a simple substitution and then decrypted by means of a confrontation of the observed and the probability-based frequency orderings, a fragment of it reads as follows:

```
i v e s t h e c e o t m s n e r c s g p t i d i w g h a r c i d d e c t
e l a t s e a r m s g i f e x p e s n e o c e r e v e o t h e i p e o d
n t e s a t m s e r h i y r t h n r t h e r e e x p e s n e o c e r o i
s u a d d g r c a t t e s e l c a o b e c i o c e o t s a t e l n o t i
a f e y u a x n u r f i s c s g p t i w s a p h n c y i s k n o p a s t
```

It cannot be read fluently, and only after fixing an irritating 3-cycle between /i/, /o/, and /n/, and a 2-cycle between /r/ and /s/, can it serve as a rough decrypt. The true text (see the beginning of Sect. 11.2) is:

```
o v e r t h e c e n t u r i e s c r y p t o l o g y h a s c o l l e c t
e d a t r e a s u r y o f e x p e r i e n c e s e v e n t h e o p e n l
i t e r a t u r e s h o w s t h i s t h e s e e x p e r i e n c e s n o
r m a l l y s c a t t e r e d c a n b e c o n c e n t r a t e d i n t o
a f e w m a x i m s f o r c r y p t o g r a p h i c w o r k i n p a r t
```

15.5.2 Cliques. Rather than working with a frequency ordering of characters, it is preferable to work with an ordering of 'equifrequency' cliques of characters that are hard to separate on account of their frequencies.

For the English language there is a decomposition of the set of letters into cliques, essentially given by Laurence Dwight Smith in 1943:

$$\{\text{etaoin}\} \qquad \{\text{srh}\}\ \{\text{ld}\}\ \{\text{cumfpgwyb}\} \qquad \{\text{vk}\} \quad \{\text{xjqz}\}\ ,$$

or somewhat more finely decomposed,

$$\{\text{e}\}\ \{\text{t}\}\ \{\text{aoin}\} \quad \{\text{srh}\}\ \{\text{ld}\}\ \{\text{cumf}\}\ \{\text{pgwyb}\} \qquad \{\text{vk}\} \quad \{\text{xjqz}\}\ ,$$

which can be further decomposed for long and 'normal' texts into

$$\{\text{e}\}\ \{\text{t}\}\ \{\text{ao}\}\ \{\text{in}\}\ \{\text{srh}\}\ \{\text{ld}\}\ \{\text{cu}\}\ \{\text{mf}\}\ \{\text{p}\}\ \{\text{gwy}\}\ \{\text{b}\}\ \{\text{v}\}\ \{\text{k}\}\ \{\text{xjqz}\}\ .$$

For the German language too there exists a decomposition of the set of letters into cliques, essentially given by André Lange and E.-A. Soudart in 1935:

$$\{\text{e}\}\ \{\text{nirsatdhu}\} \qquad \{\text{lgocmbfwkz}\} \qquad \{\text{pvjyxq}\}\ ,$$

or somewhat more finely decomposed,

$$\{\text{e}\}\ \{\text{n}\}\ \{\text{irsat}\}\ \{\text{dhu}\} \quad \{\text{lgocm}\}\ \ \{\text{bfwkz}\}\ \ \{\text{pv}\}\ \{\text{jyxq}\}\ ,$$

which can be further decomposed for long and 'normal' texts into

$$\{\text{e}\}\ \{\text{n}\}\ \{\text{ir}\}\ \{\text{sat}\}\ \{\text{dhu}\}\ \{\text{lgo}\}\ \{\text{cm}\}\ \{\text{bfwkz}\}\ \{\text{pv}\}\ \{\text{jyxq}\}\ .$$

With interactive computer support, an exhaustive procedure is indicated, which treats the cliques one after another. In particular, if the decomposition is fine enough to allow cliques of two or three elements, the exhaustive effort is feasible.

For the example of the long English text (Fig. 110), the cliques are confronted in Fig. 111 .

{e} {t} {a} {noi} {rs} {h} {dl} {c} {mufp} {ygw} {b} {v} {k} {xzqj}

{e} {t} {aoin} {srh} {ld} {cumf} {pgwyb} {vk} {xjqz}

Fig. 111. Confrontation of the cliques

In this case, the rough decryption was still rather good because the cliques did not overlap too much and there was a clear gap between /a/ and /n/, between /s/ and /h/, and between /c/ and /m/. In such cases only a few exhaustive trials are necessary.

15.5.3 Example. For the short cryptotext *Böll* of Sect. 15.3.1 with the frequencies in Fig. 103 no such fine decomposition into cliques will work. 54 H and 41 Q suggest H $\hat{=}$ e and Q $\hat{=}$ n , but in view of the next frequencies 36 D, 24 W, 23 L, 22 V, and 21 K, it cannot be expected that the cliques {ir} and {sat} are separated. But D is well separated and it looks tempting to set D$\hat{=}$ i . This would leave {rsat} confronted with {WLVK}, which means 4!=24 trials. Unfortunately, none of these give reasonable texts. In fact, the next two frequencies 19 U and 18 X are so close that a crossing-over into the clique {dhu} might be responsible. This would mean that 8!=40 320 trials were to be made, which is outside the reach of exhaustion.

{H} {Q} {D} {LUVWKX} {GOJFPEIN} {RZCS} {YMBAT}

{e} {n} {ir} {ast} {hud} {lgo} {cm} {bfwkz} {pv} {jyxq}

Fig. 112. Confrontation of the cliques for the cryptotext *Böll* of Sect. 15.3.1

Figure 112 shows the confrontation of the cliques. Obviously, for short texts mechanical decryption on the basis of individual letter frequencies does not work. At least, other stochastic peculiarities of language must be taken into account, like bigram frequencies. This will be studied in Sect. 15.7 .

15.5.4 Empirical frequencies. For the English language, Table 8 gives empirical relative frequencies $\mu_i = m_i/M$, the result of a count by Meyer-Matyas, based on 4 000 000 characters in a corpus of everyday English. Solomon Kullback pointed out in 1976 that the genre of communications gives rise to strong fluctuations, and diffentiates 'literary English' with a frequency for /e/ of 12.77% from 'telegraphic English' with a frequency for /e/ of 13.19% .

For the German language, the text basis SZ3-92 with a total of M = 681 972 characters gives results which are likewise tabulated in Table 8.

The numerical values in Table 8 may serve as a hypothetical probability distribution of a stochastic source.

character	English	German	character	English	German
a	8.04%	6.47%	n	7.09%	9.84%
b	1.54%	1.93%	o	7.60%	2.98%
c	3.06%	2.68%	p	2.00%	0.96%
d	3.99%	4.83%	q	0.11%	0.02%
e	12.51%	17.48%	r	6.12%	7.54%
f	2.30%	1.65%	s	6.54%	6.83%
g	1.96%	3.06%	t	9.25%	6.13%
h	5.49%	4.23%	u	2.71%	4.17%
i	7.26%	7.73%	v	0.99%	0.94%
j	0.16%	0.27%	w	1.92%	1.48%
k	0.67%	1.46%	x	0.19%	0.04%
l	4.14%	3.49%	y	1.73%	0.08%
m	2.53%	2.58%	z	0.09%	1.14%

Table 8. Hypothetical character probabilities of a stochastic source in English and in German

For the German language, the frequency of /e/ is skewed by the cryptographic custom of decomposing /ä/, /ö/, /ü/ into /ae/, /oe/, /ue/.

George K. Zipf and Benoît Mandelbrot have published empirical formulas for the relative frequency of the k-th letter which fit many languages astonishingly well, namely

$$p(k) \propto 1/k \qquad \text{and} \qquad p(k) \propto 1/(k+c)^m \quad \text{for suitable positive } c, m \; .$$

The actual values for the English language are shown graphically in Fig. 113. A convincing theoretical explanation has not been given.

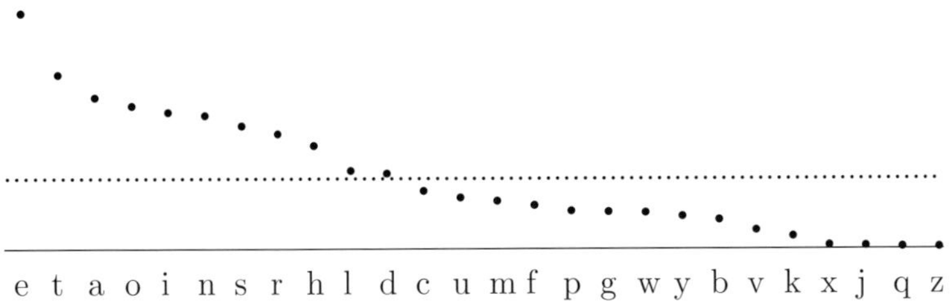

Fig. 113. Relative frequencies of characters in the English language (Meyer-Matyas count)

15.6 Optimal Matching

15.6.1 Squared distance. The frequency deviation between a given text T with M characters and an expected text T^Q of equal length from a stochastic source Q can be measured by the squared distance $d(T, T^Q)$,

$$d(T, T^Q) = \sum_{i=1}^{N} (m_i - M \cdot p_i)^2 \quad .$$

Here p_i denotes the probability for the appearance of the i-th character χ_i $(i = 1 \ldots N)$, m_i the frequency of χ_i in the text T, where $\sum_{i=1}^{N} m_i = M$.

Let σ be the decrypting substitution carrying the cryptotext characters into plaintext characters. The value of

$$d_\sigma = d(T, T^{\sigma(Q)}) = \sum_{i=1}^{N} (m_i - M \cdot p_{\sigma(i)})^2$$

measures the concordance between the cryptotext T with the observed frequencies m_i and the expected cryptotext $T^{\sigma(Q)}$ of the source $\sigma(Q)$;

$$\min_\sigma d_\sigma = \min_\sigma \sum_{i=1}^{N} (m_i - M \cdot p_{\sigma(i)})^2$$

characterizes the substitutions achieving optimal concordance, which therefore are candidates for the decryption. Because of fluctuations, substitutions bringing d_σ close to the minimum are also candidates, but with increasing d_σ they become less and less interesting.

15.6.2 Minimization. Obviously $\sum_{i=1}^{N} p_{\sigma(i)}^2 = \sum_{i=1}^{N} p_i^2$ and thus

$$\min_\sigma \sum_{i=1}^{N} (m_i - M \cdot p_{\sigma(i)})^2 = \sum_{i=1}^{N} m_i^2 + M^2 \sum_{i=1}^{N} p_i^2 - 2M \max_\sigma \sum_{i=1}^{N} m_i \cdot p_{\sigma(i)} \; .$$

To find candidates for decryption, it therefore suffices to consider the following maximum:

$$\max_\sigma \sum_{i=1}^{N} m_i \cdot p_{\sigma(i)}$$

Theorem: Assume $m_i \geq m_{i+1}$ for all i .
$\sum_{i=1}^{N} m_i \cdot p_{\sigma(i)}$ is maximal, if and only if $p_{\sigma(i)} \geq p_{\sigma(i+1)}$ for all i .

Proof: Since every permutation can be expressed as a chain of swaps of two elements, it suffices to investigate the contribution of a swap of two elements χ_j, χ_k to the sum. Then

$$m_j \cdot p_{\sigma(j)} + m_k \cdot p_{\sigma(k)} \geq m_j \cdot p_{\sigma(k)} + m_k \cdot p_{\sigma(j)} \quad \text{if and only if}$$
$$(m_j - m_k) \cdot (p_{\sigma(j)} - p_{\sigma(k)}) \geq 0 \; , \quad \text{i.e., if and only if} \quad p_{\sigma(j)} \geq p_{\sigma(k)} \; . \quad \bowtie$$

The expected result, that the optimal concordance is reached by matching in the frequency ordering, is supplemented by the strategy of finding other candidates by swaps of pairs of characters χ_j, χ_k such that each time

$$(m_j - m_k) \cdot (p_{\sigma(j)} - p_{\sigma(k)})$$

is minimal.

Example:

We assume the probabilities p_i of Table 8 and investigate the cryptotext *Böll* of Sect. 15.3.1 (the frequencies m_i are found in Fig. 103). Using a decryption σ according to the frequency order of Table 8, there results for $\sum_{i=1}^{N} m_i \cdot p_{\sigma(i)}$ a value of

$$2634.56\% = M \cdot 7.5489\%\,.$$

If D$\hat{=}$i , W$\hat{=}$r is swapped to D$\hat{=}$ r , W$\hat{=}$i, which corresponds to the frequency order of Kaeding; then the value is slightly diminished by

$(36 - 24) \cdot (7.73\% - 7.54\%) = 2.28\% = M \cdot 0.0065$ to become

$$2632.28\% = M \cdot 7.5424\%\,.$$

For the correct decryption, we obtain a value of

$$2585.80\% = M \cdot 7.4092\%\,.$$

The value of $\sum_{i=1}^{N} m_i^2$ in this example is $9347 = M^2 \cdot 7.67\%$, and the value of $\sum_{i=1}^{N} p_i^2$ according to Table 8 is 7.62%.

	.a	.b	.c	.d	.e	.f	.g	.h	.i	.j	.k	.l	.m	.n	.o	.p	.q	.r	.s	.t	.u	.v	.w	.x	.y	.z
a.	1	32	39	15		10	18		16		10	77	18	172	2	31	1	101	67	124	12	24	7		27	1
b.	8				58				6	2		21	1		11			6	5		25				19	
c.	44		12		55	1		46	15		8	16			59	1		7	1	38	16		1			
d.	45	18	4	10	39	12	2	3	57	1		7	9	5	37	7	1	10	32	39	8	4	9		6	
e.	65	11	64	107	39	23	20	15	40	1	2	46	43	120	46	32	14	154	145	80	7	16	41	17	17	
f.	21	2	9	1	25	14	1	6	21	1		10	3	2	38	3		4	8	42	11	1	4		1	
g.	11	2	1	1	32	3	1	16	10			4	1	3	23	1		21	7	13	8		2		1	
h.	84	1	2	1	251	2		5	72			3	1	2	46	1		8	3	22	2		7		1	
i.	18	7	55	16	37	27	10				8	39	32	169	63	3		21	106	88		14	1	1		4
j.					2										4						4					
k.					28				8					3	3				2	1			3		3	
l.	34	7	8	28	72	5	1		57	1	3	55	4	1	28	2	2	2	12	19	8	2	5		47	
m.	56	9	1	2	48			1	26				5	3	28	16			6	6	13		2		3	
n.	54	7	31	118	64	8	75	9	37	3	3	10	7	9	65	7		5	51	110	12	4	15	1	14	
o.	9	18	18	16	3	94	3	3	13		5	17	44	145	23	29		113	37	53	96	13	36		4	2
p.	21	1			40			7	8			29			28	26	42	3	14	7		1		2		
q.																					20					
r.	57	4	14	16	148	6	6	3	77	1	11	12	15	12	54	8		18	39	63	6	5	10		17	
s.	75	13	21	6	84	13	6	30	42		2	6	14	19	71	24	2	6	41	121	30	2	27		4	
t.	56	14	6	9	94	5	1	315	128			12	14	8	111	8		30	32	53	22	4	16		21	
u.	18	5	17	11	11	1	12	2	5			28	9	33	2	17		49	42	45				1	1	1
v.	15				53				19						6											
w.	32		3	4	30	1		48	37			4	1	10	17	2		1	3	6	1	1	2			
x.	3		5		1				4						1	4				1	1					
y.	11	11	10	4	12	3	5	5	18			6	4	3	28	7		5	17	21	1	3	14			
z.					5				2			1														1

Table 9. Bigram frequencies (in %%) in English (after O. Phelps Meaker)

15.7 Frequency of Multigrams

Even more than frequencies of individual characters, multigram frequencies imprint a language. Their importance is illustrated by

> **Invariance Theorem $3^{(n)}$:** For all monoalphabetic, functional simple substitutions, especially for all monoalphabetic linear simple substitutions (including CAESAR additions and reversals),
> *partitions of n-grams within the text are invariant.*

15.7.1 Frequency tables. According to the theorem, the frequency of n-grams in a cryptotext can be used for decryption, too. However, for $N=26$ there are already 676 bigrams and 17 576 trigrams; only in rather long cryptotexts will enough bigrams and trigrams be found, and in short texts even bigrams are quite rare, therefore the influence of fluctuations is substantial. Cryptanalysis of monographic encryptions on the basis of bigrams alone instead of single characters does not bring great advantage.

The frequencies of bigrams (as indicated by Tables 9 and 10) and trigrams are even more unbalanced than those of single characters. The 19 most frequent

	.a	.b	.c	.d	.e	.f	.g	.h	.i	.j	.k	.l	.m	.n	.o	.p	.q	.r	.s	.t	.u	.v	.w	.x	.y	.z
a.	8	31	27	11	64	15	30	20	5	1	7	59	28	102		4		51	53	46	75	2	3		1	2
b.	16	1		1	101		3	1	12		1	9		1	8			9	6	4	14		1		1	1
c.	2			2	1			242	1		14	1			2				1							
d.	54	3	1	13	227	3	4	2	93	1	3	5	4	6	9	3		10	11	6	16	3	4			3
e.	26	45	25	51	23	26	50	57	193	3	19	63	55	400	6	13	1	409	140	55	36	14	23	2	1	11
f.	19	2		9	25	12	3	1	7		1	5	1	2	9	1		18	4	20	24	1	1			1
g.	20	3		12	147	2	3	3	19	1	3	9	3	5	6	1		14	18	18	11	4	3			3
h.	70	4	1	14	102	2	4	3	23	1	3	25	11	19	18	1		37	11	47	11	4	9			3
i.	7	7	76	20	163	5	38	12	1	1	12	25	27	168	20	2		17	79	78	3	5	1			5
j.	9				9										2						5					
k.	26	1		2	26	1	1	1	7		1	10	1	1	24	1		13	5	14	9	1	1			1
l.	45	7	2	14	65	5	6	2	61	1	7	42	3	4	14	2		2	22	27	13	3	2			3
m.	40	6	1	8	50	4	4	3	44	2	3	4	23	3	15	7		2	10	8	14	4	3			2
n.	68	23	5	187	122	19	94	17	65	5	25	10	23	43	18	10		10	74	59	33	18	29			25
o.	3	8	15	7	25	6	5	9	1	1	3	31	17	64	1	6		50	19	9	3	3	7		1	6
p.	16			3	10	6		2	4			4			11	5		23	1	3	4					
q.																					2					
r.	80	25	9	67	112	18	27	19	52	4	23	18	20	31	30	9		15	54	49	48	12	17			14
s.	36	10	89	20	99	7	13	9	65	2	11	9	12	7	28	22		8	76	116	15	9	10		2	7
t.	57	8	1	35	185	5	10	14	59	2	4	11	9	9	15	3		31	50	23	26	8	21		1	26
u.	3	8	16	5	78	27	8	4	2		3	7	21	119		5		33	48	23	1	3	2			1
v.	3				37				9						43											
w.	34				48				36	1				1	17				1		9					
x.									1						1					1						
y.					1							1	1						1							
z.	4	1		1	28		1		11		1	2	1		2				1	7	43	1	9			1

Table 10. Bigram frequencies (in %%) in German (text basis SZ3-92)

bigrams in English and the 18 most frequent bigrams in German (they comprise 92.93% of all bigrams) are presented in Tables 11 and 12; the 98 most frequent trigrams in English and the 112 most frequent trigrams in German (they comprise only 52.11% of all trigrams) are presented in Tables 13 and 14. Comparing values published in the literature, it is important to know whether the word spacings are taken into consideration; sometimes (for example in Fletcher Pratt 1939) only bigrams and trigrams within words are counted. The counts show even more fluctuations than those of individual characters, as can be seen from Tables 11 and 12. Frequency tables for several other Indo-Germanic languages have been published by Gaines and Eyraud.

Tables 9 and 10 show at a glance that the matrix of bigram frequencies is not symmetric. Common bigrams with rare reverses (German *Dreher*) are

/th/, /he/, /ea/, /nd/, /nt/, /ha/, /ou/, /ng/, /hi/, /eo/, /ft/, /sc/, /rs/ ;

they are useful for the dissolution of cliques. On the other hand, the following pairs of bigrams show roughly the same frequency:

/er/ - /re/, /es/ - /se/, /an/ - /na/, /ti/ - /it/, /on/ - /no/, /in/ - /ni/,
/en/ - /ne/, /at/ - /ta/, /te/ - /et/, /or/ - /ro/, /to/ - /ot/, /ar/ - /ra/,
/st/ - /ts/, /is/- /si/, /ed/ - /de/, /of/ - /fo/ .

15.7.2 Word frequencies. Quite interesting are the frequencies of words, i.e., of multigrams with a space at the beginning and the end. The order of the most frequent words is

in English: the of and to a in that it is I for as with was his
he be not by but have you which are on or her ,

in German: die der und den am in zu ist daß es ,

in French: de il le et que je la ne on les en ce se son
mon pas lui me au une des sa qui est du ,

in Italian: la di che il non si le una lo in per un mi
io piu del ma se ,

in Spanish: de la el que en no con un se su las los es
me al lo si mi una di por sus muy hay mas .

The only one-letter-words in English are a and I ; two-letter-words are

an at as he be in is it on or to of do go no so my .

The most frequent words by far in the Indo-Germanic languages are the non-content words[1] (French *mots vides*, German *Formwörter, inhaltsleere Wörter*), namely articles, prepositions, conjunctions, and other auxiliary particles, in contrast to conceptual words (German *Begriffswörter*) like substantives, adjectives, and verbs. The 70 most frequent words of the English language are non-content words, and among the 100 most frequent ones are only ten conceptual words.

[1] In English, non-content words are the only ones not capitalized in headlines.

	Table 9	Kullback	Sinkov	Eyraud
th	315	156	270	330
he	251	40	257	270
an	172	128	152	167
in	169	150	194	202
er	154	174	179	191
re	148	196	160	169
on	145	154	154	134
es	145	108	115	149
ti	128	90	108	126
at	124	94	127	127
st	121	126	103	116
en	120	222	129	146
or	113	128	108	91
nd	118	104	95	122
to	111	100	95	79
nt	110	164	93	124
ed	107	120	111	125
is	106	70	93	79
ar	101	88	96	83

Table 11. The nineteen most frequent bigrams in English (frequencies in %%)

	Table 10	Bauer-Goos	Valerio	Eyraud
er	409	340	337	375
en	400	447	480	443
ch	242	280	266	280
de	227	214	231	233
ei	193	226	187	242
nd	187	258	258	208
te	185	178	222	178
in	168	204		197
ie	163	176	222	188
ge	147	168	160	196
es	140	181		168
ne	122	117		143
un	119	173	169	139
st	116	124		118
re	112	107	213	124
he	102	117		124
an	102	92		82
be	101	96		104

Table 12. The eighteen most frequent bigrams in German (frequencies in %%)

the	353	hat	55	man	40	ant	32	rom	28	str	25	nte	23
ing	111	ers	54	red	40	hou	31	ven	28	tic	25	rat	23
and	102	his	52	thi	40	men	30	ard	28	ame	24	tur	23
ion	75	res	50	ive	38	was	30	ear	28	com	24	ica	23
tio	75	ill	47	rea	38	oun	30	din	27	our	24	ich	23
ent	73	are	47	wit	37	pro	30	sti	27	wer	24	nde	23
ere	69	con	46	ons	37	sta	30	not	27	ome	24	pre	23
her	68	nce	45	ess	36	ine	29	ort	27	een	24	enc	22
ate	66	all	44	ave	34	whi	28	tho	26	lar	24	has	22
ver	64	eve	44	per	34	ove	28	day	26	les	24	whe	22
ter	63	ith	44	ect	33	tin	28	ore	26	san	24	wil	22
tha	62	ted	44	one	33	ast	28	but	26	ste	24	era	22
ati	59	ain	43	und	33	der	28	out	25	any	23	lin	22
for	59	est	42	int	32	ous	28	ure	25	art	23	tra	22

Table 13. The 98 most frequent trigrams in English (frequencies in %%)

ein	122	das	47	erd	33	ese	27	eni	23	ner	20	hei	18
ich	111	hen	47	enu	33	auf	26	ige	23	nds	20	lei	18
nde	89	ind	46	nen	32	ben	26	aen	22	nst	20	nei	18
die	87	enw	45	rau	32	ber	26	era	22	run	20	nau	18
und	87	ens	44	ist	31	eit	26	ern	22	sic	20	sge	18
der	86	ies	44	nic	31	ent	26	rde	22	enn	19	tte	18
che	75	ste	44	sen	31	est	26	ren	22	ins	19	wei	18
end	75	ten	44	ene	30	sei	26	tun	22	mer	19	abe	17
gen	71	ere	43	nda	30	and	25	ing	21	rei	19	chd	17
sch	66	lic	42	ter	30	ess	25	sta	21	eig	18	des	17
cht	61	ach	41	ass	29	ann	24	sie	21	eng	18	nte	17
den	57	ndi	41	ena	29	esi	24	uer	21	erg	18	rge	17
ine	53	sse	39	ver	29	ges	24	ege	20	ert	18	tes	17
nge	52	aus	36	wir	29	nsc	24	eck	20	erz	18	uns	17
nun	48	ers	36	wie	28	nwi	24	eru	20	fra	18	vor	17
ung	48	ebe	35	ede	27	tei	24	mme	20	hre	18	dem	17

Table 14. The 112 most frequent trigrams in German (frequencies in %%)

15.7.3 Positions. The frequencies of a letter depend very often on its position within a word. For example, the letter /e/ in German stands

in first position	7.7%
in second position	21.7%
in third position	16.5%
⋮	
in third from last position	8.8%
in second from last position	7.7%
in last position	15.0%

.

15.7.4 Average word length. Although in cryptography word spacing is suppressed professionally, the average word length is an important characteristic of a language (Table 15). In the German language, word lengths are distributed as follows:

1	0.05%	6	11.66%	11	3.24%	16	0.32%
2	8.20%	7	6.04%	12	2.06%	17	0.38%
3	28.71%	8	4.43%	13	1.40%	18	0.16%
4	13.49%	9	3.67%	14	0.59%	19	0.10%
5	11.55%	10	2.64%	15	0.65%		

15.7.5 Word formation. Vowels and consonants usually alternate. Vowels provide the singable sound pattern of any language. In French they can occur accumulated: *ouïe, aïeul.* Consonants in Arabian languages form the backbone of writing and occur accumulated in Slavic languages as well: *czyszczenie* (Polish), *cvŕčak* (Serbo-Croatian), *nebezpečenství* (Czech). Welsh shows strange patterns: *Llanfairpwllgwyngyllgogerychwyrndrobwllllantysiliogogogoch* is the name of a railroad station in Wales, in *rhy ddrwg* ("too bad"), *y* and *w* denote vowels. In English, words with a 4-consonant sequence like *sixths* are very rare, in German *Schlacht, schlecht, schlicht, Schlucht* are 8-letter words with one vowel only, and words like *Erstschlag* with a 7-consonant sequence can be obtained by composition. Vowel distances show typical frequences, too: in German (without spaces)

1	20.77%	5	2.63%
2	25,06%	6	1.03%
3	35.95%	7	0.15%
4	14.75%	8	0.03%

Table 15 gives a comparison of the average word length, the vowel frequency, the frequency of the five dominant consonants {l n r s t}, and the infrequent letters for five important languages with a Latin alphabet and for Russian.

	average word length	vowel frequency	{l n r s t} frequency	rare letters					
English	4.5	40%	33%	j		q		x		z
French	4.4	45%	34%		k		w			
German	5.9	39%	34%	j		q		x	y	
Italian	4.5	48%	30%	j	k		w	x	y	
Spanish	4.4	47%	31%		k		w			
Russian	6.3	45%								

Table 15. Characteristics of word formation

15.7.6 Spacing. For *informal ciphers* that preserve word spacing and possibly also punctuation, bigram tables contain also frequencies for letters at the beginning and the end of a word, and trigram tables contain frequencies for bigrams at the beginning and the end of a word and for one-letter words.

If word spacings (and possibly also punctuation marks) are not suppressed, they should be included in the encryption. Thus, the space may become the most frequent character. In German, spaces are about as frequent as /e/, while in English spaces are markedly more frequent than /e/ .

If, as in 'aristocrats', spaces are preserved, they are encrypted by themselves and thus one character is decrypted from the outset. This simplifies an entry considerably. Experienced amateur cryptologists may sometimes read such informal ciphers at first glance. "Not infrequently, the cryptogram which retains its word-divisions can be read at sight ... and this regardless of how short it may be" (Helen Fouché Gaines, 1939).

In professional cryptology, there are good reasons for using formal ciphers, as we have done. If for technical reasons, as in teletype communication and with an ASCII code, special control symbols are available, they should be used with discretion and not be mixed with the cryptographic process—a well-trained and responsible crypto clerk will know this.

15.8 The Combined Method of Frequency Matching

In an attempt to mechanize the decryption of monoalphabetic simple substitutions, particularly in the case of short texts, it may be wise to combine the information on frequencies of individual characters, bigrams, and possibly trigrams, in the sense that bigram frequencies are taken into account as soon as a clique of characters cannot be separated by monogram frequencies, and trigram frequencies as soon as even bigram frequencies do not separate the clique. More than trigrams are unlikely to be useful. In case no probable words are utilized, this is also a cryptotext-only attack, using nothing more than an assumption as to the underlying natural language.

15.8.1 Example. For the cryptotext of 280 characters (Kahn 1967)

```
G J X X N   G G O T Z   N U C O T   W M O H Y   J T K T A   M T X O B
Y N F G O   G I N U G   J F N Z V   Q H Y N G   N E A J F   H Y O T W
G O T H Y   N A F Z N   F T U I N   Z A N F G   N L N F U   T X N X U
F N E J C   I N H Y A   Z G A E U   T U C Q G   O G O T H   J O H O A
T C J X K   H Y N U V   O C O H Q   U H C N U   G H H A F   N U Z H Y
N C U T W   J U W N A   E H Y N A   F O W O T   U C H N P   H O G L N
F Q Z N G   O F U V C   N Z J H T   A H N G G   N T H O U   C G J X Y
O G H T N   A B N T O   T W G N T   H N T X N   A E B U F   K N F Y O
H H G I U   T J U C E   A F H Y N   G A C J H   O A T A E   I O C O H
U F Q X O   B Y N F G
```

a frequency count of the letters results in

17	4	13	0	7	17	23	26	5	12	3	2	2	36	25	1	5	0	0	23	20	3	6	9	13	8
A	B	C	D	E	F	G	H	I	J	K	L	M	N	O	P	Q	R	S	T	U	V	W	X	Y	Z

There is no indication of a shifted frequency profile and a CAESAR addition can be excluded by a short exhaustive test. The frequency order is:

36	26	25	23	23	20	17	17	13	13	12	9	8	7	6	5	5	4	3	3	2	2	1	0	0	0
N	H	O	G	T	U	A	F	C	Y	J	X	Z	E	W	I	Q	B	K	V	L	M	P	D	R	S

Assume that in the circumstances it is an English text. A confrontation with the frequency order of the English language, as given by Kahn in 1967,

e t a o n i r s h d l u c m p f y w g b v j k q x z

suggests in view of the marked decrease from 17 to 13 the cliques {e} {t} {aonirs}. Thus

$$\text{N} \mathrel{\hat{=}} \text{e} , \quad \text{H} \mathrel{\hat{=}} \text{t} \quad \text{and } \{\text{OGTUAF}\} \mathrel{\hat{=}} \{\text{aonirs}\} \ .$$

Bigram frequencies can be used to separate this clique. Table 16 shows the relevant segment of a bigram table for the English language (10 000 characters), Table 17 the bigram count for the present text.

	.e	.t	.a	.o	.n	.i	.r	.s	.h	.l	.d	.u	.c
e.	39	80	131	46	120	40	154	145	15	46	107	7	64
t.	94	53	56	111	8	128	30	32	315	12	9	22	6
a.	–	124	1	2	172	16	101	67	–	77	15	12	39
o.	3	53	9	23	145	13	113	37	3	17	16	96	18
n.	64	110	54	65	9	37	5	51	9	10	118	12	31
i.	37	88	18	63	169	–	21	106	–	39	16	–	55
r.	148	63	57	54	12	77	18	39	3	12	16	6	14
s.	84	121	75	71	19	42	18	41	30	6	6	30	21
h.	251	22	84	46	2	72	8	3	5	3	1	2	2
l.	72	19	34	28	1	57	2	12	–	55	28	8	8
d.	39	39	45	37	5	57	10	32	3	7	10	8	4
u.	11	45	18	2	33	5	49	42	2	28	11	–	17
c.	55	38	44	59	–	15	7	1	46	16	–	16	12

Table 16. Bigram table for the thirteen most frequent letters in English

	.N	.H	.O	.G	.T	.U	.A	.F	.C	.Y	.J	.X	.Z
N.	–	1	–	5	4	5	5	7	1	–	–	1	3
H.	3	2	4	1	2	1	1	–	1	9	1	–	–
O.	–	4	–	4	7	1	2	1	2	–	–	–	–
G.	4	2	6	2	–	–	2	–	–	–	3	–	–
T.	1	4	1	–	–	3	3	–	1	–	1	3	1
U.	–	1	–	2	4	–	–	3	5	–	–	–	1
A.	1	1	–	–	2	–	–	4	1	–	1	–	1
F.	3	2	1	2	1	2	–	–	–	1	–	–	1
C.	2	1	4	1	–	1	–	–	–	–	2	–	–
Y.	7	–	3	–	–	–	1	–	–	–	1	–	–
J.	–	2	2	–	1	2	–	2	1	–	–	3	–
X.	3	–	2	–	–	1	–	–	–	1	–	1	–
Z.	3	1	–	1	–	–	1	–	–	–	1	–	–

Table 17. Bigram count in appropriate order for cryptotext of Sect. 15.8.1

In Table 16 it can be seen that /a/, /i/ and /o/ avoid contact among themselves (they have no 'affinity'), except for the bigram /io/ . /oi/ is rare. In Table 17, O, U and A also avoid contact; OA occurs twice while AO does not occur. This indicates that

$$\text{O} \hat{=} \text{i}\,, \quad \text{A} \hat{=} \text{o} \quad \text{and therefore also} \quad \text{U} \hat{=} \text{a}\,.$$

It fits well that OU becomes /ia/ , which occurs a few times. Moreover, NU, which would represent /ea/, is frequent, while UN, which is missing, would represent the rare /ae/. Thus, the large clique is broken, and there remains only the smaller clique $\{\text{GTF}\} \hat{=} \{\text{nrs}\}$, which could even be exhausted.

The argument about contact was based on vowel contact. This is a peculiarity of the English language. The term 'vowel-solution method' to be found in the English literature (Helen Fouché Gaines, 1939) is accidental and does not describe a general method. In other languages, vowels have no tendency at all to avoid contact.

In English (and elsewhere) the consonant /n/ also has contact preferences: /n/ is regularly preceded by a vowel. This makes T rather than G, F and something from the next clique {C, Y} a candidate. It can be assumed that

$$\text{T} \hat{=} \text{n} \quad \text{and} \quad \{\text{GF}\} \hat{=} \{\text{rs}\}\,.$$

Another handle gives /h/: /th/ is very frequent and /he/ and /ha/ are frequent. The remaining G, F and C show no suitable contacts; this suggests that $\text{Y} \hat{=} \text{h}\,.$

Indeed, HY (for /th/) is very frequent, YN (for /he/) and YO (for /ha/) are frequent.

So far, seven of the ten most frequent characters are tentatively determined (and one test on $\{\text{GF}\} \hat{=} \{\text{rs}\}$ would decide about two more):

N	H	U	A	T	O	*	*	Y	*	*	*	*	*	*	*	*	*	*	*	*	*	*	*	*	*
e	t	a	o	n	i	r	s	h	d	l	u	c	m	p	f	y	w	g	b	v	j	k	q	x	z .

15.8.2 Continuation of the example.

After this entry, which was a walk, the solution should move to a trot. Indeed, the partial decryption

GJXXe	GGinZ	eaCin	WMith	JnKno	MnXiB
heFGi	GIeaG	JFeZV	QtheG	eEoJF	thinW
Ginth	eoFZe	FnaIe	ZoeFG	eLeFa	nXeXa
FeEJC	Ietho	ZGoEa	naCQG	iGint	Jitio
nCJXK	theaV	iCitQ	atCea	GttoF	eaZth
eCanW	JaWeo	Etheo	FiWin	aCteP	tiGLe
FQZeG	iFaVC	eZJtn	oteGG	entia	CGJXh
iGtne	oBeni	nWGen	tenXe	oEBaF	KeFhi
ttGIa	nJaCE	oFthe	GoCJt	ionoE	IiCit
aFQXi	BheFG				

suggests a series of improvements: in the first line Mith means with , whence $\text{M} \hat{=} \text{w}$, and JnKnown means unknown , whence $\text{J} \hat{=} \text{u}\,,\ \text{K} \hat{=} \text{k}\,.$

thinW in the second line means thing , whence W $\hat{=}$ g ; in the fourth line IethoZ means method , whence I $\hat{=}$ m , Z $\hat{=}$ d ; the word intuition fits. And there are more fragments that can help to find the remaining letters from the clique {hdlcwum}, namely /l/ and /c/ .

However, the choice between (G, F)$\hat{=}$(r, s) and (G, F)$\hat{=}$(s, r) should be made good first. The occurrence of FG in the second line and the fact, that the bigram /sr/ is very rare, gives

G $\hat{=}$ s , F $\hat{=}$ r

a very good chance. We now have the partial decryption

```
suXXe ssind eaCin gwith unkno wnXiB
hersi smeas uredV Qthes eEour thing
sinth eorde rname doers eLera nXeXa
reEuC metho dsoEa naCQs isint uitio
nCuXk theaV iCitQ atCea sttor eadth
eCang uageo Etheo rigin aCteP tisLe
rQdes iraVC edutn otess entia CsuXh
istne oBeni ngsen tenXe oEBar kerhi
ttsma nuaCE orthe soCut ionoE miCit
arQXi Bhers
```

Now in the first line suXXess means 'success', whence X $\hat{=}$ c, and deaCing means 'dealing', whence C $\hat{=}$ l .

Altogether, we now know all but a few rare letters:

N	H	U	A	T	O	F	G	Y	Z	C	J	X	I	*	*	*	M	W	*	*	*	K	*	*	*
e	t	a	o	n	i	r	s	h	d	l	u	c	m	p	f	y	w	g	b	v	j	k	q	x	z .

The resulting text can be read fluently:

```
succe ssind ealin gwith unkno wnciB
hersi smeas uredV Qthes eEour thing
sinth eorde rname doers eLera nceca
reEul metho dsoEa nalQs isint uitio
nluck theaV ilitQ atlea sttor eadth
elang uageo Etheo rigin alteP tisLe
rQdes iraVl edutn otess entia lsuch
istne oBeni ngsen tence oEBar kerhi
ttsma nualE orthe solut ionoE milit
arQci Bhers
```

and our procedure gallops along to bring almost by itself B $\hat{=}$ p , V $\hat{=}$ b , Q $\hat{=}$ y , E $\hat{=}$ f . In the third line /rname doers eLera nceca/ causes a stumble, but then in the sixth-seventh line we read /rigin altex tisve rydes/ , i.e., P $\hat{=}$ x , L $\hat{=}$ v . All actually occurring letters are determined and only /j/, /q/, and /z/ remain open.

During this gallop, we find three encryption errors:

in the third line, the fourth group should read ZBNFG ;
in the seventh line, the third group should read NVJHT ; and
in the eighth line, the first group should read OGHYN .

15.8.3 Final result. If this is not enough to convince a doubtful reader, then we can also reconstruct the password for the substitution: apart from the three missing letters, alphabetic ordering of the plaintext letters gives

a	b	c	d	e	f	g	h	i	j	k	l	m	n	o	p	q	r	s	t	u	v	w	x	y	z
U	V	X	Z	N	E	W	Y	O	*	K	C	I	T	A	B	*	F	G	H	J	L	M	P	Q	* .

The password NEWYORKCITY cannot be overlooked; it also yields for the non-occuring cryptotext characters R $\hat{=}$ j , D $\hat{=}$ q , S $\hat{=}$ z .

The message in readable form, freed from three encryption errors (whose positions are marked by underlining), is worth consideration:

> "Success in dealing with unknown ciphers is measured by these four things in the order named: perseverance, careful methods of analysis, intuition, luck. The ability at least to read the language of the original text is very desirable, but not essential." Such is the opening sentence of Parker Hitt's *Manual for the Solution of Military Ciphers*.

Colonel Parker Hitt (1877–1971) published in 1916 one of the first serious books in the U.S.A. on cryptology and dealt in this book for the first time with the systematic decryption of a PLAYFAIR encryption (Sect. 4.2.1). Hitt later became vice-president of AT&T and president of its cryptological offspring International Communication Laboratories. Hitt's sentence states that semantic support is not decisive for the success of unauthorized decryption and has been understood as encouraging mechanized solution of the laborious part (pure cryptanalysis).

Note that this decryption was carried out solely with frequency considerations, i.e., on the basis of Theorem 3 and Theorem $3^{(2)}$. Other aids like pattern finding and probable words were not given a place. More on mixed methods in Sect. 15.9 .

15.8.4 Matching a posteriori. The correct decryption shows a matching of the observed bigram frequencies and the expected frequencies based on the bigram probabilities. This is detailed in Table 18 for the thirteen most frequent letters after appropriate permutations. It can be interpreted as a way to break up the monogram cliques and leads to a combinatorial problem. A simple mechanical procedure for efficient execution of this optimal matching process is not known.

15.8.5 A different approach. Instead of looking for the plaintext belonging to a cryptotext, it is sometimes easier to reconstruct the encryption alphabet directly, provided it has been generated by a password along the method of Sect. 3.2.5 . This reconstruction could have been started in the example above as soon as the first nine letters were found:

	e	t	a	o	n	i	r	s	h	l	d	u	c
e	1.1	2.2	3.7	1.3	3.4	1.1	4.3	4.1	0.4	1.3	3.0	0.2	1.8
t	2.6	1.5	1.6	3.1	0.2	3.6	0.8	0.9	8.8	0.3	0.3	0.6	0.2
a	–	3.5	–	0.1	4.8	0.4	2.8	1.9	–	2.2	0.4	0.3	1.1
o	0.1	1.5	0.3	0.6	4.1	0.4	3.2	1.0	0.1	0.5	0.5	2.7	0.5
n	1.8	3.1	1.5	1.8	0.3	1.0	0.1	1.4	0.3	0.3	3.3	0.3	0.9
i	1.0	2.4	0.5	1.8	4.7	–	0.6	3.0	–	1.1	0.4	–	1.5
r	4.1	1.8	1.6	1.5	0.3	2.2	0.5	1.1	0.1	0.3	0.4	0.2	0.4
s	2.4	3.4	4.4	2.0	0.5	1.2	0.5	1.1	0.8	0.2	0.2	0.8	0.6
h	7.0	0.6	2.4	1.3	0.1	2.0	0.2	0.1	0.1	0.1	–	0.1	0.1
l	2.0	0.5	1.0	0.8	–	1.6	–	0.3	–	1.5	0.8	0.2	0.2
d	1.1	1.1	1.3	1.0	0.1	1.6	0.3	0.9	0.1	0.2	0.3	0.2	0.1
u	0.3	1.3	0.5	0.1	0.9	0.1	1.4	1.2	0.1	0.8	0.3	–	0.5
c	1.5	1.1	1.2	1.7	–	0.4	0.2	–	1.3	0.4	–	0.4	0.3

	N	H	U	A	T	O	F	G	Y	C	Z	J	X
N	–	1	5	5	4	–	7	5	–	1	3	–	1
H	3	2	1	1	2	4	–	1	9	1	–	1	–
U	–	1	–	–	4	–	3	2	–	5	1	–	–
A	1	1	–	–	2	–	4	–	–	1	1	1	–
T	1	4	1	–	–	3	3	–	1	–	1	3	1
O	–	4	1	2	7	–	1	4	–	2	–	–	–
F	3	2	2	–	1	1	–	2	1	–	1	–	–
G	4	2	–	2	–	6	–	2	–	–	–	3	–
Y	7	–	–	1	–	3	–	–	–	–	–	1	–
C	2	1	1	–	–	4	–	1	–	–	–	2	–
Z	3	1	–	1	–	–	–	1	–	–	–	1	–
J	–	2	2	–	1	2	2	–	–	1	–	–	3
X	3	–	1	–	–	2	–	–	1	–	–	–	1

Table 18. Expected bigram frequencies and observed ones after suitable character concordance

U	*	*	*	N	*	*	Y	O	*	*	*	*	T	A	*	*	F	G	H	*	*	*	*	*	*
a	b	c	d	e	f	g	h	i	j	k	l	m	n	o	p	q	r	s	t	u	v	w	x	y	z

There is a gap of two letters between A and F and two letters from {B, C, D, E} could be squeezed in, while two others would build up the password. This leads to six cases to be treated exhaustively, the attempt that henceforth succeeds being

U	*	*	*	N	E	*	Y	O	*	*	C	*	T	A	B	D	F	G	H	*	*	*	*	*	*
a	b	c	d	e	f	g	h	i	j	k	l	m	n	o	p	q	r	s	t	u	v	w	x	y	z

This is a highly speculative method, which nevertheless is intellectually challenging.

15.9 Frequency Matching for Polygraphic Substitutions

Polygraphic substitutions can be treated like simple substitutions if the m-grams are understood as individual characters. Nevertheless, this results in a large alphabet of N^m characters. But from the 676 bigrams of standard English normally only some hundred show up (Table 9), from the 17 576 trigrams not many more. m-grams have a markedly biased frequency distribution, facilitating an unauthorized entry.

15.9.1 A reducible case. Special bigram substitutions have peculiar methods to solve them. A trivial case is encryption with the standard matrix and permutations for column and row entries:

	a	m	e	r	i	c	...
e	AA	AB	AC	AD	AE	AF	...
q	BA	BB	BC	BD	BE	BF	...
u	CA	CB	CC	CD	CE	CF	...
a	DA	DB	DC	DD	DE	DF	...
l	EA	EB	EC	ED	EE	EF	...
i	FA	FB	FC	FD	FE	FF	...
⋮	⋮	⋮	⋮	⋮	⋮	⋮	

It can be reduced (Sect. 4.1.2) to a monographic 2-alphabetic encryption with period 2, which is treated in Chapter 17 .

15.9.2 Using a hidden symmetry. PLAYFAIR encryption (Sect. 4.2.1), once favored even by the British and German armies and also by amateurs, is not only of limited complexity, but has also a hidden torus symmetry. This has the following consequence: if a plaintext bigram contains the letter X , then the two letters of its encryption are selected among only eight letters, namely those in the row or in the column of X :

P	A	**L**	**M**	**E**		**L**	**M**	**E**	**P**	A		U	H	**I**	K	Q
R	**S**	T	O	N		T	O	N	R	**S**		Z	V	**W**	X	Y
B	**C**	D	F	G		D	F	G	B	**C**		**E**	**P**	A	**L**	**M**
H	**I**	K	Q	U		K	Q	U	H	**I**		N	R	**S**	T	O
V	**W**	X	Y	Z		X	Y	Z	V	**W**		G	B	**C**	D	F

Furthermore, the encryption of a reversed bigram is frequently the reversed encrypted bigram—in fact in all cases where a 'crossing step' was applied.

While the bigram frequencies are preserved under PLAYFAIR encryption, the individual character frequencies are not: there is a tendency for a larger clique of more frequent and a larger clique of less frequent characters to develop.

Intuitive attacks against PLAYFAIR are based on the bigram frequencies in connection with the peculiarities just mentioned. In practice, probable words are also used. Systematic treatments were first begun in 1916 by Colonel Parker Hitt, in 1918 by André Langie and in 1922 by W. W. Smith. In the Second World War, wherever PLAYFAIR was used it was routine solved; the modified PLAYFAIR (Sect. 4.2.2), used for example as a field cipher by the German *Afrika Korps*, fared no better.

The unauthorized decryptor of a polygraphic encryption normally has good reasons to assume that he knows the position of the multigram hiatus. That can be an error: If a message encrypted in PLAYFAIR is decorated with an odd number of initial nulls, the message is out of phase. It is not so much more difficult to try the two cases, but first one has to have the right idea.

15.10 Free-Style Methods

A clear separation of the methods, as made in this book, serves mainly the understanding and is indispensible if computer support is to be programmed. But an interplay of these methods, be it by a human or by a machine, can increase the efficiency of the attack. Inevitably, experienced cryptanalysts working 'manually' will combine available methods. The literature includes some pertinent reports by people like Bazeries, Hitt, Friedman—including amateurs like Babbage who were gifted with imagination.

15.10.1 A famous cryptogram. A particularly nice example has entered the world literature. In 1843, Edgar Allan Poe (1809–1849) wrote a short mystery story, "The Gold-Bug", containing an encrypted message and its solution. The alphabet is a funny hodgepodge made from figures and other symbols available to the printer—Poe was an *homme de lettres*. The cryptotext of 203 letters was like this:[2]

5 3 ‡ ‡ † 3 0 5)) 6 ∗ ; 4 8 2 6) 4 ‡ .) 4 ‡) ; 8 0 6 ∗ ; 4 8 † 8 ¶
6 0)) 8 5 ; 1 ‡ (; : ‡ ∗ 8 † 8 3 (8 8) 5 ∗ † ; 4 6 (; 8 8 ∗ 9 6 ∗
? ; 8) ∗ ‡ (; 4 8 5) ; 5 ∗ † 2 : ∗ ‡ (; 4 9 5 6 ∗ 2 (5 ∗ – 4) 8 ¶
8 ∗ ; 4 0 6 9 2 8 5) ;) 6 † 8) 4 ‡ ‡ ; 1 (‡ 9 ; 4 8 0 8 1 ; 8 : 8 ‡
1 ; 4 8 † 8 5 ; 4) 4 8 5 † 5 2 8 8 0 6 ∗ 8 1 (‡ 9 ; 4 8 ; (8 8 ; 4 (
‡ ? 3 4 ; 4 8) 4 ‡ ; 1 6 1 ; : 1 8 8 ; ‡ ? ;

Poe allows Legrand, the hero of the story, to begin with the remark that the crypto system (he calls it 'cryptograph') was adequate to the mental power of Captain Kidd, the bad guy of the story, thus impenetrable for a simple sailor, although it was 'a simple species'. Legrand, who boasts about having solved secret messages a thousand times more complicated, concludes that according to the geographic circumstances French or Spanish would come into consideration, but that fortunately the signature 'Kidd' clearly points to English. He also notes the lack of word-division spaces and complains that this makes the task more difficult. He therefore starts with a table of individual character frequencies:

33	26	19	16	16	13	12	11	10	8	8	6	5	5	4	4	3	2	1	1
8	;	4	‡	)	∗	5	6	(	†	1	0	9	2	:	3	?	¶	–	.

[2] The numerous reprints and translations are abundant in typographic errors within these six lines. It can be seen how difficult the work of a printer is, if he lacks the feedback control of semantics.

His first assumption is 8 ≐ e , which is backed by the frequent occurrence of a double /e/ in English—an argument on bigrams. Then he looks for the most frequent trigram /the/ , a pattern *123* with 8 at the end. He finds seven occurrences of ;48 , and therefore assumes ; ≙ t , 4 ≙ h .

"Thus, a great step has been taken." The entry is achieved. The last but one line, partly decrypted, reads:

```
1the†e5th)he5†52ee06*e1(‡9thet(eeth(    .
```

thet(ee towards the end of the line reminds Legrand immediately of (≙ r . This gives him thetreethr‡?3hthe and suggests /thetreethroughthe/ . Therefore ‡ ≙ o , ? ≙ u , 3 ≙ g . Next, in the second line he finds †83(88, i.e., †egree which suggests /degree/ and † ≙ d , and four characters later ;46(;88*, i.e., th6rtee* , to be read /thirteen/, whence 6 ≙ i and * ≙ n .

Now almost all of the frequent characters (except a and s) are determined. The partly decrypted text is:

```
5goodg05))inthe2i)ho.)ho)te0inthede¶
i0))e5t1ort:onedegree)5ndthirteen9in
ute)northe5)t5nd2:north95in2r5n-h)e¶
enth0i92e5)t)ide)hoot1ro9the0e1te:eo
1thede5th)he5d52ee0ine1r09thetreethr
oughthe)hot1i1t:1eetout
```

Instantly Legrand finds 5 ≙ a ,) ≙ s , as well as

0 ≙ l , 2 ≙ b , . ≙ p , ¶ ≙ v , 1 ≙ f , : ≙ y , 9 ≙ m , - ≙ c .

The (monoalphabetic) encryption step is the mapping involving 20 letters

8	;	4	‡	)	*	5	6	(	†	1	0	9	2	:	3	?	¶	-	.
e	t	h	o	s	n	a	i	r	d	f	l	m	b	y	g	u	v	c	p

and the plaintext in more readable form gives the clou:

"A good glass in the Bishop's hostel in the Devil's seat—forty-one degrees and thirteen minutes—northeast and by north—main branch seventh limb east side—shoot from the left eye of the death's-head—a bee-line from the tree through the shot fifty feet out."

15.10.2 Remark. Typically, there was not a word to say that it could not have been a polyalphabetic encryption. Poe was monoalphabetically minded.

15.11 Unicity Distance Revisited

Knowledge of the probability of an n-gram helps to understand how exhaustion in Sect. 12.7 accomplishes the sorting out of the 'right' plaintext and why a unicity distance exists. An unlikely sequence of characters is hardly a 'right' message, but we can hope that a sequence of characters with a probability near 1 may be 'right'. The unicity distance is the smallest length of a text that has probability near 1 for one of the possible decryptions and probability near 0 for all other possible decryptions.

	length 1	length 2	length 3	length 4	length 5
V F K R Q	0.76	0.02			
W G L S R	2.03	0.04			
X H M T S	0.01	0.01			
Y I N U T	0.01	0.06	0.06		
Z J O V U	1.21	0.03	0.05		
A K P W V	5.96	1.88	0.01		
B L Q X W	1.77	2.35			
C M R Y X	3.17	0.03			
D N S Z Y	5.22	1.44	0.01		
E O T A Z	17.98	1.58	0.11	1.27	
F P U B A	1.23	0.19	0.03		
G Q V C B	3.25				
H R W D C	4.61	9.54	0.45		
I S X E D	7.97	20.30	0.01		
J T Y F E	0.06				
K U Z G F	1.12	2.34			
L V A H G	3.19	0.71	0.11		
M W B I H	2.47	0.86			
N X C J I	11.06	0.03			
O Y D K J	2.00	0.14	0.01		
P Z E L K	0.59	0.05			
Q A F M L	0.01				
R B G N M	6.42	6.38	0.05		
S C H O N	7.48	22.84	90.51	98.73	100.00
T D I P O	5.55	9.09	8.56		
U E J Q P	4.87	20.09	0.03		
	100.00	100.00	100.00	100.00	100.00

Table 19. Step by step sorting out of the 'right' plaintext according to n-gram probabilities (in %)

A human sorts out the 'right' plaintext ('running down the list') by an optical and cerebral perception process, but this can be simulated by statistical analysis.

For the example of Table 5, beginning with the sixth columns, this is shown in Table 19. The multigram probabilities have been determined by the text basis SZ3-92 and are normalized to 100%, empty fields mean probabilities below 0.005%. The unicity length is in this example clearly 5.

This exhaustion, however, has its limits if it goes into the ten thousands of trials, and is inappropriate for full monoalphabetic encryption if no further information can be used.

No monoalphabetic substitution can
maintain security in heavy traffic.

David Kahn 1967

16 Kappa and Chi

Riverbank Publication No. 22,
written in 1920 when Friedman was 28,
must be regarded as the most important
single publication in cryptology.

David Kahn 1967

Astonishingly, given a monoalphabetically encrypted cryptotext, it is easier to say whether it is in English, French, or German, than to decrypt it. This is also true for plaintext: there is a reliable method to test a sufficiently long text for its membership of a known language, without 'taking notice' of it—without regarding its grammar and semantics—and there is a related test to decide whether two texts belong to the same language, without closely inspecting them.

Indeed, there exists a particular invariant of a text under monoalphabetic encryption, which is discussed in the following, and a related invariant of a pair of texts which is even invariant under a polyalphabetic encryption of both texts with the same key. And these invariants have peculiar values which differentiate between most of the common Indo-Germanic languages.

16.1 Definition and Invariance of Kappa

Given a pair of texts $T = (t_1, t_2, t_3, \ldots t_M)$, $T' = (t'_1, t'_2, t'_3, \ldots t'_M)$ of equal length $M > 1$ over the same vocabulary Z_N .

The relative frequency of finding in the two texts the same character at the same position (the character coincidence, marked by $*$) is called the *Kappa* of the pair of texts (William F. Friedman 1925, 'index of coincidence', often abbreviated *I.C.*). Thus

$$Kappa(T, T') = \sum_{\mu=1}^{M} \delta(t_\mu, t'_\mu)/M$$

with the indicator function ('delta function')

$$\delta(x, y) = \begin{cases} 1 & \text{if } x = y \\ 0 & \text{otherwise} \end{cases} .$$

Example 1 ($M = 180$)

```
T:   t h e p r e c e d i n g c h a p t e r h a s i n d i c a t e d h o w a m
T':  w o u l d s e e m t h a t o n e w a y t o o b t a i n g r e a t e r s e
                   *                                   *       *

     o n o a l p h a b e t i c c i p h e r c a n b e s o l v e d e v e n i f
     c u r i t y w o u l d b e t o u s e m o r e t h a n o n e a l p h a b e
                                       *                     *

     t h e o r i g i n a l w o r d l e n g t h s a r e c o n c e a l e d a n
     t i n e n c i p h e r i n g a m e s s a g e t h e g e n e r a l s y s t
     *                               *               *     *     * *

     d t h e s u b s t i t u t i o n a l p h a b e t i s r a n d o m i t i s
     e m c o u l d b e o n e t h a t u s e s a n u m b e r o f d i f f e r e
                             *               *           *     *

     p o s s i b l e t o f i n d a s o l u t i o n b y u s i n g f r e q u e
     n t a l p h a b e t s f o r e n c i p h e r m e n t w i t h a n u n d e
                                                           *               *
```

Example 2 ($M = 180$)

```
T:   e s t a u c h t v o n z e i t z u z e i t i m m e r w i e d e r e i n m
T':  u n t e r s c h w e i z e r p o l i t i k e r n w a e c h s t d i e a n
         *                 * *             *

     a l a u f u m k u r z d a r a u f e i l f e r t i g d e m e n t i e r t
     g s t d e n n a e c h s t e n z u g r i c h t u n g e g z u v e r p a s
                                                       *

     z u w e r d e n d a s g e r u e c h t d a s s s i c h d i e o e l e x p
     s e n a u s s e n m i n i s t e r r e n e f e l b e r s a h s i c h j e
                                   *

     o r t i e r e n d e n l a e n d e r v o m d o l l a r l o e s e n w o l
     t z t u e b e r r a s c h e n d e i n e r f o r d e r u n g a u s d e m
         *   *   *             * * * *           *       *

     l e n z u v e r d e n k e n w a e r e e s i h n e n f r e i l i c h n i
     s t a e n d e r a t a u s g e s e t z t e i n b e i t r i t t s g e s u
                 * *                 *         *     *     *
```

In example 1 (English) there results $Kappa(T, T') = 17/180 = 9.44\%$; in example 2 (German) $Kappa(T, T') = 21/180 = 11.67\%$.

16.1.1 Obviously

$$Kappa(T, T') \leq 1 \,, \quad \text{where}$$

$$Kappa(T, T') = 1 \quad \text{if and only if} \quad T \doteq T' \,.$$

There is the empirical result that sufficiently long texts $T \in \mathcal{S}, T' \in \mathcal{S}$ from one and the same language $\mathcal{S}$ (or rather from the same genre of this language) have values $Kappa(T, T')$ close to some $\kappa_{\mathcal{S}}$, while $\kappa_{\mathcal{S}}$ varies from language to language. In the literature the following values of $\kappa_{\mathcal{S}}$ are given:

$\mathcal{S}$	N	Kullback 1976	Eyraud 1953
English	26	6.61%	6.75%
German	26	7.62%	8.20%
French	26	7.78%	8.00%
Italian	26	7.38%	7.54%
Spanish	26	7.75%	7.69%
Japanese (Romaji)	26	8.19%	
Russian	32	5.29%	4.70%

The values in the literature fluctuate: 6.5 – 6.9% for English, 7.5 – 8.3% for German. From the text corpus mentioned in Sect. 15.5.4 , there results for the English language a value of $\kappa_e = 6.58\%$, from the text basis SZ3-92 for the German language a value of $\kappa_d = 7.62\%$, in good agreement with the values of Kullback. French and Spanish, for example, come very close.

The values for $\kappa_{\mathcal{S}}$, the empirical *Kappa* of a language $\mathcal{S}$, seem to reflect somewhat the redundancy of the languages: The translation of the Gospel of St. Mark, with 29 000 syllables in English (according to H. L. Mencken), needs in the Teutonic languages on average 32 650 syllables, in the Romance languages on average 40 200 syllables (36 000 in French), in the Slavic languages on average 36 500 syllables. But there is no strict connection.

16.1.2 Two results stand out:

Invariance Theorem 5: For all *poly*alphabetic, functional simple substitutions, especially for all polyalphabetic linear simple substitutions (including VIGENÈRE additions and BEAUFORT subtractions),
the Kappa of two texts of equal length, encrypted with the same key, is invariant.

Invariance Theorem 6: For all transpositions,
the Kappa of two texts of equal length, encrypted with the same key, is invariant.

16.1.3 The expectation value for the *Kappa* of two texts of equal length M over the same vocabulary Z_N is calculated from the probabilities p_i, p'_i of the appearance of the i-th character in the 'stochastic sources' Q, Q' of the texts: The expectation value for the appearance of the character χ_i in the μ-th position of the texts is $p_i \cdot p'_i$; which gives the expectation value for $Kappa(T, T')$

$$\langle Kappa(T,T')\rangle_{QQ'} = \sum_{i=1}^{N} p_i \cdot p'_i \; .$$

If the two sources are identical, $Q' = Q$, then $p'_i = p_i$ and

$$(*) \qquad \langle Kappa(T,T')\rangle_{Q} = \sum_{i=1}^{N} p_i^2 \; .$$

This equation relates the definition of *Kappa* with the classical urn experiment of probability theory.

Theorem: For identical sources $Q' = Q$,

$$\tfrac{1}{N} \leq \langle Kappa(T,T') \rangle_Q \leq 1 \; ;$$

the left bound is attained for the case of equal distribution: $Q_\mathrm{R} : p_i = \frac{1}{N}$, and only for this case; the right bound is attained for every deterministic distribution $Q_j : p_j = 1,\ p_i = 0$ for $i \neq j$, and for no other distribution.

As said above, from the hypothetical probability distribution in Sect. 15.5.4, Table 8,

$$\langle Kappa(T,T') \rangle_\mathrm{English} = 0.06577 = \kappa_\mathrm{e} \; ,$$

$$\langle Kappa(T,T') \rangle_\mathrm{German} = 0.07619 = \kappa_\mathrm{d} \; .$$

For the source with equal distribution Q_R $(N = 26)$,

$$\langle Kappa(T,T') \rangle_\mathrm{R} = \kappa_\mathrm{R} = 0.03846 = \tfrac{1}{26} \; .$$

Thus, the *Kappa* test differentiates English and German sources clearly from a source with equal distribution:

$\kappa_\mathrm{d}/\kappa_\mathrm{R} = N \cdot \kappa_\mathrm{d} = 1.98$, $\kappa_\mathrm{e}/\kappa_\mathrm{R} = N \cdot \kappa_\mathrm{e} = 1.71$.

A rule of thumb for the common languages is:

The ratio $\langle Kappa(T,T') \rangle_\mathrm{S} / \langle Kappa(T,T') \rangle_\mathrm{R}$ *is close to two.*

16.2 Definition and Invariance of Chi

Given again two texts of equal length M over the same vocabulary of N characters, $T = (t_1, t_2, t_3, \ldots t_M)$, $T' = (t'_1, t'_2, t'_3, \ldots t'_M)$. Let m_i, m'_i denote the frequency of the appearance of the character χ_i in the texts T, T', respectively; then $\sum_{i=1}^{N} m_i = M$, $\sum_{i=1}^{N} m'_i = M$.

Chi denotes the 'cross-product sum' (Solomon Kullback, 1935)

$$Chi(T,T') = \Big(\sum_{i=1}^{N} m_i \cdot m'_i\Big) / M^2 \; .$$

Written homogeneously, the definition is

$$Chi(T,T') = \Big(\sum_{i=1}^{N} m_i \cdot m'_i\Big) \Big/ \Big(\Big(\sum_{i=1}^{N} m_i\Big) \cdot \Big(\sum_{i=1}^{N} m'_i\Big) \Big) \; .$$

The two texts in Sect. 16.1, Example 1, have the following frequencies

	a	b	c	d	e	f	g	h	i	j	k	l	m	n	o	p	q	r	s	t	u	v	w	x	y	z
T	15	6	8	9	21	3	4	10	17	0	0	8	2	14	13	6	1	8	11	14	5	2	2	0	1	0
T'	15	6	4	5	30	4	4	9	8	0	0	6	6	15	12	4	0	10	10	17	8	0	4	0	3	0

This results in $Chi(T,T') = \frac{2151}{180 \cdot 180} = 6.64\%$.

For the two texts in Sect. 16.1, Example 2, the frequencies are

	a	b	c	d	e	f	g	h	i	j	k	l	m	n	o	p	q	r	s	t	u	v	w	x	y	z
T	10	0	4	11	35	4	2	5	15	0	2	10	6	14	7	1	0	15	7	9	9	3	4	1	0	6
T'	11	3	6	6	33	2	7	7	12	1	1	2	2	16	2	2	0	15	18	16	10	1	2	0	0	5

This results in $Chi(T,T') = \frac{2492}{180 \cdot 180} = 7.69\%$.

16.2.1 In analogy to Sect. 16.1.1, for simple geometric reasons

$$Chi(T,T') \leq 1, \quad \text{where}$$
$$Chi(T,T') = 1 \quad \text{if and only if } T \text{ and } T' \text{ are built from one and the same character.}$$

If all m_i are equal, $m_i = M/N$, then (for arbitrary m'_i)

$$Chi(T,T') = \tfrac{1}{N} = \kappa_R .$$

Empirically, one finds again that for sufficiently long texts from one and the same language $\mathcal{S}$ (or from the same genre of this language) not only are values of *Chi* rather close to some value typical for the language, but also this value is close to the value of *Kappa* for this language. This will be clarified in Sect. 16.3.

16.2.2 There is the important special case $T' = T$, $m'_i = m_i$. Let

$$Psi(T) = Chi(T,T) = \sum_{i=1}^{N} m_i^2/M^2 .$$

From Steiner's theorem, $\sum_{i=1}^{N}(m_i - \frac{M}{N})^2/M^2 = \sum_{i=1}^{N} m_i^2/M^2 - \frac{1}{N}$,

$$\tfrac{1}{N} \leq Psi(T) \leq 1, \quad \text{where}$$
$$Psi(T) = 1 \quad \text{if and only if } T \text{ is built from one and the same character,}$$
$$Psi(T) = \tfrac{1}{N} = \kappa_R \quad \text{if and only if all } m_i \text{ are equal.}$$

For the extreme case $M \leq N$ even

$$\tfrac{1}{M} \leq Psi(T), \quad \text{where} \quad Psi(T) = \tfrac{1}{M} \text{ if and only if } m_i \in \{0,1\}.$$

16.2.3 *Chi* and *Psi* have invariance properties, too. In contrast to *Kappa*, there is a weaker statement:

Invariance Theorem 7: For all *mono*alphabetic, functional simple substitutions, especially for all monoalphabetic linear simple substitutions (including VIGENÈRE additions and BEAUFORT subtractions),
the Chi of two texts of equal length, encrypted with the same key, as well as the Psi of a text, are invariant.

Invariance Theorem 8: For all transpositions,
the Chi of two texts of equal length, encrypted with the same key, as well as the Psi of a text, are invariant.

In so far as *Kappa*, *Chi* or *Psi* are characteristic for a language, the language can be determined from the cryptotext.

16.2.4 The expectation value for the *Chi* of two texts $T \in \mathcal{S}, T' \in \mathcal{S}$ of equal length M over the same vocabulary Z_N is calculated from the probabilities p_i, p'_i of the appearance of the i-th character in the 'stochastic sources' Q, Q' of the texts: The expectation value for the multitude of the character χ_i in T is $p_i \cdot M$, in T' is $p'_i \cdot M$, giving the expectation value for $Chi(T,T')$

$$\langle Chi(T,T')\rangle_{QQ'} = \sum_{i=1}^{N} p_i \cdot p'_i \ .$$

If the two sources are identical, $Q' = Q$, then $p'_i = p_i$ and

$$(*) \qquad \langle Chi(T,T')\rangle_{Q} = \sum_{i=1}^{N} p_i^2 \ .$$

In particular,

$$\langle Psi(T)\rangle_{Q} = \sum_{i=1}^{N} p_i^2 \ .$$

Theorem: For identical sources $Q' = Q$,

$$\tfrac{1}{N} \leq \langle Chi(T,T')\rangle_{Q} \leq 1 \ ,$$

and in particular

$$\tfrac{1}{N} \leq \langle Psi(T)\rangle_{Q} \leq 1 \ .$$

The left bound is attained for the case of equal distribution: $Q_{\mathrm{R}} : p_i = \frac{1}{N}$, and only for this case; the right bound for every deterministic distribution $Q_j : p_j = 1$, $p_i = 0$ for $i \neq j$, and for no other distribution.

Amazingly, the expectation values marked by (*) $\langle Kappa(T,T')\rangle_{Q}$ (see 16.1.3) and $\langle Chi(T,T')\rangle_{Q}$ coincide. It will turn out that there is a relation even between $Kappa(T,T')$ and $Chi(T,T')$.

16.3 The Kappa-Chi Theorem

For the following, we need two auxiliary functions $g_{i,\mu}$, $g'_{i,\mu}$.

Let $g_{i,\mu} = \begin{cases} 1 & \text{if } t_\mu \text{, the } \mu\text{-th character of } T \text{, equals } \chi_i \\ 0 & \text{otherwise} \end{cases}$;

let $g'_{i,\mu}$ for T' be defined correspondingly. Then

$$\delta(t_\mu, t'_\nu) = \sum_{i=1}^{N} g_{i,\mu} \cdot g'_{i,\nu} \quad \text{and} \quad m_i = \sum_{\mu=1}^{M} g_{i,\mu} \ , \quad m'_i = \sum_{\nu=1}^{M} g'_{i,\nu} \ .$$

16.3.1 Let $T^{(r)}$ be the text T shifted cyclically by r positions to the right. Then the number of coincidences between $T^{(r)}$ and T' is

$$Kappa(T^{(r)}, T') = \sum_{\mu=1}^{M} \delta(t_{(\mu-r-1) \bmod M+1}, t'_{\mu})/M \ .$$

In particular, $Kappa(T^{(0)}, T') = Kappa(T, T')$.

16.3.2 We now formulate the

Kappa-Chi Theorem: $$\frac{1}{M} \sum_{\rho=0}^{M-1} Kappa(T^{(\rho)}, T') = Chi(T, T') \ .$$

Thus, $Chi(T, T')$ is the arithmetic mean of all $Kappa(T^{(r)}, T')$.

Corollary : $$\frac{1}{M} \sum_{\rho=0}^{M-1} Kappa(T^{(\rho)}, T) = Psi(T) \ .$$

Proof:

$\frac{1}{M} \sum_{\rho=0}^{M-1} Kappa(T^{(\rho)}, T') =$
$\frac{1}{M} \cdot \frac{1}{M} \cdot \sum_{\rho=0}^{M-1} \sum_{\mu=1}^{M} \delta(t_{(\mu-\rho-1) \bmod M+1}, t'_{\mu}) =$
$\frac{1}{M} \cdot \frac{1}{M} \cdot \sum_{\nu=1}^{M} \sum_{\mu=1}^{M} \delta(t_{\mu}, t'_{\nu}) =$
$\frac{1}{M} \cdot \frac{1}{M} \cdot \sum_{\nu=1}^{M} \sum_{\mu=1}^{M} \sum_{i=1}^{N} g_{i,\mu} \cdot g'_{i,\nu} =$
$\frac{1}{M} \cdot \frac{1}{M} \cdot \sum_{i=1}^{N} \sum_{\nu=1}^{M} \sum_{\mu=1}^{M} g_{i,\mu} \cdot g'_{i,\nu} =$
$\frac{1}{M} \cdot \frac{1}{M} \cdot \sum_{i=1}^{N} (\sum_{\nu=1}^{M} g'_{i,\nu}) \cdot (\sum_{\mu=1}^{M} g_{i,\mu}) =$
$\frac{1}{M} \cdot \frac{1}{M} \cdot \sum_{i=1}^{N} m'_i \cdot m_i =$
$Chi(T, T')$. ⋈

It now becomes evident that in Sect. 16.1 the values for *Kappa* with 9.44% and 11.67% (accidentally) turned out rather high compared with the average values 6.64% and 7.69% in Sect. 16.2 .

16.4 The Kappa-Phi Theorem

The case $T' = T$ shows the peculiarity that $Kappa(T^{(0)}, T) = 1$, while for $r \neq 0$ essentially smaller 'normal' values of $Kappa(T^{(r)}, T)$ are found. Thus, the case $r = 0$ is untypical in the averaging process, and it would be more natural to extend the mean over the remaining $m-1$ cases only:

$$\frac{1}{M-1} \sum_{\rho=1}^{M-1} Kappa(T^{(\rho)}, T) \ .$$

16.4.1 Now
$\frac{1}{M-1} \cdot \sum_{\rho=1}^{M-1} Kappa(T^{(\rho)}, T) =$
$\frac{1}{M-1} \cdot (\sum_{\rho=0}^{M-1} Kappa(T^{(\rho)}, T) - 1) =$
$\frac{1}{M-1} \cdot (M \cdot Psi(T) \ - 1) =$
$\frac{1}{M-1} \cdot (\sum_{i=1}^{N} (m_i^2/M \ - 1)) =$
$\frac{1}{M-1} \cdot \frac{1}{M} \cdot (\sum_{i=1}^{N} (m_i^2 \ - M)) =$
$\frac{1}{M-1} \cdot \frac{1}{M} \cdot (\sum_{i=1}^{N} (m_i^2 \ - m_i)) =$
$\frac{1}{M-1} \cdot \frac{1}{M} \cdot (\sum_{i=1}^{N} m_i \cdot (m_i - 1))$.

Thus, we define a new quantity

$$Phi(T) = (\sum_{i=1}^{N} m_i \cdot (m_i - 1))/(M \cdot (M-1))$$

and state the

Kappa-Phi Theorem: $$\frac{1}{M-1} \sum_{\rho=1}^{M-1} Kappa(T^{(\rho)}, T)) = Phi(T) .$$

The calculation of $Phi(T)$ presents the small advantage, compared with $Psi(T)$, that not only for the case $m_i = 0$ but also for $m_i = 1$ nothing is contributed to the sum. This is useful for the rare letters in short texts. Note that $Phi(T) = 0$ holds if and only if all $m_i \in \{0, 1\}$. But there is another reason why people in the field work predominantly with *Phi* instead of *Psi*: it was Solomon Kullback who, using suitable stochastic arguments, first proposed the test for *Phi* (apart from the test for *Chi*).

Example 3: For the cryptotext T ($M = 280$) of Sect. 15.8.1, with the frequencies stated there,

$280^2 \cdot Psi(T) =$ 289+16+169+0+49+289+ 529+676+25+144+9+4+4+ 1296+625+1+25+0+0+529+400+9+36+81+169+64 = 5438 ,

$280 \cdot 279 \cdot Phi(T) =$ 272+12+156+0+42+272+ 506+650+20+132+6+2+2+ 1260+600+0+20+0+0+506+380+6+30+72+156+56 = 5158 , thus

$Psi(T) = 5438/78400 = 6.936\%$; $Phi(T) = 5158/78120 = 6.603\%$.

Moreover, with the frequencies of bigrams, i.e., with the text $T^{**} = T \times T^{(1)}$, one obtains

$Psi(T^{**}) = 871/77841 = 1.119\%$; $Phi(T^{**}) = 592/77562 = 0.763\%$.

16.4.2 $Phi(T)$ is not very different from $Psi(T)$:

$Psi(T) = \frac{M-1}{M} Phi(T) + \frac{1}{M} = Phi(T) + \frac{1}{M}(1 - Phi(T)),$
$Phi(T) = \frac{M}{M-1} Psi(T) - \frac{1}{M-1} = Psi(T) + (\frac{M}{M-1} - 1) Psi(T) - \frac{1}{M-1}$ and

$Psi(T) - Phi(T) = \frac{1-Phi(T)}{M} = \frac{1-Psi(T)}{M-1}$, thus $Phi(T) \leq Psi(T)$.

16.4.3 *Phi* has the same invariance properties as *Psi*:

Invariance Theorem $7^{(\mathrm{phi})}$: For all *mono*alphabetic, functional simple substitutions, especially for all monoalphabetic linear simple substitutions (including VIGENÈRE additions and BEAUFORT subtractions),
the Phi of a text is invariant.

Invariance Theorem $8^{(\mathrm{phi})}$**:** For all transpositions,
the Phi of a text is invariant.

16.4.4 The expectation value for the *Phi* of a text T of length M is calculated likewise from the probabilities p_i of the appearance of the i-th character in the 'stochastic source' Q of the text: The expectation value for $Phi(T)$ depends on M, too:

$\langle Phi(T)\rangle_Q^{(M)} = \frac{M}{M-1}\cdot\left(\sum_{i=1}^{N} p_i\cdot(p_i-\frac{1}{M})\right)$. Thus

$$\langle Phi(T)\rangle_Q^{(M)} \geq \begin{cases} \frac{M}{M-1}\cdot(\frac{1}{N}-\frac{1}{M}) = \frac{1}{N}\cdot\frac{M-N}{M-1} & \text{if } M \geq N \\ 0 & \text{if } M \leq N \end{cases};$$

equality holds if and only if all m_i are equal.

As M gets larger and larger, the expectation value for $Phi(T)$ approaches the expectation value for $Psi(T)$, namely

$$\langle Phi(T)\rangle_Q^{(\infty)} = \sum_{i=1}^{N} p_i^2 .$$

16.5 Symmetric Functions of Character Frequencies

The invariance stated in Theorems 7 and 8 for *Psi* holds for all symmetric functions of the character frequencies m_i. The simplest nonconstant polynomial function is indeed $\sum_{i=1}^{N} m_i^2$. It is a member of the following interesting family:[1]

$$Psi_a(T) = \begin{cases} \left(\sum_{i=1}^{N}(m_i/M)^a\right)^{1/(a-1)} & \text{if } 1 < a < \infty \\ \exp\left(\sum_{i=1}^{N}(m_i/M)\cdot\ln(m_i/M)\right) & \text{if } a = 1 \\ \max_{i=1}^{N}(m_i/M) & \text{if } a = \infty \end{cases}$$

with the normalization $\sum_{i=1}^{N}(m_i/M) = 1$. Psi_2 is Psi. Generalizing the result of Sect. 16.2.2,

For all a in the domain $1 \leq a \leq \infty$,

$$Psi_a(T) = \frac{1}{N} = \kappa_{\mathrm{R}} \qquad \text{if and only if all } m_i \text{ are equal.}$$

The interesting functions are Psi_1 and Psi_∞, which are the continuous limit functions of the family. Psi_1 has also the representation

$$Psi_1(T) = \prod_{i=1}^{N}(m_i/M)^{m_i/M} .$$

[1] With $x\cdot\ln x \nearrow 0$ for $x \searrow 0$; $x^x \nearrow 1$ for $x \searrow 0$.

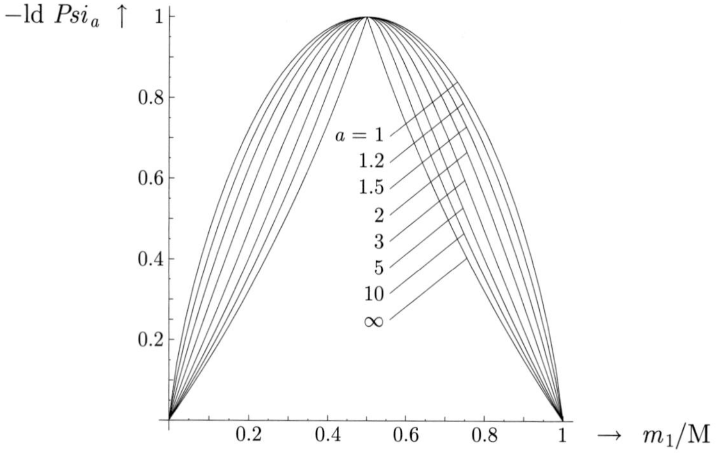

Fig. 114. Graph of the *Renyi* a-entropy for $N = 2$

The logarithmic quantity $-\mathrm{ld}\, Psi_a(T)$ is called the Renyi a-entropy of T (Rényi, 1960)[2]; the family has the following representation:

$$-\mathrm{ld}\, Psi_a(T) = \begin{cases} -\frac{1}{a-1} \cdot \mathrm{ld}\,(\sum_{i=1}^{N} (m_i/M)^a) & \text{if } 1 < a < \infty \\ -(\sum_{i=1}^{N} (m_i/M) \cdot \mathrm{ld}\,(m_i/M)) & \text{if } a = 1 \\ -\max_{i=1}^{N} \mathrm{ld}\,(m_i/M) & \text{if } a = \infty \end{cases}$$

Renyi 1-entropy $-\mathrm{ld}\, Psi_1$ is the Shannon entropy (Shannon 1945)[3].
Renyi 2-entropy $-\mathrm{ld}\, Psi_2$ could be named the Kullback entropy. Figure 114 shows the graph of $-\mathrm{ld}\, Psi_a$ for $N = 2$ and for some values of a.

For the English text T ($M = 280$) of Sect. 15.8.1, for single characters,

$$\begin{array}{lll} Psi_1(T) = 5.852\% & -\mathrm{ld}\, Psi_1(T) = 4.095 & \\ Psi_2(T) = 6.936\% & -\mathrm{ld}\, Psi_2(T) = 3.850 & \text{(Sect. 16.4.1)} \\ Psi_\infty(T) = 12.857\% & -\mathrm{ld}\, Psi_\infty(T) = 2.959\ . & \end{array}$$

For bigrams, the entropy values are slightly smaller,

$$\begin{array}{lll} \sqrt{Psi_1(T \times T^{(1)})} = 9.37\% & -\frac{1}{2}\mathrm{ld}\, Psi_1(T \times T^{(1)}) = 3.42 & \\ \sqrt{Psi_2(T \times T^{(1)})} = 10.58\% & -\frac{1}{2}\mathrm{ld}\, Psi_2(T \times T^{(1)}) = 3.24 & \text{(Sect. 16.4.1)} \\ \sqrt{Psi_\infty(T \times T^{(1)})} = 17.96\% & -\frac{1}{2}\mathrm{ld}\, Psi_\infty(T \times T^{(1)}) = 2.48\ . & \end{array}$$

---

[2] Alfréd Rényi (1921–1970), Hungarian mathematician.

[3] Claude E. Shannon (1916–2001), American mathematician, engineer, and computer scientist, first became famous in 1937 with a publication on relay circuits and Boolean algebra (*A Symbolic Analysis of Relay and Switching Circuits. Trans. AIEE 57*, 713-723, 1938). In 1941, at Bell Laboratories, he worked on mathematical problems in the communication of noisy and secret messages. This led him into information theory (*A Mathematical Theory of Communication, Bell System Technical Journal*, July 1948, p. 379, Oct. 1948, p. 623 and, together with Warren Weaver, *Mathematical Theory of Communication.* Univ. of Illinois Press, Urbana 1949).

17 Periodicity Examination

> It may be laid down as a principle that it is never worth the trouble of trying any inscrutable cypher unless its author has himself deciphered some very difficult cypher.
>
> *Charles Babbage* 1854

> The Babbage rule would have deprived cryptologists of some of the most important features of modern cryptography, such as the Vernam mechanism, the rotor, the Hagelin machine.
>
> *David Kahn* 1967

Even if a multitude of independent alphabets is used, periodic polyalphabetic encryption contains one element which is difficult to hide: the number of keys in the period of the encryption. This is based on the following stationariness property of stochastic sources: If P is a plaintext (of length M) from a source Q, then $P^{(s)}$, the plaintext P shifted cyclically by any number s of positions, is from the same source.

Theorem 1: Let p_i be the probability for the appearance of the i-th character in the source Q. Let d be the period of a periodic, polyalphabetic, functional, simple and monopartite encryption (for simplicity we assume that $d|M$). Then the encryption C of a plaintext P and $C^{(k \cdot d)}$, shifted cyclically by $k \cdot d$ positions, are from the same source, therefore

$$\langle Kappa(C^{(k\cdot d)}, C)\rangle_Q = \sum_{i=1}^{N} p_i^2 \quad \text{for all } k .$$

Proof: The encryption of $P^{(k\cdot d)}$, the cyclically shifted P, coincides with $C^{(k\cdot d)}$, the cyclically shifted encryption of P. According to Sect. 16.1.3 (*), $\langle Kappa(C^{(k\cdot d)}, C)\rangle_Q = \langle Kappa(P^{(k\cdot d)}, P)\rangle_Q = \sum_{i=1}^{N} p_i^2$. ⋈

On the other hand, such a statement cannot be made on $\langle Kappa(C^{(u)}, C)\rangle_Q$, where $d \not| u$. As a rule, $C^{(u)} = C'$ and C come from stochastic sources Q' and Q which are independent of each other. Thus, if u is not a multiple of d,

$$\langle Kappa(C^{(u)}, C)\rangle_{Q'Q} = \sum_{i=1}^{N} p_i' p_i \text{ may fluctuate around } \tfrac{1}{N} .$$

At least this is so if there are enough alphabets, and if they are chosen such that they achieve a thorough mixing of the character probabilities.

G E I E I	A S G D X	V Z I J Q	L M W L A	A M X Z Y	Z M L W H
F Z E K E	J L V D X	W K W K E	T X L B R	A T Q H L	B M X A A
N U B A I	V S M U K	H S S P W	N V L W K	A G H G N	U M K W D
L N R W E	Q J N X X	V V O A E	G E U W B	Z W M Q Y	M O M L W
X N B X M	W A L P N	F D C F P	X H W Z K	E X H S S	F X K I Y
A H U L M	K N U M Y	E X D M W	B X Z S B	C H V W Z	X P H W L
G N A M I	U K				

Fig. 115. Cryptotext of G. W. Kulp

17.1 The Kappa Test of Friedman

17.1.1 William F. Friedman proposed plotting $Kappa(C^{(u)}, C)$, the index of coincidence between $C^{(u)}$ and C. For the cryptotext in Fig. 115, this plot is shown in Fig. 116 (without $u = 0$, which is outside the frame). For multiples of 12, high values are obtained, indicating that 12 might be a period.

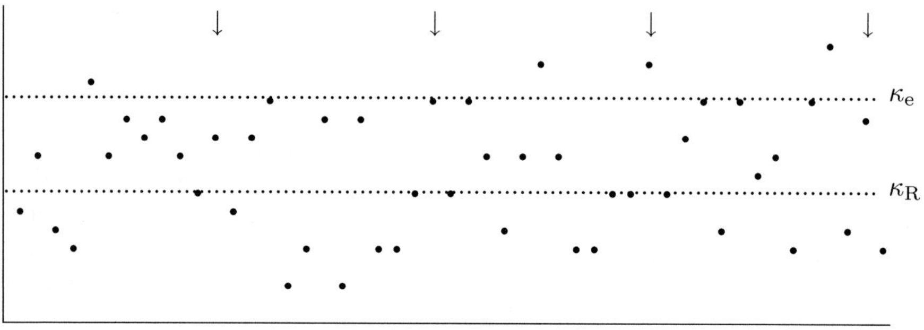

Fig. 116. *Kappa* plot for the (English) cryptotext of G. W. Kulp

17.1.2 The cryptotext of Fig. 115 has a history. It was submitted by a Mr. G. W. Kulp to a newspaper in Philadelphia, *Alexander's Weekly Messenger*, following a request by Edgar Allan Poe to send in monoalphabetically encrypted texts with word divisions preserved. It was published February 26, 1840 (Fig. 118). Poe demonstrated in a later issue that the alleged crypto was not following the rules—he did so by reducing any monoalphabetic substitution of proper English words leading to /mw/, /laam/ and /mlw/ to a contradiction, and stated that the crypto, "a jargon of random characters having no meaning whatsoever" was "an imposition". A glance at the frequency distribution, shown in Fig. 117, indeed reveals its balanced nature. Thus, frequency matching could not work.

A	B	C	D	E	F	G	H	I	J	K	L	M	N	O	P	Q	R	S	T	U	V	W	X	Y	Z
12	7	2	5	10	4	6	9	6	3	10	12	14	9	2	4	4	2	7	2	7	7	16	15	4	8

Fig. 117. Frequency distribution in the cryptotext of G. W. Kulp

17.1.3 This suggests that the encryption might be polyalphabetic. In fact, this gives a value of $\frac{1586}{187\cdot186} = 4.56\%$ for *Phi*, or $\frac{1773}{187^2} = 5.07\%$ for *Psi*, close to $\kappa_{\mathrm{R}} = \frac{1}{N}$ and too low for monoalphabetic encryption of an English text. Bigram substitution was also excluded by the rules, and PLAYFAIR was only invented in 1854. Anyhow, as already said, Poe was strictly monoalphabetically minded.

"Ge Jeasgdxv,
Zij gl mw, laam, xzy zmlwhfzek
ejlvdxw kwke tx lbr atgh lbmx aanu
bai Vsmukkss pwn vlwk agh gnumk
wdlnzweg jnbxvv oaeg enwb zwmgy
mo mlw wnbx mw al pnfdcfpkh wzkex
hssf xkiyahul. Mk num yexdm wbxy
sbc hv wyx Phwkgnamcuk?"

Fig. 118. Facsimile of the cryptotext of G. W. Kulp (1840) (it was found out later that the printer made several errors, e.g., reading q as g, and also suppressed one letter)

17.1.4 Figure 116 shows that a few values of *Kappa* come close to κ_{e} but most of them are slightly above or below κ_{R}. Large values of *Kappa* caused by a period should also show large values for all multiples; this rather excludes 5 and 15, while 12 cannot be dismissed. For the cryptotext of Kulp, a monographic encryption, polyalphabetic with a period of twelve is a promising hypothesis, but no more.

The cryptotext of Kulp, with its 187 characters, is rather short; for longer texts the multiples of a period stand out much better. This can be seen in Fig. 119 for a text of 300 characters and in Fig. 120 for a text of 3 000 characters, where the period catches the eye.

17.2 Kappa Test for Multigrams

The *Kappa* plot is not limited to single characters. Bigrams and more generally multigrams can be understood as characters, which however enlarges the vocabulary considerably.

For bigrams, $\kappa_{\mathrm{R}}^{**} = \frac{1}{N^2} = 14.8\%\%$, for trigrams $\kappa_{\mathrm{R}}^{***} = \frac{1}{N^3} = 0.569\%\%$. It is only important, how much κ_S^{**} is bigger than κ_{R}^{**}, and it turns out that the factor of about 2 in the monographic case is replaced by a factor 4.5 – 7.5 for bigrams (Fig. 121): For English, according to Kullback, κ_{e}^{**} is close to 69%% , for German, according to Kullback and Bauer, κ_{d}^{**} is close to 112%% . This means a clearer separation of the levels. For trigrams, the factor is about 40, but even with 3 000 characters the fluctuation is remarkable (Fig. 122). There is a limit to everything.

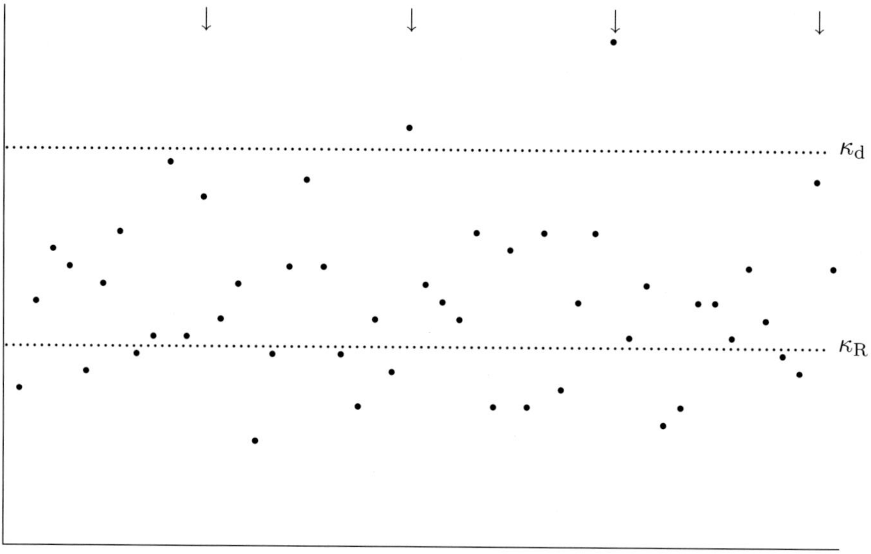

Fig. 119. *Kappa* plot for a (German) text of 300 characters

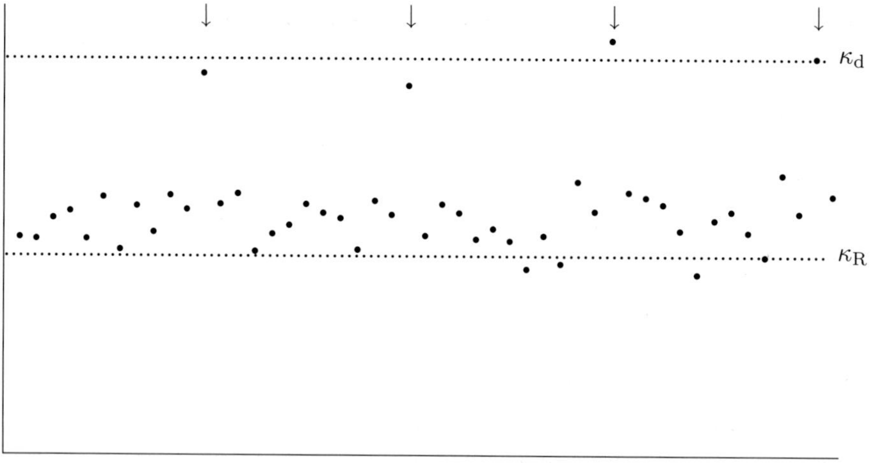

Fig. 120. *Kappa* plot for a (German) text of 3 000 characters

17.3 Cryptanalysis by Machines

17.3.1 Use of punch cards. It can be safely assumed that in the U.S.A. the methods of Friedman and Kullback were applied during the Second World War, and this by machines. As early as in 1932, Thomas H. Dyer of the U.S. Navy had used IBM punched card accounting machines for speeding up the work, the U.S. Army followed in 1936. In 1941, the year of Pearl Harbor, SIS, the Signal Intelligence Service of the U.S. Army, had 13 accounting

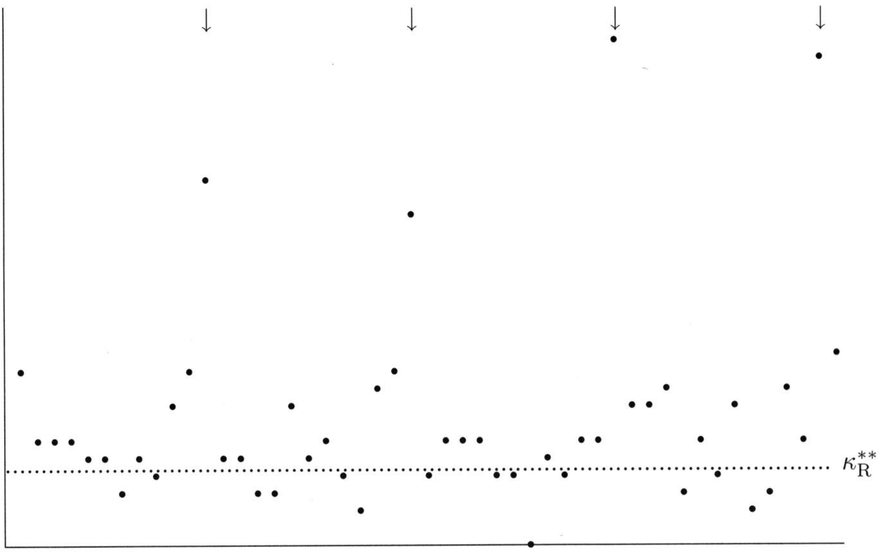

Fig. 121. *Kappa* plot for bigrams (German text of 3 000 characters)

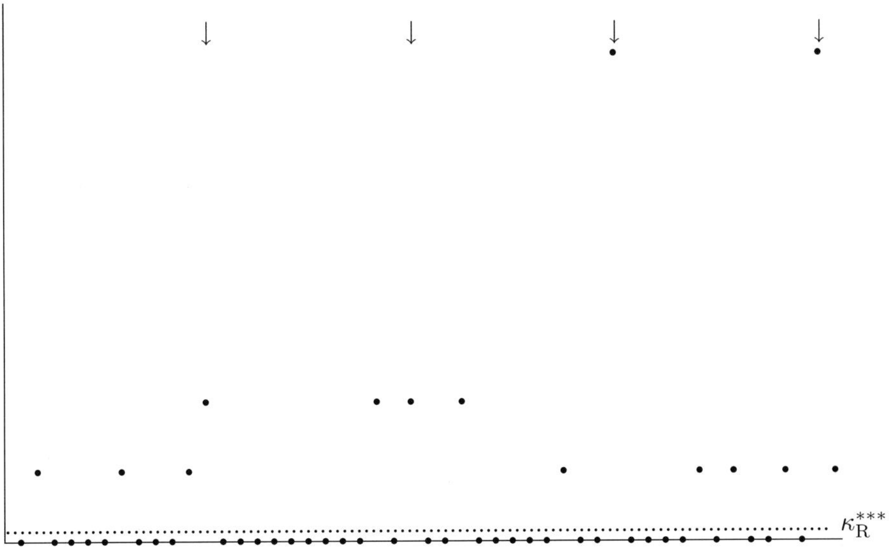

Fig. 122. *Kappa* plot for trigrams (German text of 3 000 characters)

machines at work, in 1945, at Arlington Hall, the number was 407 machines. IBM received \$750 000 per year for rent.

In Germany, accounting machines were used, too. They were, as elsewhere, particularly needed (see also Sect. 18.6.3) for stripping superencryption from codes. But they were also helpful for performing a *Kappa* test. For this purpose, they were also used (according to Kahn and Takagi) by the Japanese.

Forerunners of such automatic processing by punch card machinery were perforated sheets of paper ('overlay sheets'), used in England and elsewhere. If used for coincidence counts, the cryptotext was recorded in a binary 1-out-of-26 code by punched holes. While for a single coincidence count it suffices to write the texts one below the other, as done in Sect. 16.1, for a *Kappa* test the count is to be done repeatedly for shifted texts. Then the extra effort in preparing the overlay sheets is worthwhile, since the coincidences are seen 'at a glance' by light shining through the hole. This coincidence determination is not only much quicker, it is also more reliable. Figure 123a shows such sheets as they were used in Bletchley Park, where they were called 'Banbury sheets', because the punching was done in Banbury, a nearby small town. Single character coincidences as well as multigram coincidences can be detected and counted this way (Fig. 123b).

Punching the Banbury sheets can be done by hand. Using some simple machinery, a more refined coding, which saves paper, can be made, e.g., a 2-out-of-5 code for use in encoding decimal digits, or a 2-out-of-10 code, enough to encode both letters and digits, as was used in the German OKW by Willi Jensen. However, these codings, including the teletype 5-bit code, require more complicated means for automatic detection and registration of coincidences.

17.3.2 Saw-buck. In the Cipher Branch of the German OKW, in jargon dubbed Chi, Group IV *Analytische Kryptanalyse*, headed by Erich Hüttenhain, had a special device built by Willi Jensen for the determination of coincidences ('Doppler') and distances. Called the *Perioden- und Phasensuchgerät* (Fig. 124), it worked with two identical 5-channel teletype punched tapes, closed into a loop. One of the loops contained an additional blank punch. With each completed cycle through a pair of scanners, the phase between the two messages was shifted by one position. This 'saw-buck' principle seems to have been well known and used at several other places, too. The scanners were photoelectric and used for comparison a relay circuitry (*Zeichenvergleichslabyrinth*). The recording was done mechanically, and for a given shift the length of a stroke was proportional to the number of coincidences. After a completed cycle the recording unit moved forward one position.

Moreover, a second recording unit counted only the bigram coincidences, a third one the trigram coincidences, and so on up to 10-gram coincidences ('parallels'). The devices automatically gave a *Kappa* plot for single characters, bigrams, etc. With a scanning speed of 50 characters per second this took two hours for a text of 600 characters and was about a hundred times faster than work by hand. The device was destroyed at the end of the war.

The available material on the work of Jensen contains no references to Friedman, but it can be safely assumed that at least his early, published work

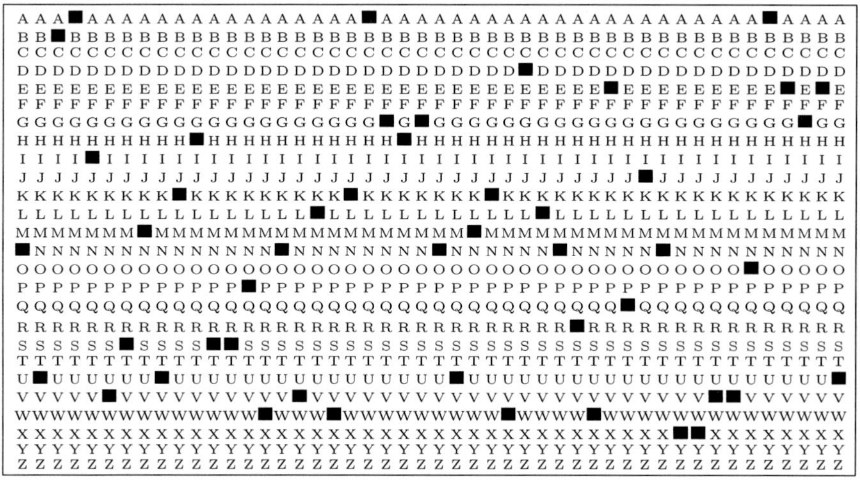

Fig. 123a. Perforated sheet with segment of the cryptotext in Fig. 115
NUBAIVSMUKHSSPWNVLWKAGHGNUMKWDLNRWEQJNXXVVOAEGEU

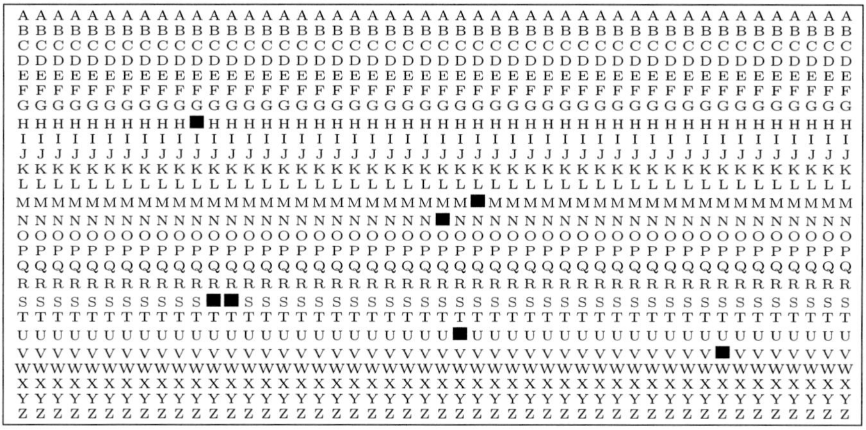

Fig. 123b. Overlay of two perforated sheets with segments of the cryptotext in Fig. 115, shifted by 72 characters

NUBA I VSMUKHSSPWNVLWKAGHGNUMKWDLN RWEQJNXXV VOAEG E U
CFPXHWZKEXHSSF X K I Y A HULMKNUMY E XDMW B XZSBCHVWZXPHWL
*** *** *

was known to Hüttenhain. However, he could have known about Friedman's main work[1] of 1938–41, which was classified, only by intelligence.

[1] William F. Friedman, Military Cryptanalysis, War Department, Office of the Chief Signal Officer. Washington, D.C.: U.S. Government Printing Office. Vol. I: *Monoalphabetic Substitution Systems* 1938, 1942. Vol. II: *Simpler Varieties of Polyalphabetic Substitution Systems* 1938, 1943. Vol. III: *Simpler Varieties of Aperiodic Substitution Systems* 1938, 1939. Vol. IV: Transposition and Fractionating Systems 1941. A copy is in the University of Pennsylvania Library, Philadelphia, PA.

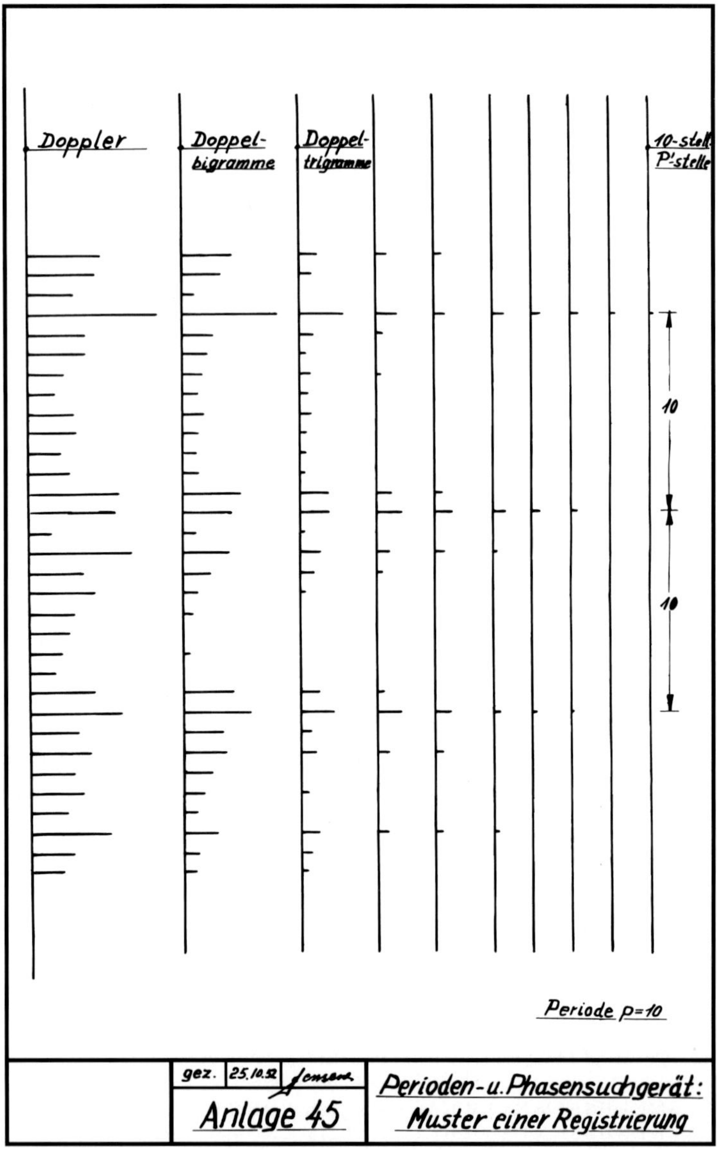

Fig. 124. Registration of coincidences in a *Perioden- und Phasensuchgerät* (Willi Jensen, *Hilfsgeräte der Kryptographie.* Draft of a Thesis, 1953)

17.3.3 The Robinsons. In Great Britain, the manual work with perforated sheets was mechanized by the HEATH ROBINSON[2], ready in May

[2] W. Heath Robinson was a British cartoonist who drew magnificent and lovely but impractical machines for all possible and impossible tasks. There were copies of HEATH ROBINSON called PETER ROBINSON and ROBINSON AND CLEAVER—names of London department stores, Altogther, by the end of 1943 12 ROBINSONs were ordered.

1943. Designed by C. E. Wynn-Williams, it had comparator and counting circuits and could read photoelectrically two loops of 5-channel teletype punched tapes with up to 2 000 characters per second, thanks to some electronic circuitry for fast counting. According to Donald Michie, HEATH ROBINSON used the saw-buck principle, too, and should have served well for coincidence examination and the finding of repetitions. It was flexible enough to serve also for stripping superencipherment and forming difference tables (Sect. 19.3). It has been reported that after W. M. Tutte had explored the internal structure of the cipher teletype machine SZ 40 (Sect. 19.2.6) it was used mainly for the $\mathbb{Z}_2$ addition of a key text to the cryptotext, shifted until the right phase was met. SUPER ROBINSON (finished May 1945) had four tapes; DRAGON (for 'dragging text through') had similar objectives. While the Bletchley Park version was electronic, the American DRAGON was made with relays.

An improvement was COLOSSUS, which had one loop stored internally and thus was able to process 5 000 characters per second without mishap. But COLOSSUS was specialized for cipher teletype machine decryption, with more complex internal operations that were performed electronically.

17.3.4 The comparator. More became known recently on American special devices for periodicity examination by *Kappa* test ('I.C.') through the work of Colin Burke. Vannevar Bush (1890–1974), well known already for his pioneering work on analog computers (the Differential Analyser) for solving differential equations, started in 1937 to build a device for counting coincidences, named COMPARATOR, for OP-20-G, the decryption branch of the U.S. Navy, following specifications of its head, Joseph N. Wenger. It worked in a 1-out-of-26 code, too. But unlike the British, who at first used approved engineering technologies, Bush had high-flying plans for very fast photoelectric scanning and electronic counters (in a 1-out-of-10 code). In 1937 it was risky to make an electronic device with more than 100 vacuum tubes working together. The project failed, also for organizational reasons. Nevertheless, it was continued, which can be explained perhaps with the role Bush played as director of the *National Defense Research Committee* (later *Office of Scientific Research and Development*) during the war. Its slow and insufficient progress not only robbed Admiral Stanford Caldwell Hooper, Chief of Naval Communications and his aide Wenger, the proponent of *pure cryptanalysis* that runs without intuitive guidance, of their immediate success, but also delayed the Navy's OP-20-GY in its use of cryptanalytic machinery. Correspondingly, in 1941, if not earlier, the U.S.A. was surpassed in its cryptanalytic potential against machine-encrypted communication channels by Great Britain. It was not until 1946 that things turned around.

Pure cryptanalysis, however, had its strong advocates among the mathematically minded cryptologists. A small group under the experienced Agnes Meyer Driscoll, with the support of the mathematician Howard T. Engstrom, attacked the ENIGMA with methods of pure cryptanalysis. For this task the microfilm machine HYPO (the 'Hypothetical Machine') was built in 1942, and was operational late in 1943. It was directed against German *offizier* messages, i.e. superencrypted ENIGMA messages. Specially oriented against the Japanese rotor machines were the relay machines VIPER and PYTHON (designed about 1943) and descendants like the electronic RATTLER.

17.3.5 RAM. Next to *Kappa*, *Chi* can also be used in periodicity examination, as we shall show in Sect. 17.5. The *Chi* test, proposed by Kullback in 1935, was not favored in 1937, because it involved not only counting but also additions and even multiplications; however it was adopted in 1940, and in the RAM machines ('Rapid Analytical Machines') in 1944 it achieved the success it deserved. In principle, the COLOSSUS machines were able to perform a Kullback examination, but whether this was actually done is not clear.

As soon as electronic universal computers were ripe, they were used in cryptanalytic work. The first special-purpose, dedicated computer models DEMON, OMALLEY, HECATE, WARLOCK were in use by the end of the 1940s, then an advanced COMPARATOR, GOLDBERG[3], and the ATLAS I and ATLAS II computers became operational at the beginning of the 1950s. More and more of the effort in cryptanalysis was transferred into the programming of universal computers with fast special and often secret additional circuitry. This process culminates at present in the architecture of the CRAY supercomputers, developed since 1976 (Plate Q).

17.4 Kasiski Examination

As a meager limit case of the *Kappa* plot for multigrams, only long multigrams that occur repeatedly are determined, and the distances between the repetitions are recorded. This search for 'parallels' (*Parallelstellensuche*) was published in 1863 by F. W. Kasiski; before the age of Friedman and Kullback, it was the preferred systematic means of attack by professional decryptors against polyalphabetic encryption, and shattered the widespread belief that this encryption is unbreakable (Sect. 8.4.2) at least in the periodic case.

17.4.1 Early steps. Unsystematic attacks on polyalphabetic encryptions started shortly after they were invented. Porta was sometimes lucky: *OMNIA VINCIT AMOR* was the (too) short and not at all esoteric key once used by an incompetent clerk, and it took Porta only a few minutes to guess it and break into the encryption. He himself used only long keys and advocated the use of keys far from daily use. And Giovanni Batista Argenti, given by his lord Iacomo Boncampagni, Duke of Sora—nephew of Pope Gregor XIII—the following cryptogram to test his ability,

Q A E T E P E E E A C S Z M D D F I C T Z A D Q G B P L E A Q T A I U I

solved it quickly, as he wrote, on October 8, 1581; he was guessing the key *INPRINCIPIOERATVERBUM* and relying on the fact that the Duke had

[3] Said to be named after Rube Goldberg, American counterpart to Heath Robinson. Possibly an allusion to Emanuel Goldberg, an inventor of photoelectric sensing.

always used ten involutory alphabets of the kind Porta had described in 1563 (Sect. 7.4.4, Fig. 53)—why should the Duke have invented something on his own? The plaintext was the beginning of the Æneid of Vergil

arma virumque cano troiæ qui primus ab oris .

Porta had early on a methodical idea: If with an Alberti disk the crypt alphabet is shifted at every step by one position, then certain frequently occurring bigrams like /ab/, /hi/, /op/ and trigrams like /def/ (in *deficio*) or /stu/ (in *studium*) generate letter repetitions in the cryptotext. Porta found MMM and 51 positions later MMM again, and he concluded that the key should have period 17 and be repeated three times, since the period 51 would be too long and the period 3 too short for a clever cipher clerk.

Porta came within a hair's breadth of finding Kasiski's method. All it needed was to understand that the pattern *111* did not matter, but just the repetition itself of some cryptotext fragment, caused by a coincidence of a frequent plaintext fragment with one and the same piece of the repeated key, which should normally happen only in a distance which is a multiple of the period. Had Porta noticed this and published, polyalphabetic encryption would not have been invulnerable still at the time of Edgar Allen Poe.

The following simplified example by Kahn may illustrate the Kasiski examination: Assume a VIGENÈRE method in $\mathbb{Z}_{26}$ works with a key *RUN* of the (too) small length 3:

t o b e o r n o t t o b e t h a t i s t h e q u e s t i o n
R U N R U N R U N R U N R U N R U N R U N R U N R U N R U N
K I O V I E E I G K I O V N U R N V J N U V K H V M G Z I A

Then the key fragment *RUNR* meets the plaintext fragment /tobe/ twice in a distance 9, which results in the repeated fragment KIOV, moreover the key fragment *UN* meets the plaintext fragment /th/ twice in a distance 6, which results in the repeated fragment NU . The distances 9 and 6 must be multiples of the period, which can only be 3 (or 1).

A similar example with a key *COMET* of length 5 is

t h e r e i s a n o t h e r f a m o u s p i a n o p l a y
C O M E T C O M E T C O M E T C O M E T C O M E T C O M E
V V Q V X K G M R H V V Q V Y C A A Y L R W M R H R Z M C

In this example the distances of the repeated fragments are 10 and 15, so the period can only be 5 (or 1).

17.4.2 Babbage on decryption. Ten years before Kasiski, Charles Babbage may have had an inkling of the importance of repetitions. Not only did he like to read the monoalphabetically encrypted messages in the agony columns of the Victorian London gazettes, he also liked to look inside polyalphabetics with word division. His dealing with linear simple encryption steps led him in 1846 to a description of VIGENÈRE and BEAUFORT encryption steps by mathematical equations (Sect. 7.4.1), and thus he could find solutions by using probable words in the plaintext as well as in the key. Via such successes, as the Babbage papers in the British Museum show, he developed an understanding for the subtleties of periodic encryptions, although even if he found out about the importance of Kasiski repetitions, as Ole Immanuel Franksen suggested in 1984, he did not write about this.

Thus, the honor of first finding a systematic means of attack against polyalphabetic encryption, not even limited to linear substitutions, and hence founding modern cryptology, goes to a retired Prussian infantry major.

Friedrich W. Kasiski was born November 29, 1805 in Schlochau, West Prussia (now Czluchow, Poland). In 1822 he entered the East Prussian 33rd *Füsilier-Regiment Graf Roon*, where he served until 1852. In his leisure time he turned to cryptography. In 1863 his 95-page booklet *Die Geheimschriften und die Dechiffrirkunst* was published by the respected Mittler & Sohn in Berlin.

At first, his publication caused no sensation, and Kasiski turned disappointed to natural history, where he won local fame. The revolution in cryptology he initiated took place after his death on May 22, 1881. Kerckhoffs commended Kasiski's work in an important paper of 1883, and the books of de Viaris in 1893 and Delastelle in 1902 were based on this. Around the turn of the century the revolution was under way, and the vulnerability of periodic polyalphabetic encryption was generally accepted among professionals.

In the light of William F. Friedman's discovery in 1925 of the index of coincidence, the Kasiski examination appears to be a rough method. Bigram repetitions are neglected, 'because they are so frequent', and single character repetitions anyway—while *Kappa* counts all repetitions and asks only whether there are more than average. Ignoring bigram repetitions was also justified by the fact that in rare cases they can come about accidentally. Even with trigrams this occurs, and it disturbs the analysis—while the index of coincidence, because of its stochastic nature, is unaffected.

The Kasiski examination establishes as the period the greatest common divisor of the distances of the recorded repetitions, excluding pragmatically those that are considered annoying and presumed to be nothing but accidental repetitions. In this respect, the Kasiski examination is intuitive and unfriendly to mechanization. Moreover, the Kasiski examination needs longer texts to be conclusive about a period than the Friedman examination.

Accidental repetitions are frequently observed and easily explained in case of linear substitutions. The reason is the commutative law that holds for addition *modulo* N : /anton/ with the key *BERTA* and /berta/ with the key *ANTON* give the same. This effect occurs with more than average frequency if both plaintext and key are from the same natural language, particularly from the same genre. Repetitions in the *keytext* can also lead to 'wrong repetitions' in the cryptotext—keys with words like *DANSEUSECANCAN*, *VIERUNDVIERZIG* can irritate the unauthorized decryptor. We shall come back to this in Sect. 18.5 .

17.4.3 An example. The model examples in the literature for a Kasiski examination almost always show window-dressing: they present more repetitions than can be expected on average. For the following example (Kahn) this cannot be said. The plaintext turns out to be a worthwhile recommendation from Albert J. Myer (1866), U.S. Signal Corps officer. The cryptotext reads:

```
ANYVG YSTYN RPLWH RDTKX RNYPV QTGHP
HZKFE YUMUS AYWVK ZYEZM EZUDL JKTUL
JLKQB JUQVU ECKBN RCTHP KESXM AZOEN
SXGOL PGNLE EBMMT GCSSV MRSEZ MXHLP
KJEJH TUPZU EDWKN NNRWA GEEXS LKZUD
LJKFI XHTKP IAZMX FACWC TQIDU WBRRL
TTKVN AJWVB REAWT NSEZM OECSS VMRSL
JMLEE BMMTG AYVIY GHPEM YFARW AOAEL
UPIUA YYMGE EMJQK SFCGU GYBPJ BPZYP
JASNN FSTUS STYVG YS
```

The character frequency count is shown in Fig. 125; it is too uniform to be explained by a monoalphabetic substitution or by a transposition. Thus, suspicion falls on a polyalphabetic substitution. This is supported by the wealth of repetitions seen when a Kasiski examination is made. There are nine repetitions of length 3 or more, among which some are very long, like LEEBMMTG and CSSVMRS. Their distances are listed in Fig. 126 together with their prime factor decompositions. The greatest common divisor is 2, but this very small apparent period is presumably caused by accidental repetitions.

14	8	7	5	22	6	12	8	5	11	14	13	16	13	4	13	5	11	18	15	14	10	9	7	16	11
A	B	C	D	E	F	G	H	I	J	K	L	M	N	O	P	Q	R	S	T	U	V	W	X	Y	Z

Fig. 125. Frequency count in the cryptotext of Kahn

Fragment	Distance	Prime factor decomposition
YVGYS	280	$2^3 \cdot 5 \cdot 7$
STY	274	$2 \cdot 137$
GHP	198	$2 \cdot 3^2 \cdot 11$
ZUDLJK	96	$2^5 \cdot 3$
LEEBMMTG	114	$2 \cdot 3 \cdot 19$
CSSVMRS	96	$2^5 \cdot 3$
SEZM	84	$2^2 \cdot 3 \cdot 7$
ZMX	48	$2^4 \cdot 3$
GEE	108	$2^2 \cdot 3^3$

Fig. 126. Prime factor decomposition of distances of Kasiski repetitions

Following Kasiski verbatim, the distances are to be decomposed into factors, the factor most frequently found being the period. The literature interprets this, following M. E. Ohaver, usually that all factors (i.e., not only the prime factors) are to be listed (Fig. 127). This leads to the possibility of two factors occurring equally often, in which case the larger one will be taken if it is a multiple of the smaller one—otherwise it might be better to follow two possibilities. In Fig. 127, apart from the factor 2 which we have dismissed as too small, the factors 3 and 6 both occur 7 times. This would make the factor 6 our candidate. As it turns out this is right, but we should put it down to luck. The correct rule, to take the greatest common divisor of all causal repetitions, suffers from the defect that we will only know afterwards which ones were causal. Intuitively we are inclined to omit those 'annoying' repetitions whose distance does not contain an otherwise most-frequent prime factor—in Fig. 126 both YVGYS and STY do not contain the otherwise frequent factor 3. Since YVGYS is rather long, it is hard to believe that it is accidental, but otherwise 2 would be the period, which is even harder to believe. If both GHP and LEEBMMTG were omitted, 12 would be a candidate for the period. But since LEEBMMTG is very long, this is unlikely, too. Thus, one has to live with the suspicion that 6 is the period.

Fragment	Distance	2	3	4	5	6	7	8	9	10	11	12	14	16	18	19	20	21	22	24	
YVGYS	280	√		√	√		√	√		√			√				√				(?)
STY	274	√																			(?)
GHP	198	√	√			√			√		√				√				√		
ZUDLJK	96	√	√	√		√		√				√		√						√	
LEEBMMTG	114	√	√			√										√					
CSSVMRS	96	√	√	√		√		√				√		√						√	
SEZM	84	√	√	√		√	√					√	√					√			
ZMX	48	√	√	√		√		√				√		√						√	
GEE	108	√	√	√		√			√			√			√						

Fig. 127. Factors of distances of Kasiski repetitions

No doubt, this shows another weak side of the Kasiski examination. A too-small value of the alleged period, caused by an accidental repetition, ruins the subsequent process of reconstruction of the alphabets. On the other hand, it can happen that the greatest common divisor of all the distances is a multiple of the genuine period. This not only causes an increase in the subsequent work load, but also makes the reconstruction of the alphabets less safe.

Under all circumstances, the Friedman examination is more reliable than the Kasiski examination.

Anticipating the later decryption (Sect. 18.1), we note that STY and YVGYS will turn out to be accidental repetitions, originating from a linear substitution over $\mathbb{Z}_{26}$. The key is *SIGNAL*, and YVGYS originates the first time from /signa/+*GNALS*, the second time from /gnals/+*SIGNA*; STY comes the first time from /als/+*SIG*, the second time from /sig/+*ALS*. This is effected by the use of a key word *SIGNAL* out of the genre of the plaintext. It complicates unauthorized decryption and is desirable for the cryptograph. By the way, it also blunts the aggressiveness of the *Kappa* and *Chi* tests.

Accidental repetitions can also occur without being the result of commutativity. The great French cryptologist Étienne Bazeries once had no luck with a BEAUFORT encryption: In 1898, in a telegram from the insurgent Duke of Orléans,

GNJLN RBEOR PFCLS OKYNX TNDBI LJNZE OIGSS HBFZN ETNDB .....

he found a Kasiski repetition TNDB of length 4 with a distance 21, but it was accidentally produced by *ERVE* –/lesd/ and by *IERV* –/prou/ (the key actually was: *VENDREDIDIXSEPTFEVRIER*). The two further repetitions EO and AQ of length 2 occurred with distances 22 and 13 . What to do? Bazeries was more confident of the longer repetition TNDB and assumed the period 21, but this was a dead end that cost him much time. In the end, it turned out that only the short repetition EO of length 2 with the period 22 was causal. Bazeries remarked bitterly *"en cryptographie, aucune règle n'est absolue."* [in cryptography, hardly any rule is absolute.]

17.4.4 Machines. Despite its weakness the Kasiski examination was still used as an auxiliary in the Second World War. The Cipher Branch of the German OKW developed and employed a special *Parallelstellensuchgerät* (Willi Jensen)—apart from the *Perioden- und Phasensuchgerät* (Sect. 17.3.2), usable for a Friedman examination. The cryptotext was punched in a 2-out-of-10 code (Sect. 17.3.1) on film tape in two copies. One copy (A) was closed into a loop and ran continuously through a scanner, while the second one (B) advanced one position in its scanner for every finished loop of (A). In case of coincidence, two holes met at the scanner, which could be discriminated by a photoelectric cell. With the help of a diaphragm of varying breadth, it was possible to detect in turn bigram, trigram, ... repetitions; within the available accuracy of measurement repetitions with up to 10 characters could be searched for. The registration was done by a spark on an aluminum plate movable in two directions, one for (A) and one for (B). The device was rather fast: to run a bundle of texts of 10 000 characters altogether, requiring 10^8 comparisons, took less than 3 hours. It served mainly to obtain quick information on texts in the bundle that were encrypted with the same key, then for detailed investigation the *Perioden- und Phasensuchgerät* was used. The device was destroyed at the end of the war, before it had been very long in practical use.

In the U.S.A., Bush built for OP-20-G in 1943 TETRA (nicknamed ICKY, TESSIE, see also 18.6.3), that could find long repetitions, or patterns of identical subgroups, and allowed a flexible selection of combinations through a plugboard. Around mid-1944, under Friedman, development of a universal cryptanalytic machine using microfilm, the Eastman 5202, started.

17.4.5 Photoelectric sensing, as used by the German, U.S. and British cryptanalysts, goes back to early attempts to use it for document retrieval. In 1927, Michael Maul of Berlin received patents which were assigned to IBM as U.S. Patents 2 000 403 and 2 000 404 . The work of Emanuel Goldberg ('Statistical machine', U.S. Patent 1 838 389, Dec. 29, 1931; filed April 5, 1928) preceded Bush's 1937 plans both for the COMPARATOR and for the first RAPID SELECTOR (later to become 'Memex'), a document retrieval system.

17.5 Building a Depth and Phi Test of Kullback

With a guess at the period d of a polyalphabetic encryption, a simple manual process for determining the number of coincidences for shifts by $k{\cdot}d$ positions consists of writing the cryptotext in lines of length d, thus forming d columns T_1, T_2, T_3, ... T_d . In the jargon of cryptanalysts, this is called 'writing out a depth' or 'building up a depth'.

	G	E	I	E	I	A	S	G	D	X	V	Z
	I	J	Q	L	M	W	L	A	A	M	X	Z
	Y	Z	M	L	W	H	F	Z	E	K	E	J
	L	V	D	X	W	K	W	K	E	T	X	L
	B	R	A	T	Q	H	L	B	M	X	A	A
	N	U	B	A	I	V	S	M	U	K	H	S
	S	P	W	N	V	L	W	K	A	G	H	G
	N	U	M	K	W	D	L	N	R	W	E	Q
	J	N	X	X	V	V	O	A	E	G	E	U
	W	B	Z	W	M	Q	Y	M	O	M	L	W
	X	N	B	X	M	W	A	L	P	N	F	D
	C	F	P	X	H	W	Z	K	E	X	H	S
	S	F	X	K	I	Y	A	H	U	L	M	K
	N	U	M	Y	E	X	D	M	W	B	X	Z
	S	B	C	H	V	W	Z	X	P	H	W	L
	G	N	A	M	I	U	K					
ϕ_ρ	14	16	12	16	30	16	14	14	18	12	18	10 Σ=190

Fig. 128. Cryptotext of G. W. Kulp, in twelve columns

Figure 128 shows the result for the cryptotext of G. W. Kulp with the guessed period $d = 12$. Coincident single characters (*dopplers*) with the minimal distance d catch the eye immediately, e.g., Z in column 12, in the first and second lines. But also repetitions with a distance $k \cdot d$ are easily seen, e.g., another Z in column 12, in the last line but two. Bigram repetitions (bigram *dopplers*) show up, too, e.g., the bigram WK in the fourth and seventh lines, the bigram MW in the second and eleventh lines, the bigram WZ in the twelfth and last but one lines, the bigram NU in the sixth, eighth, and last but two lines.

17.5.1 Forming the columns. The minor effort of arranging the cryptotext in u columns allows more than finding some Kasiski repetitions. If the guess as to the period is correct, then and only then is each column T_ρ encrypted monoalphabetically. This should be tested by forming the $Phi(T_\rho)$ for $\rho = 1, 2, \ldots, u-1, u$. In the positive case, values close to κ_S and much larger than $\kappa_R = \frac{1}{N}$ should be expected for all $Phi(T_\rho)$; in the negative case, they should fluctuate. This is the *Phi* test of Kullback, a very sharp criterion for the examination of the period.

17.5.2 Phi test is better than Kappa test. It may be appropriate to form a mean of the values $Phi(T_\rho)$ for $\rho = 1, 2, \ldots, u-1, u$. Thus, with

$$\phi_\rho = \sum_{i=1}^{N} m_i^{(\rho)} \cdot (m_i^{(\rho)} - 1) \quad \text{for } \rho = 1, 2, \ldots, u-1, u, \quad \text{where}$$

$m_i^{(\rho)}$ is the frequency of the i-th character in the ρ-th column,

we have to consider, properly normalized,

$$Phi^{(u)}(T) = u \cdot \sum_{\rho=1}^{u} \phi_\rho / M \cdot (M-1) \ .$$

Provided $u|M$, there is as in Sect. 16.4.1 the

Kappa-Phi$^{(u)}$ Theorem: $$\frac{1}{M-1} \sum_{\rho=1}^{M-1} Kappa(T^{(u \cdot \rho)}, T) = Phi^{(u)}(T) \ .$$

Thus, $Phi^{(u)}(T)$ is the arithmetic mean of all $Kappa(T^{(u\cdot\rho)}, T)$, i.e., of all coincidences at distances that are a multiple of u . It turns out to be a very sharp instrument.

17.5.3 Example. For $u = 12$ there are twelve alphabets, from each one there are 16 or 15 characters in a column. Calculation of ϕ_ρ gives (Fig. 128):
for the first column with three S , three N , two G :
$\phi_1 = 6 + 6 + 2 = 14$;
for the second column with three N , three U , two B , two F :
$\phi_2 = 6 + 6 + 2 + 2 = 16$;
for the third column with three M , two A , two B , two X :
$\phi_3 = 6 + 2 + 2 + 2 = 12$;
for the fourth column with four X , two K , two L :
$\phi_4 = 12 + 2 + 2 = 16$;
for the fifth column with four I , three M , three V , three W :
$\phi_5 = 12 + 6 + 6 + 6 = 30$;
for the sixth column with four W , two H , two V :
$\phi_6 = 12 + 2 + 2 = 16$;
and so on. Thus, $\sum_{\rho=1}^{12} \phi_\rho = 190$, $Phi^{(12)} = 12 \cdot 190/(187 \cdot 186) = 6.56\%$.

Note that $\kappa_e = 6.58\%$. Thus, $u = 12$ could well be the period.

	G	E	I	E	I	A	S	G	D	X	V	
	Z	I	J	Q	L	M	W	L	A	A	M	
	X	Z	Y	Z	M	L	W	H	F	Z	E	
	K	E	J	L	V	D	X	W	K	W	K	
	E	T	X	L	B	R	A	T	Q	H	L	
	B	M	X	A	A	N	U	B	A	I	V	
	S	M	U	K	H	S	S	P	W	N	V	
	L	W	K	A	G	H	G	N	U	M	K	
	W	D	L	N	R	W	E	Q	J	N	X	
	X	V	V	O	A	E	G	E	U	W	B	
	Z	W	M	Q	Y	M	O	M	L	W	X	
	N	B	X	M	W	A	L	P	N	F	D	
	C	F	P	X	H	W	Z	K	E	X	H	
	S	S	F	X	K	I	Y	A	H	U	L	
	M	K	N	U	M	Y	E	X	D	M	W	
	B	X	Z	S	B	C	H	V	W	Z	X	
	P	H	W	L	G	N	A	M	I	U	K	
ϕ_ρ	8	6	8	12	10	8	10	4	8	16	20	Σ=110

Fig. 129. Cryptotext of G. W. Kulp, in eleven columns

	G	E	I	E	I	A	S	G	D	X	V	Z	I	
	J	Q	L	M	W	L	A	A	M	X	Z	Y	Z	
	M	L	W	H	F	Z	E	K	E	J	L	V	D	
	X	W	K	W	K	E	T	X	L	B	R	A	T	
	Q	H	L	B	M	X	A	A	N	U	B	A	I	
	V	S	M	U	K	H	S	S	P	W	N	V	L	
	W	K	A	G	H	G	N	U	M	K	W	D	L	
	N	R	W	E	Q	J	N	X	X	V	V	O	A	
	E	G	E	U	W	B	Z	W	M	Q	Y	M	O	
	M	L	W	X	N	B	X	M	W	A	L	P	N	
	F	D	C	F	P	X	H	W	Z	K	E	X	H	
	S	S	F	X	K	I	Y	A	H	U	L	M	K	
	N	U	M	Y	E	X	D	M	W	B	X	Z	S	
	B	C	H	V	W	Z	X	P	H	W	L	G	N	
	A	M	I	U	K									
ϕ_ρ	4	4	12	10	18	10	8	12	10	10	14	8	6	Σ=126

Fig. 130. Cryptotext of G. W. Kulp, in thirteen columns

For $u = 11$ there are eleven alphabets, and from each there are 17 characters in a column. Figure 129 shows, there is a *doppler* W with distance 11 in the second and third lines, and a *doppler* V in the first, sixth and seventh lines of the eleventh column. There are no *bigram doppler*s any longer.

Calculation of ϕ_ρ gives (Fig. 129):
$\sum_{\rho=1}^{11} \phi_\rho = 110$, $Phi^{(11)} = 11 \cdot 110/(187 \cdot 186) = 3.48\%$.

Note that $\kappa_R = 3.85\%$. $Phi^{(11)}$ is remarkably smaller than $Phi^{(12)}$.
$u = 11$ has little chance to be the period.

For $u = 13$ there are thirteen alphabets, and from each there are 14 or 15 characters in a column. Figure 130 shows that calculation of ϕ_ρ gives
$\sum_{\rho=1}^{13} \phi_\rho = 126$, $Phi^{(13)} = 13 \cdot 126/(187 \cdot 186) = 4.71\%$. Thus, $u = 13$ also is not a candidate for the period.

In this way $Phi^{(u)}$ can be calculated for $u = 2, 3, 4, \ldots$ and plotted (Fig. 131). The value for $u = 12$ is much more conspicuous than in Fig. 116. It can be seen that the Kullback examination is a finer instrument than the Friedman examination.

There is a further peak at $u = 24$, which should be expected. But there are also peaks, although slightly smaller, at $u = 6$ and even at $u = 18$. It therefore cannot be excluded that $u = 6$ is the period.

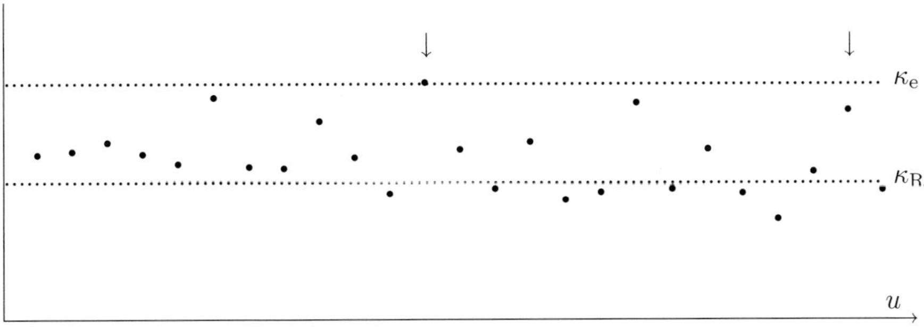

Fig. 131. $Phi^{(u)}$ plot for the cryptotext of G. W. Kulp

17.6 Estimating the Period Length

From the *Kappa-Phi* theorem of Sect. 16.4.1 we obtain for the expectation values

$$\langle Phi(T) \rangle_Q^{(M)} = \tfrac{1}{M-1} \sum_{\rho=1}^{M-1} \langle Kappa(T, T^\rho) \rangle_Q \ .$$

From a remark at the beginning of this chapter, we may deduce that

$$\langle Kappa(T, T^{k \cdot d}) \rangle_Q = \kappa_S \ ,$$

moreover, we may find

$$\langle Kappa(T, T^u) \rangle_Q \approx \kappa_R = \tfrac{1}{N} \qquad \text{for } d \nmid u.$$

Assuming furthermore for simplicity that M is a multiple of the period d, then in the sum κ_S appears $\frac{M}{d} - 1$ times, κ_R appears $M - \frac{M}{d}$ times; thus $\langle Phi(T)\rangle_Q^{(M)}$ is a mean of κ_S and κ_R,

$$(M-1) \cdot \langle Phi(T)\rangle_Q^{(M)} \approx (\tfrac{M}{d} - 1) \cdot \kappa_S + ((M-1) - (\tfrac{M}{d} - 1)) \cdot \kappa_R \,.$$

Assuming that the observed $Phi(T)$ approximates the expectation value, we find (Abraham Sinkov, around 1935):

$$(M-1) \cdot Phi(T) \quad \approx \quad (\tfrac{M}{d} - 1) \cdot \kappa_S + ((M-1) - (\tfrac{M}{d} - 1)) \cdot \kappa_R \,.$$

Although only an estimation, this fundamental relation shows qualitatively how with constant stochastic source, but increasing period of a polyalphabetic encryption, the value of Phi changes.

For large M and $d \ll M$ one can work with

$$Phi(T) \approx \frac{1}{d} \cdot \kappa_S + (1 - \frac{1}{d}) \cdot \kappa_R$$

almost as well.

Sinkov's relation can be solved for d:

$$(\frac{M}{d} - 1) \approx \frac{(M-1) \cdot (Phi(T) - \kappa_R)}{\kappa_S - \kappa_R} \text{, i.e.,}$$

$$d \approx \frac{\kappa_S - \kappa_R}{(\kappa_S - Phi(T))/M + (Phi(T) - \kappa_R)} \,.$$

For large M and $d \ll M$

$$d \approx \frac{\kappa_S - \kappa_R}{Phi(T) - \kappa_R} \,.$$

can also be used.

For example, for the cryptotext of G. W. Kulp with $M = 187$, according to Sect. 17.1.3, $Phi = 4.56\%$, so with $\kappa_S = \kappa_e = 6.58\%$, $\kappa_R = \frac{1}{N} = 3.85\%$ we obtain

$$d \approx \frac{2.73\%}{(2.02\%/187) + 0.71\%} = 3.79 \quad \text{or simplified}$$

$$d \approx \frac{2.73\%}{0.71\%} = 3.85 \,.$$

This value is low compared with a presumptive period $d = 12$, and would fit better with $d = 6$. But the estimate is rather unstable and should not be taken too seriously. The Sinkov estimate can only give support to a serious Kasiski, Friedman, or Kullback examination if the period is small.

18 Alignment of Accompanying Alphabets

Provided the period d of a polyalphabetic encrypted text is determined sufficiently reliably, and by building a depth can be reduced to solving d monoalphabetic encryptions, one can try to reduce the accompanying alphabets—if possible—to a primary alphabet. In case of VIGENÈRE encryption, an exhaustion of all accompanying standard alphabets by matching profiles (Sect. 18.1) is easy enough. The same aligning can be done in case of ALBERTI encryption, if one of the standard alphabets is known or has been found out (Sect. 18.2). In general, however, a mutual aligning of all alphabets (Sect. 18.3) is needed to reconstruct the unknown primary alphabet (Sect. 18.4). For this purpose, a Kullback examination will work wonders. The case of unknown unrelated alphabets, where each one is to be determined by itself, cannot be treated this way.

18.1 Matching the Profile

In view of the wide acceptance VIGENÈRE encryption has found, it may often be worthwhile to try this entry, which does not need much effort. Thus, if d is the period, d profiles are to be plotted. The strip method for exhaustion (Chapter 12) and pattern finding (Chapter 13) for the individual monoalphabetic CAESAR additions do not work, since the texts are torn to pieces (German *zerrissen*).

2	0	0	0	1	0	2	3	0	6	3	3	3	0	0	0	0	0	6	1	5	2	5	0	2	4
A	B	C	D	E	F	G	H	I	J	K	L	M	N	O	P	Q	R	S	T	U	V	W	X	Y	Z

Fig. 132. Frequency distribution in the first column of the cryptotext of Kahn

18.1.1 Using a depth. Thus, for the example of Myer's text (Sect. 17.4.3) we recommend building a depth and counting the frequencies for each of the six columns. For the first column, i.e., for the subtext consisting of the 1st, 7th, 13th, ... characters, the result can be seen in Fig. 132. Even without plotting the English profile is immediately recognizable: NOPQR is the v-w-x-y-z lowland, to its left JKLM is the r-s-t-u ridge. Then DEFGH is at the right distance to be the l-m-n-o ridge, which does not show clearly. But the cryptanalyst has to be prepared for such fluctuations, in particular if the depth is not large. This remark applies also to the observation that W does not have the frequency one would expect for the e-peak.

With S: S $\hat{=}$ a the first column is aligned, and it can be expected that the whole is a VIGENÈRE system. The first key letter S is found. With the other columns a similar procedure is carried through, to give us step by step the key

$$SIGNAL\ ,$$

which is confirmed by subsequent decryption of the whole cryptotext. With fragments leading to causal repetitions underlined, the plaintext is

```
i f s i g   n a l s a   r e t o b   e d i s p   l a y e d   i n t h e
p r e s e   n c e o f   a n e n e   m y t h e   y m u s t   b e g u a
r d e d b   y c i p h   e r s t h   e c i p h   e r s m u   s t b e c
a p a b l   e o f f r   e q u e n   t c h a n   g e s t h   e r u l e
s b y w h   i c h t h   e s e c h   a n g e s   a r e m a   d e m u s
t b e s i   m p l e c   i p h e r   s a r e u   n d i s c   o v e r a
b l e i n   p r o p o   r t i o n   a s t h e   i r c h a   n g e s a
r e f r e   q u e n t   a n d a s   t h e m e   s s a g e   s i n e a
c h c h a   n g e a r   e b r i e   f f r o m   a l b e r   t j m y e
r s m a n   u a l o f   s i g n a   l s
```

Now we can even see how the causal repetitions were accomplished: the longest, LEEBMMTG, originates from a repeated combination of /frequent/ with *GNALSIGN*; another one, ZUDLJK, from a repeated combination of /mustbe/ with *NALSIG*. CSSVMRS comes from /changes/ with *ALSIGNA*. Strangely, the repeated occurrence of /cipher/ in the plaintext did not lead to a repetition. SEZM, GHP, ZMX, GEE are caused by /sthe/, /the/, /her/, /are/ meeting *ALSI*, *NAL*, *SIG*, *GNA*. YVGYS and STY are accidental repetitions.

18.1.2 Plotting the profiles. In case of relatively long keys the depth is small, and it may be difficult to recognize frequency differences. Then there is still the simple possibility of a graphical plot. For the cryptotext of G. W. Kulp (Fig. 115) the preparatory work of forming the columns for $d = 12$ is already done in Sect. 17.4, and the alignment can immediately follow the Kullback examination of the period. From Fig. 128 the twelve profiles in Fig. 133 are derived. In Fig. 134 they are aligned to match somewhat the frequency profile of English. The trial was successful here, too.

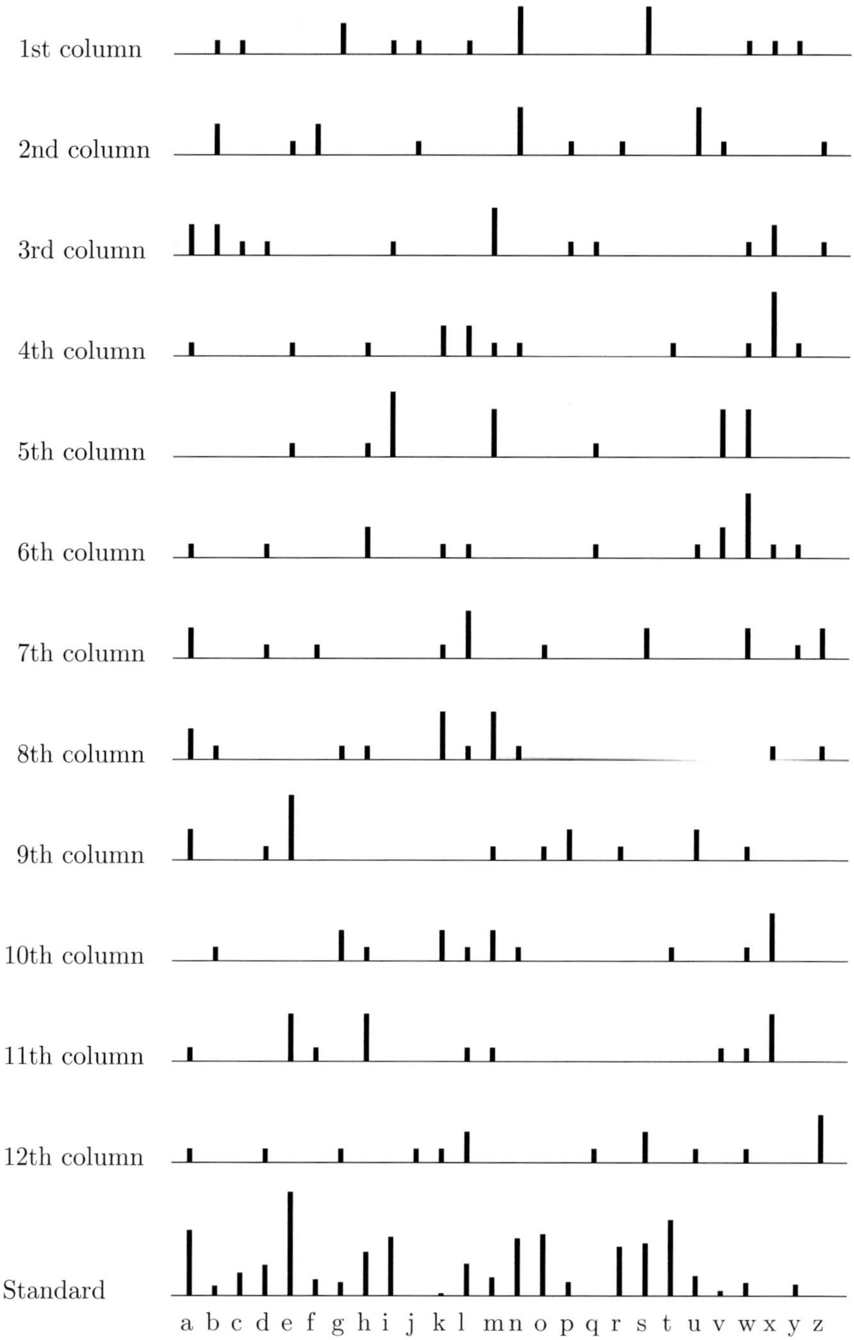

Fig. 133. Profiles for the cryptotext of G. W. Kulp

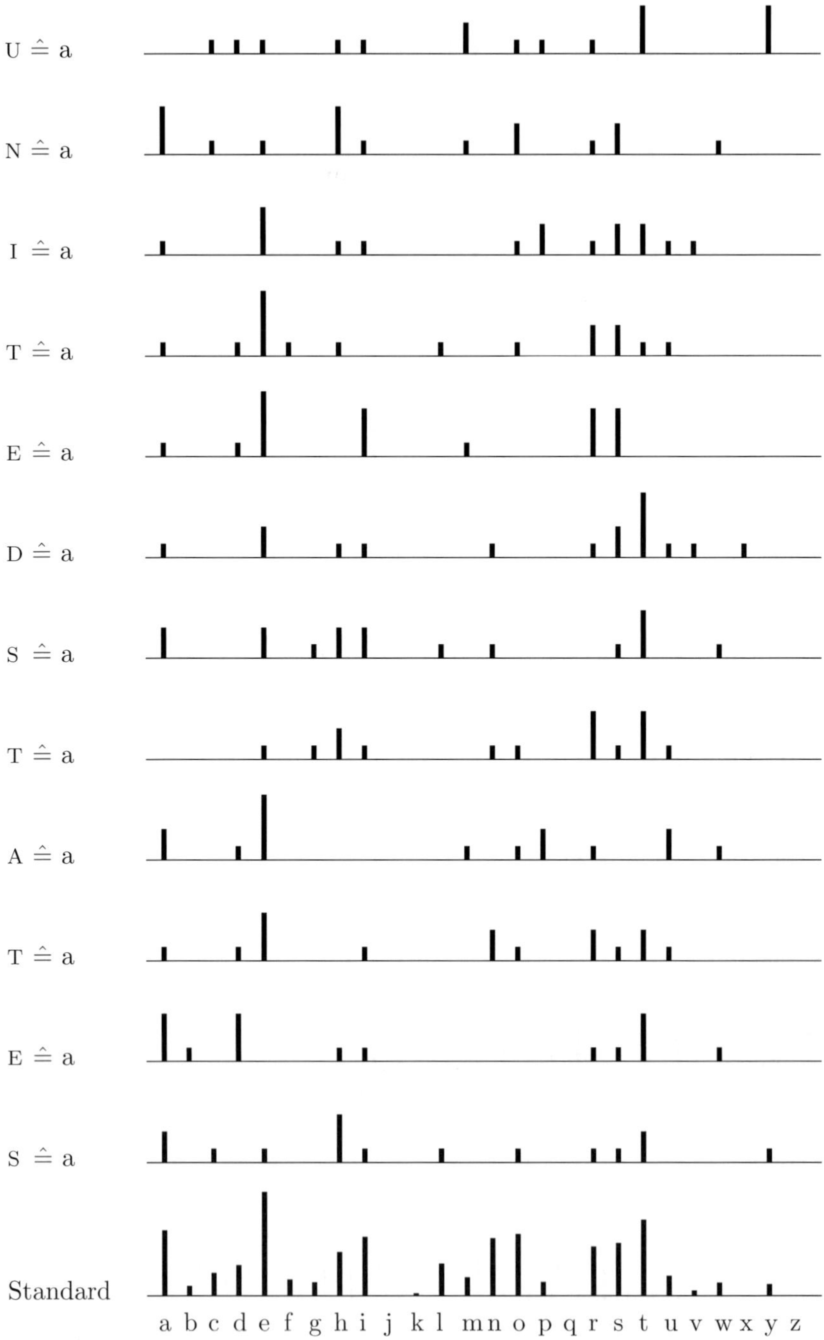

Fig. 134. Aligned profiles for the cryptotext of G. W. Kulp

The 11th column is misleading: /e/, the most frequent character, does not occur. Quite generally it can be said that in case of small sets of characters it is preferable to align first the rare characters, particularly the rarest ones, the missing ones. Then some of the /e/—the ones in the third, fourth, fifth, ninth and tenth column—give a good clue.

In this way, it turns out that the plaintext /a/ corresponds in the first alphabet to U, in the second alphabet N, in the third alphabet I, in the fourth, eighth and tenth alphabets T, in the fifth and eleventh alphabets E, in the sixth alphabet D, in the seventh and twelfth alphabets S. In the ninth alphabet A corresponds to plaintext /a/, therefore this substitution is the identity. To suppress it in the case of a VIGENÈRE would be a technical error, since this would open the possibility of a non-coincidence attack (Sect. 14.1).

The alignment is certainly facilitated here by the fact that one alphabet occurs three times. Such patterns in the key are technical errors, too.

The key word is now revealed as

UNITEDSTATES .

This makes sense in the circumstances of Philadelphia in 1840. In the sense of Rohrbach's maxim, the decryption, with the key period $d=12$, is completely convincing. Thus, Edgar Allen Poe was a little bit unfair in saying the text was an imposition. For the curious, the decrypted text is given in Fig. 135. The decryption was successfully done by Brian J. Winkel in 1975 and published in Martin Gardner's column in *Scientific American*, August 1977.

m r a l e	x a n d e	r h o w i	s i t t h	a t t h e	m e s s e
n g e r a	r r i v e	s h e r e	a t t h e	s a m e t	i m e w i
t h t h e	s a t u r	d a y c o	u r i e r	a n d o t	h e r s a
t u r d a	y p a p e	r s w h e	n a c c o	r d i n g	t o t h e
d a t e i	t i s p u	b l i s h	e d t h r	e e d a y	s p r e v
i o u s i	s t h e f	a u l t w	i t h y o	u o r t h	e p o s t
m a s t e	r s				

Fig. 135. Plaintext of the message of G. W. Kulp

18.2 Aligning Against Known Alphabet

Aligning the alphabets with the naked eye may seem difficult in Fig. 133, say with the first or with the eighth column.

18.2.1 Using Chi. Computational means turn out to be a sharper instrument. A natural idea is to determine the alignment shift by calculating the *Chi* between the frequencies of the alphabet in question and the primary alphabet. This is shown in Fig. 136 for the unshifted standard alphabet and in Fig. 137 for the suitably shifted alphabet with /a/ corresponding to U.

In the first case (a $\hat{=}$ A) the value 64.81/16 %= 4.05% results for *Chi*; in the second case (a $\hat{=}$ U) the markedly larger value 86.03/16 % = 5.38%. Table 20 lists the *Chi* values for all shifts. It turns out that besides a $\hat{=}$ U also a $\hat{=}$ J and a $\hat{=}$ F are distinguished. These three choices need further, exhaustive treatment.

0	1	1	0	0	0	2	0	1	1	0	1	0	
A	B	C	D	E	F	G	H	I	J	K	L	M	
8.04	1.54	3.06	3.99	12.51	2.30	1.96	5.49	7.26	0.16	0.67	4.14	2.53	
a	b	c	d	e	f	g	h	i	j	k	l	m	
3	0	0	0	0	3	0	0	0	1	1	1	0	
N	O	P	Q	R	S	T	U	V	W	X	Y	Z	
7.09	7.60	2.00	0.11	6.12	6.54	9.25	2.71	0.99	1.92	0.19	1.73	0.09	64.81
n	o	p	q	r	s	t	u	v	w	x	y	z	

Fig. 136. *Chi* for standard alphabet against first column

0	0	1	1	1	0	0	1	1	0	0	0	2	
U	V	W	X	Y	Z	A	B	C	D	E	F	G	
8.04	1.54	3.06	3.99	12.51	2.30	1.96	5.49	7.26	0.16	0.67	4.14	2.53	
a	b	c	d	e	f	g	h	i	j	k	l	m	
0	1	1	0	1	0	3	0	0	0	0	3	0	
H	I	J	K	L	M	N	O	P	Q	R	S	T	
7.09	7.60	2.00	0.11	6.12	6.54	9.25	2.71	0.99	1.92	0.19	1.73	0.09	86.03
n	o	p	q	r	s	t	u	v	w	x	y	z	

Fig. 137. *Chi* for standard alphabet against first column, shifted a $\hat{=}$ U

This shows that a periodic VIGENÈRE system can be decrypted mechanically under reasonably fortunate circumstances. For a text as long as that in the case Kulp vs. Poe this is not only successful, but also feasible with the support of a personal computer.

18.2.2 Strip method. The basic idea of aligning against a primary alphabet—in case of a VIGENÈRE system the standard alphabet, in case of an ALBERTI system a mixed alphabet fallen into unauthorized hands—albeit without calculation of the *Chi*, is found rather early in the literature. A common version uses strips as in Sect. 12.8.1 for the primary alphabet, with the most frequent characters (in English the nine characters e t a o n i r s h) printed in boldface or in red color—in mechanical solutions in the Second World War (Ernst Witt, Hans Rohrbach) using semitranslucent paper—and with the rarest characters (in English the five characters j k q x z) missing. Using a column of the cryptotext as line, the corresponding plaintext—which is torn, however—is to be found in some other line, and it is plausible to take a line with a maximum of boldface characters, provided that line has no or only a few missing characters. This is shown in Fig. 138 for the first column

G I Y L B N S N J W X C S N S G

of the example of G. W. Kulp, Fig. 128. The line marked by a $\hat{=}$ U clearly stands out. The line marked by a $\hat{=}$ F has two boldface characters more, but also one handicap : . As indicated by the values in Table 20, the problem

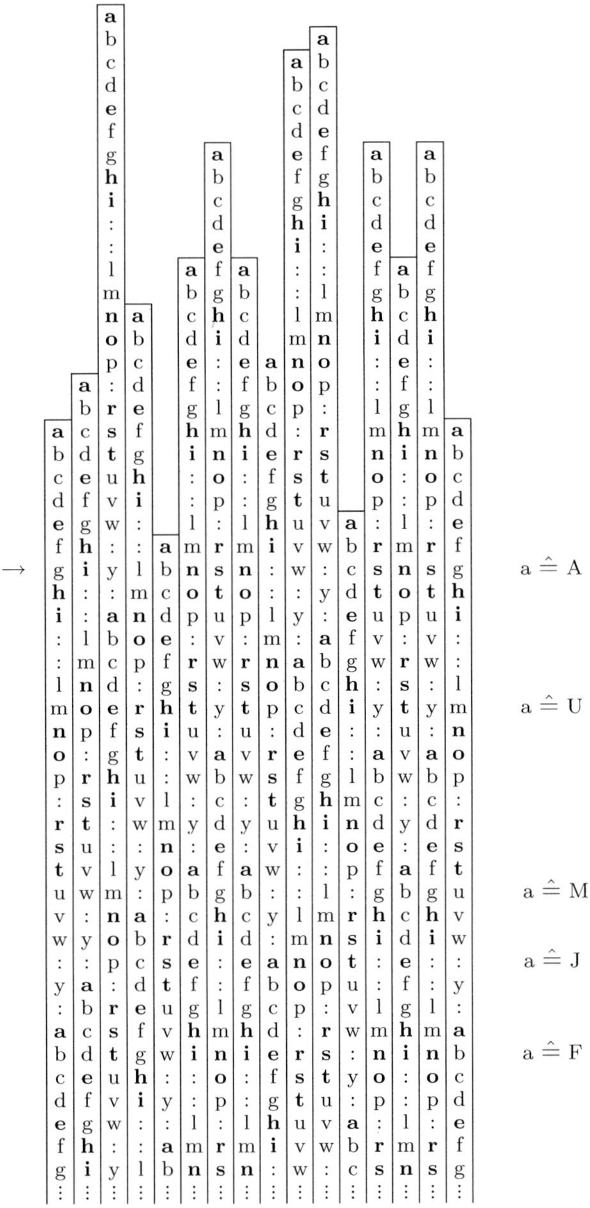

Fig. 138. Strip method: Search for the 'boldest' line (for first column in Fig. 128)

emerges of deciding between these two keys *U* and *F* for the first column. The remaining lines, e.g., a ≙ J, are inferior. a ≙ M shows the worst case.

18.2.3 Additional help. As soon as the shift is determined for a second column, too, we can hope that bigram frequencies will help in such decisions, in our case between the competing keys *U* and *F*. The determination of the individual key letters is thus mutually supporting.

Alignment	*Chi*	
a ≙ A	4.05%	
a ≙ B	3.54%	
a ≙ C	3.70%	
a ≙ D	2.64%	
a ≙ E	4.38%	
a ≙ F	5.54%	⟵
a ≙ G	4.07%	
a ≙ H	2.97%	
a ≙ I	2.98%	
a ≙ J	5.13%	⟵
a ≙ K	4.43%	
a ≙ L	3.60%	
a ≙ M	1.72%	
a ≙ N	4.30%	
a ≙ O	4.85%	
a ≙ P	4.11%	
a ≙ Q	3.01%	
a ≙ R	2.77%	
a ≙ S	4.71%	
a ≙ T	3.59%	
a ≙ U	5.38%	⟵
a ≙ V	3.71%	
a ≙ W	3.37%	
a ≙ X	2.65%	
a ≙ Y	4.18%	
a ≙ Z	4.62%	

Table 20.
Calculated values of *Chi*
for standard alphabet against first column

18.2.4 Slide method. A related method uses a slide or a disk, corresponding exactly to the original, carrying the standard alphabet or a mixed alphabet fallen into unauthorized hands. The most frequent letters on the plaintext side are again specially marked, say in boldface type, and the rarest ones omitted. On the cryptotext side, the observed frequencies are marked, say by strokes. For the letters of the column G I Y L B N S N J W X C S N S G, the frequencies are marked as follows (Fig. 139):

$$\mathrm{A}\ \overline{\mathrm{B}}\ \overline{\mathrm{C}}\ \mathrm{D}\ \mathrm{E}\ \mathrm{F}\ \overline{\overline{\mathrm{G}}}\ \mathrm{H}\ \overline{\mathrm{I}}\ \overline{\mathrm{J}}\ \mathrm{K}\ \overline{\mathrm{L}}\ \mathrm{M}\ \overline{\overline{\overline{\mathrm{N}}}}\ \mathrm{O}\ \mathrm{P}\ \mathrm{Q}\ \mathrm{R}\ \overline{\overline{\overline{\mathrm{S}}}}\ \mathrm{T}\ \mathrm{U}\ \mathrm{V}\ \overline{\mathrm{W}}\ \overline{\mathrm{X}}\ \overline{\mathrm{Y}}\ \mathrm{Z} \quad .$$

Then the two slides or disks are moved against each other until the 'boldest' confrontation is found. Clearly, this is only a variant of the method above.

18.2.5 Reach of the method. This principle of 'controlled exhaustion' can be used in all cases where it is known how to obtain from a primary substitution all other substitutions, among others also in case of ROTOR encryption steps. A depth of at least 6–9 is aimed at, i.e., normally 6–9 cryptotext letters per key letter are required for success.

a b c d **e** f g **h** **i** . . l m **n** o p . **r** **s** **t** u v w . y . **a** b c d **e** f g **h** **i** . . l m **n** o p . **r** **s** **t** u v w . y .

a $\hat{=}$ U   A $\bar{B}$ $\bar{C}$ D E F $\bar{\bar{G}}$ H $\bar{I}$ $\bar{J}$ K $\bar{L}$ M $\bar{\bar{\bar{N}}}$ O P Q R $\bar{\bar{\bar{S}}}$ T U V $\bar{W}$ $\bar{X}$ $\bar{Y}$ Z

a b c d **e** f g **h** **i** . . l m **n** o p . **r** **s** **t** u v w . y . **a** b c d **e** f g **h** **i** . . l m **n** o p . **r** **s** **t** u v w . y .

a $\hat{=}$ F   A $\bar{B}$ $\bar{C}$ D E F $\bar{\bar{G}}$ H $\bar{I}$ $\bar{J}$ K $\bar{L}$ M $\bar{\bar{\bar{N}}}$ O P Q R $\bar{\bar{\bar{S}}}$ T U V $\bar{W}$ $\bar{X}$ $\bar{Y}$ Z

a b c d **e** f g **h** **i** . . l m **n** o p . **r** **s** **t** u v w . y . **a** b c d **e** f g **h** **i** . . l m **n** o p . **r** **s** **t** u v w . y .

a $\hat{=}$ M   A $\bar{B}$ $\bar{C}$ D E F $\bar{\bar{G}}$ H $\bar{I}$ $\bar{J}$ K $\bar{L}$ M $\bar{\bar{\bar{N}}}$ O P Q R $\bar{\bar{\bar{S}}}$ T U V $\bar{W}$ $\bar{X}$ $\bar{Y}$ Z

Fig. 139. Slide for decryption of a column of torn text
a $\hat{=}$ U: a good match   a $\hat{=}$ F: a good match, too   a $\hat{=}$ M: a bad match

In other words, to prevent this attack, the plaintext should not be longer than six times the key length.

Moreover, the method can also be used for polyalphabetic encryption with arbitrary unrelated alphabets that have fallen into unauthorized hands—say a whole cylinder M-94 or a whole strip device CSP 642, or that have already been grouped in families (Sect. 14.3.6). Then no more is needed than to test a column of the torn plaintext against every single alphabet. Again, use of a personal computer will be sufficient.

However, the depth of the columns will frequently be too short to succeed with this method. Normally, at least 40 or 50 cryptotext letters per key letter will be required for success.

18.3 Chi Test: Mutual Alignment of Accompanying Alphabets

If the primary alphabet is not known, it is still possible to align mutually the individual accompanying alphabets and again to replace the polyalphabetically encrypted cryptotext by a monoalphabetically encrypted intermediary cryptotext, which can be treated with the methods of Chapters 12–15. This procedure is also useful if the primary alphabet is the standard alphabet, but this fact has not been recognized, say because the cryptotext is rather short.

18.3.1 Example. The following cryptotext (Abraham Sinkov 1968) of 303 characters has a well-balanced frequency distribution shown in Fig. 140, which does not suggest monoalphabetic encryption.

11	8	12	9	12	8	3	4	10	21	5	7	19	9	20	10	8	20	12	4	8	15	22	13	7	26
A	B	C	D	E	F	G	H	I	J	K	L	M	N	O	P	Q	R	S	T	U	V	W	X	Y	Z

Fig. 140. Frequency distribution in the cryptotext of Sect. 18.3.1

SWWJR GPRDN FMWJE XEWGR ZJQDN VJZRV
SZXOJ VWWRO VBHRM MOFDL IPAXV EZWUT
CZOZA AQQJL UPKZZ XUMJA PCZOE BAWZR
ZYKZI POFOL UOCRE NYKRI CAMOX IOORR
ZJKOL VWWJN VPKZA AFOCA MZOMR CJZDY
EJXEL XRFQI ZJCMA RJVWI DSWZX ASOTR
BJBZO QPXMI PDJVZ ZXHGQ SZFDQ FJZJR
BMWIC EZMWL MECVY VWZOX TWHSR UUBMT
NSJDW SSOOW CUNJY VJEWI PPFSL MOQVY
CVWRI SMMHW XMEJY NUZMV MXWCR NBRDE
SNB

The value of *Phi* is 4.58% and confirms the conjecture. There are nine repetitions of length 3, but no longer ones, and the distances are:

WWJ	:	$125 = 5 \cdot 5 \cdot 5$
RZJ	:	$100 = 2 \cdot 2 \cdot 5 \cdot 5$
JVW	:	$132 = 2 \cdot 2 \cdot 3 \cdot 11$
VWW	:	$90 = 2 \cdot 3 \cdot 3 \cdot 5$
CZO	:	$21 = 3 \cdot 7$
ZAA	:	$70 = 2 \cdot 5 \cdot 7$
PKZ	:	$60 = 2 \cdot 2 \cdot 3 \cdot 5$
ZZX	:	$121 = 11 \cdot 11$
CAM	:	$28 = 2 \cdot 2 \cdot 7$.

The Kasiski examination is unable to differentiate between the two possible periods 5 and 7. However, this can be done with a Kullback examination. Writing a depth in 7 columns yields the low value $Phi^{(7)} = 4.44\%$, while for a depth in 5 columns the calculation of the values of ϕ_ρ as shown in Fig. 141a yields $Phi^{(5)}{=}5{\cdot}(196{+}236{+}258{+}262{+}240)/(303\cdot 302){=}6.51\%$, a much higher value in the expected range. Since also the frequencies are rather unbalanced, each column could indeed be monoalphabetically encrypted. However, none of the columns seems to have a frequency distribution with a shifted profile belonging to English, German, or French, as a glance ahead at Fig. 141b shows. Thus, not a VIGENÈRE system, but more generally an ALBERTI system seems likely, and we have to determine its primary alphabet.

18.3.2 Obtaining an intermediary cryptotext. If the text were ten times as long, we could treat every column separately and in the end perhaps be surprised to find that all the alphabets have a common primary alphabet. But with 61 or 60 characters in a column, the text basis is too small to do this. Therefore, we can only hope that *under the assumption* of an ALBERTI system the five alphabets can be mutually aligned such that a monoalphabetically encrypted intermediary cryptotext of 303 characters is obtained, enough to use standard methods.

1st column

A	B	C	D	E	F	G	H	I	J	K	L	M	N	O	P	Q	R	S	T	U	V	W	X	Y	Z
3	3	5	1	3	2	1	0	2	0	0	0	5	4	0	4	1	1	6	1	3	7	0	4	0	5

$\phi_1 = 196$

2nd column

A	B	C	D	E	F	G	H	I	J	K	L	M	N	O	P	Q	R	S	T	U	V	W	X	Y	Z
2	2	1	1	2	1	0	0	0	10	0	0	4	1	5	6	1	1	4	0	4	1	5	2	2	6

$\phi_2 = 236$

3rd column

A	B	C	D	E	F	G	H	I	J	K	L	M	N	O	P	Q	R	S	T	U	V	W	X	Y	Z
1	3	3	0	2	5	0	3	0	2	5	0	4	1	6	0	3	2	0	0	0	1	11	3	0	6

$\phi_3 = 258$

4th column

A	B	C	D	E	F	G	H	I	J	K	L	M	N	O	P	Q	R	S	T	U	V	W	X	Y	Z
0	0	2	7	1	0	2	1	1	8	0	0	5	0	7	0	1	7	2	1	1	3	3	1	0	7

$\phi_4 = 262$

5th column

A	B	C	D	E	F	G	H	I	J	K	L	M	N	O	P	Q	R	S	T	U	V	W	X	Y	Z
5	0	1	0	4	0	0	0	7	1	0	7	1	3	2	0	2	9	0	2	0	3	3	3	5	2

$\phi_5 = 240$

Fig. 141a. Frequency distribution for five columns and values of ϕ_ρ

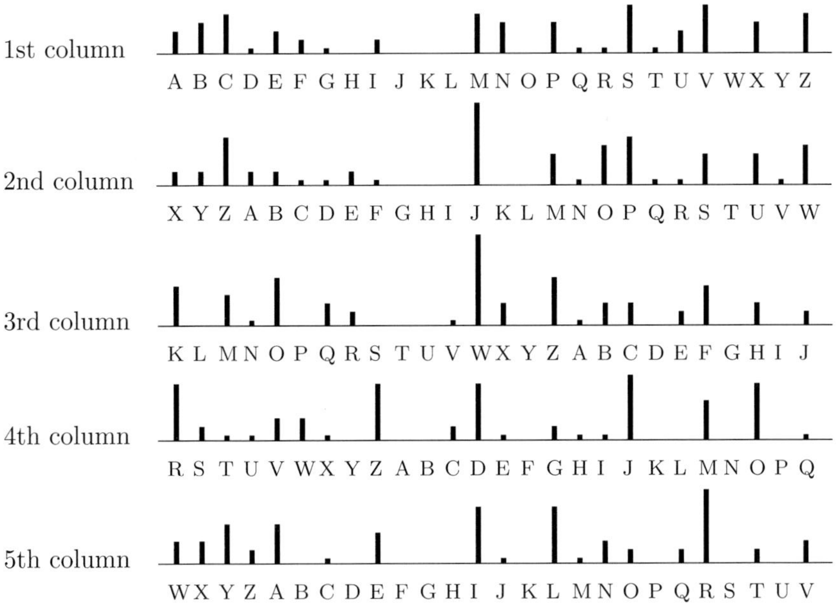

Fig. 141b. Profiles for five columns, aligned

To obtain the mutual alignment of the i-th and the k-th columns, *Chi* is calculated for the i-th column, shifted cyclically by q positions, and the k-th column, for $q = 0 \ldots N{-}1$. Normally, in this sequence all the values fluctuate around κ_R with one exception, which should be in the neighborhood of κ_S, and the corresponding q giving the alignment shift. Table 21 shows this for the first and second column, with the resulting alignment $A^{(1)} \hat{=} X^{(2)}$.

Alignment	*Chi*	
$A^{(1)} \hat{=} A^{(2)}$	$157/61^2 = 4.22\%$	
$A^{(1)} \hat{=} B^{(2)}$	$133/61^2 = 3.57\%$	
$A^{(1)} \hat{=} C^{(2)}$	$162/61^2 = 4.35\%$	
$A^{(1)} \hat{=} D^{(2)}$	$122/61^2 = 3.28\%$	
$A^{(1)} \hat{=} E^{(2)}$	$144/61^2 = 3.87\%$	
$A^{(1)} \hat{=} F^{(2)}$	$138/61^2 = 3.71\%$	
$A^{(1)} \hat{=} G^{(2)}$	$102/61^2 = 2.74\%$	
$A^{(1)} \hat{=} H^{(2)}$	$170/61^2 = 4.57\%$	
$A^{(1)} \hat{=} I^{(2)}$	$119/61^2 = 3.20\%$	
$A^{(1)} \hat{=} J^{(2)}$	$126/61^2 = 3.39\%$	
$A^{(1)} \hat{=} K^{(2)}$	$188/61^2 = 5.05\%$	
$A^{(1)} \hat{=} L^{(2)}$	$83/61^2 = 2.23\%$	
$A^{(1)} \hat{=} M^{(2)}$	$160/61^2 = 4.30\%$	
$A^{(1)} \hat{=} N^{(2)}$	$133/61^2 = 3.57\%$	
$A^{(1)} \hat{=} O^{(2)}$	$165/61^2 = 4.43\%$	
$A^{(1)} \hat{=} P^{(2)}$	$137/61^2 = 3.68\%$	
$A^{(1)} \hat{=} Q^{(2)}$	$106/61^2 = 2.85\%$	
$A^{(1)} \hat{=} R^{(2)}$	$172/61^2 = 4.62\%$	
$A^{(1)} \hat{=} S^{(2)}$	$130/61^2 = 3.49\%$	
$A^{(1)} \hat{=} T^{(2)}$	$123/61^2 = 3.31\%$	
$A^{(1)} \hat{=} U^{(2)}$	$190/61^2 = 5.11\%$	
$A^{(1)} \hat{=} V^{(2)}$	$132/61^2 = 3.55\%$	
$A^{(1)} \hat{=} W^{(2)}$	$148/61^2 = 3.98\%$	
$A^{(1)} \hat{=} X^{(2)}$	$236/61^2 = 6.34\%$	⟵
$A^{(1)} \hat{=} Y^{(2)}$	$91/61^2 = 2.45\%$	
$A^{(1)} \hat{=} Z^{(2)}$	$154/61^2 = 4.14\%$	

Table 21. Calculated values of *Chi* for first column against second column

Abraham Sinkov indicates different alignment strategies: a chaining one with an alignment of the 2nd column against the 1st, the 3rd column against the 2nd, the 4th column against the 3rd, the 5th column against the 4th, and so on, possibly cyclically closed; a starlike one with an alignment of the 2nd column against the 1st, the 3rd column against the 1st, the 4th column against the 1st, the 5th column against the 1st, and so on. The chain has the disadvantage that the results cannot be better than the effect of the weakest alignment in the chain. The star has the disadvantage that the common reference may be ill-chosen. In general, bypasses are sometimes necessary.

A weak alignment occurs if more than one value of *Chi* is raised above the background. This happens in our example for the calculation of *Chi* between the 3rd and the 4th column. As Table 22 shows, there is almost no difference between $A^{(3)} \hat{=} H^{(4)}$ and $A^{(3)} \hat{=} N^{(4)}$. It is possible to follow up several cases, and also possible to bypass weak alignments. In our example it turns out that the chain as well as the star work well with the alignment $A^{(3)} \hat{=} H^{(4)}$. This in the end gives the alignment shown in Fig. 141b.

Alignment	*Chi*	
$A^{(3)} \hat{=} A^{(4)}$	$187/61 \cdot 60 = 5.11\%$	
$A^{(3)} \hat{=} B^{(4)}$	$86/61 \cdot 60 = 2.35\%$	
$A^{(3)} \hat{=} C^{(4)}$	$148/61 \cdot 60 = 4.04\%$	
$A^{(3)} \hat{=} D^{(4)}$	$164/61 \cdot 60 = 4.48\%$	
$A^{(3)} \hat{=} E^{(4)}$	$165/61 \cdot 60 = 4.51\%$	
$A^{(3)} \hat{=} F^{(4)}$	$117/61 \cdot 60 = 3.20\%$	
$A^{(3)} \hat{=} G^{(4)}$	$82/61 \cdot 60 = 2.24\%$	
$A^{(3)} \hat{=} H^{(4)}$	$231/61 \cdot 60 = 6.31\%$	$\longleftarrow$
$A^{(3)} \hat{=} I^{(4)}$	$122/61 \cdot 60 = 3.33\%$	
$A^{(3)} \hat{=} J^{(4)}$	$110/61 \cdot 60 = 3.01\%$	
$A^{(3)} \hat{=} K^{(4)}$	$143/61 \cdot 60 = 3.91\%$	
$A^{(3)} \hat{=} L^{(4)}$	$109/61 \cdot 60 = 2.98\%$	
$A^{(3)} \hat{=} M^{(4)}$	$150/61 \cdot 60 = 4.10\%$	
$A^{(3)} \hat{=} N^{(4)}$	$229/61 \cdot 60 = 6.26\%$	$\longleftarrow$
$A^{(3)} \hat{=} O^{(4)}$	$53/61 \cdot 60 = 1.45\%$	
$A^{(3)} \hat{=} P^{(4)}$	$180/61 \cdot 60 = 4.92\%$	
$A^{(3)} \hat{=} Q^{(4)}$	$146/61 \cdot 60 = 3.99\%$	
$A^{(3)} \hat{=} R^{(4)}$	$103/61 \cdot 60 = 2.77\%$	
$A^{(3)} \hat{=} S^{(4)}$	$204/61 \cdot 60 = 5.57\%$	
$A^{(3)} \hat{=} T^{(4)}$	$108/61 \cdot 60 = 2.95\%$	
$A^{(3)} \hat{=} U^{(4)}$	$126/61 \cdot 60 = 3.44\%$	
$A^{(3)} \hat{=} V^{(4)}$	$190/61 \cdot 60 = 5.19\%$	
$A^{(3)} \hat{=} W^{(4)}$	$114/61 \cdot 60 = 3.11\%$	
$A^{(3)} \hat{=} X^{(4)}$	$124/61 \cdot 60 = 3.39\%$	
$A^{(3)} \hat{=} Y^{(4)}$	$145/61 \cdot 60 = 3.96\%$	
$A^{(3)} \hat{=} Z^{(4)}$	$124/61 \cdot 60 = 3.39\%$	

Table 22. Calculated values of *Chi* for third column against fourth column

Thus, the example shows how a periodic ALBERTI encryption under reasonably fortunate circumstances can be reduced mechanically to an intermediary cryptotext which is most likely monoalphabetically encrypted. In this example from Sinkov, 60 cryptotext letters per key letter were more than sufficient. The support obtainable from a personal computer is enough.

18.3.3 A side result. In Fig. 141b it may be noticed that the letters, read vertically, which produce AXKRW, BYLSX, CZMTY, DANUZ and so on, give among others ROBIN. This is presumably the 5-letter key word.

18.4 Reconstruction of the Primary Alphabet

The monoalphabetically encrypted intermediary text is produced by systematic change of letters, as indicated by Fig. 141b: the fragment SWWJR is treated as follows:

$$\mathrm{SWWJR} = \mathrm{S}^{(1)}\mathrm{W}^{(2)}\mathrm{W}^{(3)}\mathrm{J}^{(4)}\mathrm{R}^{(5)} = \mathrm{S}^{(1)}\mathrm{Z}^{(1)}\mathrm{M}^{(1)}\mathrm{S}^{(1)}\mathrm{V}^{(1)} = \mathrm{SZMSV}^{(1)}$$

Altogether there is the following intermediary cryptotext, expressed in the alphabet $^{(1)}$ of the first column:

SZMSV	GSHMR	FPMSI	XHMPV	ZMGMR	VMPAZ
SCNXN	VZMAS	VEXAQ	MRVMP	ISQGZ	ECMDX
CCEIE	ATGSP	USAID	XXCSE	PFPXI	BDMIV
ZBAIM	PRVXP	URSAI	NBAAM	CDCXB	IREAV
ZMAXP	VZMSR	VSAIE	AIELE	MCEVV	CMPMC
EMNNP	XUVZM	ZMSVE	RMLFM	DVMIB	AVECV
BMRIS	QSNVM	PGZED	ZAXPU	SCVMU	FMPSV
BPMRG	ECCFP	MHSEC	VZPXB	TZXBV	UXRVX
NVZMA	SVEXA	CXDSC	VMUFM	PSVBP	MRGEC
CYMAM	SPCQA	XPUSC	NXPVZ	MAMLV	NEHMI
SQR					

Its frequency distribution with respect to the alphabet $^{(1)}$ is

A	B	C	D	E	F	G	H	I	J	K	L	M	N	O	P	Q	R	S	T	U	V	W	X	Y	Z
20	10	21	7	19	6	7	4	14	0	0	3	40	9	0	23	5	13	24	2	8	31	0	20	1	16

The entry is therefore

$\mathrm{M}^{(1)} \hat{=}$ e , $\mathrm{V}^{(1)} \hat{=}$ t , $\mathrm{S}^{(1)} \hat{=}$ a .

From the frequently occurring trigram VZM $\hat{=}$ tZe one obtains

$\mathrm{Z}^{(1)} \hat{=}$ h .

Working freestyle, one can from the occurrence of /heat/ tentatively conjecture that /temperature/ occurs, indeed the repetition VMUFMPSVBP $\hat{=}$ teUFePatBP occurring towards the end of the seventh line and again in the ninth has the requested pattern. This gives already

$\mathrm{U}^{(1)} \hat{=}$ m , $\mathrm{F}^{(1)} \hat{=}$ p , $\mathrm{P}^{(1)} \hat{=}$ r , $\mathrm{B}^{(1)} \hat{=}$ u .

At the beginning of the fifth line there is

VZMAXPVZMSRV $\hat{=}$ theAXrtheaRt $\hat{=}$ thenortheast . Thus

$\mathrm{A}^{(1)} \hat{=}$ n , $\mathrm{X}^{(1)} \hat{=}$ o , $\mathrm{R}^{(1)} \hat{=}$ s .

The decryption now moves to a gallop, since e t a o n r s h and some rare letters are already determined: from the fragmentary decryption

aheat	GaHes	preaI	oHert	heGes	ternh
aCNoN	thena	tEonQ	ester	IaQGh	ECeDo
CCEIE	nTGar	manID	ooCaE	rproI	uDeIt

```
hunIe rstor msanI Nunne CDCou IsEnt
henor theas tanIE nIELE eCEtt CereC
EeNNr omthe heatE seLpe DteIu ntECt
uesIa QaNte rGhED hnorm aCtem perat
uresG ECCpr eHaEC throu Thout mosto
Nthen atEon CoDaC tempe ratur esGEC
CYene arCQn ormaC North eneLt NEHeI
aQs
```

one deduces step by step

$\mathrm{G}^{(1)} \hat{=}\ \mathrm{w}$, $\mathrm{H}^{(1)} \hat{=}\ \mathrm{v}$, $\mathrm{I}^{(1)} \hat{=}\ \mathrm{d}$, $\mathrm{C}^{(1)} \hat{=}\ \mathrm{l}$, $\mathrm{N}^{(1)} \hat{=}\ \mathrm{f}$, $\mathrm{E}^{(1)} \hat{=}\ \mathrm{i}$,
$\mathrm{Q}^{(1)} \hat{=}\ \mathrm{y}$, $\mathrm{D}^{(1)} \hat{=}\ \mathrm{c}$, $\mathrm{T}^{(1)} \hat{=}\ \mathrm{g}$, $\mathrm{L}^{(1)} \hat{=}\ \mathrm{x}$, $\mathrm{Y}^{(1)} \hat{=}\ \mathrm{b}$,

and ends up with a plaintext that obviously makes sense:

```
aheat waves pread overt hewes ternh
alfof thena tiony ester daywh ileco
llidi ngwar mandc oolai rprod ucedt
hunde rstor msand funne lclou dsint
henor theas tandi ndixi elitt lerel
ieffr omthe heati sexpe ctedu ntilt
uesda yafte rwhic hnorm altem perat
uresw illpr evail throu ghout mosto
fthen ation local tempe ratur eswil
lbene arlyn ormal forth enext fived
ays
```

At that moment the primary alphabet is reconstructed up to an arbitrary shift. Nothing is known about /j/, /k/, /q/, /z/ , which do not show up. Assuming that *ROBIN* indeed was the key word, a primary alphabet results that belongs to the key letter *A* and reordered for decryption reads:

	A	B	C	D	E	F	G	H	I	J	K	L	M	N	O	P	Q	R	S	T	U	V	W	X	Y	Z
A	s	a	g	m	t	*	o	b	h	n	u	l	c	i	p	w	v	d	*	*	x	e	f	*	r	y .

Now even the 'second' key, the password for the construction of the alphabet, becomes transparent: if the alphabet sequence is written in five columns

```
s o l v e
a b c d f
g h i * *
m n p * r
t u w x y
*
```

the password /solve/ suddenly appears. The primary alphabet is constructed according to the method discussed in Sect. 3.2.5 . In this way, we can even complete the alphabet:

s o l v e
a b c d f
g h i j k
m n p q r
t u w x y
z

The complete primary alphabet belonging to the key letter A is

	a	b	c	d	e	f	g	h	i	j	k	l	m	n	o	p	q	r	s	t	u	v	w	x	y	z
A	B	H	M	R	V	W	C	I	N	S	X	L	D	J	G	O	T	Y	A	E	K	Q	P	U	Z	F .

With this result, the decryption is perfect in the sense of Rohrbach.

Causal repetitions (see 18.3.1) are RZJ and VWW, which both become /the/; PKZ is decrypted /and/ ; WWJ is /hea/ in hea(t) and in (nort)hea(st) , ZAA is /din/ in (colli)din(g) and in (an)d_in . The quite unlikely occurrence of the four accidental repetitions JVW, CZO, ZZX, CAM is confirmed; they have distances $2 \cdot 2 \cdot 3 \cdot 11$, $3 \cdot 7$, $11 \cdot 11$, $2 \cdot 2 \cdot 7$, all missing the factor 5.

18.5 Kerckhoffs' Symmetry of Position

In Sect. 18.4, for methodical reasons, a free-style frequency analysis was performed. Frequently, there are clues for probable words leading to a pattern analysis, which may also give decryption of rare characters in some of the accompanying alphabets. In 1883, Auguste Kerckhoffs detected that it is possible in suitable cases to infer from such a decryption of characters of some column the decryption of characters of some other, separate column. He called the corresponding property of the accompanying alphabets, based (Chapter 5) on the commutativity of addition in $\mathbb{Z}_N$, *symétrie de position* (symmetry of position). The method, generally known under Kerckhoffs' name, is explained in the following using an original example from Kerckhoffs.

18.5.1 Example. Let the cryptotext be

```
RBNBJ  JHGTS  PTABG  JXZBG  JICEM  QAMUW
IVGAG  NEIMW  REZKZ  SUABR  RBPBJ  CGYBG
JJMHE  NPMUZ  CHGWO  UDCKO  JKKBC  PVPMJ
NPGKW  PWADW  CPBVM  RBZBH  JWZDN  MEUAO
JFBMN  KEXHZ  AWMWK  AQMTG  LVGHC  QBMWE
```

and assume that a Kasiski examination of the bigram repetitions RB, BJ, BG, RE, MJ, PQ has raised suspicion of a polyalphabetic encryption of period 5, possibly with accompanying shifted alphabets. As *mots probables* of the telegram dated from September 2, 1882 and sent from London to the Agence Havas in Le Caire (Cairo) are listed *Arabie, Wolseley*[1], *Suez, Ismaïlia, canal, général, soldats*. Kerckhoffs first makes a frequency analysis of the five columns and finds that in the first column $\text{J}^{(1)} \hat{=} \text{e}$, in the second and fourth

[1] Lord Garnet J. Wolseley, Commander-in-Chief of the British Army.

$\mathrm{B}^{(2)} \mathrel{\hat{=}} \mathrm{B}^{(4)} \mathrel{\hat{=}} \mathrm{e}$, in the third $\mathrm{M}^{(3)} \mathrel{\hat{=}} \mathrm{e}$, in the fifth $\mathrm{Z}^{(5)} \mathrel{\hat{=}} \mathrm{e}$ is a reasonable guess. This gives him a partial decryption

RBNBJ JHGTS PTABG JXZBG JICEM QAMUW
* e * e * e * * * * * * * e * e * * e * e * * * * * * e * *

and he tries in the circumstances the hypothesis that this should be completed to *le général Wolseley .....* . This gives an entry

RBNBJ JHGTS PTABG JXZBG JICEM QAMUW
legen eralw olsel ey*e* e**** **e**

Thus $\mathrm{G}^{(5)} \mathrel{\hat{=}} \mathrm{l}$; then he tries a continuation with *télégraphie*:

RBNBJ JHGTS PTABG JXZBG JICEM QAMUW
legen eralw olsel eytel egrap hie**

This is the prehistory; the specific method starts here. So far we have

	a	b	c	d	e	f	g	h	i	j	k	l	m	n	o	p	q	r	s	t	u	v	w	x	y	z
(1)					J			Q				R			P											
(2)					B		I		A			T						H							X	
(3)	G				M		N											C	A	Z						
(4)	E				B							T														
(5)					Z							G		J		M							S			

The symmetry of position is now introduced: To start with, the second and the fourth lines must be identical (because of $\mathrm{B}^{(2)} \mathrel{\hat{=}} \mathrm{B}^{(4)} \mathrel{\hat{=}} \mathrm{e}$, $\mathrm{T}^{(2)} \mathrel{\hat{=}} \mathrm{T}^{(4)} \mathrel{\hat{=}} \mathrm{l}$), which when supplemented gives

	a	b	c	d	e	f	g	h	i	j	k	l	m	n	o	p	q	r	s	t	u	v	w	x	y	z
(2)	E				B		I		A			T						H							X	
(4)	E				B		I		A			T						H							X	

Since J occurs in the first and in the fifth line and $\mathrm{J}^{(1)} \mathrel{\hat{=}} \mathrm{e}$, $\mathrm{J}^{(5)} \mathrel{\hat{=}} \mathrm{n} = \mathrm{e}+9$ we conclude that the whole fifth line is shifted against the first line by nine positions to the right. This gives a determination of eight cryptotext characters:

	a	b	c	d	e	f	g	h	i	j	k	l	m	n	o	p	q	r	s	t	u	v	w	x	y	z
(1)			G		J		M	Q				R		S	P							Z				
(5)					Z							G		J		M	Q				R		S	P		

But the third and the fifth lines are also connected, among others by $\mathrm{M}^{(3)} \mathrel{\hat{=}} \mathrm{e}$, $\mathrm{M}^{(5)} \mathrel{\hat{=}} \mathrm{p} = \mathrm{e}+11$. This leads to the following fixation of eleven cryptotext characters:

	a	b	c	d	e	f	g	h	i	j	k	l	m	n	o	p	q	r	s	t	u	v	w	x	y	z
(1)			G		J		M	Q	N			R		S	P					C	A	Z				
(3)	G		J		M	Q	N			R		S	P					C	A	Z						
(5)			C	A	Z							G		J		M	Q	N			R		S	P		

Finally, the second and the third lines are connected by one letter, namely A: $\mathrm{A}^{(2)} \mathrel{\hat{=}} \mathrm{i}$ and $\mathrm{A}^{(3)} \mathrel{\hat{=}} \mathrm{s} = \mathrm{i}+11$. This now gives connections with all five alphabets and a fixation of seventeen cryptotext characters:

	a	b	c	d	e	f	g	h	i	j	k	l	m	n	o	p	q	r	s	t	u	v	w	x	y	z
(1)			G	H	J		M	Q	N		X	R	E	S	P		B		I	C	A	Z		T		
(2)	E	S	P		B		I	C	A	Z		T					G	H	J		M	Q	N		X	R
(3)	G	H	J		M	Q	N		X	R	E	S	P		B		I	C	A	Z		T				
(4)	E	S	P		B		I	C	A	Z		T					G	H	J		M	Q	N		X	R
(5)		I	C	A	Z		T					G	H	J		M	Q	N		X	R	E	S	P		B

Decryptions are still missing for the nine cryptotext characters D, F, K, L, O, U, V, W, Y . However, it can be expected that with 17 out of a total of 26 characters the further decryption is trifling. Indeed, the fragmentary decryption of the first three lines of the telegram

```
RBNBJ  JHGTS  PTABG  JXZBG  JICEM  QAMUW
legen  eralw  olsel  eytel  egrap  hie**

IVGAG  NEIMW  REZKZ  SUABR  RBPBJ  CGYBG
s*ail  iaqu*  latte  n*seu  lemen  tq*el

JJMHE  NPMUZ  CHGWO  UDCKO  JKKBC  PVPMJ
eserv  ice*e  tra**  **r**  e**ec  ommun
```

provides V, W with the probable word *Ismaïlia.* Obvious filling of gaps gives U, Y and if in the third line *transports* is recognized, then D, K, O are given. F and L both occur only once (in the fifth line of the telegram) and are harder.

But there is already a better way to bring the decryption to an end: a password used in the formation of the alphabet has emerged and is obviously RESPUBLICA . Thus, the five alphabets used can be completed to give

	a	b	c	d	e	f	g	h	i	j	k	l	m	n	o	p	q	r	s	t	u	v	w	x	y	z
(1)	D	F	G	H	J	K	M	Q	N	O	X	R	E	S	P	U	B	L	I	C	A	Z	Y	T	V	W
(2)	E	S	P	U	B	L	I	C	A	Z	Y	T	V	W	D	F	G	H	J	K	M	Q	N	O	X	R
(3)	G	H	J	K	M	Q	N	O	X	R	E	S	P	U	B	L	I	C	A	Z	Y	T	V	W	D	F
(4)	E	S	P	U	B	L	I	C	A	Z	Y	T	V	W	D	F	G	H	J	K	M	Q	N	O	X	R
(5)	L	I	C	A	Z	Y	T	V	W	D	F	G	H	J	K	M	Q	N	O	X	R	E	S	P	U	B

In the completed column under plaintext /a/ appears DEGEL (French *dégel*, thaw) as the key word, which makes sense.

Finally, the whole encryption table (*tabula recta*) is shown as Table 23. It remains open which line should be the first; following Kerckhoffs also here, we choose the line with the password at the end. If then the lines, i.e., the alphabets, are numbered from A to Z, the key word of length 5 is $FRHRW$. But the 'true' key DEGEL uses the column under plaintext /a/ .

The plaintext reads: "*Le général Wolseley télégraphie d'Ismaïlia qu'il attend seulement que le service de transports et de communication soit complètement organisé pour faire une nouvelle marche en v...*".

Summarizing, the *symétrie de position* allows one "to extort more plaintext from a paucity of ciphertext" (David Kahn).

	a	b	c	d	e	f	g	h	i	j	k	l	m	n	o	p	q	r	s	t	u	v	w	x	y	z
A	Z	Y	T	V	W	D	F	G	H	J	K	M	Q	N	O	X	R	E	S	P	U	B	L	I	C	A
B	Y	T	V	W	D	F	G	H	J	K	M	Q	N	O	X	R	E	S	P	U	B	L	I	C	A	Z
C	T	V	W	D	F	G	H	J	K	M	Q	N	O	X	R	E	S	P	U	B	L	I	C	A	Z	Y
D	V	W	D	F	G	H	J	K	M	Q	N	O	X	R	E	S	P	U	B	L	I	C	A	Z	Y	T
E	W	D	F	G	H	J	K	M	Q	N	O	X	R	E	S	P	U	B	L	I	C	A	Z	Y	T	V
F	D	F	G	H	J	K	M	Q	N	O	X	R	E	S	P	U	B	L	I	C	A	Z	Y	T	V	W
G	F	G	H	J	K	M	Q	N	O	X	R	E	S	P	U	B	L	I	C	A	Z	Y	T	V	W	D
H	G	H	J	K	M	Q	N	O	X	R	E	S	P	U	B	L	I	C	A	Z	Y	T	V	W	D	F
I	H	J	K	M	Q	N	O	X	R	E	S	P	U	B	L	I	C	A	Z	Y	T	V	W	D	F	G
J	J	K	M	Q	N	O	X	R	E	S	P	U	B	L	I	C	A	Z	Y	T	V	W	D	F	G	H
K	K	M	Q	N	O	X	R	E	S	P	U	B	L	I	C	A	Z	Y	T	V	W	D	F	G	H	J
L	M	Q	N	O	X	R	E	S	P	U	B	L	I	C	A	Z	Y	T	V	W	D	F	G	H	J	K
M	Q	N	O	X	R	E	S	P	U	B	L	I	C	A	Z	Y	T	V	W	D	F	G	H	J	K	M
N	N	O	X	R	E	S	P	U	B	L	I	C	A	Z	Y	T	V	W	D	F	G	H	J	K	M	Q
O	O	X	R	E	S	P	U	B	L	I	C	A	Z	Y	T	V	W	D	F	G	H	J	K	M	Q	N
P	X	R	E	S	P	U	B	L	I	C	A	Z	Y	T	V	W	D	F	G	H	J	K	M	Q	N	O
Q	R	E	S	P	U	B	L	I	C	A	Z	Y	T	V	W	D	F	G	H	J	K	M	Q	N	O	X
R	E	S	P	U	B	L	I	C	A	Z	Y	T	V	W	D	F	G	H	J	K	M	Q	N	O	X	R
S	S	P	U	B	L	I	C	A	Z	Y	T	V	W	D	F	G	H	J	K	M	Q	N	O	X	R	E
T	P	U	B	L	I	C	A	Z	Y	T	V	W	D	F	G	H	J	K	M	Q	N	O	X	R	E	S
U	U	B	L	I	C	A	Z	Y	T	V	W	D	F	G	H	J	K	M	Q	N	O	X	R	E	S	P
V	B	L	I	C	A	Z	Y	T	V	W	D	F	G	H	J	K	M	Q	N	O	X	R	E	S	P	U
W	L	I	C	A	Z	Y	T	V	W	D	F	G	H	J	K	M	Q	N	O	X	R	E	S	P	U	B
X	I	C	A	Z	Y	T	V	W	D	F	G	H	J	K	M	Q	N	O	X	R	E	S	P	U	B	L
Y	C	A	Z	Y	T	V	W	D	F	G	H	J	K	M	Q	N	O	X	R	E	S	P	U	B	L	I
Z	A	Z	Y	T	V	W	D	F	G	H	J	K	M	Q	N	O	X	R	E	S	P	U	B	L	I	C

Table 23. Table of alphabets (*tabula recta*) for the example of Kerckhoffs

18.5.2 Volapük. The Fleming Auguste Kerckhoffs (the complete list of his given names is Jean-Guillaume-Hubert-Victor-François-Alexandre-Auguste, his nobility name was *von Nieuwenhof*) was born January 19, 1835 in Nuth in the duchy of Limburg (now in Belgium). He went to school near Aachen, studied after a stay in England at the university of Luik (*Liège*, *Lüttich*), became a high school teacher in modern languages and worked as a traveling secretary, to find finally a position in Melun, south-east of Paris. He was somewhat eccentric as a teacher, but very active in learned societies. In 1873, he became a French citizen, and in 1873–1876 he studied at the universities of Bonn and Tübingen and became *Docteur ès lettres.* In 1878 he was given a chair for German Language at the École des Hautes Études Commerciales and at the École Arago in Paris. His first contribution to cryptology was in 1882, when he wrote—for unknown reasons—*La cryptographie militaire.* This 64-page article in the *Journal des Sciences militaires*, January and February 1883, and Kasiski's work of 1863 are the foundation stones of scientific cryptology in the 19th century.

However, for most people Kerckhoffs' fame stems from his ardent and tragic support for the international, universal language Volapük, proposed in 1879 by Johann Martin Schleyer. Kerckhoffs was appointed in 1887 *Dilekel* (director) of the International Volapük Academy. Like Esperanto (1887) and other later proposals, Volapük was unable to establish itself. Kerckhoffs lived long enough to see the decline of Volapük and died broken-hearted in 1903.

18.5.3 An example with a surprise. The symmetry of position is also useful in dealing with a VIGENÈRE system, of course. We shall show this for the cryptotext of G. W. Kulp (Fig. 115), first assuming only that it is an ALBERTI system, and that there are reasons to expect a period of 12. Then we start by writing a depth of 12 columns as in Fig. 128:

(1)	(2)	(3)	(4)	(5)	(6)	(7)	(8)	(9)	(10)	(11)	(12)
G	E	I	E	I	A	S	G	D	X	V	Z
I	J	Q	L	M	W	L	A	A	M	X	Z
Y	Z	M	L	W	H	F	Z	E	K	E	J
L	V	D	X	W	K	W	K	E	T	X	L
B	R	A	T	Q	H	L	B	M	X	A	A
N	U	B	A	I	V	S	M	U	K	H	S
S	P	W	N	V	L	W	K	A	G	H	G
N	U	M	K	W	D	L	N	R	W	E	Q
J	N	X	X	V	V	O	A	E	G	E	U
W	B	Z	W	M	Q	Y	M	O	M	L	W
X	N	B	X	M	W	A	L	P	N	F	D
C	F	P	X	H	W	Z	K	E	X	H	S
S	F	X	K	I	Y	A	H	U	L	M	K
N	U	M	Y	E	X	D	M	W	B	X	Z
S	B	C	H	V	W	Z	X	P	H	W	L
G	N	A	M	I	U	K					

and find that cryptotext characters in the same column frequently show distances 4 or 7 or 11. In fact, we find six triples with these distances:

	(3)	(4)	(6)	(7)	(10)	(12)
	B	M	W	L	M	L
+7						
	I	T	D	S	T	S
+4						
	M	X	H	W	X	W

This finding also shows that the 4th and the 10th columns, as well as the 7th and the 12th columns are subordinate to the same key.

Using a systematic procedure we form for every column the pairwise differences between the cryptotext characters and list them in a table. This 'difference method' is only a variant of the *symétrie de position.* It yields in our case a predominance of 4, 7, 11 (and of the complements 22, 19, 15) and thus some more cases of the appearance of the differences 4 or 7 or 11:

	(1)	(2)	(3)	(4)	(5)	(6)	(7)	(8)	(9)	(10)	(11)	(12)
	N		B	M		W	L	M		M	X	L
+7												
		N	I	T	E	D	S		A	T	E	S
+4												
	Y	R	M	X	I	H	W	X	E	X		W
Shift	*0*	*19*	*14*	*25*	*10*	*9*	*24*	*25*	*6*	*25*	*10*	*24*

Note that the word UNITEDSTATES shines through in the middle line. The shifts needed to align the columns with the first column are given in a footline. They produce a key; decryption with respect to this key reduces the cryptotext to a monoalphabetically encrypted intermediary cryptotext. Subtraction of the respective key letters gives the beginning of this text

G	L	U	F	Y	R	U	H	X	Y	L	B
I	Q	C	M	C	N	N	B	U	N	N	B
Y	G	Y	M	M	Y	H	A	Y	L	U	L
L	C	P	Y	M	B	Y	L	Y	U	N	N .

Decryption by means of a frequency analysis offers no problems, and the intermediary encryption turns out to be a CAESAR addition—we did not use this at all—with U$\hat{=}$ a; decryption is performed by counting forward six places in the alphabet order. The beginning of the plaintext is therefore (compare Sect. 18.1.2):

m	r	a	l	e	x	a	n	d	e	r	h
o	w	i	s	i	t	t	h	a	t	t	h
e	m	e	s	s	e	n	g	e	r	a	r
r	i	v	e	s	h	e	r	e	a	t	t .

The 'difference method' which sprang up here will again be found useful for stripping off superencrypted code in the next section, since it is totally free of frequency analysis. If in the symmetry of position method frequency considerations were included, one could guess that in the above triples the last line, which is more densely populated than the other two, corresponds to /e/ (the most frequent letter) and thus the first line to /t/ and the second line to /a/. Since /a/ is the zero element in $\mathbb{Z}_{26}$, this observation also clarifies why the key word *UNITEDSTATES* shone through. (There are also 'wrong' differences, e.g., in the third column M and X with difference +11).

18.6 Stripping off Superencryption: Difference Method

We resume the discussion of Sect. 18.3 . The (mutual) alignment of accompanying alphabets does not refer to the plaintext, which is regained only from the monoalphabetically encrypted theoretical intermediary text (Sect. 18.4). This reflects the fact that ALBERTI steps are compositions: monoalphabetic functional substitution, followed by polyalphabetic VIGENÈRE addition.

18.6.1 A strip. Thus, the technique of Sects. 18.3, 18.5 is also applicable for superencrypted code, i.e., for a composition (Sect. 9.2.2) of coding and following VIGENÈRE over $\mathbb{Z}_{26}$ (literal code) or $\mathbb{Z}_{10}$ (numeral code). The latter case is called in English jargon 'stripping off a numerical additive from enciphered code', French *libeller par soustraction de l'additive*, German *Subtraktion einer Überschlüsselungszahl*. Enciphered code is 'encicode' for short, and the code taken from the codebook is 'plain code' or 'placode' for short.

Assuming a certain width of the placode (normally known, e.g., 5) and a certain period, say 15, columns of equally superencrypted placode words (columns of encicode groups) are formed. Frequently occurring plaintext words or phrases lead to frequency differences in each column of encicode groups. If the material is voluminous enough, this may allow the calculation of the mutual *Chi* of two columns and thus help alignment, but frequently the material will not be rich enough to establish by mutual alignment a reference encicode.

18.6.2 Symmetry of position. But there is still the *symétrie de position.* Two encicode groups that are prominent in two columns belong to the same placode if and only if their difference equals the difference of the additives that belong to the columns. Note that differences, according to Shannon, are the 'residue classes' of linear polygraphic substitutions.

To give an example, assume there are three columns and in each one are found three prominent encicode groups,

(1)	(2)	(3)
47965	60597	27904
69451	34689	41537
11057	10056	26443

If 47965 from the first and 60597 from the second column of encicode groups belong to the same placode, then 11057 from the first and 34689 from the second column of encicode groups belong also to one and the same placode, since

$$47965 - 11057 = 60597 - 34689 = \mathbf{36918}\ .$$

To find such coincidences systematically, for each column of encicode groups all mutual differences are calculated, in our example a three-by-three matrix for each column of encicode groups:

	(1)			(2)			(3)	
00000	88514	**36918**	00000	**36918**	50541	00000	**86477**	01561
22596	00000	58404	**74192**	00000	**24633**	**24633**	00000	25194
74192	52606	00000	50569	**86477**	00000	09549	85916	00000

With this information, we also find relations between encicode groups in different columns:

Within the first and within the second column, we have

$$47965 - 11057 = 60597 - 34689 = 36918\ .$$

Therefore, between the first and the second columns, there is

$$47965 - 60597 = 11057 - 34689 = 87478\ .$$

Within the second and within the third column, we have

$$34689 - 10056 = 41537 - 27904 = 24633\ .$$

Therefore, between the second and the third columns, there is

$$34689 - 41537 = 10056 - 27904 = 93152\ .$$

Reducing the second column of encicode groups (2) relative to the first one (1) by adding everywhere in (2) the difference 87478, and reducing the third column of encicode groups (3) relative to the first one (1) by adding everywhere in (3) the difference 87478 (for (2) against (1)) and then the difference 93152 (for (3) against (2)); altogether therefore the difference 70520, we obtain:

	(1')	(2')	(3')
	47965	47965	97424
	69451	11057	11057
	11057	97424	96963
Shift	*0*	*87478*	*70520*

The next step is to look in the first column of encicode groups for another occurrence—though more rare—of 97424, likewise in the (reduced) third column of encicode groups for another occurrence—though more rare–of 47965; moreover for occurrences of 69451, 97424, 96963. With luck, further commonly occurring placodes can be discovered.

18.6.3 Use of machines. This procedure, although logically simple, requires cumbersome calculations and it is unsurprising that in the 1920s cryptanalysts sought mechanical support for the alignment. Punch card equipment was available and was suited for the task. Particularly in the Second World War, British (J. Tiltman, Sept. 1939), Americans (R. J. Fabian, Nov. 1940), and Germans relied upon such help.

Next, special devices were built. In the Cipher Branch of the German *Oberkommando der Wehrmacht*, a *Differenzenrechengerät* was designed that processed encicode groups punched on tape with the help of mechanical scanners and relay circuitry. It provided seven differences of five-digit groups per second and recorded the output on a typewriter, and was thus 10 to 15 times faster than a human calculator at top speed.

In contrast to these digital methods, analog devices with photoelectric measurement for tetragrams were used to determine the most frequently occurring placodes, both in the Chi Branch of the O.K.W. and in the *Sonderdienst Dahlem* unit of the German Foreign Ministry. For the reduction, i.e., the subtraction of a difference from a column of encicode groups with known relative basis, special optical analog devices were designed by the mathematician Ernst Witt.

Less is known about special devices used by the Allies of the Second World War. The HEATH ROBINSON and COLOSSUS machines built in Bletchley Park handled binary additives and were oriented mainly against teletype cipher machines like the Lorenz *Schlüsselzusatz* SZ 42 (Sect. 19.2.6). In the U.S.A., the COPPERHEAD machines of 1943, technologically on the level of HEATH ROBINSON, worked with optical scanning, too, and were used against Japanese superencrypted codes. Comparable, perhaps, with German developments was the TESSIE machine of 1942, a device built by the Eastman company for the Navy, working with photoelectric measurement, which was used to find four-digit encicode groups needed for stripping off a superencryption. It was directed both against the Japanese high level fleet code and the German 'brief signals manual' for U-boats that was used to flash location messages and tapped by the Allies for cribs.

18.7 Decryption of Code

At the very end, after stripping off the superencryption, there remains the decryption of the intermediary encicode, the reconstruction of the code book ('book-building'). The intermediary encicode is shifted against the placode by a constant, but this is totally irrelevant for the work to be done, which is mainly linguistic in nature. Systematically, this work belongs rather in Chapter 15. It is much simplified if the code is a one-part code (Sect. 4.4.2); then a codegroup lying between two groups with already known plaintext equivalents has an in-between plaintext equivalent. At this point, imagination and vision, association and combination find ample scope for application. Book-building is that part of cryptanalysis where mathematics alone is helpless. A systematic treatment of the linguistic side of cryptanalysis was first attempted (in 1892) by Paul Louis Eugène Valério.

18.8 Reconstruction of the Password

The advantage offered by accompanying alphabets is that they result from a single primary alphabet. This can only help the hurried cryptographer if he can easily remember or construct the primary alphabet. For this purpose, passwords (Sect. 3.2.5) are very popular. Reconstructing them not only gives the unauthorized decryptor additional security but can also be used methodically. The use of meaningful passwords therefore creates a weakness.

18.8.1 Friedman. At first sight, what William F. Friedman presented in 1917 looks like a conjurer's trick: Let the primary alphabet be

a b c d e f g h i j k l m n o p q r s t u v w x y z
N T U V P W X J F Y Z D K Q C A B O G R L I S H M E

It turns out that it is monocyclic with the cycle

(a n q b t r o c u l d v i f w s g x h j y m k z e p) .

Now, starting from an arbitrary character, say /a/ , the substitution is iterated and the results are written down with distances of 1, of 3, of 5:

N Q B T R O C U L D V I F W S G X ... ,
N * * Q * * B * * T * * R * * O * ... ,
N * * * * Q * * * * B * * * * T * ... ,

and so on. This gives altogether

1 N Q B T R O C U L D V I F W S G X H J Y M K Z E P A
3 N D J Q V Y B I M T F K R W Z O S E C G P U X A L H
5 N K X I C Q Z H F U B E J W L T P Y S D R A M G V O
7 N G R Y L E F Q X O M D P W B H C K V A S T J U Z I
9 **N T** C D F G J K P Q R U V W X Y Z **A B O L I S H M E**
⋮ ⋮

and produces with the distance 9 a sequence that contains a meaningful password: ABOLISHMENT. Now taking on the plaintext side the same sequence but shifted 9 places to the right, the following substitution is obtained:

  a b o l i s h m e n t c d f g j k p q r u v w x y z
9 N T C D F G J K P Q R U V W X Y Z A B O L I S H M E

Reordering yields the initial alphabet, whose construction from the keyword is now clarified: it is the ninth power of the cycle

(a b o l i s h m e n t c d f g j k p q r u v w x y z)

with the password a b o l i s h m e n t.

The conjurer's trick becomes better understood if one realizes that for *every* distance an alphabet is obtained, from which by reordering the initial alphabet is regained, e.g., for the distance 7:

  a s t j u z i n g r y l e f q x o m d p w b h c k v
7 N G R Y L E F Q X O M D P W B H C K V A S T J U Z I

although this one does not produce a meaningful password. Or with the distance 1 one obtains, of course,

  a n q b t r o c u l d v i f w s g x h j y m k z e p
1 N Q B T R O C U L D V I F W S G X H J Y M K Z E P A

In fact, the third power of this alphabet reconstructs the password, since 3 times 9 equals 1 *modulo* 26.

18.8.2 Friedman again. William F. Friedman also gave in 1918 a process for the reconstruction of passwords in the very general case (Sect. 3.2.5) of an ALBERTI system with passwords both for the plaintext side and the cryptotext side. We shall come back to this in Sect. 19.5.3.

19 Compromises

The quality of a machine
depends largely on its use.
Boris Hagelin

Among the cryptographic faults listed in Chapter 11, the compromises are worst, because they open methodical lines of attack. Next to the plaintext-cryptotext compromise, discussed in Sect. 14.6, we deal in this chapter with plaintext-plaintext and cryptotext-cryptotext compromises.

19.1 Kerckhoffs' Superimposition

Polyalphabetic encryption with periodic keytext, even with unknown and unrelated alphabets, provides no security against unauthorized decryption. Once the period is determined (Chapter 17), building of a depth (Chapter 18) leads to a monoalphabetically encrypted plaintext. However, the plaintext is torn, which makes the decryption of very short texts difficult or impossible (Sects. 18.2.5, 18.3.2).

But even if the key is not periodic or comparable in length to the plaintext, the methods of Chapter 18 can be applied whenever a number of plaintexts are encrypted with the same key. Provided the cryptotexts can be adjusted to be in phase with the keytext, this plaintext-plaintext compromise of the key likewise allows one to build a depth, i.e., to build columns of cryptocharacters or of encicode groups, each one consisting of monoalphabetically encrypted (but still torn) plaintext. Auguste Kerckhoffs also discussed this situation in his 1883 paper. The in-phase adjustment of several texts is called superimposition (German *Überlagerung*). Superimposition will only work if plenty of plaintexts are available and can be adjusted, but because of the logistic problems of key assignment, this is very likely if encryption machines are used which have the same or only slightly varying starting position of the (mechanically generated) key. In this way, John H. Tiltman[1] succeeded in breaking the older ENIGMA without plugboard (British codename *rocket*), used by the German *Reichsbahn* in the Second World War for the transmission of timetables for transport trains.

---

[1] "... he was charming and intelligent and though he looked military he certainly didn't behave like a stuffed shirt." (Robin Denniston)

It is evident that the periodic use of a not-too-long key, which leads to some repetitions, also means a plaintext-plaintext compromise—but since this was common practice, it was not called so. This observation has the consequence that all methods usable for the determination of a key period can also be used to test whether the different plaintexts are in phase, and if they are not, to adjust them. In this case, the *Kappa* test or the *Chi* test will be just right.

19.1.1 Example. The following superimposition example given by Kerckhoffs assumes that the plaintexts are in phase. With altogether 13 of them, it gives a conveniently simple exercise (typographic errors are corrected):

	1 2 3 4 5	6 7 8 9 10	11 12 13 14 15	16 17 18 19 20	21 22 23 24 25	26 27 28 29
(i)	U H Y B R	J I M B C	F A M M F	J H D M R	I Q	
(ii)	U H W P R	B Q L K I	B L W R E	J R B K L	H I X B Q	E X H M
(iii)	I E W H C	H Q K Q M	T M V G J	J E D Z V	A	
(iv)	U W V R R	H I K M C	W W R G H	D C X S R	Q H	
(v)	U H S H A	H K S V C	J W Z V X	J Y N D M	Q Q N	
(vi)	Y H V H M	A G Q K C	W X P V I	H H W L Z	V L T H V	
(vii)	L H V H A	A G R L P	F M S O H	I P W Z Z	J E L Q R	B W
(viii)	S W U I R	X I C J U	F S H G W	R S Z B A	A L	
(ix)	U H W H V	A Y U L C	J W O U K	D E B K Q		
(x)	Y W X H Y	H B A L G	B V P S W	I W W J R	R H	
(xi)	W Q R E X	B I E N H	M V Y M H	S I Y M		
(xii)	S W U H D	H P J J C	K X G M H	L		
(xiii)	G Q V Q R	V O T Q Q	S P W R			

Kerckhoffs begins with the statement that for frequency considerations presumably $H^{(2)} \hat{=}$ e, $H^{(4)} \hat{=}$ e, $R^{(5)} \hat{=}$ e, $H^{(6)} \hat{=}$ e, $I^{(7)} \hat{=}$ e, $L^{(9)} \hat{=}$ e, $C^{(10)} \hat{=}$ e, and that because of the many coincidences the second, the fourth, and the sixth positions fall under the same key (wisely he does not assume $U^{(1)} \hat{=}$ e). Cryptotext (iv) would then be decrypted (iv) ****eee**..... , which suggests looking for a word that ends with ée ; l' armée would be suitable: (iv) larmeee**..... . Cryptotext (v) with (v) le*e*e***..... suggests (v) legeneral..... . Cryptotext (vi) offers the solution (vi) *ere*v***....., this leaves a choice between (vi) serezvous..... or (vi) ferezvous..... . Cryptotext (vii) with (vii) *erenvo*e*..... is interpreted by Kerckhoffs somewhat convincingly as (vii) nerenvoyez..... . He then continues with the remaining cryptotexts. But he has already made an entry.

Superimposition as a methodical idea is also suited for the case of unrelated alphabets. Provided that not too many different alphabets are used, say not more than two dozen in the case of monographic alphabets, and that the cryptotext is long enough that most of these alphabets are used at least a few times, then the effective depth of the material is correspondingly multiplied as soon as the identity of these alphabets is established. If the key is formulated in German, then on average every sixth key character is an E and thus every sixth alphabet is the same. In English, things are only slightly better.

19.1.2 Symétrie de position. In the present case the further decryption would be cumbersome, detailed work, had not Kerckhoffs made the assumption that we have accompanying alphabets which are simply shifted and thus the *symétrie de position*, the climax of his work, can be used. Then everything goes like clockwork. Some of the decrypted messages are (the genre is French North Africa):

(i)	leprefetdepoliceestici	*'le préfet de police est ici'*
(ii)	lespertesdelennemisontgrandes	*'les pertes de l' ennemi sont grandes'*
(iii)	onsemetsurladefensive	*'on se met sur la défensive'*
(iv)	larmeeestentreeaucaire	*'l' armée est entrée au Caire'*
(v)	legeneralestaalexandrie	*'le général est à Alexandrie'*
(vi)	serezvousenetatderesister	*'serez vous en état de résister'*
(vii)	nerenvoyezpaslesprisonniers	*'ne renvoyez pas les prisonniers'* .

It turns out that Kerckhoffs used the same ALBERTI steps as in Sect. 18.5.1. The key is periodic; it can be reconstructed with the encryption table of Table 23 and runs thus

$$JEMEMETSSURLADEFENSIVE \; .$$

19.2 Superimposition for Encryptions with a Key Group

Under favorable circumstances even the extreme case (not treated by Auguste Kerckhoffs) of a superimposition of only two cryptotexts encrypted with the same key is not hopeless, provided the alphabets are known.

19.2.1 Pure encryption. We assume in this section that the crypto system is not only, as usual, injective and definal, i.e., for every encryption step $\chi_s : V^{(n)} \dashrightarrow W^{(m)}$ there exists a decryption step $\chi_s^{-1} : W^{(m)} \dashrightarrow V^{(n)}$:

(*) $$\chi_s^{-1}(\chi_s(p)) = p \quad \text{for all} \quad p \in V^{(n)},$$

but also that it is functional and surjective (Sect. 2.6.2):

(**) $$\chi_s(\chi_s^{-1}(c)) = c \quad \text{for all} \quad c \in W^{(m)} \quad .$$

Then $|V^{(n)}| = |W^{(m)}|$. In this case, it is convenient to identify plaintext characters and cryptotext characters, thus $n = m$, $V \doteq W$, and the endomorphic case $\chi_s : V^n \longleftrightarrow V^n$ is assumed. Thus, let $M \subseteq V^n \times V^n$ be the crypto system, $|V| = N$. M is the key space.

The important assumption is now that the crypto system M is a pure crypto system (Sect. 9.1.1), it is closed under composition: The composition of two encryption steps $\chi_s \in M$, $\chi_t \in M$ belongs to the set M of encryption steps: $\chi_s(\chi_t(p)) = \chi_{s \bullet t}(p)$, whereby $s \bullet t$ is uniquely defined.

The composition is associative: $\chi_{r \bullet s}(\chi_t(p)) = \chi_r(\chi_{s \bullet t}(p))$. Since we have assumed that every encryption step $\chi_s \in M$ has an inverse $\chi_s^{-1} \in M$, the encryption steps build a group under composition, the key group M. $\chi_{s^{-1}}(p)$ is defined by $\chi_s^{-1}(p)$.

Trivially, the key group can be a singleton $M = \{\mathrm{id}\}$; or it can have N^n elements, say $M = \{\mathrm{id}, \chi, \chi^2, \chi^3, \ldots \chi^{N^n-1}, \}$, where χ is monocyclic; or it can have maximally $(N^n)!$ elements, $M \doteq V^n \longleftrightarrow V^n$.

Now let $c' = (c'_1, c'_2, c'_3, \ldots)$ and $c'' = (c''_1, c''_2, c''_3, \ldots)$ be two cryptotexts that are encryptions with the same key $k = (k_1, k_2, k_3, \ldots)$ of the two plaintexts $p' = (p'_1, p'_2, p'_3, \ldots)$ and $p'' = (p''_1, p''_2, p''_3, \ldots)$:

$$c'_i = \chi_{k_i}(p'_i) \ , \quad c''_i = \chi_{k_i}(p''_i) \ .$$

Furthermore, we assume that the crypto system is transitive (Sect. 14.3.4). Then the key group M is a transitive permutation group in the classical sense that there exists a character $a \in V^n$ such that for each character $y \in V^n$ there exists an encryption step $\chi_t \in M$ such that $y = \chi_t(a)$. This implies that the number of keys is greater than or equal to the power of the alphabet V^n, $|M| \geq N^n$, and every character of V^n can be related injectively to a key. In other words, the characters are equivalence classes of the keys.

We may also assume a Shannon crypto system (Sect. 2.6.4), where the key k_i is uniquely determined by a pair consisting of plaintext character p_i and cryptotext character c_i, which implies $|M| \leq N^n$. In general, the key k_i does not need to be uniquely determined by p_i and c_i.[2]

Thus, we have a pure transitive Shannon crypto system with $|M| = N^n$, belonging to a Latin square. The relation between characters and keys is one-to-one; identification of keys and characters according to $s = \chi_s(a)$ results in

$s \bullet t = \chi_{s \bullet t}(a) = \chi_s(\chi_t(a)) = \chi_s(t)$ and thus $\chi_s(p) = s \bullet p$, $\chi_{s \bullet t}(p) = \chi_{\chi_s(t)}(p)$.

Furthermore,

$$\chi_{s^{-1}}(c) = s^{-1} \bullet c = \chi_s^{-1}(c) \ .$$

Now it makes sense to speak of $\chi_{c'_i}^{-1}(c''_i)$, the cryptotext character c''_i decrypted with the cryptotext character c'_i as in-phase key. A simple calculation shows the important result that under the given conditions the key in $\chi_{c'_i}^{-1}(c''_i)$ is canceled out, or more precisely,

$$\chi_{c'_i}^{-1}(c''_i) = \chi_{p'_i}^{-1}(p''_i) \ .$$

19.2.2 Differences. For (endomorphic) Shannon crypto systems with a transitive key group, we form the difference $d_i \overset{\text{def}}{=} \chi_{c'_i}^{-1}(c''_i)$ of the two observed cryptotexts and look for two plaintexts p'_i, p''_i such that their difference $\chi_{p'_i}^{-1}(p''_i)$ equals d_i. This can be attempted in a zig-zag way much like in

[2] Only if $(N^n)! = N^n$, i.e., for $N^n = 1$ or $N^n = 2$, is the key necessarily determined. This includes as interesting case only $V \doteq Z_2$, $n = 1$; then there are only the two encryption steps identity O and involution L (Sect. 8.3.1); $p_i = c_i$ gives $k_i \doteq O$, $p_i \neq c_i$ gives $k_i \doteq L$.

Sect. 14.4; to solve it uniquely two plaintexts are needed such that the sum of their redundancies (Sect. 12.6, footnote 4) is at least 100%.

If the encryption steps even form a commutative group with respect to composition, then we have Kerckhoffs' *symétrie de position*,

$$\chi_s(t) = \chi_t(s) \quad .$$

For (endomorphic) crypto systems with a commutative key group, the key k is uniquely determined both by p' and c' and by p'' and c'', since with $c'_i = \chi_{k_i}(p'_i)$ also $c'_i = \chi_{p'_i}(k_i)$ and thus $k_i = \chi^{-1}_{p'_i}(c'_i)$. They are necessarily Shannon crypto systems; if $|M| > N^n$, then the key group is not commutative. We shall find such key groups in Sect. 19.2.4.

Moreover, in the commutative case, since $\chi_{p'_i}(d_i) = p''_i$ holds, $\chi_{d_i}(p'_i) = p''_i$ holds too. Thus p''_i results from p'_i by encryption with d_i as key.

For every key group there is a set of *dual encryption steps* $\{\breve{\chi}_s\}$ with

$$\breve{\chi}_s(p) = s \bullet p^{-1} \;, \quad \breve{\chi}^{-1}_s(c) = c^{-1} \bullet s \quad \text{and} \quad \breve{\chi}_{s \bullet t^{-1}}(p) = \breve{\chi}_{\breve{\chi}_s(t)}(p) \quad .$$

Now $\breve{\chi}^{-1}_{c'_i}(c''_i) = \breve{\chi}_{p''_i}(p'_i)$. For the case of a commutative key group the dual encryption is self-reciprocal: $\breve{\chi}^{-1}_s(t) = \breve{\chi}_s(t)$.

19.2.3 Cyclic key groups. If (for $n = 1$) accompanying alphabets are constructed by a cyclic shift of a primary alphabet of N characters, then the number of keys coincides with the number of characters; in fact the key group is commutative and is the cyclic group of order N. Thus, VIGENÈRE encryption has this group as key group, with addition modulo N as a model, whereas BEAUFORT encryption (as used in the Hagelin M-209) is the dual of VIGENÈRE encryption. Here, the primary alphabets are known anyway.

ALBERTI encryption can be treated as well, if the difference is modified. In the encryption table of Kerkhoffs' example, Table 23, with ρ as generating cycle and the primary alphabet

P:	a	b	c	d	e	f	g	h	i	j	k	l	m	n	o	p	q	r	s	t	u	v	w	x	y	z
	Z	Y	T	V	W	D	F	G	H	J	K	M	Q	N	O	X	R	E	S	P	U	B	L	I	C	A

there is $A = \rho^0 P$, $B = \rho^1 P$, $C = \rho^2 P$, $D = \rho^3 P$, ... , $Z = \rho^{25} P$.

We may assume that P is known already, say because an ALBERTI disk has fallen into the wrong hands.

We now take up again the cryptotexts (i) and (ii) of Sect. 19.1.1. Forming from $c'' \stackrel{\text{def}}{=}$ (ii) and $c' \stackrel{\text{def}}{=}$ (i) the P-modified difference $d_i = \chi^{-1}_{P^{-1}c'_i}(P^{-1}c''_i)$, we find that $d_i \mathrel{\hat{=}} \rho^{\delta_i} P$ if and only if in the known permuted alphabet P one has to go δ_i steps in order to get from c''_i to c'_i. The result is

	1	2	3	4	5	6	7	8	9	10	11	12	13	14	15	16	17	18	19	20	21	22
c''	U	H	W	P	R	B	Q	L	K	I	B	L	W	R	E	J	R	B	K	L	H	I
c'	U	H	Y	B	R	J	I	M	B	C	F	A	M	M	F	J	H	D	M	R	I	Q
d	a	a	x	c	a	o	l	p	l	b	l	d	h	v	p	a	s	k	b	u	p	p
δ	0	0	23	2	0	14	11	15	11	1	11	3	7	21	15	0	18	10	1	20	15	15

Now $d_i = \chi^{-1}_{P^{-1}c'_i}(P^{-1}c''_i) = \chi^{-1}_{P^{-1}p'_i}(P^{-1}p''_i)$; the known d can be interpreted as cryptotext, obtained under χ^{-1} with $P^{-1}p'$ as (unknown) key from $P^{-1}p''$. This *swapping of roles* opens up all possible attacks for the determination of the key, using key patterns and key letter frequencies.

For example, $d_i =$ a (which occurs for $i = 1, 2, 5, 16$) means identity of p'_i and p''_i . In French, this happens with a frequency of about 30% for $p'_i = p''_i =$ /e/ , while $p'_i = p''_i =$ /a/ and $p'_i = p''_i =$ /s/ each occur with a frequency of only 10%. (In fact, the bold assumption /e/ would be fulfilled for $i = 2, 5, 16$, while /l/ occurs for $i = 1$). In view of the information given about the genre, the method of the probable word is to be recommended. Thus, assuming we are dealing with the French language and in the circumstances the probable word *ennemi*, it remains to check exhaustively whether there corresponds to one of the possible positions of *ennemi* in p'' a meaningful French word in p' (or vice versa). The following successive trials

	1	2	3	4	5	6	7	8	9
p''	e	n	n	e	m	i	*	*	*
d	a	a	x	c	a	o	l	p	l
δ	0	0	23	2	0	14	11	15	11
p'	e	n	k	g	m	w	*	*	*

	1	2	3	4	5	6	7	8	9
p''	*	e	n	n	e	m	i	*	*
d	a	a	x	c	a	o	l	p	l
δ	0	0	23	2	0	14	11	15	11
p'	*	e	k	p	e	a	t	*	*

	1	2	3	4	5	6	7	8	9
p''	*	*	e	n	n	e	m	i	*
d	a	a	x	c	a	o	l	p	l
δ	0	0	23	2	0	14	11	15	11
p'	*	*	b	p	n	s	x	x	*

	1	2	3	4	5	6	7	8	9
p''	*	*	*	e	n	n	e	m	i
d	a	a	x	c	a	o	l	p	l
δ	0	0	23	2	0	14	11	15	11
p'	*	*	*	g	n	b	p	b	t

are unsuccessful, but in the 13th position (and nowhere else)

	1	2	3	4	5	6	7	8	9	10	11	12	13	14	15	16	17	18	19	20	21	22
p''	*	*	*	*	*	*	*	*	*	*	*	*	e	n	n	e	m	i	*	*	*	*
d	a	a	x	c	a	o	l	p	l	b	l	d	h	v	p	a	s	k	b	u	p	p
δ	0	0	23	2	0	14	11	15	11	1	11	3	7	21	15	0	18	10	1	20	15	15
p'	*	*	*	*	*	*	*	*	*	*	*	*	l	i	c	e	e	s	*	*	*	*

emerges with /licees/ a fragment, that can be enlarged reasonably into /police_est/ . (In p' , /ennemi/ is a flop.) Now interchanging the roles of p' and p'' gives

	1	2	3	4	5	6	7	8	9	10	11	12	13	14	15	16	17	18	19	20	21	22
p''	*	*	*	*	*	*	*	*	*	*	e	l	e	n	n	e	m	i	s	*	*	*
d	a	a	x	c	a	o	l	p	l	b	l	d	h	v	p	a	s	k	b	u	p	p
δ	0	0	23	2	0	14	11	15	11	1	11	3	7	21	15	0	18	10	1	20	15	15
p'	*	*	*	*	*	*	*	*	*	*	p	o	l	i	c	e	e	s	t	*	*	*

suggesting enlargement to /de l' ennemi sont/ . Backwards again we get

	1	2	3	4	5	6	7	8	9	10	11	12	13	14	15	16	17	18	19	20	21	22
p''	*	*	*	*	*	*	*	*	*	d	e	l	e	n	n	e	m	i	s	o	n	t
d	a	a	x	c	a	o	l	p	l	b	l	d	h	v	p	a	s	k	b	u	p	p
δ	0	0	23	2	0	14	11	15	11	1	11	3	7	21	15	0	18	10	1	20	15	15
p'	*	*	*	*	*	*	*	*	*	e	p	o	l	i	c	e	e	s	t	i	c	i

In this way, a probable word can serve as a seed which in a zig-zag manner grows to the right and to the left in both texts. The method allows especially the use of non-content words, endings and prefixes, which are not rare; in English /and/, /the/, /that/, /which/, /under/, /tion/, in French /les/, /que/, /ion/, in German /und/, /ein/, /ung/, /bar/, /heit/, /unter/ . In our example a new seed is successful, and with /les/ in

	1	2	3	4	5	6	7	8	9	10	11	12	13	14	15	16	17	18	19	20	21	22
p''	l	e	s	*	*	*	*	*	*	d	e	l	e	n	n	e	m	i	s	o	n	t
d	a	a	x	c	a	o	l	p	l	b	l	d	h	v	p	a	s	k	b	u	p	p
δ	0	0	23	2	0	14	11	15	11	1	11	3	7	21	15	0	18	10	1	20	15	15
p'	l	e	p	*	*	*	*	*	*	e	p	o	l	i	c	e	e	s	t	i	c	i

we get with a little bit of luck perhaps /leprefetd/ in p' and a confirmation by supplementing /lespertes/ in p'' :

	1	2	3	4	5	6	7	8	9	10	11	12	13	14	15	16	17	18	19	20	21	22
p''	l	e	s	p	e	r	t	e	s	d	e	l	e	n	n	e	m	i	s	o	n	t
d	a	a	x	c	a	o	l	p	l	b	l	d	h	v	p	a	s	k	b	u	p	p
δ	0	0	23	2	0	14	11	15	11	1	11	3	7	21	15	0	18	10	1	20	15	15
p'	l	e	p	r	e	f	e	t	d	e	p	o	l	i	c	e	e	s	t	i	c	i

This ends the decryption of the two cryptotexts. Only the decryption of the shorter one of two such texts can be obtained in this way, of course, but there is still the key that has not been used so far. It can be reconstructed now: Juxtaposition of p' and c' gives according to Table 23, which was assumed to be known,

	1	2	3	4	5	6	7	8	9	10	11	12	13	14	15	16	17	18	19	20	21	22
c'	U	H	Y	B	R	J	I	M	B	C	F	A	M	M	F	J	H	D	M	R	I	Q
p'	l	e	p	r	e	f	e	t	d	e	p	o	l	i	c	e	e	s	t	i	c	i
k	*J*	*E*	*M*	*E*	*M*	*E*	*T*	*S*	*S*	*U*	*R*	*L*	*A*	*D*	*E*	*F*	*E*	*N*	*S*	*I*	*V*	*E*

and again confirmation by a 'meaningful' key sentence. For the success of this zig-zag method, it was sufficient that both plaintexts had distinctly more than 50% redundancy.

19.2.4 Other key groups. The key group we just dealt with, typical for ALBERTI encryption and especially for VIGENÈRE (dually: BEAUFORT) encryption, is as said above the cyclic group $\mathcal{C}_N$ of order N, where we have

$V = W = \mathbb{Z}_N$. It is only one example for groups of prescribed order. For $\mathbb{Z}_{26}$, there is besides $\mathcal{C}_{26}$ another commutative group: the direct product $\mathcal{C}_{13} \times \mathcal{C}_2$ of the cyclic group of order 13 and the cyclic group of order 2. It is the group generated by 13 PORTA encryptions which is obtained by an intermediate coding $\mathbb{Z}_{26} \longrightarrow \mathbb{Z}_{13} \times \mathbb{Z}_2$. There is also a non-commutative group of order 26, the Dieder group $\mathcal{D}_{13}$, with the generators S and T; $S^{13} = T^2 = (ST)^2 = I$ which it seems has so far no relevance in cryptology.

For $\mathbb{Z}_{25}$ there is next to the cyclic group $\mathcal{C}_{25}$ a further commutative group: the direct product $\mathcal{C}_5^2 \doteq \mathcal{C}_5 \times \mathcal{C}_5$ of two cyclic groups of order 5, obtained by the intermediate Polybios coding $\mathbb{Z}_{25} \longrightarrow \mathbb{Z}_5 \times \mathbb{Z}_5$. There is no non-commutative group of order 25. For $\mathbb{Z}_{10}$ there is next to the cyclic group $\mathcal{C}_{10}$ the commutative group $\mathcal{C}_2 \times \mathcal{C}_5$ reached by the intermediate biquinary coding $\mathbb{Z}_{10} \longrightarrow \mathbb{Z}_2 \times \mathbb{Z}_5$. There is also the non-commutative Dieder group $\mathcal{D}_5$ with the generators S and T; $S^5 = T^2 = (ST)^2 = I$.

In view of binary coding, particularly interesting groups are those of order 2^n. For arbitrary j the commutative groups $\mathcal{C}_{2^j}$ and $\mathcal{C}_2^j \doteq \mathcal{C}_2 \times \mathcal{C}_2 \times \ldots \times \mathcal{C}_2$ are most prominent. For $j = 2$ we have the cyclic group of order 4 and Klein's *Vierergruppe*. For $j = 3$ there are additionally the non-commutative quaternion group $\mathcal{Q}$ and the non-commutative Dieder group $\mathcal{D}_4$ with the generators S and T; $S^4 = T^2 = (ST)^2 = I$, both without cryptological relevance. This remark also applies to a handful of non-commutative groups for $j = 4, 5$.

For the difference between $\mathbb{Z}_{2^n}$ and $\mathbb{Z}_2^n$ (as well as for the difference between $\mathbb{Z}_{10^n}$ and $\mathbb{Z}_{10}^n$), namely the lack of the carry mechanism, see Sect. 8.3.3.

0	t	4	o	2	h	n	m	5	l	r	g	i	p	c	v	e	z	d	b	s	y	f	x	a	w	j	3	u	q	k	1	
O	**O**	**O**	**O**	**O**	**O**	**O**	**O**	**O**	**O**	**O**	**O**	**O**	**O**	**O**	**O**	**L**	**L**	**L**	**L**	**L**	**L**	**L**	**L**	**L**	**L**	**L**	**L**	**L**	**L**	**L**	**L**	16
O	**O**	**O**	**O**	**O**	**O**	**O**	**O**	**L**	**L**	**L**	**L**	**L**	**L**	**L**	**L**	**O**	**O**	**O**	**O**	**O**	**O**	**O**	**O**	**L**	**L**	**L**	**L**	**L**	**L**	**L**	**L**	8
•	•	•	•	•	•	•	•	•	•	•	•	•	•	•	•	•	•	•	•	•	•	•	•	•	•	•	•	•	•	•	•	
O	**O**	**O**	**O**	**L**	**L**	**L**	**L**	**O**	**O**	**O**	**O**	**L**	**L**	**L**	**L**	**O**	**O**	**O**	**O**	**L**	**L**	**L**	**L**	**O**	**O**	**O**	**O**	**L**	**L**	**L**	**L**	4
O	**O**	**L**	**L**	**O**	**O**	**L**	**L**	**O**	**O**	**L**	**L**	**O**	**O**	**L**	**L**	**O**	**O**	**L**	**L**	**O**	**O**	**L**	**L**	**O**	**O**	**L**	**L**	**O**	**O**	**L**	**L**	2
O	**L**	**O**	**L**	**O**	**L**	**O**	**L**	**O**	**L**	**O**	**L**	**O**	**L**	**O**	**L**	**O**	**L**	**O**	**L**	**O**	**L**	**O**	**L**	**O**	**L**	**O**	**L**	**O**	**L**	**O**	**L**	1
0	1	2	3	4	5	6	7	8	9	10	11	12	13	14	15	16	17	18	19	20	21	22	23	24	25	26	27	28	29	30	31	

Table 24. Binary coding of the International Teletype Alphabet No. 2 (CCITT 2)
0: Void, 1: Letter Shift, 2: Word Space, 3: Figure Shift, 4: Carriage Return, 5: Line Feed
(/ 8 9 5 3 4 in B.P.)

19.2.5 The special case $\mathcal{C}_2^5$.

A representation of $\mathbb{Z}_{25}$ is the vocabulary of the International Teletype Alphabet No. 2 (CCITT 2) from 1929, going back to Donald Murray in 1900. The 5-channel representation suggests an encryption by addition mod. 2, $c_i = p_i \oplus k_i$, the key group of which is $\mathcal{C}_2^5 \doteq \mathcal{C}_2 \times \mathcal{C}_2 \times \mathcal{C}_2 \times \mathcal{C}_2 \times \mathcal{C}_2$ (and not $\mathcal{C}_{25}$), namely an encryption with 32 alphabets, generated by substitutions O or L (Sect. 8.3.1) of the five binary characters. The actual coding $\mathbb{Z}_{32} \longrightarrow \mathbb{Z}_2^5$ of CCITT 2 is shown[3] in Table 24. Apart from

---

[3] The following correspondence, known from the typewriter keyboard, holds between figures and letters: 1 ≐ q, 2 ≐ w, 3 ≐ e, 4 ≐ r, 5 ≐ t, 6 ≐ y, 7 ≐ u, 8 ≐ i, 9 ≐ o, 0 ≐ p.

26 (lowercase) letters there are six control characters of the teletype machine, whose function is meaningless on the cryptographic line; we designate them with 0, 1, 2, 3, 4, 5 and use 2 as a word separator (actually, 12 was used; this was a weakness Beurling exploited). Denoting the keys correspondingly by *0, A, B, C,..., Z, 2, 3, 4, 5, 1* , Table 25 shows the natural encryption table which one can presume to be known.

	0	a	b	c	d	e	f	g	h	i	j	k	l	m	n	o	p	q	r	s	t	u	v	w	x	y	z	2	3	4	5	1
0	0	A	B	C	D	E	F	G	H	I	J	K	L	M	N	O	P	Q	R	S	T	U	V	W	X	Y	Z	2	3	4	5	1
A	A	0	G	F	R	5	C	B	Q	S	4	N	Z	1	K	3	Y	H	D	I	W	2	X	T	V	P	L	U	O	J	E	M
B	B	G	0	Q	T	O	H	A	F	1	L	P	J	S	Y	E	K	C	W	M	D	V	U	R	2	N	4	X	5	Z	3	I
C	C	F	Q	0	U	K	A	H	G	4	S	E	M	L	5	P	O	B	2	J	V	D	T	X	W	3	1	R	Y	I	N	Z
D	D	R	T	U	0	4	2	W	X	K	5	I	3	Y	S	Z	1	V	A	N	B	C	Q	G	H	M	O	F	L	E	J	P
E	E	5	O	K	4	0	N	3	Y	U	R	C	W	X	F	B	Q	P	J	2	Z	I	1	L	M	H	T	S	G	D	A	V
F	F	C	H	A	2	N	0	Q	B	J	I	5	1	Z	E	Y	3	G	U	4	X	R	W	V	T	O	M	D	P	S	K	L
G	G	B	A	H	W	3	Q	0	C	M	Z	Y	4	I	P	5	N	F	T	1	R	X	2	D	U	K	J	V	E	L	O	S
H	H	Q	F	G	X	Y	B	C	0	L	1	3	I	4	O	N	5	A	V	Z	2	W	R	U	D	E	S	T	K	M	P	J
I	I	S	1	4	K	U	J	M	L	0	F	D	H	G	R	V	T	Z	N	A	P	E	O	Y	3	W	Q	5	X	C	2	B
J	J	4	L	S	5	R	I	Z	1	F	0	2	B	Q	U	W	X	M	E	C	3	N	Y	O	P	V	G	K	T	A	D	H
K	K	N	P	E	I	C	5	Y	3	D	2	0	X	W	A	Q	B	O	S	R	1	4	Z	M	L	G	V	J	H	U	F	T
L	L	Z	J	M	3	W	1	4	I	H	B	X	0	C	V	R	2	S	O	Q	5	Y	N	E	K	U	A	P	D	G	T	F
M	M	1	S	L	Y	X	Z	I	4	G	Q	W	C	0	T	2	R	J	P	B	N	3	5	K	E	D	F	O	U	H	V	A
N	N	K	Y	5	S	F	E	P	O	R	U	A	V	T	0	H	G	3	I	D	M	J	L	1	Z	B	X	4	Q	2	C	W
O	O	3	E	P	Z	B	Y	5	N	V	W	Q	R	2	H	0	C	K	L	X	4	1	I	J	S	F	D	M	A	T	G	U
P	P	Y	K	O	1	Q	3	N	5	T	X	B	2	R	G	C	0	E	M	W	I	Z	4	S	J	A	U	L	F	V	H	D
Q	Q	H	C	B	V	P	G	F	A	Z	M	O	S	J	3	K	E	0	X	L	U	T	D	2	R	5	I	W	N	1	Y	4
R	R	D	W	2	A	J	U	T	V	N	E	S	O	P	I	L	M	X	0	K	G	F	H	B	Q	1	3	C	Z	5	4	Y
S	S	I	M	J	N	2	4	1	Z	A	C	R	Q	B	D	X	W	L	K	0	Y	5	3	P	O	T	H	E	V	F	U	G
T	T	W	D	V	B	Z	X	R	2	P	3	1	5	N	M	4	I	U	G	Y	0	Q	C	A	F	S	E	H	J	O	L	K
U	U	2	V	D	C	I	R	X	W	E	N	4	Y	3	J	1	Z	T	F	5	Q	0	B	H	G	L	P	A	M	K	S	O
V	V	X	U	T	Q	1	W	2	R	O	Y	Z	N	5	L	I	4	D	H	3	C	B	0	F	A	J	K	G	S	P	M	E
W	W	T	R	X	G	L	V	D	U	Y	O	M	E	K	1	J	S	2	B	P	A	H	F	0	C	I	5	Q	4	3	Z	N
X	X	V	2	W	H	M	T	U	D	3	P	L	K	E	Z	S	J	R	Q	O	F	G	A	C	0	4	N	B	I	Y	1	5
Y	Y	P	N	3	M	H	O	K	E	W	V	G	U	D	B	F	A	5	1	T	S	L	J	I	4	0	2	Z	C	X	Q	R
Z	Z	L	4	1	O	T	M	J	S	Q	G	V	A	F	X	D	U	I	3	H	E	P	K	5	N	2	0	Y	R	B	W	C
2	2	U	X	R	F	S	D	V	T	5	K	J	P	O	4	M	L	W	C	E	H	A	G	Q	B	Z	Y	0	1	N	I	3
3	3	O	5	Y	L	G	P	E	K	X	T	H	D	U	Q	A	F	N	Z	V	J	M	S	4	I	C	R	1	0	W	B	2
4	4	J	Z	I	E	D	S	L	M	C	A	U	G	H	2	T	V	1	5	F	O	K	P	3	Y	X	B	N	W	0	R	Q
5	5	E	3	N	J	A	K	O	P	2	D	F	T	V	C	G	H	Y	4	U	L	S	M	Z	1	Q	W	I	B	R	0	X
1	1	M	I	Z	P	V	L	S	J	B	H	T	F	A	W	U	D	4	Y	G	K	O	E	N	5	R	C	3	2	Q	X	0

Table 25. Encryption table (Latin square) for teletype symbols: addition modulo 2 in $\mathbb{Z}_2^5$

The teletype coding was widely known from the turn of the century, and via Vernam the professional cryptologists were also familiar with it; the obvious natural key group $\mathcal{C}_2^5$ was concretely known. Thus, all preconditions for an attack as in Sect. 19.2.3 were fulfilled, and in particular the key character could be reconstructed (because of the commutativity of the key group) from plaintext character and cryptotext character.

A fictitious example of a break may have gone like this: Two cryptotexts of roughly the same length near 4000, picked up by the British at the time of the German attack on Crete in mid-May 1941 on a *Wehrmacht* line Vienna-Athens, contained after coinciding preambles, presumably in phase, the fragments (in B. P. called a 'depth of two')

c''	2 W H N R	G 1 A T U	A P L B V	R W O U F	Y P B S X	Z N R 4 J	S R
c'	L 0 G 2 A	W G H 2 Z	K B V Z V	Q Z W Y K	Y W J I 0	K T 5 A Z	2 K

The British formed the difference $d = c'' \oplus c'$ (i.e., performed addition mod. 2)

	1 2 3 4 5	6 7 8 9 10	11 12 13 14 15	16 17 18 19 20	21 22 23 24 25	26 27 28 29 30	31 32
c''	2 W H N R	G 1 A T U	A P L B V	R W O U F	Y P B S X	Z N R 4 J	S R
c'	L 0 G 2 A	W G H 2 Z	K B V Z V	Q Z W Y K	Y W J I 0	K T 5 A Z	2 K
d	f w c w d	d v q k p	n k n 4 0	x 5 j 1 5	0 s l a x	v m 4 j g	g s

and used the probable word /2kreta2/ to find a reasonable counterpart. They discovered at the fourth position

	1 2 3 4 5	6 7 8 9 10	11 12 13 14 15	16 17 18 19 20	21 22 23 24 25	26 27 28 29 30	31 32
p''	* * * 2 k	r e t a 2	* * * * *	* * * * *	* * * * *	* * * * *	* *
d	f w c w d	d v q k p	n k n 4 0	x 5 j 1 5	0 s l a x	v m 4 j g	g s
p'	* * * n i	a 2 u n d	* * * * *	* * * * *	* * * * *	* * * * *	* *

A short look at the map of Greece suggests for p'' a supplementation to /chania/ and gives:

	1 2 3 4 5	6 7 8 9 10	11 12 13 14 15	16 17 18 19 20	21 22 23 24 25	26 27 28 29 30	31 32
p''	a u f 2 k	r e t a 2	* * * * *	* * * * *	* * * * *	* * * * *	* *
d	f w c w d	d v q k p	n k n 4 0	x 5 j 1 5	0 s l a x	v m 4 j g	g s
p'	c h a n i	a 2 u n d	* * * * *	* * * * *	* * * * *	* * * * *	* *

Now some more geographic names, following the /und/ , could be tried. Another way takes a further probable word, say /2angriff2/. The people at Bletchley Park had success in position 19 of p'' with:

	1 2 3 4 5	6 7 8 9 10	11 12 13 14 15	16 17 18 19 20	21 22 23 24 25	26 27 28 29 30	31 32
p''	a u f 2 k	r e t a 2	* * * * *	* * * 2 a	n g r i f	f 2 * * *	* *
d	f w c w d	d v q k p	n k n 4 0	x 5 j 1 5	0 s l a x	v m 4 j g	g s
p'	c h a n i	a 2 u n d	* * * * *	* * * f e	n 2 o s t	w a e * *	* *

Now it is almost finished: the missing piece in p' should read /2die2haefen/ , followed by /ostwaerts2/. This gives:

	1 2 3 4 5	6 7 8 9 10	11 12 13 14 15	16 17 18 19 20	21 22 23 24 25	26 27 28 29 30	31 32
p''	a u f 2 k	r e t a 2	w i r d 2	d e r 2 a	n g r i f	f 2 d e r	2 g
d	f w c w d	d v q k p	n k n 4 0	x 5 j 1 5	0 s l a x	v m 4 j g	g s
p'	c h a n i	a 2 u n d	2 d i e 2	h a e f e	n 2 o s t	w a e r t	s 2

In readable form:

'auf_kreta_wird_der_angriff_der_g .....'
'chania_und_die_haefen_ostwaerts_ .....'

The cryptanalysts could proceed similarly with other fragments of the text. The exacting cross-ruff was certainly fascinating, too.

The British could now also reconstruct the key. However, since in $\mathcal{C}_2^n$ subtraction and addition coincide, first the right pairing of plaintexts and cryptotexts had to be found. The two possible juxtapositions result in:

	1	2	3	4	5	6	7	8	9	10	11	12	13	14	15	16	17	18	19	20	21	22	23	24	25	26	27	28	29	30	31	32
p'	c	h	a	n	i	a	2	u	n	d	2	d	i	e	2	h	a	e	f	e	n	2	o	s	t	w	a	e	r	t	s	2
k_1	M	H	B	W	S	T	S	W	W	O	T	T	O	T	E	A	L	L	O	C	B	N	W	A	T	M	W	A	D	E	G	T
c'	L	0	G	2	A	W	G	H	2	Z	K	B	V	Z	V	Q	Z	W	Y	K	Y	W	J	I	0	K	T	5	A	Z	2	K

	1	2	3	4	5	6	7	8	9	10	11	12	13	14	15	16	17	18	19	20	21	22	23	24	25	26	27	28	29	30	31	32
p''	a	u	f	2	k	r	e	t	a	2	w	i	r	d	2	d	e	r	2	a	n	g	r	i	f	f	2	d	e	r	2	g
k_2	Z	U	Q	0	N	B	3	6	M	C	M	2	H	O	E	V	T	B	R	N	B	D	E	0	F	5	K	J	5	3	0	Y
c'	L	0	G	2	A	W	G	H	2	Z	K	B	V	Z	V	Q	Z	W	Y	K	Y	W	J	I	0	K	T	5	A	Z	2	K

The cryptanalysts may well have decided for a less irregular key text, like k_1, assuming that the machine had a reasonably uncomplicated mechanism showing some local regularity.

19.2.6 Tunny. The use at the highest level of the German *Wehrmacht-Nachrichtenverbindungen* of the cipher teletype machines SZ 40, SZ 42 (*Schlüsselzusatz*) built by the Lorenz company, which generated their key by a semiregular movement, was thus utterly risky if a plaintext-plaintext compromise could not be excluded. Moreover, P. W. Filby has reported that in 1932 some 'Mr. Lorenz' offered to the Foreign Office a cipher machine and disclosed it. In addition, the competing product, the cipher teletype machine T 52 (*Geheimschreiber*) built by the Siemens company, was openly described in the German Patent No. 615016 by August Jipp and Ehrhard Rossberg (Sect. 9.1.3); the U.S. Patent No. 1 912 983 for Jipp, Rossberg, and Eberhard Hettler was granted June 6, 1933. Thus it was not too difficult for the British in Bletchley Park to judge the situation realistically. Luckily for the British, the *Schlüsselzusatz* used only the 32 encryption steps originating by substitutions O or L of the five binary characters, while the *Geheimschreiber*, as in the patent explained, used also transpositions of the bits (Sect. 9.1.3) and thus had a key set with many more than 32 keys.

The disaster for the Germans indeed developed much earlier than one would expect. As reported in 1993 by Jack Good (first allusions were made in 1978 by Brian Johnson, in 1983 by Andrew Hodges), a plaintext-plaintext compromise happened even before the *Schlüsselzusatz* came into regular use, when during tests on the newly established *Wehrmacht Hellschreiber* radio line between Vienna and Athens, in consequence of a mistake by a German telegraphist, two rather long messages p', p'' were sent with the same initial position of the key wheels, i.e., encrypted in phase with the same key. Somebody in Bletchley Park developed a suspicion; the compromise allowed Colonel (later Brigadier) John H. Tiltman, then chief cryptanalyst in B. P., in painstaking work over two weeks in the fall of 1941, to deduce the two plaintexts from the difference d of the two recorded cryptotexts, a 'depth of two'.

As is known now, the accident happened on August 30, 1941: With the same indicator HQIBPEXEZMUG two isologs of roughly 4000 characters, coinciding in the first seven characters, were recorded; the characters #51 – #120 are quoted below together with the diffences that Tiltman formed:

	51 52 53 54 55	56 57 58 59 60	61 62 63 64 65	66 67 68 69 70	71 72 73 74 75	76 77 78 79 80	81 82 83 84 85
c''	U B 2 3 R	5 W E V G	Q I 2 4 5	G R J M L	C Y 5 0 H	K A S 1 I	S 5 X U N
c'	Y U H V H	3 H E E 0	T G 2 H H	1 Q J X V	K 1 B J M	K 2 O M Z	Y V I N 3
d	l v t s v	b u 0 1 g	u m 0 m p	s x 0 e n	e r 3 j 4	0 u x a q	t m 3 j q

	86 87 88 89 90	91 92 93 94 95	96 97 98 99 00	01 02 03 04 05	06 07 08 09 10	11 12 13 14 15	16 17 18 19 20
c''	S R Z Z B	D B B 1 C	L S Q H H	U H 5 X D	0 F N 3 J	3 V O C A	D J C D N
c'	H M C 3 D	U Q 3 4 Z	R 2 M R M	O H * J Q	P W U E Y	C D R G 1	L D A T I
d	z p 1 r t	c c 5 q 1	o e j v 4	1 0 * p v	p v j g v	y q l h m	3 5 f b r

If Tiltman tried as probable word the very frequent /geheim2/, he would have succeeded twice in finding an intelligible counterpart,

	61 62 63 64 65	66 67 68 69 70	71 72 73 74 75	76 77 78 79 80	81 82 83 84 85	86 87 88 89 90	91 92 93 94 95
p''	* * * * g	e h e i m	2 * * * *	* * * * *	* * g e h	e i m 2 *	* * * * *
d	u m 0 m p	s x 0 e n	e r e j 4	0 u x a q	t m 3 j q	z p 1 r t	c c 5 q 1
p'	* * * * n	2 d e u t	s * * * *	* * * * *	* * e r a	t t a c *	* * * * *

/n2deuts/ can easily be supplemented to /an2deutsch/, and /erattac/ leads to /2militaerattache2/; the gap is filled by /an2deutschen/. Thus, there results already a fragment for p' of 29 characters

	61 62 63 64 65	66 67 68 69 70	71 72 73 74 75	76 77 78 79 80	81 82 83 84 85	86 87 88 89 90	91 92 93 94 95
p''	* * * * g	e h e i m	2 * * * *	* * * * *	* * g e h	e i m 2 *	* * * * *
d	u m 0 m p	s x 0 e n	e r e j 4	0 u x a q	t m 3 j q	z p 1 r t	c c 5 q 1
p'	* * * a n	2 d e u t	s c h e n	2 m i l i	t a e r a	t t a c h	e 2 * * *

and produces for p'' 29 consecutive characters as well

	61 62 63 64 65	66 67 68 69 70	71 72 73 74 75	76 77 78 79 80	81 82 83 84 85	86 87 88 89 90	91 92 93 94 95
p''	* * * 1 g	e h e i m	2 2 k r 2	2 3 3 z z	0 1 g e h	e i m 2 2	k r * * *
d	u m 0 m p	s x 0 e n	e r e j 4	0 u x a q	t m 3 j q	z p 1 r t	c c 5 q 1
p'	* * * a n	2 d e u t	s c h e n	2 m i l i	t a e r a	t t a c h	e 2 * * *

This shows that, following a pet mistake, /geheim/ was doubled; the doubling of the complete group /1geheim22kr2233zz/ leads to an extension, which up to two mistakes in typing makes sense. At the latest now the suspicion arises that the rest of the message p'' is simply shifted against p', and this by 39 positions, since /an2deutschen2militaerattache2/ in the 103rd position makes sense again, namely /lage11nr33mwoou211g/ (in readable form *lage nr.2997_g*):

	86 87 88 89 90	91 92 93 94 95	96 97 98 99 00	01 02 03 04 05	06 07 08 09 10	11 12 13 14 15	16 17 18 19 20
p''	e i m 2 2	k r 2 2 3	3 z z 1 2	* * a n 2	d e u t s	c h e n 2	m i l i t
d	z p 1 r t	c c 5 q 1	o e j v 4	1 0 * p v	p v j g v	y q l h m	3 5 f b r
p'	t t a c h	e 2 i w 2	a t g e n	* l a g e	1 1 n r 3	3 m w o o	u 2 1 1 g

Thus, the continuation can now be produced automatically through a look-ahead by 39 characters—quite similar to the autokey fallacy Shannon described (Sect. 8.7.2). It is reported that Tiltman finished the deciphering in ten days, which may have been in consequence of further mistakes in typing more difficult than a retrospective analysis shows.

That the enciphering was made by addition modulo 2, with the consequence that subtraction coincided with addition, was irrelevant for the method and only facilitated clerical work. However, the enciphering was self-reciprocal, which was considered advantageous in practice, without being properly self-reciprocal with the drawback connected to this.

The two messages were most likely of little value. What was important was that a fragment of about 4000 characters of the key of the hitherto unknown machine, which the British gave the codename *tunny*, was exposed, namely by the calculation $k = c' + p' \bmod 2$; it started with (cf. Table 24)

	61 62 63 64 65	66 67 68 69 70	71 72 73 74 75	76 77 78 79 80	81 82 83 84 85	86 87 88 89 90	
k	* * * *QO*	*3VRGC*	*RZFRT*	*JOVCQ*	*SXUIO*	*2NFYX*	
	LO	**LOOOO**	**OLLOO**	**LOOOL**	**LLLOO**	**OOLLL**	16
	LO	**LLLLL**	**LOOLO**	**LOLLL**	**OOLLO**	**OOOOO**	8
	LO	**OLOOL**	**OOLOO**	**OOLLL**	**LLLLO**	**LLLLL**	4
	OL	**LLLLL**	**LOLLO**	**LLLLO**	**OLOOL**	**OLLOL**	2
	LL	**LLOLO**	**OLOOL**	**OLLOL**	**OLOOL**	**OOOLL**	1

	91 92 93 94 95	96 97 98 99 00	01 02 03 04 05	06 07 08 09 10	11 12 13 14 15	16 17 18 19 20	
k	*IWX2Y*	*DH4JT*	*MIEZP*	*DNJIC*	*YRB5U*	*YFMKM*	
	OLLOL	**LOOLO**	**OOLLO**	**LOLLO**	**LOLOL**	**LLOLO**	16
	LLOOO	**OOOLO**	**OLOOL**	**OOLLL**	**OLOLL**	**OOOLO**	8
	LOLLL	**OLOOO**	**LLOOL**	**OLOOL**	**LOOOL**	**LLLLL**	4
	OOLOO	**LOLLO**	**LOOOO**	**LLLLL**	**OLLOO**	**OLLLL**	2
	OLLOL	**LLOOL**	**LOOLL**	**OOOOO**	**LOLOO**	**LOLOL**	1

With the reconstructed key fragment of the hypothetical *tunny* machine it was possible to analyze the key generator of the German machine. First, it was necessary to find the periods of the individual keying wheels (whose existence in analogy to the Siemens cipher teletype machine T 52 could be guessed). From the indicator, it could be vaguely guessed that there were a total of 12 keying wheels. Since none of the channels k_1 until k_5 of the key k had a period of length below 100, it was to be assumed that each channel was enciphered by a composition of (at least) two keying wheels. The British called them χ_i und ψ_i, where first the *Chi*-wheels and second the *Psi*-wheels were applied. By periodicity examination they found first the periods of the *Chi* keying wheels, in particular 41 for χ_1, as is shown in Fig. 142. Next, the periods of the *Psi* keying wheels were determined (43 for ψ_1) and the manner of working of the remaining two motor wheels (s. 9.1.4) was found out.

This was accomplished mainly by the young mathematician William Thomas Tutte from Trinity College in Cambridge, who later became well known in graph theory. By February 1942, the whole structure of the machine that had been used for the HQIBPEXEZMUG message (nicknamed afterwards ZMUG) was clarified; after the end of the war the captured machines confirmed this. Jack Good proudly stated "We did not capture a German *tunny* until the last days of the war in Europe." It was indeed 'pure cryptanalysis'.

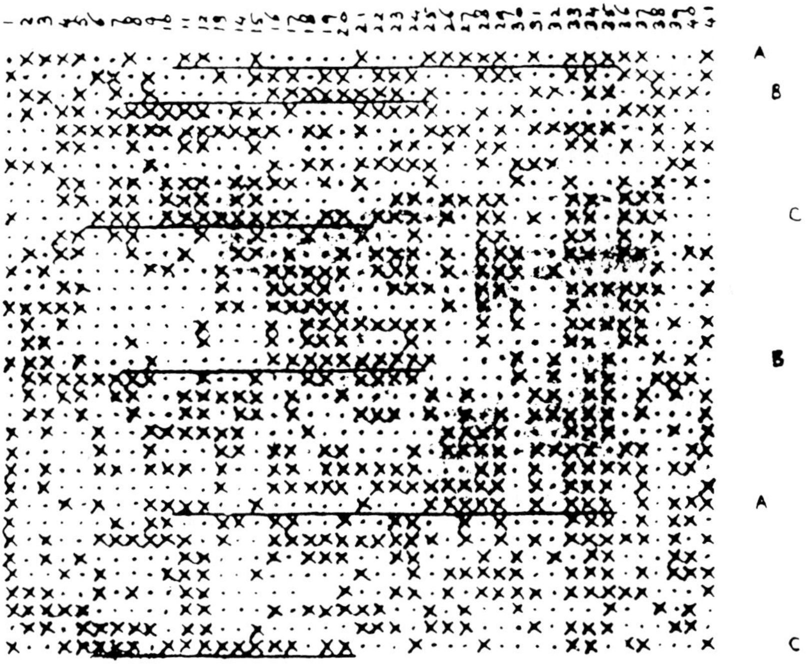

Fig. 142. Periodicity examination of the keying wheel χ_1. Kasiski repetitions: A, B, C. In Bletchley Park, a dot was used to represent **O**, and a cross to represent **L**

The first practical task was to find for a whole day the right keying sequence, periodic **O-L** sequences caused by the cams on the keying wheels (‘*wheel breaking*’)—possibly by ‘dragging’ probable words along the ciphertext. Then, the remaining practical task was to find for each message the right initial setting of the keying wheels (‘*wheel setting*’)—when this had led to a success, a replica of the *tunny* machine printed the decrypted plaintext.

This work was guided by Maxwell H. A. Newman, who was convinced of *pure cryptanalysis*. There was a keying text and the observed cryptotext, and they had to be brought in phase, presumably by some sort of *Kappa* test. The problem could be treated functionally by the ‘saw-buck’ principle (Sect. 17.3.2). In May 1943, a first model of a fast machine, HEATH ROBINSON (discussed already in Sect. 17.3.3), went into operation. Its two loops, one for the cryptotext, one for the key, were punched on tape. Keeping the two tapes synchronized led to problems of mechanical wear. Therefore, Tom H. Flowers, supported by Sid W. Broadhurst, W. W. Chandler and A. W. M. Coombs, developed an improved version, where the key function was internally generated and the other tape was read photoelectrically with a speed of 5 000 characters per second. Most important, the new machine used electronic switching circuits and was correspondingly fast. In December 1943, the prototype model, dubbed COLOSSUS Mark I, was ready. By February 1944 it was operational against *tunny*. It was the first functioning electronic com-

Fig. 143. Partial view of COLOSSUS (presumably Mark II). The closed loop tape for the cryptotext, the plugboard field, and an array of tubes, presumably of type *Mullard EF36*, are clearly visible

puter in the world. On June 1, 1944, just in time for the D-Day landings in Normandy, the improved COLOSSUS Mark II came into use (Fig. 143).

The internal electronic circuits with about 1 500 tubes worked with ring counters[4] in a 1-out-of-n code. COLOSSUS allowed a flexible plugboard programming of elementary Boolean operations and binary (biquinary) arithmetic in 5-fold parallelization. The improved COLOSSUS Mark II with about 2 000 tubes was also able to perform conditional branching, and it had a 'logic switching panel' for presetting and manually changing Boolean operations between tracks during the run. Altogether, 10 COLOSSUS machines were built.

It is no denigration of the British to state that the COLOSSUS protocomputers were mainly oriented towards primitive comparison operations, on computing only in a very special sense, and that their control was at the same level as the machines of Konrad Zuse (1910–1995), which were loop-controlled; more precisely, following the saw-buck principle, they had a loop within a loop. Whether they were used for tasks other than 'wheel-setting'—they could well have been—remains open.

In the U.S.A., a successful cryptanalytical electronic development comparable the British COLOSSUS did not exist, according to Colin Burke, until the end of the Second World War, because of a number of failures in the attempts

---

4 Johnson was misinformed in 1978 to assume that COLOSSUS was oriented against the Siemens machine T 52 and therefore mentions 10 ring counters instead of 12, corresponding to the 12 key wheels of the SZ 40.

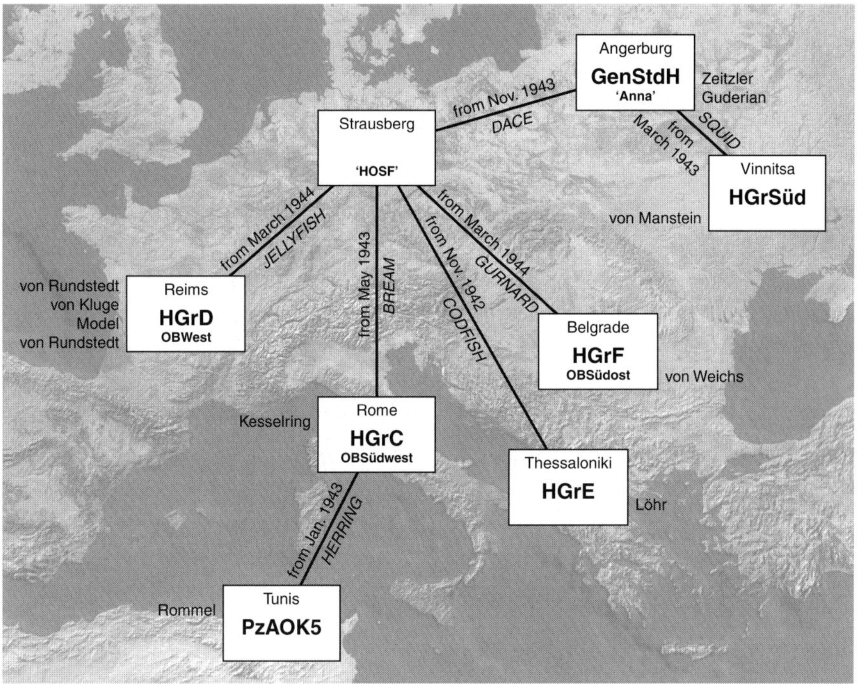

Fig. 144. Some wireless teletype connections with SZ 40, SZ 42, broken November 1942–July 1944 by Bletchley Park COLOSSUS (Hinsley)

of Vannevar Bush to build an electronic COMPARATOR. But "by the time Japan surrendered, the Americans were building electronic machines using twice as many tubes as the British Colossus" (Burke)—the Eastman 5202.

The British after 1943 supplemented with their successful breaking of the German *Funkfernschreibverbindungen* (wireless teletype connections), which mainly used SZ 40 and SZ 42, the ENIGMA stream of messages that was mainly fed by the German Air Force. Although the decryption, due to the higher cryptological level, sometimes took time (normally 4 days), the strategic intelligence received was worth the effort. The breaks (Fig. 144) included from November 1942 the line between Berlin and Army Group E in Saloniki, and from January 1943 the line between Army Group C in Rome and Field Marshal Erwin Rommel's *Panzerarmee* in Tunis. From May 1943 on, the line between the Berlin center in Straussberg and Army Group C (Field Marshal Albert Kesselring) had a leak, and from March 1943 also the line between the *Führerhauptquartier* message center ANNA and Army Group *Süd* in Winniza. As one of the consequences, the German attack against Kursk in July 1943 turned into disaster. These early breaks were in 1944 supplemented by many more, among them one in the line between Berlin and the *Oberbefehlshaber West*, Field Marshal Gerd von Rundstedt. The Germans were unsuspecting. They changed the *Chi*-wheels every 3 months, the *Psi*-wheels every 6 months, later every month; and the motor wheels daily.

The British had bad luck, too. On June 10, 1944, four days after the D-Day landings and ten days after starting COLOSSUS Mark II, they lost their entry into the line from Berlin to von Rundstedt and in July also into the line from Berlin to Kesselring; only in September 1944 did they catch up again. This was caused by a radical addition to the the Lorenz machine, similar to the *Klartextfunktion* on the Siemens machine: the British called it 'Plaintext Bit 5 Two steps back'.

Now more and more COLOSSUS machines came into use. Their growing success against the SZ 42 came just in time to compensate for the growing difficulties in the ENIGMA decryption. The British successes reached a culmination in March 1945; from then on the collapsing *Wehrmacht* no longer provided enough work even for Bletchley Park.

19.2.7 Sturgeon. The British efforts were less concerned with the *Schlüsselfernschreibmaschine* T 52 of Siemens (*Geheimschreiber*), although two machines were captured in North Africa by units of the British Eighth Army.[5] One reason was that T 52a and T 52c was exclusively used on wire lines, partly because synchronization of early T 52 was not stable enough on noisy wireless channels. Thus there was much less signal material available. Moreover, since the German Air Force, which used mainly T 52c and T 52e, had such a low standard of cryptanalytic security for their ENIGMA lines, it was not so urgent to listen also to their T 52 traffic. On the other hand, breaking into the disciplined ENIGMA traffic of the German Army was more difficult (Sect. 19.7). However, thanks to carelessness on the side of the German operators, some *Geheimschreiber* encryptions (the first ones in 1942 on the line Sicily-Libya and on a line from the Aegean to Sicily, codenamed 'Maquerel') were broken, too, despite the more difficult cryptanalytic situation. The T 52a and T 52d used besides 32 substitutions also 32 transpositions of the five teletype bits, 30 of which were different; the wiring that led to these transpositions was known from the German and U.S. patents. The methods of attack that had proved useful against SZ 42 could be extrapolated (with a grain of salt), although the reconstruction of the key was somewhat more elaborate. The construction of the key wheels was also basically known. The changes in T 52a that led to the T 52c were minor; the irregular movement of T 52d and T 52e was not quite as irregular as it could have been. Encryptions *mit Klartextfunktion* (Sect. 8.7.3), involving an autokeying, were very obstinate for the unauthorized decryptor—and for the authorized recipient, too, if the radio channel was noisy. They were practically nonexistent.

In May and June 1940, the Swedish mathematician Arne Beurling—a genius like Turing—working for *Försvarets Radioanstalt* (FRA), broke by manual work into a German T 52a teletype line to Oslo running over Swedish territory. He exploited the German stereotyped use of a 'letter shift' character before a 'word space' character and their abundance: He observed that not

---

[5] The fact that COLOSSUS was mainly directed against SZ 42 and not, as was previously assumed, against T 52, was revealed in 1980 by Rex Malik.

Fig. 145. *Geheimschreiber* replica built by the Swedish cryptanalytic bureau FRA

only plain 1 and 2 have one bit in common and four different, but that this had also to be the case if they are enciphered both at the same key position ('differential cryptanalysis'). In this way, he was able to reconstruct the spaces between words. Some details on how he further proceeded he took with him in the grave. Anyhow, from a message intercepted on May 25, 1940 and confirmed by a message two days later, he was able to reconstruct finally the T52a/b machine completely (June 12, 1940) and (Fig. 145) had replicas—called 'apps'—built, 32 by the end of 1942. On June 17, 1942, the Germans were warned by Finnish sources, but did not react appropriately. In July 1942, the Swedes could even penetrate the T52c traffic. The break ended in May 1943 when the Germans changed the indicator procedures.

19.3 In-Phase Superimposition of Superencrypted Code

The examples so far invited immediate superimposition, since the cryptotexts were already in phase—as they were in the operation of the Swiss army ENIGMA, which used the same initial setting for all messages of one day. If two cryptotexts are encrypted with different initial settings of a (mechanically generated) key, they must be mutually adjusted to allow in-phase superimposition. This can be achieved as in Sect. 17.1 by a *Kappa* examination. The cryptotexts are presumably in phase as soon as their mutual *Kappa* becomes maximal and is close to κ_S —provided the keys overlap at all.

19.3.1 Use of the indicator. Sometimes texts can be adjusted more simply. If for the superencryption of code a frequently changing key is to be used, it may be recommended to start each message at a different key position of one and the same key sequence. Avoiding a prearranged procedure, it is common practice to indicate at the beginning of the message the starting position. This so-called indicator (German *Spruchschlüssel*, not to be confused with discriminant, German *Kenngruppe*, which indicates the system to be used) may mean anything from the page and line in a book used for the key text to the initial setting of the wheels of an encryption machine. This hides the key, of course, but it does not prevent adjustment. It seems that this has been frequently overlooked. As far as superimposition is concerned, it is only a *complication illusoire*, if from the indicators of two messages their phase difference can be calculated or somehow determined. In professional encryption, the indicator was therefore encrypted itself, as for example in the ENIGMA traffic. If the indicator simply shows page and line in a book, only the number of lines per page has to be found out to calculate the phase shift—and even this number is commonly roughly known. This is the situation in the following example by Kahn of a 4-digit code with 4-digit numbers as additives, the discriminator in front of the message comprising 2 digits for the page and 2 digits for the line. If five messages even use the same page

(i)	6218	6260	7532	8291	2661	6863	2281	7135	5406	7046	9128	.....
(ii)	6216	3964	3043	1169	5729	3392	1952	7572	2754	7891	6290	.....
(iii)	6218	4061	6509	4513	1881	0398	3402	8671	4326	8267	6810	.....
(iv)	6218	5480	9325	3811	4083	5373	4882	8664	8891	6337	5914	.....
(v)	6217	7260	8931	8100	5787	6807	2471	0480	9892	1199	8426	.....

they can be adjusted immediately:

			1	2	3	4	5	6	7	8	9	10	
(i)			6260	7532	8291	2661	6863	2281	7135	5406	7046	9128	.....
(ii)	3964	3043	1169	5729	3392	1952	7572	2754	7891	6290	6719	7529	.....
(iii)			4061	6509	4513	1881	0398	3402	8671	4326	8267	6810	.....
(iv)			5480	9325	3811	4083	5373	4882	8664	8891	6337	5914	.....
(v)		7260	8931	8100	5787	6807	2471	0480	9892	1199	8426	1710	..... .

For a frequency analysis of the columns of encicode groups there is usually not enough material. In case of linear substitution, in particular for an additive superencryption in $\mathbb{Z}_{10}^4$, as should be assumed in the present case, the *symétrie de position* introduced in Sect. 18.6.2 can help. Thus, the difference method forms for every column a difference table, of which two examples (for the first and for the fifth column) are given in Fig. 146.

1					5				
0000	5101	2209	1880	8339	0000	9391	6575	1590	4492
5909	0000	7108	6789	3238	1719	0000	7284	2209	5101
8801	3902	0000	9681	6130	4535	3826	0000	5025	8927
9220	4321	1429	0000	7559	9510	**8801**	5085	0000	3902
2771	7872	4970	3551	0000	6618	5909	2183	7108	0000

Fig. 146. Two examples of difference tables

In view of the large size of 8 difference tables, each with 20 essential entries, it is preferable to order all entries for finding multiple occurrences. Figure 147 shows a suitable segment of such a table.

	difference in $\mathbb{Z}_{10}^4$	column	message
	⋮	⋮	⋮
	8209 = 0480 − 2281	6	(v)−(i)
→	8801 = 4061 − 6260	1	(iii)−(i)
→	8801 = 5373 − 7572	5	(iv)−(ii)
	9077 = 6509 − 7532	2	(iii)−(i)
	9106 = 5914 − 6810	10	(iv)−(iii)
→	9220 = 5480 − 6260	1	(iv)−(i)
→	9220 = 1881 − 2661	4	(iii)−(i)
	9308 = 3811 − 4513	3	(iv)−(iii)
→	9391 = 1952 − 2661	4	(ii)−(i)
→	9391 = 6337 − 7046	9	(iv)−(i)
→	9391 = 6810 − 7529	10	(iii)−(ii)
	9510 = 5373 − 6863	5	(iv)−(i)
	⋮	⋮	⋮

Fig. 147. Differences occurring, ordered

If now a difference occurs repeatedly, then the subtrahends are to be subtracted in the respective columns. As shown in Fig. 148, for the difference 8801, *6260* is subtracted in column 1, *7572* in column 5; likewise for the difference 9391, *2661* is subtracted in column 4, *7046* in column 9, *7529* in column 10. Here these two subtractions cover already the ones originating from the difference 9220, which is a confirmation that the phases were adjusted correctly.

	1'	2	3	4'	5'	6	7	8	9'	10'	
(i)	**0000**	7532	8291	**0000**	**9391**	2281	7135	5406	**0000**	2609	....
(ii)	**5909**	5729	3392	**9391**	**0000**	2754	7891	6290	9773	**0000**	....
(iii)	**8801**	6509	4513	**9220**	3826	3402	8671	4326	1221	**9391**	....
(iv)	**9220**	9325	3811	2422	**8801**	4882	8664	8891	**9391**	8495	....
(v)	2771	8100	5787	4246	**5909**	0480	9892	1199	1480	4291	....
	6260			*2661*	*7572*				*7046*	*7529*	

Fig. 148. Partially reduced messages

In the reduced columns in Fig. 148, the placode groups **0000**, **9391**, **5909**, **8801**, **9220** occur repeatedly. The groups in slanted typeface below the columns give the relative key. Further reductions bring all five messages into a monoalphabetically encrypted intermediary text, into a relative placode, which can be treated as in Sect. 18.3.2 .

The difference method does not always work as well as it may seem from this example. Frequently, only islands of interconnected groups are found at first, and further material is needed to join them into archipelagos. If the depth of the messages is insufficient, it may still happen that only partial solutions can be reached. Moreover, wrong coincidences of differences may occur. In our example, the difference 1480 originates not only from the column 9: 1480 = 8426 − 7046, but also from the column 6: 1480 = 4882 − 3402 . This would indicate a reduction of column 6 by *3402* . This, however, would introduce wrong placode groups, as one would find out later.

19.3.2 Kunze. The experts in the cryptanalytic service of the German *Auswärtiges Amt*, Paschke, Kunze, Schauffler, and Langlotz, had a long standing in the profession. They all joined the office in 1918 or 1919, first headed by Kurt Selchow. Adolf Paschke was the nominal head of the linguistic section. Dr. Werner Kunze, the mathematician (at that time a rarity in the cryptanalytic service) started attacking a French superencrypted code in 1921 and finally reconstructed it in 1923; he resumed this work in 1927. Thus, he had long years of experience in stripping off superencryption of code. The unit was first camouflaged as Z section of Division I, Personnel and Budget; in 1936, a reorganization changed its name into Pers Z.

The cumbersome stripping work was mechanically supported and semiautomated. Hans-Georg Krug built such 'robots', as Kahn called them, partly from punch card equipment, partly from standard telecommunications components. With this help, Hans-Kurt Müller, Asta Friedrichs and others succeeded in decrypting the diplomatic code of the U.S.A., effective August 1941 till the summer of 1943. Allen W. Dulles, who was then the U.S. Secret Service boss in Europe, suspected nothing, until warned by Hans Bernd Gisevius from the German *Widerstand*. Similar ideas were followed at *Chi*, the cipher branch of the OKW, by the engineers Wilhelm Rotscheidt and Willi Jensen, as mentioned in Sect. 18.6.3 . And the *B-Dienst* of the *Kriegsmarine* succeeded in solving the British Naval Cyphers.

The Allies and some bureaus of the neutrals used the same techniques. "The single most common cryptanalytic procedure of the war [was] the stripping of a numerical additive from enciphered code" (David Kahn).

In 1936, Kunze did fine work, too, in solving the Japanese ORANGE rotor machine (Sect. 8.5.7) and later the RED machine. In 1997, Otto Leiberich reported that during the war, the *Auswärtiges Amt* supported twelve linguists (among them Cort Rave, who kept up a liaison with Erich Hüttenhain at the cooperating OKW *Chi*) solving day by day the PURPLE signals of the Japanese ambassador Hiroshi Oshima. Foreign Minister Joachim von Ribbentrop considered Pers Z as a special weapon in his struggle with his rivals "Reichsmarschall" Hermann Göring and "Reichsführer SS" Heinrich Himmler.

19.4 Cryptotext-Cryptotext Compromises

A difficult situation occurs in practice if a message is to be repeated only insignificantly changed, for example by correcting a typing mistake. If the corrected message is sent again with the *same* key, a plaintext-plaintext compromise starts from the position of the mistake, with all the bad effects discussed so far. If the corrected message is sent again with a *different* key, a cryptotext-cryptotext compromise occurs up to the position of the mistake.

19.4.1 Cryptotext-cryptotext compromise of the keys. It happens quite generally if the same message or at least a large part of it is encrypted twice or more (each time with a different key). A dangerous risk of a break exists if this is done in the same system; if the resulting cryptotexts are 'isomorphic'(Sect. 2.6.3) then they have equal length, which is very conspicuous. A classical example of such an attack was offered in December 1938 and January 1939 by a pair of radio signals from the Rumanian military attaché in Paris to his Foreign Ministry, which differed in length by only two 5-letter groups. According to Hüttenhain, *Chi* succeeded in decrypting the signals, it turned out that only the plaintext fragment /Heft 17/ of the first signal was replaced by the plaintext fragment /Heft 15 statt 17/ in the second signal.

Cryptotext-cryptotext compromise of the keys is inherent in message network systems if a circular message is to be sent out, possibly to dozens or hundreds of recipients, each with their own key. It seems that German cryptologists underrated this danger, that their signal officers were not warned enough, and that this negligence continued till the end of the war. Rear Admiral Ludwig Stummel, responsible for the cryptanalytical security of the radio signals of the German Navy, introduced in 1943 a great number of key nets and in mid-1944 gave each U-boat its own key. It was thought this would give the adversary so much individual work that it would contribute to Germany's cryptanalytical security. But it was a self-defeating complication: "... was actually helpful, because the same message would often appear in several keys, sometimes on different days" (Rolf Noskwith). Even Stummel could not manage to formulate general orders for each key net or even for each boat individually.

When in 1942 the 4-rotor ENIGMA (see 11.1.11) was introduced only for the submarines, compromises were frequently caused by transmitting general orders for the other ships encrypted with the 3-rotor ENIGMA. Here the 1914 warning of Sir Alfred Ewing, head of Room 40, would have been appropriate: "It is never wise to mix your ciphers. Like mixing your drinks, it may lead to self-betrayal." But *Grossadmiral* Karl Dönitz's staff did it anyway.

We cannot properly speak of a cryptotext-cryptotext compromise of the keys if they are public. But note that only the encryption keys are public, not the decryption keys, and they are what we mean—in a symmetric encryption method, there is no difference. In fact, the risk of cryptotext-cryptotext compromise is inherent in public key systems.

In the jargon of Bletchley Park, a cryptotext-cryptotext compromise of the keys was called a 'kiss'. One could not have better expressed the joy at such a stroke of luck. Fortunately for the British, the smaller boats of the German Navy did not have ENIGMAs, but had to use a simple bigram encryption (*Werftschlüssel*). The large ships did not have this key or did not like to use it. If now certain messages—warnings about floating mines—had to be transmitted quickly, nobody took the trouble to reformulate the plaintexts. The British occasionally provoked such situations with the aim of establishing cryptotext-cryptotext compromises of the difficult 4-rotor ENIGMA with the easily breakable simple bigram substitution. With British humor, they called this 'gardening'. In fact, this technique meant a transition to a classical plaintext-cryptotext compromise situation, where the decrypted message furnishes a 'crib' of not only probably, but certainly contained words.

On May 7, 1941, the German weather ship *München* was captured by the British Navy and the *Wetterkurzschlüssel* fell into British hands. From the weather reports of the U-boats, a flow of kisses originated, filling the cribs of Bletchley Park for the ENIGMAs, and this continued until 1944. The bridge players called these 'cross-ruffs'[6]. Seizing *U-559* on October 30, 1942 even resulted in a compromise of the new 4-rotor ENIGMA by weather reports.

The lesson is that the collateral use of code superenciphered by a one-time key generated by a machine, together with repeatedly used additives, if the latter encryption is broken already, compromises the 'individual' key and leads to a reconstruction of the machine that produced it. This misadventure befall the German *Auswärtiges Amt* which, presumably because of a shortage of key material, on the line Berlin-Dublin used the FLORADORA superencryption which was already (see Sect. 9.2.1) broken by the British. The key could thus be investigated; it turned out that it was generated by a modified Lorenz SZ40, already reconstructed in Bletchley Park. Thus, the total traffic of the AA, considered unbreakable, was laid open.

19.4.2 Reduction to a plaintext-plaintext compromise. For the case of a VIGENÈRE encryption, in particular superencryption by additives, a cryptotext-cryptotext compromise can be reduced simply to a plaintext-plaintext compromise: the plaintext is regarded as keytext, the keytext as plaintext. This is just another case of the swapping of roles we saw in Sect. 19.2.3 . This means that the methods of superimposition and of *symétrie de position* are applicable. It is not required that the polyalphabetic encryption is periodic. The precondition for the swapping of roles is again that the keys are in prose and show frequency characteristics and/or patterns.

To give an example, there are five signals of equal length

	1 2 3 4 5	6 7 8 9 10	11 12 13 14 15	16 17 18 19 20	21 22 23 24 25	26 27 28 29 30
(i)	T C C V L	E S K P T	X M P V W	H Y M V G	X B O R V	C W A R F
(ii)	V L L B V	C K W F P	E H E C F	C G N Z E	K K K V I	H D D I D
(iii)	M Y Y R D	M J W M C	U I G L O	K M X L R	E W H X M	R J H A S
(iv)	B K Q T Z	B Z W K W	Z X G Z O	V T B A T	K W M G M	R J K L P
(v)	M Y Y V H	B W J D X	C P C Z O	H V T S I	V M E B S	O H R A U

[6] A cross-ruff was successfully used in July 1918 by J. Rives Childs of G.2 A.6, A.E.F. on the Mackensen telegram about withdrawal of German troops in Rumania.

	31 32 33 34 35	36 37 38 39 40	41 42 43 44
(i)	R R D Y C	T K L B L	M G L W
(ii)	S V F K Q	A J V C R	F K L K
(iii)	H B R N U	T R V G J	X J P W
(iv)	K O W H U	C B D U F	T V E F
(v)	S D A N I	T Y H F K	Z Z W G

and there is reason to expect a linear substitution. For each column, the differences over $\mathbb{Z}_{26}$ are determined. Six of these show particular coincidences in the differences 4, 7 and 11, as shown in Fig. 149.

1

0	24	**7**	18	**7**
2	0	**9**	20	**9**
19	17	0	11	0
8	6	15	0	15
19	17	0	11	0

13

0	11	9	9	13
15	0	24	24	**2**
17	**2**	0	0	**4**
17	**2**	0	0	**4**
13	24	22	22	0

18

0	25	15	11	19
1	0	16	12	20
11	10	0	22	**4**
15	14	4	0	8
7	6	22	18	0

22

0	17	5	5	15
9	0	14	14	24
21	12	0	0	10
21	12	0	0	10
11	**2**	16	16	0

27

0	19	13	13	15
7	0	20	20	22
13	6	0	0	**2**
13	6	0	0	**2**
11	**4**	24	24	0

36

0	19	0	17	0
7	0	**7**	24	**7**
0	19	0	17	0
9	**2**	**9**	0	**9**
0	19	0	17	0

Fig. 149. Six difference tables

In column 18 boldface **11** alone as sum of **4** and **7** fits, in column 1 11 is excluded, since 11 and 19, the complement of **7**, stand in the same line.

Among the further differences occurring in column 36, **2** and **9** are also found in column 1, **2** in column 13, and **9** in column 22. In the difference table of column 13, the two differences **2** in the second column are to be distinguished from the difference **2** in the fifth column; one of these differences occurs in column 22. (It will turn out that the differences **9** and **11** in the first line of 13 are accidental.)

Taking now the first column as reference and aligning the other five columns on the basis of these differences produces the following skeleton:

	1′	13′	18′	22′	27′	36′
(i)	T	G	M	M	M	M
(ii)	V	V	N	V	T	T
(iii)	M	X	X	H	Z	M
(iv)	B	X	B	H	Z	V
(v)	M	T	T	X	X	M
	0	*9*	*0*	*15*	*10*	*7*

This skeleton can be tested by the columns 5, 6, 7, 10, 12, 19, 20, 24, 30, 31, 33, 37, 38 to obtain:

	5′	6′	7′	10′	12′	19′	20′	24′	30′	31′	33′	37′	38′
(i)	X	P	T	X	B	H	Z	X	X	W	V	M	B
(ii)	H	N	L	T	W	L	X	B	V	X	A	L	L
(iii)	P	X	K	G	X	X	K	D	K	M	M	T	L
(iv)	L	M	A	A	M	M	M	M	H	P	R	D	T
(v)	T	M	X	B	E	E	B	H	M	X	V	A	X
	14	*15*	*25*	*22*	*11*	*14*	*7*	*20*	*8*	*21*	*5*	*24*	*10*

The multiple occurrence of four characters M, T, V, X indicates that we are on the right track. The other frequent characters B, L, A, G, K can be used to continue the formation of differences. This gives an alignment for 40 of the 44 columns.

	1′	2′	3′	4′	5′	6′	7′	8′	9′	10′	11′	12′	13′	14′	15′	16′	17′	18′	19′	20′	21′	22′
(i)	T	E	B	V	X	P	T	L	U	X	Z	B	G	G	B	G		M	H	Z	X	M
(ii)	V	N	K	B	H	N	L	X	K	T	G	W	V	N	K	B		N	L	X	K	V
(iii)	M	A	X	R	P	X	K	X	B	G	W	X	X	W	T	J		X	X	K	E	H
(iv)	B	M	P	T	L	M	A	X	P	A	B	M	X	K	T	U		B	M	M	K	H
(v)	M	A	X	V	T	M	X	K	I	B	E	E	T	K	T	G		T	E	B	V	X
	0	*24*	*1*	*0*	*14*	*15*	*25*	*25*	*21*	*22*	*24*	*11*	*9*	*15*	*21*	*1*		*0*	*14*	*7*	*0*	*15*

	23′	24′	25′	26′	27′	28′	29′	30′	31′	32′	33′	34′	35′	36′	37′	38′	39′	40′	41′	42′	43′	44′
(i)	O	X	K		M	B	K	X	W	H	Y	L	B	M	M	B	G	Z			A	X
(ii)	K	B	X		T	E	B	V	X	L	A	X	P	T	L	L	H	F			A	L
(iii)	H	D	B		Z	I	T	K	M	R	M	A	T	M	T	L	L	X			E	X
(iv)	M	M	B		Z	L	E	H	P	E	R	U	T	V	D	T	Z	T			T	G
(v)	E	H	H		X	W	T	M	X	T	V	A	H	M	A	X	K	Y			L	H
	0	*20*	*11*		*10*	*25*	*7*	*8*	*21*	*10*	*5*	*13*	*1*	*7*	*24*	*10*	*21*	*12*			*11*	*25*

As before, the additives written in the footlines are to be added to the intermediate text letters to obtain the original cryptotext characters. Thus, they are themselves a CAESAR encryption of the pseudo-key (the original plaintext) which reads:

1 2 3 4 5	6 7 8 9 10	11 12 13 14 15	16 17 18 19 20	21 22 23 24 25	26 27 28 29 30
A Y B A O	*P Z Z V W*	*Y L J P V*	*B ∗ A O H*	*A P A U L*	*∗ K Z H I*

31 32 33 34 35	36 37 38 39 40	41 42 43 44
V K F N B	*H Y K V M*	*∗ ∗ L Z*

Now exhaustion is indicated: among the 26 possible alignments the addition of 19 yields the following fragmentary English plaintext:

1 2 3 4 5	6 7 8 9 10	11 12 13 14 15	16 17 18 19 20	21 22 23 24 25	26 27 28 29 30
t r u t h	i s s o p	r e c i o	u ∗ t h a	t i t n e	∗ d s a b

31 32 33 34 35	36 37 38 39 40	41 42 43 44
o d y g u	a r d o f	∗ ∗ e s

In cleartext, the complete quotation, from Churchill's autobiography of 1949, reads: "In wartime, truth is so precious that she should always be attended by a bodyguard of lies" (Churchill to Roosevelt and Stalin, November 1943).

The five pseudo-plaintexts (the five original keys) can be reconstructed easily. They are taken from a 'children's book' for non-children:

"Alice was beginning to get very tired of sitting by he[r sister on the bank ...]"
" 'Curiouser and curiouser!' cried Alice (she was so much s[urprised ...])"
"They were indeed a queer-looking party that assemble[d on the bank ...]"
"It was the White Rabbit, trotting slowly back again, an[d looking ...]"
"The Caterpillar and Alice looked at each other for so[me time in silence ...]"

19.5 A Method of Sinkov

19.5.1 A direct product of keys. In the very general case of polyalphabetic encryptions with unrelated alphabets, a method can be tried, which may even work with only two encryptions of the same plaintext, provided the two keys are periodic with known periods of different length.

The following example of such a cryptotext-cryptotext compromise, first published by Sinkov in 1968, explains the procedure which is found in classified 1938 work of Friedman, declassified in 1984.

There are two messages observed on the same day, of 149 characters each:

```
(i)  WCOAK TJYVT VXBQC ZIVBL AUJNY BBTMT
     JGOEV GUGAT KDPKV GDXHE WGSFD XLTMI
     NKNLF XMGOG SZRUA LAQNV IXDXW EJTKI
     YAOSH NTLCI VQMJQ FYYPB CZOPZ VOGWZ
     KQZAY DNTSF WGOVI IKGXE GTRXL YOIP

(ii) TXHHV JXVNO MXHSC EEYFG EEYAQ DYHRK
     EHHIN OPKRO ZDVFV TQSIC SIMJK ZIHRL
     CQIBK EZKFL OZDPA OJHMF LVHRL UKHNL
     OVHTE HBNHG MQBXQ ZIAGS UXEYR XQJYC
     AIYHL ZVMQV QGUKI QDMAC QQBRB SQNI
```

Since the two cryptotexts have equal length, a suspicion arises that the plaintexts are identical. First, an examination of the period is appropriate. It turns out that the key for the first cryptotext presumably has period 6, the key for the second presumably 5. In this case, 30 is a period both (unknown) keys have in common. Then each character coincidence between the two texts should be repeated in a distance of 30 positions. Indeed the write-up above shows the XX coincidence in column 12 repeated in column 42 as DD coincidence, in column 72 as ZZ coincidence, and so on. A similar repetition holds for the columns 15, 45, 75, and so on. This observation strongly corroborates the conjecture that both cryptotexts belong to one and the same plaintext. In fact,

$$12+30i = \begin{cases} 6 \pmod{6} \\ 2 \pmod{5} \end{cases}, \qquad 15+30i = \begin{cases} 3 \pmod{6} \\ 5 \pmod{5} \end{cases}.$$

Therefore, the 6th alphabet of the first message should coincide with the 2nd alphabet of the second message and the 3rd alphabet of the first message should coincide with the 5th alphabet of the second message. A calculation of the corresponding *Chi* gives high values supporting this hypothesis.

Sinkov's method now decomposes the two messages according to the two keys used. The six alphabets used in keying the first message are named α, β, γ, δ, ϵ, ζ and the five alphabets used in keying the second message are named ι, κ, λ, μ, ν. Thus, the beginning of this decomposition reads:

	1	2	3	4	5	6	7	8	9	10	11	12	13	14	15	16	17	18	19	20	21	22	23	24	25	26	27	28	29	30
α	W						J						B						B						Y					
β		C						Y						Q						L						B				
γ			O						V						C						A						B			
δ				A						T						Z						U						T		
ϵ					K						V						I						J						M	
ζ						T						X						V						N						T
ι	T					J					M					E					E					D				
κ		X					X					X					E					E					Y			
λ			H					V					H					Y					Y					H		
μ				H					N					S					F					A					R	
ν					V					O					C					G					Q					K

More entries can be made into this diagram: Since both cryptotexts belong to the same plaintext, column 2, column 7, and column 12, all showing X for key κ , can be superimposed. In the same way, column 3, column 13, and column 28, all showing H for key λ , can be superimposed. Moreover, column 16 , and column 21, both showing E for key ι ; column 17, and column 22, both showing E for key κ ; column 18, and column 23, both showing Y for key λ , can be superimposed. This gives the following array:

	1	2	3	4	5	6	7	8	9	10	11	12	13	14	15	16	17	18	19	20	21	22	23	24	25	26	27	28	29	30
α	W	J	B				J					J	B						B						Y			B		
β		C					C	Y				C		Q						L						B				
γ			O						V				O		C	A					A						B	O		
δ			T	A						T			T			Z	U				Z	U						T		
ϵ					K						V						I	J				I	J						M	
ζ		X				T	X					X						V					V	N						T
ι	T					J					M					E					E					D				
κ		X					X					X					E					E					Y			
λ			H					V					H					Y					Y					H		
μ				H					N					S					F					A					R	
ν					V					O					C					G					Q					K

Superimposition can be done in the same way also in the lines ι, κ, λ, μ, ν: column 13, and column 19, both showing B for the key α , can be superimposed, and so on. This superimposition is extended, of course, over the full length 149 of the messages. Altogether, we obtain the following array:

	1	2	3	4	5	6	7	8	9	10	11	12	13	14	15	16	17	18	19	20	21	22	23	24	25	26	27	28	29	30
α	W	J	B	M		D	J	P		B		J	B	W	M	J		N	B	U	J		N	X	Y		L	B	J	D
β	Q	C	K	E	X		C	Y	X	K		C	K	Q	E	C		A	K	L	C		A		T	B		K	C	
γ	L	A	O	C	V	K	A	W	V	O	N	A	O	L	C	A		Z	O	G	A		Z		Q	S	B	O	A	K
δ	G	Z	T	A	F		Z		F	T		Z	T	G	A	Z	U	X	T	I	Z	U	X		L			T	Z	
ϵ	Y	M	D	R	K	F	M		K	D	V	M	D	Y	R	M	I	J	D	W	M	I	J		S			D	M	F
ζ	I	X	Q	Z	D	T	X		D	Q		X	Q	I	Z	X	E	V	Q	P	X	E	V	N	G		Y	Q	X	T
ι	T	E		U	Z	J	E	S	Z		M	E		T	U	E		C		Q	E		C	H	O	D			E	J
κ	I	X	Q	Z	D	T	X		D	Q		X	Q	I	Z	X	E	V	Q	P	X	E	V	N	G		Y	Q	X	T
λ	J	S	H	B			S	V		H	I	S	H	J	B	S		Y	H	N	S		Y		M		A	H	S	
μ	S	R	F	H	N		R	G	N	F		R	F	S	H	R		M	F	Y	R		M	A	B	Q		F	R	
ν	L	A	O	C	V	K	A	W	V	O	N	A	O	L	C	A		Z	O	G	A		Z		Q	S	B	O	A	K
$\rightarrow$	A	B	C	H	D	E	B	F	D	C	G	B	C	A	H	B	I	J	C	K	B	I	J	L	Q	M	N	C	B	E

	31	32	33	34	35	36	37	38	39	40	41	42	43	44	45	46	47	48	49	50	51	52	53	54	55	56	57	58	59	60
α	J		B			Y	U		J	B			P	B		W	B	J	H	M	P	W	Y		D		W	B	J	W
β	C	G	K			T	L	G	C	K	X	X	Y	K	X	Q	K	C		E	Y	Q	T			X	Q	K	C	Q
γ	A		O		N	Q	G		A	O	V	V	W	O	V	L	O	A		C	W	L	Q		K	V	L	O	A	L
δ	Z		T	E		L	I		Z	T	F	F		T	F	G	T	Z	E	A		G	L			F	G	T	Z	G
ϵ	M		D	P	V	S	W		M	D	K	K		D	K	Y	D	M	P	R		Y	S		F	K	Y	D	M	Y
ζ	X	H	Q			G	P	H	X	Q	D	D		Q	D	I	Q	X		Z		I	G	F	T	D	I	Q	X	I
ι	E				M	O	Q		E		Z	Z	S		Z	T		E		U	S	T	O		J	Z	T		E	T
κ	X	H	Q			G	P	H	X	Q	D	D		Q	D	I	Q	X		Z		I	G	F	T	D	I	Q	X	I
λ	S	K	H			M	N	K	S	H			V	H		J	H	S		B	V	J	M				J	H	S	J
μ	R		F	I		B	Y		R	F	N	N	G	F	N	S	F	R	I	H	G	S	B	J		N	S	F	R	S
ν	A		O		N	Q	G		A	O	V	V	W	O	V	L	O	A		C	W	L	Q		K	V	L	O	A	L
$\rightarrow$	B	O	C	P	G	Q	K	O	B	C	D	D	F	C	D	A	C	B	P	H	F	A	Q	R	E	D	A	C	B	A

	61	62	63	64	65	66	67	68	69	70	71	72	73	74	75	76	77	78	79	80	81	82	83	84	85	86	87	88	89	90
α	N	B		Y	D	J	M		B	W	Y	M	R		J	Y		B	N			N	B	J	W	M		B		W
β	A	K		T		C	E	G	K	Q	T	E		U	C	T		K	A	V		A	K	C	Q	E		K	X	Q
γ	Z	O	N	Q	K	A	C		O	L	Q	C			A	Q		O	Z	F	I	Z	O	A	L	C	J	O	V	L
δ	X	T		L		Z	A		T	G	L	A			Z	L		T	X			X	T	Z	G	A		T	F	G
ϵ	J	D	V	S	F	M	R		D	Y	S	R			M	D	A	D	J			J	D	M	Y	R		D	K	Y
ζ	V	Q		G	T	X	Z	H	Q	I	G	Z			X	G	J	Q	V			V	Q	X	I	Z	K	Q	D	I
ι	C		M	O	J	E	U			T	O	U			E	O			C	X	L	C		E	T	U			Z	T
κ	V	Q		G	T	X	Z	H	Q	I	G	Z			X	G	J	Q	V			V	Q	X	I	Z	K	Q	D	I
λ	Y	H	I	M		S	B	K	H	J	M	B	D		S	M		H	Y			Y	H	S	J	B		H		J
μ	M	F		B		R	H		F	S	B	H		P	R	B		F	M	K		M	F	R	S	H		F	N	S
ν	Z	O	N	Q	K	A	C		O	L	Q	C			A	Q		O	Z	F	I	Z	O	A	L	C	J	O	V	L
$\rightarrow$	J	C	G	Q	E	B	H	O	C	A	Q	H	S	T	B	Q	U	C	J	V	W	J	C	B	A	H	X	C	D	A

	91	92	93	94	95	96	97	98	99	00	01	02	03	04	05	06	07	08	09	10	11	12	13	14	15	16	17	18	19	20
α	Y	N	B			X	T	U	M	U		B	M		Y		W	L	P		M	J		U	Z		B	W	U	M
β	T	A	K					L	E	L		K	E	J	T	X	Q		Y	B	E	C		L		V	K	Q	L	E
γ	Q	Z	O		E			G	C	G	N	O	C		Q	V	L	B	W	S	C	A		G	R	F	O	L	G	C
δ	L	X	T	S				I	A	I		T	A		L	F	G				A	Z		I			T	G	I	A
ϵ	S	J	D		H			W	R	W	V	D	R		S	K	Y				R	M	O	W			D	Y	W	R
ζ	G	V	Q			N	B	P	Z	P		Q	Z		G	D	I	Y			Z	X		P			Q	I	P	Z
ι	O	C				H		Q	U	Q	M		U		O	Z	T		S	D	U	E		Q		X		T	Q	U
κ	G	V	Q			N	B	P	Z	P		Q	Z		G	D	I	Y			Z	X		P			Q	I	P	Z
λ	M	Y	H					N	B	N	I	H	B		M		J	A	V		B	S	E	N			H	J	N	B
μ	B	M	F	T		A		Y	H	Y		F	H	X	B	N	S		G	Q	H	R		Y		K	F	S	Y	H
ν	Q	Z	O		E			G	C	G	N	O	C		Q	V	L	B	W	S	C	A		G	R	F	O	L	G	C
$\rightarrow$	Q	J	C	Y	Z	L	1	K	H	K	G	C	H	2	Q	D	A	N	F	M	H	B	3	K	4	V	C	A	K	H

1..	21	22	23	24	25	26	27	28	29	30	31	32	33	34	35	36	37	38	39	40	41	42	43	44	45	46	47	48	49
α	K	W	N	M	W		N	Y			U	Y	O			U		Y	X	M	U	B	M	J	L	P	B	U	
β		Q	A	E	Q	X	A	T	B	X	L	T		V	G	L	X	T		E	L	K	E	C		Y	K	L	
γ		L	Z	C	L	V	Z	Q	S	V	G	Q		F	I	G	V	Q		C	G	O	C	A	B	W	O	G	
δ		G	X	A	G	F	X	L		F	I	L				I	F	L		A	I	T	A	Z			T	I	E
ϵ		Y	J	R	Y	K	J	S		K	W	S				W	K	S		R	W	D	R	M			D	W	P
ζ		I	V	Z	I	D	V	G		D	P	G			H	P	D	G	N	Z	P	Q	Z	X	Y		Q	P	
ι	A	T	C	U	T	Z	C	O	D	Z	Q	O		X		Q	Z	O	H	U	Q		U	E		S		Q	
κ		I	V	Z	I	D	V	G		D	P	G			H	P	D	G	N	Z	P	Q	Z	X	Y		Q	P	
λ		J	Y	B	J		Y	M			N	M	U		K	N		M		B	N	H	B	S	A	V	H	N	
μ		S	M	H	S	N	M	B	Q	N	Y	B		K		Y	N	B	A	H	Y	F	H	R		G	F	Y	I
ν		L	Z	C	L	V	Z	Q	S	V	G	Q		F	I	G	V	Q		C	G	O	C	A	B	W	O	G	
$\rightarrow$	5	A	J	H	A	D	J	Q	M	D	K	Q	6	V	O	K	D	Q	L	H	K	C	H	B	N	F	C	K	P

As we should have expected, the alphabets named γ and ν are identical[7], likewise the alphabets named ζ and κ. Not all fields of entries are filled, and some columns such as 8 and 17 could possibly be identical. Indeed, there are a total of 32 different columns, which are *ad hoc* abbreviated with A ... Z and 1 ... 6 and listed more or less in the order of their formation in the superposition process. 32 characters—that is more than the 26 letters of the common alphabet. In fact, nothing prohibits the adversary from using a plaintext alphabet of more than 26 characters. But most likely some columns mean the same plaintext character. Thus, we have presumably reached a monoalphabetically encrypted intermediate text, but the encryption includes homophones—a rather surprising result.

Fortunately, it will turn out that the victims of homophony are not the most frequent characters—as they often are for the purpose of equalizing the frequencies—but just the rare ones; they are too rare to provide enough material to fill the fields.

19.5.2 Intermediate encryption. It is given by the footlines of the array

A B C H D	E B F D C	G B C A H	B I J C K	B I J L Q	M N C B E
B O C P G	Q K O B C	D D F C D	A C B P H	F A Q R E	D A C B A
J C G Q E	B H O C A	Q H S T B	Q U C J V	W J C B A	H X C D A
Q J C Y Z	L 1 K H K	G C H 2 Q	D A N F M	H B 3 K 4	V C A K H
5 A J H A	D J Q M D	K Q 6 V O	K D Q L H	K C H B N	F C K P

which shows clearly the frequency distribution of English, with a peak-C and B, A, H, D, O, N of about equal frequency. Among the many ways that would give an entry, we assume that /treasurysecretary/ is a probable word that fits the pattern *123456275382314 27* of the beginning. With eight letters, a lot is achieved:

t r e a s	u r y s e	c r e t a	r I J e K	r I J L Q	M N e r u
r O e P c	Q K O r e	s s y e s	t e r P a	y t Q R u	s t e r t
J e c Q u	r a O e t	Q a S T r	Q U e J V	W J e r t	a X e s t
Q J e Y Z	L 1 K a K	c e a 2 Q	s t N y M	a r 3 K 4	V e t K a
5 t J a t	s J Q M s	K Q 6 V O	K s Q L a	K e a r N	y e K P

[7] For further manual work one will identify the corresponding lines but in programmed execution it is simpler to repeat them.

/yesterday/ in the second line catches the eye, but does not help much. A few places earlier /congress/ brings more. There are now twelve letters determined and only two from the *etaonirsh* are still missing, /i/ and /h/ :

```
t r e a s   u r y s e   c r e t a   r I J e n   r I J L o   M N e r u
r g e d c   o n g r e   s s y e s   t e r d a   y t o R u   s t e r t
J e c o u   r a g e t   o a S T r   o U e J V   W J e r t   a X e s t
o J e Y Z   L 1 n a n   c e a 2 o   s t N y M   a r 3 n 4   V e t n a
5 t J a t   s J o M s   n o 6 V g   n s o L a   n e a r N   y e n d
```

Now we get rid of some homophones: /henry/ in the first line means that I homophone with F is /y/ . For the /i/ it is more difficult, but in the last line we can read /no signs/ , provided 6 homophone with D is /s/ . Thus:

```
t r e a s   u r y s e   c r e t a   r y h e n   r y h L o   M N e r u
r g e d c   o n g r e   s s y e s   t e r d a   y t o R u   s t e r t
h e c o u   r a g e t   o a S T r   o U e h i   W h e r t   a X e s t
o h e Y Z   L 1 n a n   c e a 2 o   s t N y M   a r 3 n 4   i e t n a
5 t h a t   s h o M s   n o s i g   n s o L a   n e a r N   y e n d
```

which presents in the second line /to muster/, in the third line /approve higher taxes/ (S and T homophone), in the fourth line /help finance a costly war in vietnam/ , whence finally in the first line the name /henry h fowler/ emerges and the plaintext reads:

```
t r e a s   u r y s e   c r e t a   r y h e n   r y h f o   w l e r u
r g e d c   o n g r e   s s y e s   t e r d a   y t o m u   s t e r t
h e c o u   r a g e t   o a p p r   o v e h i   g h e r t   a x e s t
o h e l p   f i n a n   c e a c o   s t l y w   a r i n v   i e t n a
m t h a t   s h o w s   n o s i g   n s o f a   n e a r l   y e n d
```

So far, the 32 columns reduced to 21 characters; 5 letters are missing in the plaintext. The fragmentary encryption table is as follows:

	a	b	c	d	e	f	g	h	i	j	k	l	m	n	o	p	q	r	s	t	u	v	w	x	y	z
α	M				B	X		N	T			L	K	U	Y	R		J	O	W	D	Z			P	
β	E		J		K		G	A	V					L	T	U		C	X	Q			B		Y	
$\gamma=\nu$	C		N		O		I	Z	F			B		G	Q	E		A	V	L	K	R	S	J	W	
δ	A			E	T			X				S		I	L			Z	F	G					U	
ϵ	R		V	P	D			J	O					W	S	H		M	K	Y	F	A			I	
$\zeta=\kappa$	Z				Q	N	H	V	B			Y	F	P	G			X	D	I	T	J		K	E	
ι	U		M			H	L	C	X				A	Q	O			E	Z	T	J		D		S	
λ	B		I		H		K	Y	E			A		N	M	D		S	U	J					V	
μ	H		X	I	F	A		M	K			T	J	Y	B	P		R	N	S			Q		G	
	H		G	P	C	L	O	J	V			N	R	K	Q	S		B	D	A	E	U	M	X	F	
$\longrightarrow$			2				W		1			Y	5			T			6			4			I	
									3							Z										

19.5.3 Reconstruction. The complete encryption table (*tabula recta*) could be reconstructed if the alphabets were obtained by shifting a primary alphabet. This turns out to be the case, and a method developed by Friedman (mentioned in Sect. 18.8.2) allows one to reconstruct the primary alphabet. For the decryption this was not necessary, but it helps to fulfill Rohrbach's maxim.

Friedman starts from the observation that in the desired encryption table all *columns* are derived from a single one by cyclic shift. Picking out, say, the lines λ and μ, then at some unknown distance k below /t/ are the characters J and S, below /r/ the characters S and R, below /s/ U and N, then N and Y, Y and M, M and B, B and H, H and F, finally V and G —all with the same distance k. Thus, we have already three chains

J-S-R , U-N-Y-M-B-H-F V-G

with the distance k. But the characters J and S are also found below /r/ in the lines α and λ, then at the same distance k there are also the characters W and J, R and D, O and U, U and N, P and V, L and A. The chains can be extended with the new links; we now have

W-J-S-R-D , O-U-N-Y-M-B-H-F , P-V-G and L-A .

J and D have the distance $3k$; in the lines ι and α are found below /u/ the characters J and D , thus not only U and M, O and Y, but also C and N, X and T, A and K, Q and U, E and J, Z and O, T and W, S and P have distance $3k$. We now have the chains

T-E-*-W-J-S-R-D-P-V-G , Q-C-O-U-N-Y-M-B-H-F-*-X , L-A-*-*-K .

We can now close these chain fragments by observing that W and G , to be found in the lines α and δ , have the distance $7k$, thus this is also the distance of J and Z, B and T, M and A, U and I, H and E, Y and L, which closes the cycle:

T-E-K-W-J-S-R-D-P-V-G-Z-Q-C-O-U-N-Y-M-B-H-F-I-X-L-A- .

But this is not necessarily the original order. With the method of Sect. 18.8.1, the fifth power

T-S-G-U-H-A-J-V-O-B-L-W-P-C-M-X-K-D-Q-Y-I-E-R-Z-N-F-

brings success: the sequence

H A J V O B L W P C M X K D Q Y I E R Z N F T S G U

results—columnwise—from the password HOPKINS by the method described in Sect. 3.2.5:

H	O	P	K	I	N	S
A	B	C	D	E	F	G
J	L	M	Q	R	T	U
V	W	X	Y	Z		

With this sequence, a *tabula recta* is built (Table 26). To determine the headline, the following can be done: A 'rich' line like the one named with $\gamma = \nu$, reordered, reads:

		r	x	s	e	l	t	y		a			u		o		g	p	v	h	c	i		w	n	
$\gamma = \nu$	H	A	J	V	O	B	L	W	P	C	M	X	K	D	Q	Y	I	E	R	Z	N	F	T	S	G	U

The other lines considered so far drop into place and determine further plaintext characters. Altogether a fragmentary headline is obtained:

∗ r x s e l t y ∗ a f m u ∗ o ∗ g p v h c i ∗ w n d

where only the rare plaintext characters /b/, /j/, /k/, /q/, /z/ are still missing. This headline now also discloses its secret; it is built by the method described in Sect. 3.2.5 with the password /johns/ .

j	o	h	n	s
a	b	c	d	e
f	g	i	k	l
m	p	q	r	t
u	v	w	x	y
z				

.

This settles the five missing characters, too. The encryption table in Table 26 falls under the heading 'treble key' (Sect. 8.2.3). The passwords /johns/ and HOPKINS are obvious allusions to Johns Hopkins (1795–1873), American financier and philanthropist.[8] The keys are fittingly *CIPHER* and *GROUP*, as can be seen from the entries for $\alpha\beta\gamma\delta\epsilon\zeta$ and $\iota\kappa\lambda\mu\nu$ in Table 26.

		j	a	f	m	u	z	o	b	g	p	v	h	c	i	q	w	n	d	k	r	x	s	e	l	t	y
δ	*H*	H	A	J	V	O	B	L	W	P	C	M	X	K	D	Q	Y	I	E	R	Z	N	F	T	S	G	U
	A	A	J	V	O	B	L	W	P	C	M	X	K	D	Q	Y	I	E	R	Z	N	F	T	S	G	U	H
	J	J	V	O	B	L	W	P	C	M	X	K	D	Q	Y	I	E	R	Z	N	F	T	S	G	U	H	A
	V	V	O	B	L	W	P	C	M	X	K	D	Q	Y	I	E	R	Z	N	F	T	S	G	U	H	A	J
λ	*O*	O	B	L	W	P	C	M	X	K	D	Q	Y	I	E	R	Z	N	F	T	S	G	U	H	A	J	V
	B	B	L	W	P	C	M	X	K	D	Q	Y	I	E	R	Z	N	F	T	S	G	U	H	A	J	V	O
	L	L	W	P	C	M	X	K	D	Q	Y	I	E	R	Z	N	F	T	S	G	U	H	A	J	V	O	B
	W	W	P	C	M	X	K	D	Q	Y	I	E	R	Z	N	F	T	S	G	U	H	A	J	V	O	B	L
$\gamma=\nu$	*P*	P	C	M	X	K	D	Q	Y	I	E	R	Z	N	F	T	S	G	U	H	A	J	V	O	B	L	W
α	*C*	C	M	X	K	D	Q	Y	I	E	R	Z	N	F	T	S	G	U	H	A	J	V	O	B	L	W	P
	M	M	X	K	D	Q	Y	I	E	R	Z	N	F	T	S	G	U	H	A	J	V	O	B	L	W	P	C
	X	X	K	D	Q	Y	I	E	R	Z	N	F	T	S	G	U	H	A	J	V	O	B	L	W	P	C	M
	K	K	D	Q	Y	I	E	R	Z	N	F	T	S	G	U	H	A	J	V	O	B	L	W	P	C	M	X
	D	D	Q	Y	I	E	R	Z	N	F	T	S	G	U	H	A	J	V	O	B	L	W	P	C	M	X	K
	Q	Q	Y	I	E	R	Z	N	F	T	S	G	U	H	A	J	V	O	B	L	W	P	C	M	X	K	D
	Y	Y	I	E	R	Z	N	F	T	S	G	U	H	A	J	V	O	B	L	W	P	C	M	X	K	D	Q
β	*I*	I	E	R	Z	N	F	T	S	G	U	H	A	J	V	O	B	L	W	P	C	M	X	K	D	Q	Y
ϵ	*E*	E	R	Z	N	F	T	S	G	U	H	A	J	V	O	B	L	W	P	C	M	X	K	D	Q	Y	I
$\zeta=\kappa$	*R*	R	Z	N	F	T	S	G	U	H	A	J	V	O	B	L	W	P	C	M	X	K	D	Q	Y	I	E
	Z	Z	N	F	T	S	G	U	H	A	J	V	O	B	L	W	P	C	M	X	K	D	Q	Y	I	E	R
	N	N	F	T	S	G	U	H	A	J	V	O	B	L	W	P	C	M	X	K	D	Q	Y	I	E	R	Z
	F	F	T	S	G	U	H	A	J	V	O	B	L	W	P	C	M	X	K	D	Q	Y	I	E	R	Z	N
	T	T	S	G	U	H	A	J	V	O	B	L	W	P	C	M	X	K	D	Q	Y	I	E	R	Z	N	F
	S	S	G	U	H	A	J	V	O	B	L	W	P	C	M	X	K	D	Q	Y	I	E	R	Z	N	F	T
ι	*G*	G	U	H	A	J	V	O	B	L	W	P	C	M	X	K	D	Q	Y	I	E	R	Z	N	F	T	S
μ	*U*	U	H	A	J	V	O	B	L	W	P	C	M	X	K	D	Q	Y	I	E	R	Z	N	F	T	S	G

Table 26. Encryption table with permuted headline

[8] The Johns Hopkins University in Maryland, U.S.A., was the place where in the Second World War the proximity fuze was developed.

19.6 Cryptotext-Cryptotext Compromise: Indicator Doubling

The double encipherment of each text setting was a gross error.

Gordon Welchman 1982

The Poles were known for the high standard of their cryptanalytic abilities: They won their war against Russia in 1920 with the help of cryptanalysis.

As Władysław Kozaczuk disclosed in 1967,[9] a typical case of cryptotext-cryptotext compromise allowed from 1932 on the Polish *Biuro Szyfrów* under Major Gwido Langer ('Luc') with its cryptanalytic service B.S.-4 under Maksymilian Ciężki (1899–1951) and the young mathematicians Marian Rejewski, Jerzy Różycki, Henryk Zygalski to penetrate the encryption of the German *Wehrmacht*, whose practice transmissions of ENIGMA-encrypted radio signals in the Eastern provinces of Prussia gave a copious supply of cryptotexts.

It was a typical problem of machine encryption with a key sequence generated by the machine itself (Sect. 8.5). If every message is started with its own initial setting, immediate superimposition is inhibited. But it was widely considered too difficult and prone to mistakes to prearrange such new initial settings for every message. This did not hold for the ENIGMA only, but is the general problem of key negotiation and key administration. Therefore, as already mentioned in Sect. 19.3, indicators are used for the 'text setting' or 'message setting'. Of course, they should not give the setting plainly, but must be themselves somehow encrypted.

Thus, it was thought to be a clever idea to use the encryption machine itself for this purpose. One specific weakness with both the Army and Air Force ENIGMA was that the encryption of the indicator was based *exclusively* on the ENIGMA itself. However, for the Navy ENIGMA, at least a bigram superencryption was done, starting in May 1937 .

Another weakness was that (for reasons to be given below) the plain indicator was doubled before encryption—the old treacherous trick—, thus introducing a tiny cryptotext-cryptotext compromise at a fixed place, the very beginning of the message. All this was the fault not of the machine, but of the rules for operating it.

The fancy idea of doubling the indicator seems to go back to recommendations for the commercial ENIGMA of 1924.

---

[9] In his book *Bitwa o tajemnice: Służby wywiadowcze Polski i Rzeszy Niemieckiej 1922-1939*, Warsaw 1967 (in Polish), which was largely overlooked in the West (a review was published in a Göttingen journal in 1967). However, in 1968, Donald Cameron Watt (in a foreword to *Breach of Security* by David Irving, London 1968) revealed that in 1939 Great Britain "received from Polish Military Intelligence keys and machines for decoding German official military and diplomatic ciphers". This seemed unbelievable, until in 1973 the French General Gustave Bertrand, who had been involved in the deal, confirmed it.

Otherwise, the security of the ENIGMA seemed to be high. A *Tagesschlüssel*, valid for one day for all machines in a 'key net', determined, apart from the *Steckerverbindung*, the order (and later also the choice) of the three rotors (*Walzenlage*), the internal ring setting (*Ringstellung*) on each rotor, and a ground setting (basic wheel setting, *Grundstellung*) of the rotors. With the machine set accordingly, a freely chosen 3-letter group, the plain indicator, in Bletchley Park called the text setting or message setting (German *Spruchschlüssel*), was doubled and encrypted, and the resulting 6-letter group, the encrypted doubled indicator (German *chiffrierter Spruchschlüssel*), was transmitted after a preamble (in plain) containing call-sign, time of origin, and number of letters in the cryptotext. Then the message followed, encrypted with the indicator as initial setting. The authorized recipient first decrypted the encrypted indicator with his *Tagesschlüssel* and checked whether it was correctly doubled in order to find the initial setting for decryption of the cryptotext proper. This encryption procedure was valid until September 15, 1938.

Weakening the security of the cryptotext proper by using an indicator was accepted without scruples by the Germans, since the indicator was protected well by the ENIGMA, which was judged to be *indéchiffrable*. Nobody observed that this was logically a vicious circle. Anyhow, the ENIGMA seemed to have enough combinatory complexity, and when after July 15, 1928 more and more ENIGMA-encrypted radio signals of the German Army appeared in the ether, the Polish bureau, although familiar with the commercial ENIGMA, which had been on the open market since 1926, was not able to break in at first. (The internal wiring of the rotors in the military ENIGMA of 1930 was different from that in the commercial version, of course.)

The reason for doubling the plain indicator, the 'double encipherment of each text setting', was that radio signals then were frequently disturbed by noise. An encrypted indicator corrupted during transmission would cause nonsense in the authorized decryption, with all the risks of repetition. To transmit the whole cryptomessage twice for the sake of error-detection seemed out of the question. Thus, the error-detecting possibility was restricted to the plain indicator. But this led to a much more dangerous compromise which could have been avoided. And the doubling was not necessary at all: When it was discontinued on May 1, 1940, ENIGMA traffic did run smoothly.

How did the Polish bureau find out about the indicator at all? It discovered around 1930 quickly that two signals which started with the same 6-letter-group showed a higher character coincidence near κ_d, thus invited superimposition. As a consequence, the initial setting of the rotors was determined by the first 6-letter-group—in other words. it was an indicator. Since it was clear that it was not plain, it could only be somehow encrypted. But how was this done?

19.6.1 France I. It is not known whether the Poles had reasons to believe the Germans would be stupid enough to use the ENIGMA itself for the indicator encryption. In any case, they found in 1931 help from their French friends.

The spy Hans-Thilo Schmidt (1888-1943, with code name ASCHE, Asché, the French pronunciation of *H.E.*), working until 1938 in the *Chiffrier-Stelle* of the *Reichswehrministerium*, who was arrested 23.3.1943 and allegedly executed in July 1943 (suicide 19.9.1943), had since October 1931 forwarded via the secret agent 'Rex' manuals on the use of the ENIGMA, on the encryption procedure, and even *Tagesschlüssel* for September and October 1932 (including the ring settings and cross-pluggings for these two months) to the French *grand chef*, then Major, later Général Gustave Bertrand (1896–1976, *nom de guerre* 'Bolek'), who forwarded them in turn to the Polish bureau.

19.6.2 Poland I. Cię̇żki's young aide, the highly gifted[10] Marian Rejewski (1905–1980)—he knew him since 1929 and had hired him permanently in 1932—had first to find out how the French gift could be made useful. According to a report by Władysław Kozaczuk in 1984, he proceeded as follows.

19.6.2.1 He had already guessed that each signal began with the 6 letters of the encrypted doubled indicator. Let $P_1, P_2, P_3, P_4, P_5, P_6$ (Rejewski used the letters A, B, C, D, E, F) denote the permutations performed upon the 1st, 2nd, 3rd, ... 6th plaintext letter, starting with some basic wheel setting. From $\mathrm{a}P_i = \mathrm{X}$ and $\mathrm{a}P_{i+3} = \mathrm{Y}$ $(i = 1, 2, 3)$ it follows that $\mathrm{X}P_i^{-1}P_{i+3} = \mathrm{Y}$. The properly self-reciprocal character of the ENIGMA was known. Therefore, from $\mathrm{a}P_i = \mathrm{X}$ and $\mathrm{a}P_{i+3} = \mathrm{Y}$ $(i = 1, 2, 3)$ it follows that $\mathrm{X}P_iP_{i+3} = \mathrm{Y}$. The known characters X, Y standing in the 1st and 4th, or the 2nd and 5th, or the 3rd and 6th positions of the cryptotext thus impose conditions on the three products P_iP_{i+3} of the unknown permutations $P_1, P_2, P_3, P_4, P_5, P_6$.

1.	AUQ	AMN	14.	IND	JHU	27.	PVJ	FEG	40.	SJM	SPO	53.	WTM	RAO
2.	BNH	CHL	15.	JWF	MIC	28.	QGA	LYB	41.	SJM	SPO	54.	WTM	RAO
3.	BCT	CGJ	16.	JWF	MIC	29.	QGA	LYB	42.	SJM	SPO	55.	WTM	RAO
4.	CIK	BZT	17.	KHB	XJV	30.	RJL	WPX	43.	SUG	SMF	56.	WKI	RKK
5.	DDB	VDV	18.	KHB	XJV	31.	RJL	WPX	44.	SUG	SMF	57.	XRS	GNM
6.	EJP	IPS	19.	LDR	HDE	32.	RJL	WPX	45.	TMN	EBY	58.	XRS	GNM
7.	FBR	KLE	20.	LDR	HDE	33.	RJL	WPX	46.	TMN	EBY	59.	XOI	GUK
8.	GPB	ZSV	21.	MAW	UXP	34.	RFC	WQQ	47.	TAA	EXB	60.	XYW	GCP
9.	HNO	THD	22.	MAW	UXP	35.	SYX	SCW	48.	USE	NWH	61.	YPC	OSQ
10.	HNO	THD	23.	NXD	QTU	36.	SYX	SCW	49.	VII	PZK	62.	YPC	OSQ
11.	HXV	TTI	24.	NXD	QTU	37.	SYX	SCW	50.	VII	PZK	63.	ZZY	YRA
12.	IKG	JKF	25.	NLU	QFZ	38.	SYX	SCW	51.	VQZ	PVR	64.	ZEF	YOC
13.	IKG	JKF	26.	OBU	DLZ	39.	SYX	SCW	52.	VQZ	PVR	65.	ZSJ	YWG

Fig. 150. 65 observed encrypted doubled indicators to the same *Tagesschlüssel*

[10] Marian Rejewski's intuitive abilities are illustrated by the following episode: In the commercial ENIGMA, the contacts on the entry ring belonged to letters in the order QWERTZU... of the letters on the keyboard. This seemed to be different with the military ENIGMA I. Rejewski said to himself "the Germans rely upon order" and tried around New Year 1933 the alphabetic order (see Sect. 7.3.2)—and that was it. Knox, who had long racked his brain about this question, was told the solution in July 1939 by Rejewski. Penelope Fitzgerals, Knox's niece reported "Knox was furious when he learned how simple it was." Later, he was chanting "Nous avons le QWERTZU" (Peter Twinn).

Figure 150 presents 65 encrypted doubled indicators. They show that for P_1P_4 the character a goes into itself (1.), likewise
the character s goes into itself (35.), while
the character b goes into c and vice versa (2., 4.),
the character r goes into w and vice versa (30., 53.). For the remaining characters it turns out that they belong under P_1P_4 to the cycles

(d v p f k x g z y o) (5., 49., 27., 7., 17., 57., 8., 63., 61., 26.) and
(e i j m u n q l h t) (6., 12., 15., 21., 48., 23., 28., 19., 9., 45.) .

In short, P_1P_4 has two 1-cycles, two 2-cycles and two cycles of ten characters, and since this embraces all 26 characters, P_1P_4 is fully determined. The cycle determination is complete if every character occurs at least one time in the first, the second and the third position; as a rule this requires fifty to a hundred messages—this much was certainly the result of a busy manœuvre day.

For P_2P_5 and P_3P_6 the work is similar. Altogether there is the lucky result

$P_1P_4=$ (a) (s) (b c) (r w) (d v p f k x g z y o) (e i j m u n q l h t)
$P_2P_5=$ (a x t) (b l f q v e o u m) (c g y) (d) (h j p s w i z r n) (k)
$P_3P_6=$ (a b v i k t j g f c q n y) (d u z r e h l x w p s m o)

The 1-cycles ('females')[11] play a particular role: Since each one of the permutations $P_1, P_2, P_3, P_4, P_5, P_6$ because of its self-reciprocal character consists of 2-cycles ('swappings') only, $P_iP_{i+3}x = x$ implies that there exists a character y such that $P_ix = y$ and $P_{i+3}y = x$, i.e., both P_i and P_{i+3} contain the 2-cycle $(x\ y)$. In the given example, both P_1 and P_4 contain the 2-cycle (a s). A theorem of group theory about products of properly self-reciprocal permutations states that the cycles of P_iP_{i+3} occur in pairs of equal length: if

P_i contains the 2-cycles $(x_1y_1), (x_2y_2), \ldots, (x_\mu y_\mu)$, and
P_{i+3} contains the 2-cycles $(y_1x_2), (y_2x_3), \ldots, (y_\mu x_1)$, then
P_iP_{i+3} contains the μ-cycles $(x_1x_2 \ldots x_\mu)$, $(y_\mu y_{\mu-1} \ldots y_1)$.

Thus, if one of the cycles of P_iP_{i+3} is written in reversed order (←) below the other one, then the 2-cycles of P_i can be read vertically—provided the cycles are in phase. To find this phase is the problem. It could be solved by exhaustion, for P_1P_4 above in 2×10 trials, for P_2P_5 above in 3×9 trials.

19.6.2.2 But Marian Rejewski found a shortcut. He observed that the encrypted indicators actually used showed deviations from equal distribution, which probably meant that the German crypto clerks, like most people playing in the lottery, were unable to choose the text setting truly at random. Thus, Rejewski directed his interest primarily to conspicuous patterns, and he was right. In fact, the German security regulations were not too clear on this point, and a German officer who had given the order to take as text

[11] In the jargon of Bletchley Park the term 'female' was used, originating from the Polish pun *te same* ('the same') ↔ *samiczka* ('female'). Most people in Bletchley Park did not know—in fact did not have to know—the Polish origin and found their own explanations, like *female* screw for a threaded hole.

setting the end position of the rotors in the previous message could argue that he had made sure the text setting was changed after every message.

Thus, it was common practice to use even stereotyped 3-letter groups like /aaa/, /bbb/, /sss/ . When in the spring of 1933 the mere repetition of letters was explicitly forbidden, it was too late. The Poles had already made their entry into the ENIGMA. Later, the bad habit developed of using horizontally or vertically adjacent letters on the keyboard: /qwe/ , /asd/ (horizontally); /qay/ , /cde/ (vertically), etc.

Rejewski's frequency argument was that the most frequently occurring encrypted indicator SYX SCW , occurring five times (35.–39.), should correspond to a most conspicuous pattern. There were still a number of those to be tested. Assume, we test with the plain indicator aaa . This fits in P_1 with the 2-cycle (a s), in P_2 it gives the 2-cycle (a y), in P_3 the 2-cycle (a x); it fits in P_4 with the 2-cycle (a s), in P_5 it gives the 2-cycle (a c), in P_6 it gives the 2-cycle (a w) . Thus, for P_3 and P_6 the phase of the two cycles

```
     ↓
→ ( a b v i k t j g f c q n y )
← ( x l h e r z u d o m s p w )
```

is already determined; in a zig-zag the 2-cycles of P_3 and P_6 , beginning with (a x) , can be calculated:

P_3= (a x) (b l) (v h) (i e) (k r) (t z) (j u) (g d) (f o) (c m) (q s) (n p) (y w)

P_6= (x b) (l v) (h i) (e k) (r t) (z j) (u g) (d f) (o c) (m q) (s n) (p y) (w a)

P_3 contains among others the 2-cycle (q s). Thus, the plain indicator to AUQ AMN (1.) has the pattern ∗∗s ; since P_1 contains among others the 2-cycle (as) , it even has the pattern *s*∗s . If one now guesses that the plain indicator to AUQ AMN reads sss , then in P_2 , apart from (a y) , also (s u) is determined. Thus, the phase for the cycles of P_2 and P_5 is also completely fixed:

```
     ↓                   ↓
→ ( a x t ) ( b l f q v e o u m ) ( d )
← ( y g c ) ( j n h r z i w s p ) ( k ) .
```

In a zig-zag the 2-cycles of P_2 and P_5 , beginning with (a y), can be calculated:

P_2= (a y) (x g) (t c) (b j) (l n) (f h) (q r) (v z) (e i) (o w) (u s) (m p) (d k)

P_5= (y x) (g t) (c a) (j l) (n f) (h q) (r v) (z e) (i o) (w u) (s m) (p b) (k d)

Another frequently occurring encrypted indicator was RJL WPX , occurring four times (30.–33.). The corresponding plain indicator has the pattern ∗bb . P_1 can only have the 2-cycles (r b) or (r c) . In the first case with the more likely plain indicator bbb : P_1 contains the 2-cycle (b r), P_4 the 2-cycle (r c). For a pairing of the 10-cycles, another encrypted indicator may be used, say LDR HDE (19.–20.) . Since P_3 and P_6 contain (r k) and (k e) , P_2 and P_5 contain (d k) and (k d) , the pattern of the plain indicator is ∗kk . This suggests again the stereotype kkk , with the result that P_1

and P_4 contain the 2-cycles (l k) and (k h). Thus, the phase for the cycles of P_1 and P_4 is also completely fixed:

$$\begin{array}{l} \rightarrow (\ a\ )\ (\overset{\downarrow}{b}\ c)\ (d\ v\ p\ f\ \overset{\downarrow}{k}\ x\ g\ z\ y\ o) \\ \leftarrow (\ s\ )\ (r\ w)\ (i\ e\ t\ h\ l\ q\ n\ u\ m\ j) \quad \text{and thus} \end{array}$$

P_1= (a s) (b r) (c w) (d i) (v e) (p t) (f h) (k l) (x q) (g n) (z u) (y m) (o j)

P_4= (s a) (r c) (w b) (i v) (e p) (t f) (h k) (l x) (q g) (n z) (u y) (m o) (j d)

Altogether the first three permutations read in ordered form:

P_1 = (a s) (b r) (c w) (d i) (e v) (f h) (g n) (j o) (k l) (m y) (p t) (q x) (u z)

P_2 = (a y) (b j) (c t) (d k) (e i) (f h) (g x) (l n) (m p) (o w) (q r) (s u) (v z)

P_3 = (a x) (b l) (c m) (d g) (e i) (f o) (h v) (j u) (k r) (n p) (q s) (t z) (w y)

19.6.2.3 The reconstruction of all plain indicators used on this busy manœuvre day is now possible (Fig. 151). The bad habits of the ENIGMA crypto clerks are evident. First, the use of stereotypes has led to multiple use of identical indicators, something that should by no means happen. Second, a look on the keyboard of the ENIGMA (Fig. 152) is frightening: only two out of forty, namely abc and uvw , are not keyboard stereotypes; instead they are alphabet stereotypes. Neither the crypto clerks nor their signal officers would have dreamed that peaceful practice transmissions with an innocently invented combat scenario would give away so much of the secret of the ENIGMA.

AUQ AMN : sss	IKG JKF : ddd	QGA LYB : xxx	VQZ PVR : ert
BNH CHL : rfv	IND JHU : dfg	RJL WPX : bbb	WTM RAO : ccc
BCT CGJ : rtz	JWF MIC : ooo	RFC WQQ : bnm	WKI RKK : cde
CIK BZT : wer	KHB XJV : lll	SYX SCW : aaa	XRS GNM : qqq
DDB VDV : ikl	LDR HDE : kkk	SJM SPO : abc	XOI GUK : qwe
EJP IPS : vbn	MAW UXP : yyy	SUG SMF : asd	XYW GCP : qay
FBR KLE : hjk	NXD QTU : ggg	TMN EBY : ppp	YPC OSQ : mmm
GPB ZSV : nml	NLU QFZ : ghj	TAA EXB : pyx	ZZY YRA : uvw
HNO THD : fff	OBU DLZ : jjj	USE NWH : zui	ZEF YOC : uio
HXV TTI : fgh	PVJ FEG : tzu	VII PZK : eee	ZSJ YWG : uuu

Fig. 151. 40 different indicators decrypted

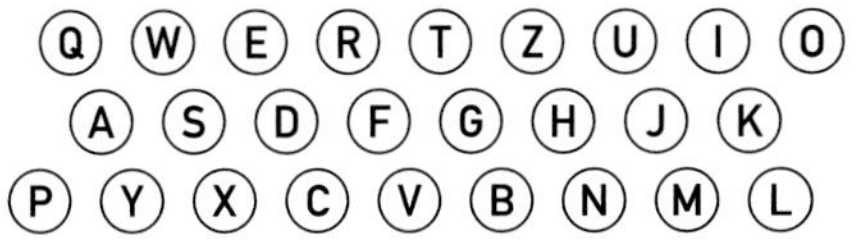

Fig. 152. Keyboard of the ENIGMA

19.6.2.4 The Polish bureau certainly learned something also from the content of the decrypted signals. But much more important was that the compromise had endangered the wiring of the rotors. Since the indicator analysis involved only the first six letters, it was mostly only the fast rotor R_N (the rightmost one) which was moved, and the two other ENIGMA rotors remained in 20 out of 26 cases at rest. This, together with the material from ASCHE, was sufficient for Rejewski in December 1932 to reconstruct the wiring of the fast rotor core, and since the rotor order at that time was changed every quarter (from 1936 every month, later every day), each rotor came finally under examination by the Polish Biuro Szyfrów. Once all the plain indicators of a day were decrypted, all the signals that day could be decrypted with the help of Polish ENIGMA replicas. But what about the next day?

19.6.2.5 Reconstruction of the basic wheel setting (German *Grundstellung*) was accomplished with the help of a theorem of group theory, which was highlighted by Deavours as "the theorem that won World War II". It says

S and TST^{-1} have the same cycle decomposition ('characteristic').

Therefore, Marian Rejewski and his co-workers made use of the fact that the cycle lengths in the three observable $P_i P_{i+3}$ are independent of the choice of the cross-plugging (and of the ring setting anyway). The number of essentially different cycle arrangments is the number of partitions of $26/2 = 13$, which is 101; three such partitions—in the example above the partitions ('characteristics') 10+2+1, 9+3+1, 13—in general characterize uniquely the $6 \times 26 \times 26 \times 26 \approx 10^5$ basic wheel settings. Rejewski, supported by Różycki and Zygalski, was now able to produce with the help of the ENIGMA replica a catalogue for every rotor order containing the partitions of the cycles for all basic wheel settings. For this purpose, an electromechanical device called the 'cyclometer' was built in the factory AVA in Stepinska Street, Warsaw. The Biuro Szyfrów finished the catalogue in 1937; to find the *Tagesschlüssel* then took no longer then 10–20 minutes. Unfortunately for the Poles, on November 1, 1937 the Germans changed the reflecting rotor.

19.6.2.6 There remained the problem of finding the 'right' ring setting on the rotor core. The exhaustive treatment could be simplified by an observation Rejewski had made in 1932, thanks to the material of ASCHE: most plaintexts started with /anx/, where /x/ replaced the word space. According to Kerckhoffs' admonition, it had to be expected that the machine was in the wrong hands, so it was pretty silly to use a stereotyped beginning. We shall resume this trivial case of a plaintext-cryptotext compromise in Sect. 19.7.

19.6.2.7 The Poles also used for a while a method they called *metoda rusztu* ('grill method', 'grid method', 'grate method'), which was "manual and tedious" as Kozaczuk says. It was usable only as long as the number of cross-pluggings was small (six up to October 1, 1936) and served to determine the ring setting of the fast rotor. Details were published in 1979 by Garlinski.

19.6.3 Poland II. All this success was only possible because of the *properly* self-reciprocal character[12] of the ENIGMA rotor encryption; the reflecting rotor of Scherbius and Korn turned out to be a grandiose illusory complication.

[12] The simply self-reciprocal encryption machines of Boris Hagelin did not suffer from this defect—nevertheless the M-209 was broken from 1942 by the Germans in North Africa.

In 1937, introduction of bigram tables stopped reading the *Marine* ENIGMA. In 1938, the situation was aggravated. The Germans changed the encryption procedure on September 15, and introduced on December 15 a fourth and a fifth rotor, giving $60 = 5\times 4\times 3$ instead of $6 = 3\times 2\times 1$ possible rotor orders.

19.6.3.1 The Poles had to find out the wiring of the new rotors quickly, and they were lucky. Among the traffic they regularly decrypted were signals from the S.D. (*Sicherheitsdienst*), the intelligence service of the Nazi Party. The S.D. did not change their encryption procedure, but introduced the new rotors in December 1938. These rotors came from time to time into the position of the fast rotor and their wiring could be reconstructed the same way as previously with the first three rotors.

The use of two methods, one of which possibly was compromised, was a grave error.

As an aside, there is a story of how B.S.-4 came to read the S.D. signals. The *Sicherheitsdienst* officers were distrustful of everybody and encoded their messages by hand before giving them to an ENIGMA operator for superencryption. The Poles, decrypting all ENIGMA traffic, obtained meaningless text and thought at first the cryptotext was encrypted in a different system. Then, one day in 1937, the three letter word /ein/ was read. This could only mean that a plaintext group by mistake was mixed with a code; probably the numeral 1 had not been transcribed and the ENIGMA operator knew no better than to send /ein/ instead. The Poles then found it easy to break the simple hand encryption.

19.6.3.2 The new encryption procedure, valid until the end of April 1940, did not use the same ground setting for all messages of the day, but for each message an arbitrary ground setting was to be chosen, which should precede the signal in plain. With this chosen ground setting (in B.P. called 'indicator setting'), as before, a randomly chosen plain indicator, still doubled, was to be encrypted and also used as message setting for the encryption of the text.

To give an example, for a signal beginning (after the plain preamble) with RTJWA HWIK..... , rtj is the ground setting, WAH WIK is the encrypted doubled indicator, encrypted with rtj . For this situation, we shall write in the sequel rtj | WAH WIK .

The authorized recipient uses the ground setting rtj to find from WAH WIK the plain indicator doublet (which has the pattern *123123*); then decrypts the cryptotext with the first three letters (the true indicator) as text setting.

As long as ring setting and rotor order had not fallen into the wrong hands, the foe could do nothing with the openly displayed ground setting. The search space still contained resp.105 456 or 1 054 560 possibilities: 26^3 ring settings, 6 rotor orders, in addition by December 1938 10 choices of 3 rotors out of 5.

19.6.3.3 The methods Rejewski and his friends had used so far did not work any longer, since they were based on the multiple use of the same ground setting for a full day. But the Germans, almost incredibly, kept the doubling

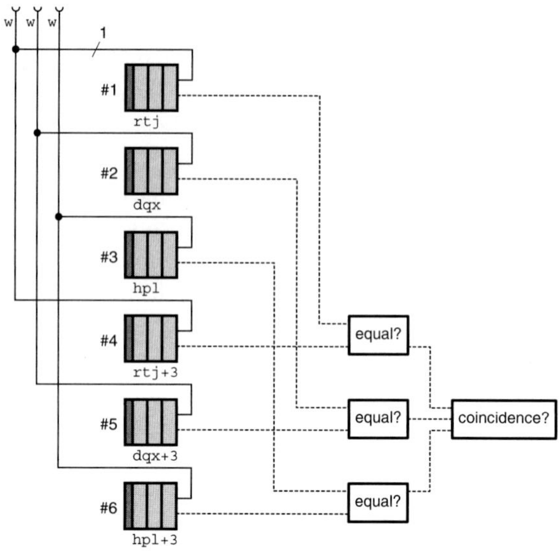

Fig. 153. Abstract function of a Polish *bomba* (October 1938)

of the plain indicator[13] and thus allowed the attack of searching for a pattern, i.e., for the pattern *123123* , at a known position. This method would have worked before, too, but in the fall of 1938, there was no choice but to go to the trouble of much more work. The Poles therefore thought of mechanisation. Rejewski ordered six machines from the AVA factory in October 1938, each one simulating one of the six rotor orders, and tested parallel on them the 17 576 positions of the rotor core (in BP called 'rod-positions'), which needed at most 110 minutes.

The 'right' ring setting on the rotor core was found using the 1-cycles in the following way. The machine was built from three pairs of ENIGMA rotor sets. In each pair the rod-positions of all rotors were shifted by three; the position of the rotor sets of the first pair was shifted by one against the position of the rotor sets of the second pair, which in turn was shifted by one against the position of the rotor sets of the third pair.

As soon as there was enough material to provide three encrypted doubled indicators such that the same character appeared once in the first and the fourth, once in the second and the fifth, and once in the third and the sixth position—like the letter W in (the example goes back to Rejewski)

rtj	WAH WIK
dqx	DWJ MWR
hpl	RAW KTW

—and thus as in Sect. 19.6.2.1 exhibited a 1-cycle ('fixpoint'), a promising attack was possible (Fig. 153). The machine was started with the three initial

[13] The double encipherment of each text setting continued, according to Peter Twinn, for the 4-rotor *Abwehr* ENIGMA until the end of the war.

settings rtj, dqx and hpl, and the test character W was input repeatedly until in each one of the three pairs the same character occurred, i.e., the pattern *123123* was found. Such a coincidence triggered a simple relay circuit to stop the whole machine, whose appearance led the Poles to call it the *bomba*[14]. Of course, sometimes there were mishits.

If in this way the core position was revealed, a comparison of the encrypted doubled indicators with the encryptions produced by the ENIGMA replica reconstructed the ring setting and the cross-plugging. In this way, all signals of the day (later of the 8-hour shift) from one and the same key net (during the war there were up to 120 such key nets) could be decrypted.

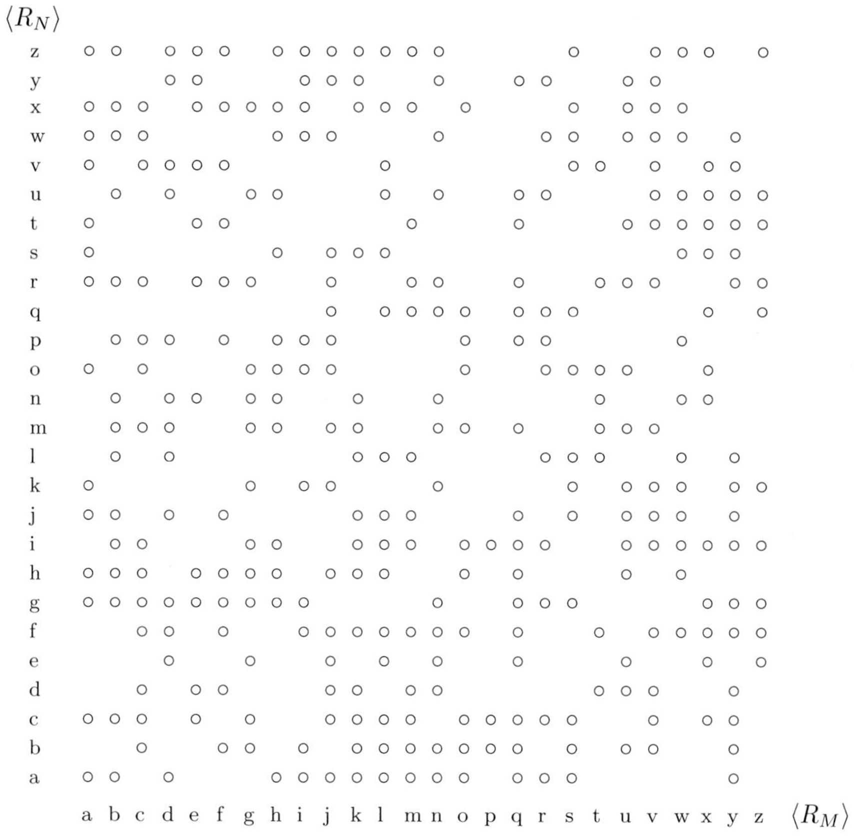

Fig. 154. Zygalski sheet K_{14}^{413} for rotor order IV-I-III, $\langle R_L \rangle = $ k ; holes show possible 1-cycles (fixpoints) of $P_1 P_4$ for ground setting $\langle R_L \rangle \langle R_M \rangle \langle R_N \rangle$

The *bomba* was sensitive to cross-pluggings, and the method worked only if the test character (in the example above the letter W) was 'unplugged'. The likelihood for this to happen was around 50%, as long as five to eight

[14] According to Tadeusz Lisicki, it was originally named by Jerzy Różycki after an ice-cream bombe. While they were eating it, the idea for the machine came to him and his friends.

plugs were used. If three different fixpoints had been allowed, as the British originally had in mind, the likelihood would have dropped to 12.5%.

19.6.3.4 For overcoming the difficulties, another kind of mechanizing was developed in autumn 1938 by Henryk Zygalski (1907–1978), namely a 'punch card catalog' for determining the *Tagesschlüssel* from about ten to twelve (arbitrary) fixpoints. For the six rotor orders it was calculated for P_1P_4, P_2P_5 or P_3P_6, whether for a ground setting $\langle R_L\rangle\langle R_M\rangle\langle R_N\rangle$ of the rotors R_L, R_M, R_N some fixpoint is possible at all. For each one of the 26 letters $\langle R_L\rangle$ this was recorded in a $\langle R_M\rangle\times\langle R_N\rangle$ matrix by a punched hole ('*female*') (Fig. 154); roughly 40% of the squares on a sheet contained holes. In fact, to allow full overlay, sheets of 51×51 fields, made by horizontal and vertical duplication, were used. By superimposing the sheets aligned according to their ground setting $\langle R_M\rangle\langle R_N\rangle$, the core position was determined, as a rule uniquely as soon as about ten to twelve fixpoints were available. Most important, the method was insensitive to the cross-plugging used and was still useful when ten plugs were used after August 19, 1939—as long as the double encipherment lasted.

The Germans had always tried by suitable encryption security not to become victims of a Kerckhoffs superimposition, and finally became victims of a trivial weakness, the doubling of the indicator.

19.6.4 Great Britain. In a meeting on January 9, 1939 in Paris, the Polish Lieutenant Colonel Gwido Langer (1894–1948) supplemented his French connections by contacts with his British colleagues. With increasing danger of war, closer cooperation was indicated. The result was a meeting on July 24–25, 1939 in Warsaw of Alfred Dillwyn ('Dilly') Knox, the leading British cryptanalyst in the Foreign Office, his boss Alastair Denniston, the head of the Government Code and Cypher School (GC&CS) and the mysterious 'Mr. Sandwich' (Commander Humphrey Sandwith) with the French Commandant Gustave Bertrand and Capitaine Henri Braquenié, and with the Polish side, represented by Ciężki, Langer, and the Grand Chef Colonel Stefan Mayer. Rejewski, Różycki, and Zygalski proudly presented all their results in Pyry, to the south of Warsaw. At this occasion, the French as well as the British also obtained Polish replicas of the ENIGMA with all its five rotors.

Since the crisis that led to the Munich Conference of September 1938 the British had looked for somewhere to evacuate their cryptological service, to insiders known as 'Room 47' of the Foreign Office, that worked from the address 56 Broadway (Whitehall), Westminster. They found it in *Bletchley Park* (BP for short, radio codename 'Station X'), geographically well located about fifty miles north of London. Before war broke out in 1939, the GC&CS was established there and reinforced. Among its many duties was decryption of the ENIGMA, and the group around Knox and Turing made good use of Polish punched sheets which they called 'canvasses' or 'Jeffreys sheets' after John Jeffreys (dec. 1940), who supervised their preparation; they were ready in January 1940. However, for the Polish *bomba* a development was necessary in order to cope with now 60 instead of formerly 6 rotor orders.

Oliver Strachey[15] had several times arranged contacts between the young Alan Mathison Turing, who already had a reputation as a logician and had been interested since childhood in cryptology, and the GC&CS. The head of the cryptanalytic service, Knox, was a classics scholar, who in 1915 had preferred Room 40 of the Admiralty to a Fellowship at King's College in Cambridge, and already had experience with the commercial ENIGMA used by the Italians. On September 4, 1939, the second day of the Second World War, Turing reported to BP. He worked on a further development of the Polish *bomba* and was joined in this by Gordon Welchman (1906–1985), who also arrived September 4. Turing had experience with relay circuitry (Sect. 5.7.3) and thus was not merely theoretically interested in cryptology. His contacts with GC&CS may have reached back to 1936.

In January 1940, BP for the first time broke an ENIGMA key, the key RED for Jan. 6 of the carelessly transmitting *Luftwaffe*, and continued to do so.

19.6.4.1 When in mid-January 1940 Turing met Rejewski, who had fled to France, in Gretz-Armainvilliers, north of Paris, he was, according to Rejewski, very interested in the Polish ideas for defeating the cross-plugging. By then, he had arrived at his own ideas (see Sect. 19.7), but of course he could not mention how far he had come. It was only natural that Turing tried to improve the Polish *bomby* to make them insensitive to cross-plugging like the Zygalski sheets. The British, like the Poles, were afraid that with fewer and fewer *self-steckered* letters their methods would soon become useless. Thus, Turing wanted to get rid of the restriction to 'self-steckered' letters.

He wanted late in 1939, as Joan Murray née Clarke (1917–1996) remembers the argument, to test all 26 letters in parallel to see what output they would have; this would allow a 'simultaneous scanning' of all 26 possibilities of the test letter. Thus, Turing thought of replacing the Scherbius rotors by 'Turing rotors', each one having both on the entry side and exit side *two* concentric rings of contacts—one for the journey towards the reflector, one for the return journey—both mimicking the same ENIGMA rotor. Likewise, the reflector would have two rings of contacts. This modification would have input and output by 26 wires in parallel, and result in a *double-ended scrambler* (Welchman, U.S. Jargon *'commutator'*), really representing a classical ENIGMA substitution $P_i = S_i U S_i^{-1}$ for $i = 1 \ldots 26^3$. In accordance with its self-reciprocal character, the scrambler had to be input-output symmetrical; this was provided for by a symmetric wiring between the contacts of the inner and outer ring of the reflector.

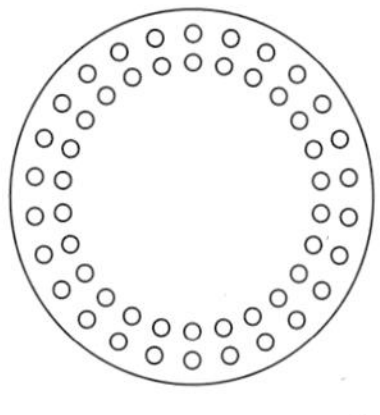

In this Turing version, the Polish *bomba* amounted to three closed cycles, each built from two double-ended scramblers; one such cycle is shown in

[15] Oliver Strachey, husband of the feminist Ray Strachey, father of the computer scientist Christopher Strachey, and brother of the writer Lytton Strachey, replaced in 1941 in the Canadian services the former U.S. Major Herbert Osborne Yardley, who had fallen into disgrace in the United States.

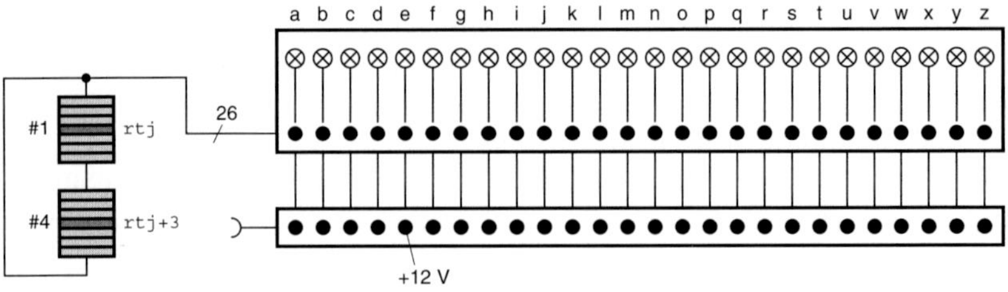

Fig. 155. Hypothetical Turing version of the Polish *bomba* (with '*simultaneous scanning*'). Diagram of one of the three cycles

Fig. 155 . Turing had thus managed to strip off the superencryption by the cross-plugging mechanically. And he recognized that the 1-cycles, the said females of the Zygalski sheets, being natural fixpoints of a mapping, could be determined by an iterative feedback process, which normally diverged and thus indicated that the rotor positions in question did not allow a fixpoint; if it did not diverge, it gave the fixpoint. The logician Turing, familiar with the *reductio ad absurdum*, thus turned to the general principle of feedback.

Technically, the distinction between the divergent and the non-divergent case was made by a 'test register' attached to the 26-line bus of the feedback (in Fig. 155 from #4 to #1). Voltage is applied to the wire belonging to the test letter (W in our example). In the divergent case all light bulbs of the test register light. In the non-divergent case the feedback cycle of the fixpoint is electrically isolated from the remaining wiring; correspondingly either exactly one light bulb (the one belonging to W) is lit, or all light bulbs but this one are, depending on whether the cross-plugging was correctly chosen or not.

A battery of double-ended scramblers was to be moved simultaneously. Thus Turing could have simulated a Polish *bomba*. But the actual development took a different, much more general path (Sect. 19.7).

In the last quarter of 1939, Turing's design had progressed far enough that Bletchley Park was allowed to ask the British Tabulating Machine Company in Letchworth to build a machine, which was also called BOMBE. Harold 'Doc' Keen, with a crew of twelve people, finished it by March 1940. Later, *Keen* was equally successful in building the 4-rotor-bombe MAMMOTH.

Welchman, by the way, late in 1939 when he was still a novice, arrived independently at similar conclusions, although at first he was not involved in the ENIGMA decryption by machines. He also reinvented the Zygalski sheets, not knowing that John Jeffreys in another building already had a production line going. Likewise, he did not know of Turing's ideas, but this was intended.

19.6.4.2 Turing may have already thought before the Pyry meeting of making use of probable words to break into ENIGMA. After this meeting, having heard about the *bomba*, he turned his thoughts to mechanizing his method. The major advantage with the device Turing had in mind was that it not only found the rotor order, like the Zygalski sheet, but it also found at least one *stecker*.

Presumably, Turing's ideas and precautions were guided by Knox's fear that the Germans would again change their encryption procedure and give up the indicator doubling. Then, having learned in the meantime from decryptions a lot on the habits and styles of the Germans, the British hoped to be able to produce the necessary feedbacks efficiently with probable words the Germans used so plentifully: /wettervorhersage biskaya/, /wettervorhersage deutsche bucht/ etc., or /obersturmbannführer/, obergruppenführer/ etc., or /keine besonderen ereignisse/. Thus, BP was prepared when on May 1, 1940, shortly before the campaign in France, the next change came: *Heer* and *Luftwaffe* dropped the indicator doubling and put *bomba* and Zygalski-Jeffreys sheets out of action. (The Kriegsmarine superencrypted the message indicator with a bigram substitution, see Sect. 4.1.2). The Turing BOMBE prototype 'Victory' was operational by March 18. Starting August 8, 1940, regular 'spider', nicknamed 'Agnes', 'Jumbo', 'Funf' followed; 'Ming' was ready by May 26, 1941. (More about the mode of operation in Sect. 19.7.) "It was he who first formulated the principle of mechanizing a search for logical consistency based on a *probable word*" (*Andrew Hodges*). Moreover, Turing had designed the BOMBE in a way that allowed universal working.

19.6.4.3 If, however, the probable word method would not work, there was still the fundamental possibility of an in-phase superimposition (Sect. 19.3). This made adjustment necessary and thus something like a search for repetitions or a coincidence count. The mechanism used for performing this was called a 'Banburism' because the long sheets of paper containing the message in a 1-out-of-26 code were produced in Banbury, a little town near Oxford. Turing developed for this purpose a particular scoring of the parallels, which means weighing the repetitions, where for example two bigram repetitions favored adjustment more than four monogram repetitions. Turing used a logarithmic unit [ban] (Chapter 12), a decimal counterpart to Shannon's binary unit of information [bit]; 1 [ban] $\hat{=}$ ${}^2\log 10$ [bit]. 1 [deciban] $\approx$ 0.332 [bit], the practical unit used in Bletchley Park, corresponds to 1 [dB].

Turing and Shannon obviously developed their ideas independently of each other, and met only around the end of 1942. "Turing and I never talked about cryptography" (Claude Shannon). It could well be that before July 1941 nobody in Bletchley Park knew the *Kappa* test of Friedman (Sect. 17.1) or the *Phi* test of Kullback (Sect. 17.5). Despite I. J. Good's attempts, it remains unclear how Turing found what he called 'repetition frequency' in his *Treatise on the Enigma* written in late summer or early autumn 1940.

Banburism was actually an elaboration of a method the Poles had already used, the 'clock method' of Jerzy Różycki (1909–1942). The adjustment of the cryptotexts resulted in a difference in the phase of the plain indicators. Thus, at least the fast rotor R_N (the initial setting of which was indicated by the third letter of the text setting) was revealed. Deavours and Kruh give the following example: From several pairs of in-phase messages, the third position of the encrypted indicator showed the following observed values:

first message	R	F	D	T	N	M	K	B	M	N
second message	F	K	Y	Y	Q	Q	O	Y	C	K
Difference of phases *modulo* 26	07	12	04	02	14	21	06	11	06	03

These data form two chains:

1	2	3	4	5	6	7	8	9	10	11	12	13	14	15	16	17	18	19	20	21	22	23	24	25	26
R				Q			F		M						C	N			K						O
D		T		Y															B						

Sliding the plaintext alphabet along the first chain, a self-reciprocal pair, namely (k n) , is found for the following shift:

a	b	c	d	e	f	g	h	i	j	k	l	m	n	o	p	q	r	s	t	u	v	w	x	y	z
	F		M						C	N			K						O	R				Q	

Using all possibilities for self-reciprocal pairs, the following hull is obtained

a	b	c	d	e	f	g	h	i	j	k	l	m	n	o	p	q	r	s	t	u	v	w	x	y	z
	F	J	M		B				C	N		D	K	T		Y	U		O	R				Q	

which is consistent with the second chain

a	b	c	d	e	f	g	h	i	j	k	l	m	n	o	p	q	r	s	t	u	v	w	x	y	z
					B							D		T		Y									

Thus, fourteen starting positions of the fast rotor are revealed.

In this way, for the rotors I ... V the position of the one notch (Sect. 8.5.3) could be determined, and thus which rotor was used as fast rotor. This helped in the Polish *bomba* and in the British BOMBE to reduce the number of rotor orders to be tested from 60 to $12 = 4 \times 3$.

19.6.5 France II. Rejewski, Zygalski and Różycki, escaping the Polish disaster, fled via Rumania to France. End of September 1939, they joined the French radio intelligence group under Commandant Bertrand in the *Château de Vignolles* near Gretz-Armainvilliers (cover name *Poste de Commandement* Bruno), 30 miles southeast of Paris. Until the German attack (‚Fall Gelb') on France in May 1940, 'Group Z' worked with Zygalski sheets made in Bletchley Park, and solved nearly the same number of signals as BP, mainly German Army administration (key net GREEN: on January 17, 1940 the messages of October 28, 1939) and *Luftwaffe* (key net RED: around January 25, 1940 the messages of January 6, 1940), then after the Norway invasion the signals of *Fliegerführer* Trondheim (key net YELLOW: starting April 10, 1940).

After the collapse of France, P.C. Bruno was first transferred on June 24, 1940 to Oran in Algeria, then in October 1940 transferred back to the *Château des Fouzes* near Uzès (cover name 'Cadix') in the unoccupied part of France. Since the Zygalski sheets had become useless by May 1, 1940, Poles and British had to rely for a while on Cillis and Herivel tips (Sect. 19.7). The Polish unit under Lieutenant-Colonel Gwido Langer, named 'Expositur 3000' by the British, was evacuated November 9, 1942, after the landing of the Allies in North Afrika and the German occupation of the rest of France; Rejewski und Zygalski were imprisoned for a while in Spain and reached London via Gibraltar on August 3, 1943. They continued with cryptanalytic work on hand ciphers; however, they were kept away from Bletchley Park with the Turing–Welchman Bombes and the Colossus machines.

19.7 Plaintext-Cryptotext Compromise: Feedback Cycle

The ENIGMA system was practically open for the British in May 1940. A probable word attack, a plaintext-cryptotext compromise, could profit from the sharpness of the Turing (and Welchman) feedback pursued since 1939. And Turing had designed the British BOMBE in a way that allowed this. Thus, a method was resumed that Rejewski had used in 1932 (see Sect. 19.6.2.6). So, systematically, this chapter belongs at the end of Chap. 14.

The British faced to be forced to prepare the menu day by day, and they found enough cribs to do so. Meanwhile, they were helped by continuing violations of even the simplest rules of crypto security on the German side. John Herivel observed in May 1940 that for the first signal of the day the ground setting was frequently very close to (if it did not actually coincide with) the position of the wheels for the ring setting of the day ('Herivel tip'). Moreover, the use of stereotyped indicators continued, which the British placed under the heading 'Cillis'[16]. There was also the abuse of taking the ground setting as indicator, called JABJAB by Dennis Babbage. When the German supervisors finally reacted, the damage was already irreparable.

Crypto security discipline was lowest in the Air Force of the pompous parvenu Göring. From May 26, 1940 on, before the Turing-Welchman-BOMBE was working, mathematicians and linguists in BP regularly managed to read the ENIGMA signals of the *Luftwaffe* (key net RED), while for the signals of the *Kriegsmarine* (key net DOLPHIN, „Heimische Gewässer", later „Hydra") they had to wait until June 1941 before they had mastered the bigram superencryption of the message keys. In December 1940, they succeeded in breaking into the radio signals of the SS (key net ORANGE). From September 1942 on, Field Marshal Rommel's ENIGMA traffic with Berlin (net CHAFFINCH) was no longer secure, and from mid-1942 on the British achieved deep and lasting breaks, above all in the heavy *Luftwaffe* traffic (key net WASP of *Fliegerkorps* IX, GADFLY of *Fliegerkorps* X, HORNET of *Fliegerkorps* IV, SCORPION of *Fliegerführer Afrika*). Most obstinate, to British judgement, was the radio communication of the German *Heer*, which was a consequence of the solid training of the operators. Before the spring of 1942, no ENIGMA traffic line of the *Heer* except one, VULTURE I in Russia (June 1941) was broken.

19.7.1 Turing BOMBE. In the general probable word attack, Turing (and in parallel Welchman) used instead of the three isolated, two-fold cycles of the *bomba* a whole system of *feedback cycles* formed by a battery of first 10 and later 12 double-ended scramblers. Such feedback cycle systems, which are *independent of the steckering*, are obtained from a juxtaposition of a probable word and a fragment of the cryptotext. Fortunately, for long enough probable words the non-coincidence exhaustion method (Sect. 14.1) allows one to exclude many positions, furthermore conspicuous probable words are

---

[16] Sometimes interpreted 'sillies'. Welchkman: 'I have no idea how the term arose'.

often at the beginning or the end of the message (unless Russian copulation has been used). Therefore, it is not unrealistic to establish a new feedback cycle system for every juxtaposition; there are not too many.

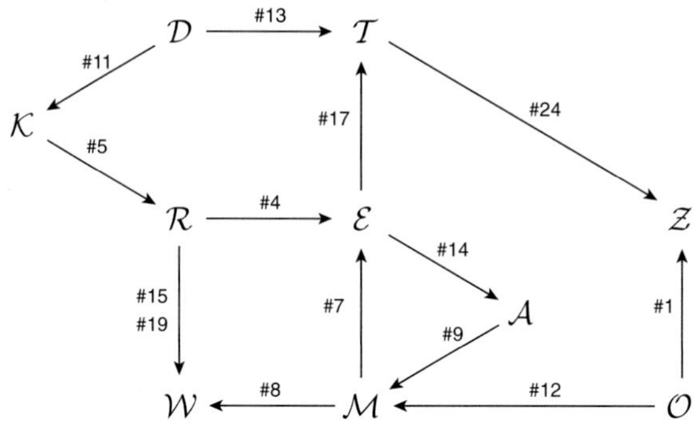

Fig. 156. Plaintext letter / ciphertext letter pairings for crib
o b e r k o m m a n d o d e r w e h r m a c h t
Z M G E R F E W M L K M T A W X T S W V U I N Z

The following example[17] goes back to C. A. Deavours and L. Kruh. Let

OVRLJ BZMGE RFEWM LKMTA WXTSW VUINZ GYOLY FMKMS GOFTU EIU...

be the cryptotext and /oberkommandoderwehrmacht/ the probable word. The third leftmost position not excluded by non-coincidence gives the 'crib'

						1	2	3	4	5	6	7	8	9	10	11	12	13	14	15	16	17	18	19	20	21	22	23	24
						o	b	e	r	k	o	m	m	a	n	d	o	d	e	r	w	e	h	r	m	a	c	h	t
O	V	R	L	J	B	Z	M	G	E	R	F	E	W	M	L	K	M	T	A	W	X	T	S	W	V	U	I	N	Z

with 24 double-ended scramblers numbered from #1 to #24 .

The plaintext letter / ciphertext letter pairings can be compressed into a directed graph, shown in Fig. 156 . The self-reciprocal character of the double-ended scramblers, reflected in the electrical connection of their inputs and outputs, means a transition to an undirected graph. From this graph, a subgraph may be selected, in jargon a 'menu'—for our example the graph with eight nodes shown in Fig. 157 at the upper right corner. Each cycle (in Turing's jargon 'closure') in this subgraph establishes a feedback in the Turing BOMBE setup. A menu with 6 letters and 4 cycles is of course more lucrative than one with 12 letters and one cycle: it reduces the danger of mishits.
Corresponding to such a subgraph with 10 transitions, 10 double-ended scramblers are now connected (with 26-line buses) and a test register is connected, say at $\mathcal{E}$ (Fig. 157). To some entry, say e, voltage is applied.

The positions 14, 9, 7 form a cycle ('closure'): denoting the internal contacts with $a, b, c, \ldots y, z$, the cross-plugging with T, and the substitution performed by scrambler #i with P_i, we obtain the relations

[17] Rotor order IV I II, reflector B . Ring setting 000, cross-plugging VO WN CR TY PJ QI . Text setting tgv .

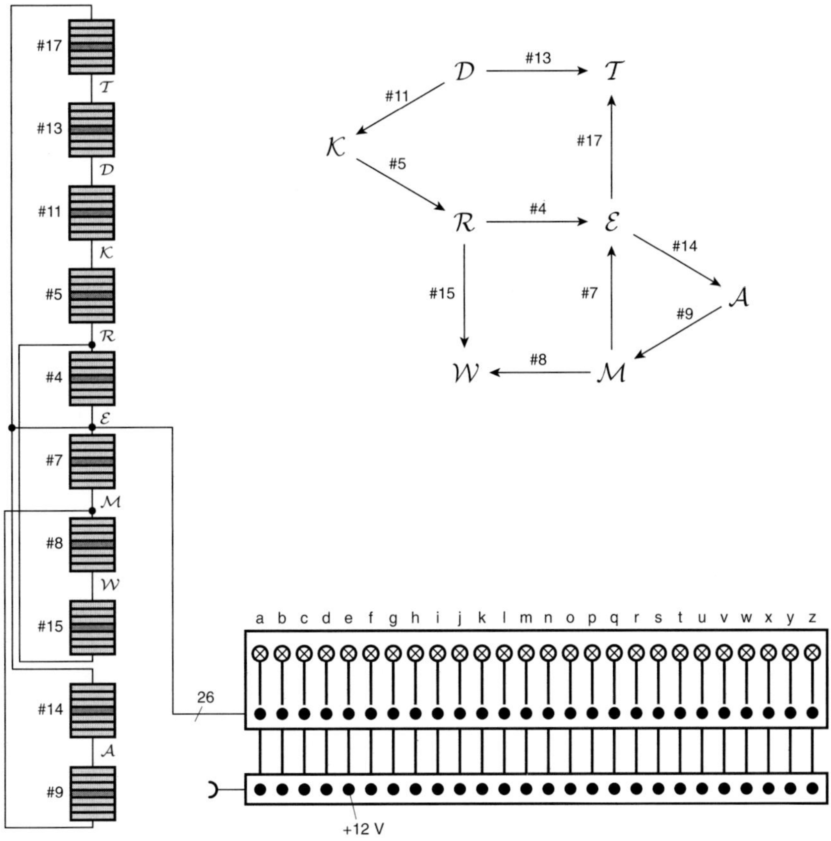

Fig. 157. Turing BOMBE setup for feedback cycle system of Fig. 156

$eT = mTP_7$, $mT = aTP_9$, $aT = eTP_{14}$, or $eT = eTP_{14}P_9P_7$.

Thus, eT is a fixpoint of $P_{14}P_9P_7$.

But the positions 4, 15, 8, 7 form also a cycle: at first we have the relations

$eT = rTP_4$, $wT = rTP_{15}$, $wT = mTP_8$, $eT = mTP_7$;

since the scrambler substitutions are self-reciprocal, we obtain

$eT = mTP_7$, $mT = wTP_8$, $wT = rTP_{15}$, $rT = eTP_4$, or
$eT = eTP_4P_{15}P_8P_7$. Thus, eT is also a fixpoint of $P_4P_{15}P_8P_7$.

Moreover, the positions 4, 5, 11, 13, 17 form a cycle: using again that the scrambler substitutions are self-reciprocal, we obtain the relations

$eT = rTP_4$, $rT = kTP_5$, $kT = dTP_{11}$, $dT = tTP_{13}$, $tT = eTP_{17}$.

Thus, $P_{17}P_{13}P_{11}P_5P_4$ also has the fixpoint eT .

Assume that the position of the scramblers is not the 'right' one. Then (normally, i.e., if enough cycles exist) the voltage spreads over the total system and all the test register light bulbs are lit. A relay circuit discovers this divergent case and moves the scramblers on to the next position.

Now assume that the position of the scramblers is the 'right' one, i.e., the one used for encryption (such that the scrambler #4 maps $rT{=}/\mathrm{r}/$ into $eT{=}\mathrm{E}$). Then there are two subcases. If the cross-plugging is correctly chosen, i.e., the entry e to which voltage is applied is /e/, then the voltage does not spread, and apart from the light bulb belonging to e no lamp is lit. If, however, the cross-plugging is not correctly chosen, then the voltage (normally, i.e., if enough cycles exist) spreads over the whole remaining system and all lamps are lit except one, which indicates the cross-plugging. In both of these convergent subcases, the machine setting and the light bulb indication can be noted down. The scrambler position determines the ring setting. It can be a mishit. This can be quickly decided by using the resulting setting to try to decrypt the surrounding text.

The possibility of Turing's feedback cycle attack was totally overlooked by Gisbert Hasenjäger (of age 23), responsible for the security of the ENIGMA in the *Referat* IVa, Security of own ciphers (Karl Stein) of the Cipher Branch OKW. This attack, as was shown above, is strongly supported by the properly self-reciprocal character of the ENIGMA encryption; however, it would also work in principle for non-selfreciprocal double-ended scramblers, although such cycles occur much less frequently. For example, the only true cycle in Fig. 156 is the cycle

7	9	14
m	a	e
E	M	A

and very long probable words would be needed to make the attack succeed, or a larger menu would be needed. For example, in the feedback cycle system of Fig. 156, the crib would allow one to adjoin a node $\mathcal{U}$ connected with $\mathcal{A}$, or a node $\mathcal{V}$ connected with $\mathcal{M}$, or a few more. There is, compared to Fig. 157, even one cycle more in the example: from $\mathcal{T}$ over $\mathcal{Z}$, $\mathcal{O}$, $\mathcal{M}$ and $\mathcal{A}$ to $\mathcal{T}$. But this would increase the number of scramblers needed in the setup.

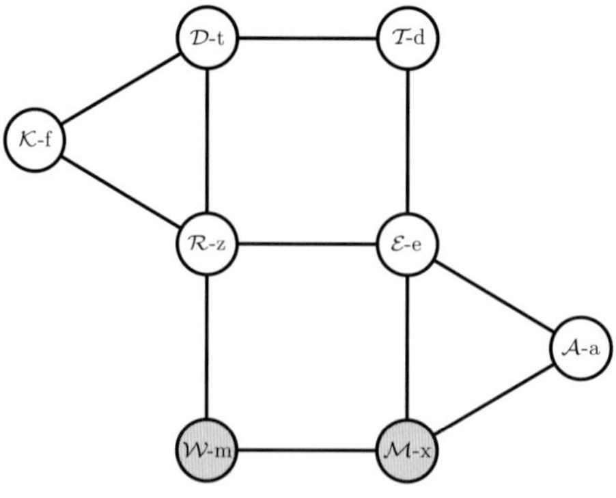

Fig. 158. Contradictory halting configuration

Gordon Welchman

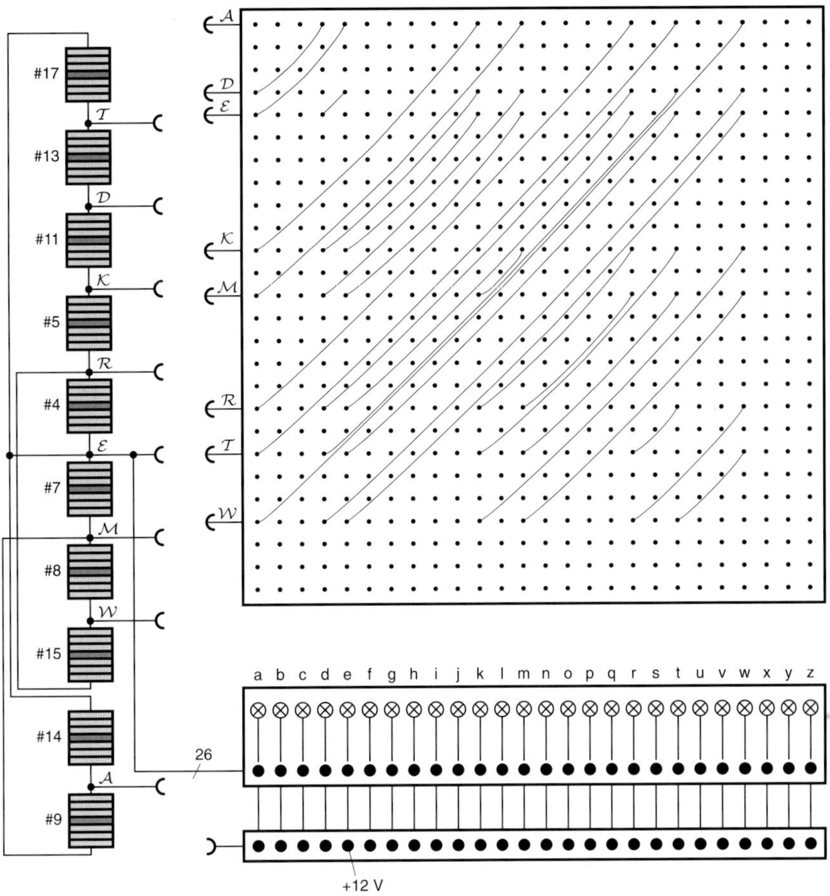

Fig. 159a. Welchman BOMBE setup for feedback cycle system of Fig. 156

19.7.2 Turing-Welchman BOMBE. Gordon Welchman improved the Turing feedback cycle attack quite decisively by taking all relations effected by the self-reciprocal character of the typical ENIGMA cross-plugging explicitly into account. Whenever Turing's BOMBE stopped, the nodes like $\mathcal{A}$, $\mathcal{D}$, $\mathcal{E}$, $\mathcal{K}$ and so on were assigned certain internal contacts. Figure 158 shows such a halting configuration, showing two 'self-steckered' interpretations $\mathcal{A}$-a, $\mathcal{E}$-e. The two interpretations $\mathcal{D}$-t, $\mathcal{T}$-d indicate a cross-plugging t-d. The two interpretations $\mathcal{W}$-m and $\mathcal{M}$-x, however, contradict the involutory character of the cross-plugging. The BOMBE should not have halted in such a configuration; the contradiction found by this reasoning should have caused divergence inside the BOMBE's electrical wiring.

Welchman found in November 1939 a simple electrical realization of such a 'forming the hull with respect to cross-plugging', the 'diagonal board' shown in Fig. 159a. Its functioning is explained in Fig. 159b:

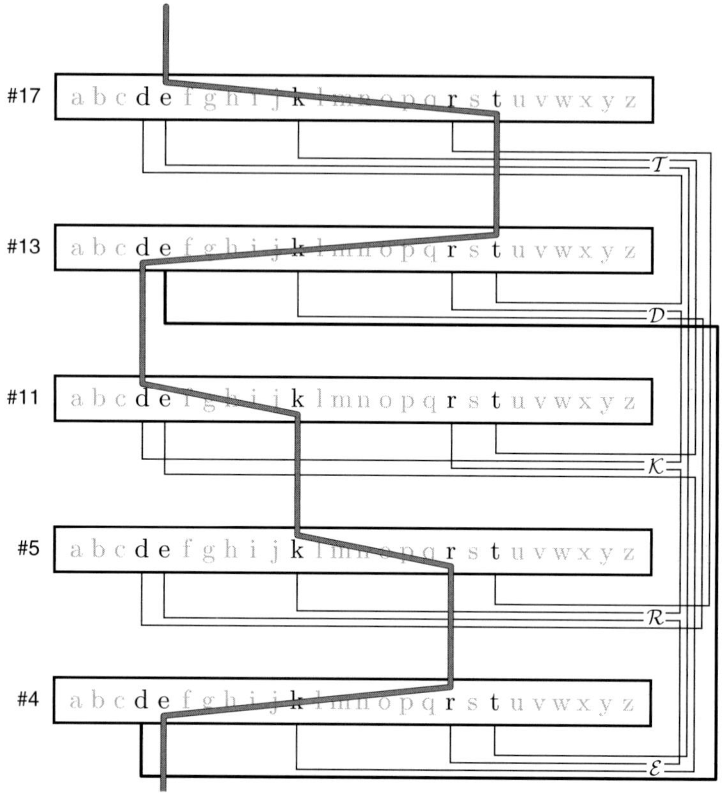

Fig. 159b. Functioning of the 'diagonal board' of Welchman for the example of Fig. 156

Assume $eT = dT\,P_{11}P_5P_4$. The bold overlay in the wiring shows how the electrical connection made by the scramblers from e in bus $\mathcal{E}$ to d in bus $\mathcal{D}$ is supplemented by the diagonal board with a fixed electrical connection from d in bus $\mathcal{E}$ to e in bus $\mathcal{D}$.

With Welchman's improvement, Turing's feedback cycle attack attained its full power and the efficiency of the BOMBE increased dramatically. Many fewer cycles were needed to fill the test register. This not only helped to save scramblers, it also allowed shorter cribs and thus increased the chance that the middle rotor remained at rest. So Welchman is the true hero of the BOMBE story in Bletchley Park. Devours and Kruh (1985) formulated it in a way that may console Hasenjäger:

> "It is doubtful that anyone else would have thought of Welchman's idea because most persons, including Turing, were initially incredulous when Welchman explained his concept."

19.7.3 More BOMBEs. People in Bletchley Park that called the aggregate of scramblers together with the test register and the diagonal board a 'bomb' were not aware of the Polish origin of name and idea. 'Agnes'[18], the first Turing–Welchman production BOMBE after the prototype 'Victory', (which did not yet have a diagonal board) was ready by mid-August 1940, it needed about 15 minutes for a complete exhaustion of one wheel order. In the spring of 1941 ('Ming': End of May 1941) there were 8 BOMBEs at work, and 12 towards the end of the year, built by the British Tabulating Machine Company in Letchworth. The number increased rapidly to 30 in August 1942, 60 in March 1943, and 200 at the end of the Second World War. B.P. cracked tens of thousands of German military messages a m onth.

In the U.S.A., both Army and Navy developed high-speed versions of the BOMBE that were in service for a few years after the end of the war.

The X-68 003 of the U.S. Army (SIS), a genuine relay machine constructed by Sam B. Williams from Western Electric, in operation since October 1943 and equipped with 144 double-ended scramblers, became known[19] as MADAME X; using stepping switches it allowed a quick change of the crib. The simulation of scramblers by relays was slow, but avoided rotating masses. Developed with the help of Bell Laboratories, it was directed against 3-rotor ENIGMAs and unwieldy against the 4-rotor ENIGMAs of the *Kriegsmarine.* Only one MADAME X was actually built, it corresponded in power to six or eight British BOMBEs. Design and construction cost a million dollars which was not thought cost-effective compared with the Navy's BOMBEs.

For the U.S. Navy Op-20-G, Joseph Desch at NCR, who had experience with rapid circuitry for elementary particle counters and thus a reputation in electronics, accepted in September 1942 the ambitious commission to build 350 BOMBEs, each one several times larger than the Turing–Welchman BOMBE. Moreover, they expected to have these machines operating by the spring of 1943. Desch and his group "thought that American technology and mass production methods could work miracles" (Burke). Desch, however, rejected the request by Joseph N. Wenger to build an electronic version: "An electronic BOMBE was an impossibility." He was wise: for a 'super-BOMBE' he had calculated 20 000 tubes, while the British needed only about 2000 tubes for COLOSSUS.

The British could do no more than help the Navy of their ally. Howard Engstrom had sent Joseph Eachus in July 1942 to Bletchley Park. Turing traveled November 7–13, 1942 in the *Queen Elizabeth* over the Atlantic Ocean and departed again from New York harbor on the night of March 23, 1943 in the *Empress of Scotland* to arrive safely back in England on March 29. He made good use of the four months in the U.S.A., meeting also Claude Elwood Shannon at Bell Labs while doing work on voice scrambling. Op-20-G was provided with money and the most able people, but security measures were tighter

[18] Turing had dubbed it originally 'Agnus Dei'.

[19] It is unclear whether the name is an allusion to Agnes Meyer Driscoll, the brave fighter for pure cryptanalysis (Sect. 17.3.4), who was referred to in Op-20-G as Madam 'X'.

Fig. 160a. British BOMBE 'Atlanta' (standard model) in Eastcote. A scrambler is formed by three vertically adjacent rotors

Abb. 160b. US Navy BOMBE for 4-Rotor-ENIGMA. A scrambler is formed by four vertically adjacent rotors

than those for the Atomic Bomb project. NCR in Dayton, Ohio provided the setting for the BOMBEs. Despite the efforts of Eachus, it took longer than expected, and by the spring of 1943 only two prototypes, ADAM and EVE, were halfway ready. Franklin Delano Roosevelt himself gave the project support and impetus. Meanwhile, the situation in the Atlantic improved for the Allies, mainly thanks to BP decryptions. Desch had particular problems with fast spinning electromechanical scramblers with brush contacts. In mid-June 1943 it was hoped to overcome the difficulties soon. When on July 26, 1943, 13 production models did not function at all, it looked like the whole project would be killed. But Desch did not give up. The mechanical difficulties were surmounted step by step and reliability increased. In September 1943, the first machines built by NCR were sent from Dayton to Washington, where they were to start work. By mid-November 50 BOMBEs were in operation and 30 more on the way. In 1944, success was certain, although it took slightly longer than optimistically projected and cost almost three times as much as planned, but after all a Desch BOMBE (the project was so secret that the machine did not even have a name) cost only \$ 45 000.

The US Navy 4-rotor BOMBE comprised 16 4-rotor-scramblers and a Welchman diagonal board and was 200 times faster than the Polish *bomba*, 20 times faster than the Turing–Welchman BOMBE (the specification had said 26 times faster). It was still 30% faster than the 1943 British Bombe attachment WALRUS ('COBRA') directed against 4-rotor-ENIGMAs of the *Kriegsmarine*. Op-20-G had caught up: by December 1943 the decryption of a TRITON ENIGMA signal took on average only 18 hours, compared to 600 in June 1943. In contrast to their British cousins, they localized the scrambler positions and controlled the whole job by digital electronics with 1500 thyratrons (gas-filled tubes). At least 100 Desch BOMBEs were built. They proved to be so reliable that by the end of 1943 all work on the decryption of the TRITON key net of the German U-boats was assigned to the U.S. Navy—a great step forward from the rivalries of mid-1942.

The BRUSA pact *Cooperation in Code/Cipher Matters* of May 1943 between the U.S.A. and the U.K., and for their Navies in particular the Holden[20] Agreement of October 1942 "began to move the two nations towards a level of unprecedented cooperation" (Burke) in cryptanalysis. But some frictions and tensions remained. "It was not until the UKUSA agreement of 1946 that the two nations forged that unique relationship of trust that was maintained throughout the Cold War" (Burke).

VIPER and PYTHON were American machines directed against the Japanese rotor cipher machines. They were built from relays and stepping switches, and little by little equipped with electronic additions.

Finally, towards the end of the war, the inevitable transition to truly electronic machines was made: Op-20-G built RATTLER against Japan's JN-157

[20] Carl F. Holden, Capt. U.S. Navy, Director of Naval Communications.

while SIS built in 1945 a successor to the relay-based AUTOSCRITCHER which was correspondingly called the SUPERSCRITCHER[21]. Op-20-G also built DUENNA, and the British built GIANT—names that did not exist until recently in the open literature. All these machines were directed against cross-plugging and reflector plugging. Design of DELILAH started in September 1943, construction in June 1944; it was just finished by May 6, 1945.

19.7.4 The advent of computers. The idea of the universal stored-program computer, which had originated in mid-1940 with Eckert and Mauchly and had been elaborated by von Neumann and Goldstine, was pretty soon, although not publicly, influencing cryptanalysis with machines. James T. Pendergrass advocated the use of universal computers. It started in 1948 with ABNER at SIS (a development that took four years) and the ATLAS efforts going back to August 1947 at Op-20-G (supplementing those already mentioned in Sects. 17.3.5 and 18.6.3). The National Security Agency, the 'super' authority, successor to both SIS and Op-20-G, gave great impact to the emerging computer field. Howard H. Campaigne, Samuel S. Snyder, and Erwin Tomash reported on the influence of U.S. cryptological organizations on the aspiring digital computer industry. Some former Navy reserve officers, Howard T. Engstrom, William C. Norris, and Ralph Meader founded a private company, Engineering Research Associates, Inc. (E.R.A.) early in 1946; they were assisted by Charles B. Tompkins and John E. Howard and cooperated with Joseph Eachus and James T. Pendergrass of Op-20-G . They developed computers in close contact with the Navy. The milestone was Task 13, renamed ATLAS in 1948, delivered in December 1950. This led to a marketing of the computer ERA 1101 (announced in December 1951). Its successor (Task 29, code-named ATLAS II, completed in 1953) was marketed as ERA 1103 and was an immediate success. E.R.A. became a Remington Rand subsidiary in 1952. By 1954, Remington Rand enjoyed a strong second position on the market with the improved 1101A (and the UNIVAC II developed by the Eckert-Mauchly group).

Their competitor IBM announced in 1951 the Defense Calculator, renamed IBM 701 and marketed in 1953; the initial delivery of the 701 to a commercial buyer took place in April 1953, of the ATLAS II to the government in October 1953. IBM's STRETCH developed from the N.S.A. HARVEST of 1962.

In the 1970s, Seymour R. Cray (1925–1996), an electrical engineer who formerly worked at E.R.A. under Engstrom, formed his own company and in 1976 designed the CRAY-1 . N.S.A. is still partly based on commercial manufacturers. The sensitive circuitry of CRAY computers hides some of the cryptanalytic algorithms N.S.A. is relying upon. The Cold War in its present, miniature form has crept into the chips.

---

[21] The expression 'scritchmus' comes from the jargon of Bletchley Park ("I cannot now recall what technique was nicknamed a scritchmus," wrote Derek Taunt) and the method was developed by Dennis Babbage. For the origin see also Sect. 14.5 . Ralph Erskine thinks, 'scritching' comes from 'scratching out contradictions'.

20 Linear Basis Analysis

> It would not be an exaggeration to state
> that abstract cryptography is identical
> with abstract mathematics.
>
> *A. Adrian Albert* 1941

20.1 Reduction of Linear Polygraphic Substitutions

In favorable cases, a linear polygraphic substitution with an encryption width n can be reduced to an exhaustion of width n, namely if a decryption of a set of n frequently occurring crypto n-grams into a set of n plain n-grams can be guessed. This is sometimes easier than it looks at first sight, if a long enough probable word can be assumed that is met in n different phases.

20.1.1 Example. To keep the computations verifiable, we limit ourselves to an example with $n = 3$. Given the cryptotext

F D Y S W   I J X N Z   N S N R E   N H U W A   W M I E I   E X W S X
I S I G Q   J N T B D   B W D P U   ....

we assume that we have in the circumstances a linear polygraphic substitution of width 3 over the standard alphabet—possibly a second trial in a series of trials with increasing encryption width. The cryptotext then reads in trigrams over $\mathbb{Z}_{26}$

5 3 24   18 22 8   9 23 13   25 13 18   **13 17 4**   13 7 20   **22 0 22**   12 8 4
8 4 23   22 18 23   4 18 8   **6 16 9**   13 19 1   3 1 22   3 15 20   ....

and we assume that the (boldface) trigrams **13 17 4** , **22 0 22** and **6 16 9** appear quite frequently in the further cryptotext. In view of the very frequent occurence of /ation/ in English, French and German, we can try the conjecture that the three plaintext trigrams /ati/, /tio/ and /ion/ are involved; in which order remains to be seen.

In $\mathbb{Z}_{26}$ the plaintext trigrams are **0 19 8** , **19 8 14** and **8 14 13** . Therefore, the matrix X of the linear substitution is determined as follows, where P is a permutation matrix unknown at the moment:

$$\begin{pmatrix} 0 & 19 & 8 \\ 19 & 8 & 14 \\ 8 & 14 & 13 \end{pmatrix} X \;=\; P \begin{pmatrix} 13 & 17 & 4 \\ 22 & 0 & 22 \\ 6 & 16 & 9 \end{pmatrix} .$$

There are six solutions in $\mathbb{Z}_{26}$ for the $6 = 3!$ permutations. We can imagine that after trying two or three of them, we found

$$\begin{pmatrix} 0 & 19 & 8 \\ 19 & 8 & 14 \\ 8 & 14 & 13 \end{pmatrix} X = \begin{pmatrix} 22 & 0 & 22 \\ 6 & 16 & 9 \\ 13 & 17 & 4 \end{pmatrix} \qquad \text{to give} \qquad X = \begin{pmatrix} 12 & 8 & 17 \\ 8 & 18 & 24 \\ 13 & 19 & 14 \end{pmatrix},$$

which seems to be the right path, since translated into the letter alphabet,

$X = \begin{pmatrix} M & I & R \\ I & S & Y \\ N & T & O \end{pmatrix}$ comes from the 'reasonable' password $MINISTRYO[F]$.

This confirms the solution in Rohrbach's sense.

20.1.2 A pitfall. But there is a complication. If we now try to decrypt the cryptotext, we might look for an inverse matrix X and surprisingly find there is none. In fact, the use of a 'reasonable' password does not guarantee that the encryption is injective, and X is not injective: the vector $(0 \;\; 13 \;\; 0)$ is annihilated by X. This means, that even the authorized decryptor has the fun of looking for the 'right' solution:

$$\begin{array}{llllll} \text{to} & 5 \;\; 3 \;\; 24 & \text{belong} & 8 \;\; 0 \;\; 5 \,\hat{=}\, \mathrm{i\,a\,f} & \text{and} & 8 \;\; 13 \;\; 5 \,\hat{=}\, \mathrm{i\,n\,f}\;\; ; \\ \text{to} & 18 \;\; 22 \;\; 8 & \text{belong} & 14 \;\; 4 \;\; 12 \,\hat{=}\, \mathrm{o\,e\,m} & \text{and} & 14 \;\; 17 \;\; 12 \,\hat{=}\, \mathrm{o\,r\,m} \quad \text{and so on.} \end{array}$$

The polyphone decryption is:

$$\mathrm{i}\;{\genfrac{}{}{0pt}{}{\mathrm{a}}{\mathrm{n}}}\;\mathrm{f}\;\;\mathrm{o}\;{\genfrac{}{}{0pt}{}{\mathrm{e}}{\mathrm{r}}}\;\mathrm{m}\;\;\mathrm{d}\;{\genfrac{}{}{0pt}{}{\mathrm{i}}{\mathrm{v}}}\;\mathrm{r}\;\;\mathrm{e}\;{\genfrac{}{}{0pt}{}{\mathrm{c}}{\mathrm{p}}}\;\mathrm{t}\;\;\mathrm{i}\;{\genfrac{}{}{0pt}{}{\mathrm{b}}{\mathrm{o}}}\;\mathrm{n}\;\;\mathrm{o}\;{\genfrac{}{}{0pt}{}{\mathrm{f}}{\mathrm{s}}}\;\mathrm{n}\;\;\mathrm{a}\;{\genfrac{}{}{0pt}{}{\mathrm{g}}{\mathrm{t}}}\;\mathrm{i}\;\;\mathrm{o}\;{\genfrac{}{}{0pt}{}{\mathrm{a}}{\mathrm{n}}}\;\mathrm{a}$$

$$\mathrm{l}\;{\genfrac{}{}{0pt}{}{\mathrm{e}}{\mathrm{r}}}\;\mathrm{a}\;\;\mathrm{d}\;{\genfrac{}{}{0pt}{}{\mathrm{i}}{\mathrm{v}}}\;\mathrm{o}\;\;\mathrm{s}\;{\genfrac{}{}{0pt}{}{\mathrm{g}}{\mathrm{t}}}\;\mathrm{a}\;\;\mathrm{t}\;{\genfrac{}{}{0pt}{}{\mathrm{i}}{\mathrm{v}}}\;\mathrm{o}\;\;\mathrm{n}\;{\genfrac{}{}{0pt}{}{\mathrm{a}}{\mathrm{n}}}\;\mathrm{b}\;\;\mathrm{o}\;{\genfrac{}{}{0pt}{}{\mathrm{h}}{\mathrm{u}}}\;\mathrm{t}\;\;\mathrm{o}\;{\genfrac{}{}{0pt}{}{\mathrm{h}}{\mathrm{u}}}\;\mathrm{r}\;\;\ldots\ldots\ldots\;.$$

The correct plaintext is easily discovered: "inform direction of national radio station about our ... ."

20.2 Reconstruction of the Key

If a quasi-nonperiodic key of a polyalphabetic linear polygraphic substitution of width n is generated by iteration of a regular $n \times n$ matrix A over $\mathbb{Z}_N$, a swapping of roles between plaintext and keytext can be made. A probable word of length k, $k \geq n^2 + n$ is shifted along the cryptotext and subtracted in every position. What remains is in favorable cases a key fragment $(s_{M+1}, s_{M+2}, \ldots s_{M+n^2+n}, s_{M+k})$ of a length $k \geq n^2 + n$. The n equations

$$\begin{aligned} (s_{M+1}, s_{M+2}, \ldots s_{M+n})\; A &= (s_{M+n+1},\; s_{M+n+2},\; \ldots s_{M+2n}) \\ (s_{M+n+1}, s_{M+n+2}, \ldots s_{M+2n})\; A &= (s_{M+2n+1}, s_{M+2n+2}, \ldots s_{M+3n}) \\ (s_{M+2n+1}, s_{M+2n+2}, \ldots s_{M+3n})\; A &= (s_{M+3n+1}, s_{M+3n+2}, \ldots s_{M+4n}) \\ &\;\;\vdots \\ (s_{M+n^2-n+1}, s_{M+n^2-n+2}, \ldots s_{M+n^2})\; A &= (s_{M+n^2+1}, s_{M+n^2+2}, \ldots s_{M+n^2+n}) \end{aligned}$$

in $\mathbb{Z}_N$ suffice to determine A; for $k > n^2+n$ we even have an overdetermined system of linear equations.

To give an example, the three pairs of numbers (1 0), (3 5), (23 22) are obtained from (1 0) by two iterations with a matrix A; A is determined by

$$(1\ 0)\,A=(3\ 5) \text{ and } (3\ 5)\,A=(23\ 22)$$

and the result in $\mathbb{Z}_{26}$ is $$A=\begin{pmatrix} 3 & 5 \\ 8 & 17 \end{pmatrix}.$$

If in the favorable case the position of the probable word fits, then the system can be solved, and for the overdetermined case some such systems are solvable and give a common solution, strongly indicating a correct solution. In the unfavorable case that the position of the probable word does not fit, the system or one of the systems may not be solvable. If it is accidentally solvable, then the keytext can be prolonged and subtracted from the crypto text, which as a rule produces nonsense text—indicating a flop. For a sufficiently long probable word, the key is normally completely revealed, and mishits should not occur.

20.3 Reconstruction of a Linear Shift Register

Linear shift registers in the wider sense fall as a special case under the kind of attack treated in Sect. 20.2. In this case, the matrix A is an $n \times n$ companion matrix (Sect. 8.6.1),

$$A = \begin{pmatrix} 0 & 0 & 0 & \ldots & 0 & \alpha_k \\ 1 & 0 & 0 & \ldots & 0 & \alpha_{k-1} \\ 0 & 1 & 0 & \ldots & 0 & \alpha_{k-2} \\ & & \vdots & & & \\ 0 & 0 & 0 & \ldots & 0 & \alpha_2 \\ 0 & 0 & 0 & \ldots & 1 & \alpha_1 \end{pmatrix}.$$

If a key is generated by iteration with such an $n \times n$ companion matrix, then a fragment of length $2n$ of the key text suffices to reconstruct the companion matrix and thus to generate the whole key.

Again, to keep the computations verifiable, we limit ourselves to an example with $n = 4$. Assume we have the following cryptotext:

C G V J F M C I H T X U F S D Y V L M R ... ...

Assume, too, that in the circumstances the encryption is a polygraphic VIGENÈRE with a quasiperiodic key sequence generated by a linear polygraphic substitution of width 4 in $\mathbb{Z}_{26}$. Among the probable words we conjecture the word /broadcast/.

We may start with the hypothesis that the probable word is right at the beginning. This leads to the situation

C	G	V	J	F	M	C	I	H	T	X	U	F	S	D	Y	V	L	M	R	...
2	6	21	9	5	12	2	8	7	19	23	20	5	18	3	24	21	11	12	17	...
b	r	o	a	d	c	a	s	t												
1	17	14	0	3	2	0	18	19												
1	*15*	*7*	*9*	*2*	*10*	*2*	*16*	*14*												

This yields the iteration equation in $\mathbb{Z}_{26}$

$$\begin{pmatrix} 1 & 15 & 7 & 9 \\ 15 & 7 & 9 & 2 \\ 7 & 9 & 2 & 10 \\ 9 & 2 & 10 & 2 \\ 2 & 10 & 2 & 16 \end{pmatrix} \begin{pmatrix} 0 & 0 & 0 & t_1 \\ 1 & 0 & 0 & t_2 \\ 0 & 1 & 0 & t_3 \\ 0 & 0 & 1 & t_4 \end{pmatrix} = \begin{pmatrix} 15 & 7 & 9 & 2 \\ 7 & 9 & 2 & 10 \\ 9 & 2 & 10 & 2 \\ 2 & 10 & 2 & 16 \\ 10 & 2 & 16 & 14 \end{pmatrix}$$

and the overdetermined linear system

$$\begin{pmatrix} 1 & 15 & 7 & 9 \\ 15 & 7 & 9 & 2 \\ 7 & 9 & 2 & 10 \\ 9 & 2 & 10 & 2 \\ 2 & 10 & 2 & 16 \end{pmatrix} \begin{pmatrix} t_1 \\ t_2 \\ t_3 \\ t_4 \end{pmatrix} = \begin{pmatrix} 2 \\ 10 \\ 2 \\ 16 \\ 14 \end{pmatrix},$$

which cannot be solved. The first four lines can be transformed by Gaussian elimination into

$$\begin{pmatrix} 1 & 15 & 7 & 9 \\ 0 & 1 & 9 & 9 \\ 0 & 0 & 1 & 17 \\ 0 & 0 & 0 & 1 \end{pmatrix} \begin{pmatrix} t_1 \\ t_2 \\ t_3 \\ t_4 \end{pmatrix} = \begin{pmatrix} 2 \\ 18 \\ 0 \\ 8 \end{pmatrix},$$

with a solution by back-substitution

$$t_4 = 8,\ t_3 = 20,\ t_2 = 0,\ t_1 = 24\,,$$

but this obviously does not fulfill the fifth equation.

The next hypothesis to be tested could be that the probable word begins at the second position of the plaintext, which leads to the situation

C	G	V	J	F	M	C	I	H	T	X	U	F	S	D	Y	V	L	M	R	...
2	6	21	9	5	12	2	8	7	19	23	20	5	18	3	24	21	11	12	17	...
	b	r	o	a	d	c	a	s	t											
	1	17	14	0	3	2	0	18	19											
	5	*4*	*21*	*5*	*9*	*0*	*8*	*15*	*0*											

and to the iteration equation in $\mathbb{Z}_{26}$

$$\begin{pmatrix} 5 & 4 & 21 & 5 \\ 4 & 21 & 5 & 9 \\ 21 & 5 & 9 & 0 \\ 5 & 9 & 0 & 8 \\ 9 & 0 & 8 & 15 \end{pmatrix} \begin{pmatrix} 0 & 0 & 0 & t_1 \\ 1 & 0 & 0 & t_2 \\ 0 & 1 & 0 & t_3 \\ 0 & 0 & 1 & t_4 \end{pmatrix} = \begin{pmatrix} 4 & 21 & 5 & 9 \\ 21 & 5 & 9 & 0 \\ 5 & 9 & 0 & 8 \\ 9 & 0 & 8 & 15 \\ 0 & 8 & 15 & 0 \end{pmatrix}.$$

This yields the overdetermined linear system

$$\begin{pmatrix} 5 & 4 & 21 & 5 \\ 4 & 21 & 5 & 9 \\ 21 & 5 & 9 & 0 \\ 5 & 9 & 0 & 8 \\ 9 & 0 & 8 & 15 \end{pmatrix} \begin{pmatrix} t_1 \\ t_2 \\ t_3 \\ t_4 \end{pmatrix} = \begin{pmatrix} 9 \\ 0 \\ 8 \\ 15 \\ 0 \end{pmatrix},$$

which can be solved: the first four lines can be transformed by Gaussian elimination into

$$\begin{pmatrix} 1 & 6 & 25 & 1 \\ 0 & 1 & 23 & 7 \\ 0 & 0 & 1 & 4 \\ 0 & 0 & 0 & 1 \end{pmatrix} \begin{pmatrix} t_1 \\ t_2 \\ t_3 \\ t_4 \end{pmatrix} = \begin{pmatrix} 7 \\ 18 \\ 23 \\ 3 \end{pmatrix},$$

with a solution by back-substitution

$$t_4 = 3, t_3 = 11, t_2 = 4, t_1 = 17,$$

which obviously does fulfill the fifth equation.

The iteration matrix for the continuation of the key text is thus in $\mathbb{Z}_{26}$

$$A = \begin{pmatrix} 0 & 0 & 0 & 17 \\ 1 & 0 & 0 & 4 \\ 0 & 1 & 0 & 11 \\ 0 & 0 & 1 & 3 \end{pmatrix} \quad \text{with the inverse} \quad A^{-1} = \begin{pmatrix} 12 & 1 & 0 & 0 \\ 7 & 0 & 1 & 0 \\ 9 & 0 & 0 & 1 \\ 23 & 0 & 0 & 0 \end{pmatrix}.$$

Therefore, the key can be supplemented to

2 5 4 21 5   9 0 8 15 0   15 7 25 4 24   23 20 9 19 3 ...

and leads to the following decryption

C	G	V	J	F	M	C	I	H	T	X	U	F	S	D	Y	V	L	M	T	...
2	6	21	9	5	12	2	8	7	19	23	20	5	18	3	24	21	11	12	17	...
2	*5*	*4*	*21*	*5*	*9*	*0*	*8*	*15*	*0*	*15*	*7*	*25*	*4*	*24*	*23*	*20*	*9*	*19*	*3*	...
0	1	17	14	0	3	2	0	18	19	8	13	6	14	5	1	1	2	19	14	...
a	b	r	o	a	d	c	a	s	t	i	n	g	o	f	b	b	c	t	o	... .

("A broadcasting of BBC tonight announced the Allied invasion to be expected within forty-eight hours.")

The iteration matrix is deduced from the password

$$\textit{(FID)DLER} \doteq (5\ 8\ 3)\ 3\ 11\ 4\ 17\ .$$

For a key sequence generated by a binary linear shift register in connection with a VIGENÈRE over $\mathbb{Z}_2$, i.e., a VERNAM, everything said above holds, too. A shift register encryption should be nonlinear to avoid this line of attack (Beth et al. 1982).

Quite generally, linear substitutions are much more vulnerable to cryptanalytic attacks than non-linear ones.

21 Anagramming

Abandonner les méthodes de substitution pour celles de transposition a été changer son cheval borgne pour un aveugle.

[Abandoning the methods of substitution for those of transposition was like changing one's one-eyed horse for a blind one.]

Étienne Bazeries 1901

Transpositions were for a while the favorites of the military, particularly in the late 18th and early 19th century in France, Germany, Austria, and elsewhere. They seemed to be suitable above all as field ciphers ('trench codes') above all for the lower ranks. Bazeries, around 1900, made fun of this and generally ascribed to transposition systems that seemed difficult at first sight a *complication illusoire*. Cryptanalysts usually loved adversaries that used simple transpositions (like the German *Abwehr* hand ciphers) because they promised to be easy prey; likewise the literature treats cryptanalysis of transpositions as relatively unsophisticated.

21.1 Transposition

Simple transposition (Sect. 6.2.1), i.e., throwing single characters about, with small encryption width n, can be treated for very small known n by systematic studies of contact in bigrams, possibly also in trigrams and tetragrams of characters. In the Second World War, the cryptanalytic services in the German *Auswärtiges Amt* (Pers Z) and *Oberkommando der Wehrmacht* (*Chi*) used special machines, called *Spezialvergleicher* and *Bigrammbewertungsgerät* (Rohrbach, Jensen) for the semiautomatic solution of simple column transposition and simple block transposition. Essentially, the exhaustive scissors-and-paste method of Sect. 12.8.2 was mechanized. A piece of cryptotext was confronted with the whole cryptotext in all relative positions and for the observed bigrams the theoretical bigram frequencies were multiplied, then positions where this product was high were singled out. The method is even helpful if the columns of a columnar transposition do not all have the same length.

For the U.S. Army, towards the end of the war SIS built FREAK, a bigram counter based on electric condensers, a substitute for the 1943 NCR-built MIKE which, according to Burke, was "a huge electromechanical contraption."

21.1.1 Example. We consider an example of this 'contact method' for the following cryptotext:

S S N K L H O N I W M M E U N T A H U L I N N A H N C I N F C I E R O
N A C B A M Z G H N K T H W C D E S I N K C A I E A N I M

Counting the frequencies of single letters results in a distribution not very different from that of the German language and allows us to consider transposition. The total number of characters is 64, which even suggests a transposition with an 8×8 or 4×16 square. Trying the 8×8 square

S	I	A	H	E	M	W	C
S	W	H	N	R	Z	C	A
N	M	U	C	O	G	D	I
K	M	L	I	N	H	E	E
L	E	I	N	A	N	S	A
H	U	N	F	C	K	I	N
O	N	N	C	B	T	N	I
N	T	A	I	A	H	K	M

we take a column (it could also be a line) that contains many frequent characters, say the 5th, ERONACBA, and confront it with the other columns. The resulting bigrams and their expected frequency (in %%) is given in the following diagram (empty entries mean a frequency below 0.5%%):

ES 140	E I 193	EA 26	EH 57	EM 55	EW 23	EC 25
RS 54	RW 17	RH 19	RN 31	RZ 14	RC 9	RA 80
ON 64	OM 17	OU 3	OC 15	OG 5	OD 7	O I 1
NK 25	NM 23	NL 10	N I 65	NH 17	NE 122	NE 122
AL 59	AE 64	A I 5	AN 102	AN 102	AS 53	AA 8
CH 242	CU	CN	CF	CK 14	C I 1	CN
BO 8	BN 1	BN 168	BC	BT 4	BN 1	B I 12
AN 102	AT 46	AA 8	A I 5	AH 20	AK 7	AM 28

The confrontation of the column ERONACBA with the column SSNKLHON shows clearly higher frequencies than the others. Multiplying the frequencies gives the value $1.41 \times 10^{14} \times 10^{-32} = 1.41 \times 10^{-18}$, while all other columns give values below $3.74 \times 10^{9} \times 10^{-32} = 3.74 \times 10^{-23}$. Because of this good result, the next column we test is the first, SSNKLHON. Now the confrontation with the other columns gives

S I 65	SA 36	SH 9	SM 12	SW 10	SC 89
SW 10	SH 9	SN 7	SZ 7	SC 89	SA 36
NM 23	NU 33	NC 5	NG 94	ND 187	N I 65
KM 1	KL 10	K I 7	KH 1	KE 26	KE 26
LE 65	L I 61	LN 4	LN 4	LS 22	LA 45
HU 11	HN 19	HF 2	HK 3	H I 23	HN 19
ON 64	ON 64	OC 15	OT 9	ON 64	O I 1
NT 59	NA 68	N I 65	NH 17	NK 25	NM 23

This time, confrontation with the column WCDESINK stands out, not as clearly as before, but with a product of $3.50 \times 10^{12} \times 10^{-32} = 3.50 \times 10^{-20}$ still indubitably, since all others are below $5.39 \times 10^{11} \times 10^{-32} = 5.39 \times 10^{-21}$. Daring to continue with the 7th column WCDESINK, the next confrontation gives a preference for the 3rd column AHULINNA. If the columns which are singled out in this way are written side by side, the result so far is

ESWA
RSCH
ONDU
NKEL
ALSI
CHIN
BONN
ANKA

Surprisingly, this is already plaintext; since the columns used so far are the 5th, the 1st, the 7th and the 3rd, it is quite likely that the transposition uses a 4×16 square. The remaining columns are simply subjected now to the same permutation, doubling the length of the columns:

MICH
ZWAN
GMIC
HMEI
NEAN
KUNF
TNIC
HTMI

The complete plaintext reads

"*es war schon dunkel als ich in bonn ankam ich zwang mich meine ankunft nicht mi*[*t der automatik* ...]" (Heinrich Böll, *Ansichten eines Clowns*, 1963).

21.1.2 Shifted columns. In the example just discussed the first column of the plaintext had more of the frequent letters than the other columns. This will usually not be the case, and not only the contact to the right, but also the contact to the left will need to be investigated. As soon as the very first or the very last column of the plaintext is reached, continuation makes sense only with columns shifted by one place. Using trigram frequencies increases the number of exhaustive steps, but may give more stable permutations.

21.1.3 Caveat. We have seen that simple transposition with fixed encryption steps of some width provides no security if the text is a few times longer than the width. Transposition with a width equal to the length of the text, as a rule, allows more than one 'meaningful' solution, even for very long texts. A smart lawyer therefore could have saved Brother Tom of Jonathan Swift (Sect. 6.3), if he had found another, harmless solution of the anagram. However, the security this kind of transposition offers rests fully on the one-time use of the permutation of the places, which means an individual key. As soon as such an encrypting transposition step is used a few times, the simple attack of Sect. 21.1.1 can be tried, and a specific method to be discussed in Sect. 21.3 .

21.1.4 Codegroup patterns. Even when code has been superencrypted by simple transposition, it can be treated in the way mentioned above if its codegroups have certain patterns. For example, this is the case if 'pronounceable' codegroups are built in a vowel-consonant pattern, like $\mathcal{CVCVC}$ in the GREEN code of the U.S. State Department (Sect. 4.4.2). Still in the Second World War, the U.S. State Department used codes of type $\mathcal{CVCVC}$ and $\mathcal{CVCCV}$, a property which also helped Pers Z (Sect. 19.3.2) to strip off an additive.

21.1.5 Illusory complication. Furthermore, the 'contact method' works also for mixed-rows columnar transposition and mixed-rows block transposition (Sect. 6.2.3), since the contact is only occasionally interrupted. It yields an intermediate cryptotext with permuted lines, for example in our 8×8 square

MICHZWAN ALSICHIN ONDUNKEL NEANKUNF
GMICHMEI ESWARSCH TNICHTMI BONNANKA .

Both Givierge and Eyraud pointed out that *transposition double* in the form of mixed-rows columnar transposition and mixed-rows block transposition, including Nihilist transposition, are not much more resistant than the simplest columnar transposition. The *double* suggests a *complication illusoire.*

21.2 Double Columnar Transposition

Double columnar transposition (Sect. 6.2.4)—except in particular cases, as in Sect. 6.2.5—is a much harder task for the unauthorized decryptor. The reason is that after the first transposition all contacts are completely torn. Eyraud treats the case in some detail, but is unable to give a complete method. Kahn writes:

"... in theory the cryptanalyst merely has to build up the columns of the second block by twos and threes so that their digraphs and trigraphs would in turn be joinable into good plaintext fragments. But this is far more easily said than done. Even a gifted cryptanalyst can accomplish it only on occasion; and even with help, such as a probable word, it is never easy."

A really powerful means of attack, if possible, is multiple anagramming.

21.3 Multiple Anagramming

For the most general case of transposition, even with a width about as large as the text, including also grilles and route transcriptions, there is a general method, requiring nothing more than that two plaintexts of the same length have been encrypted with the same encryption step, i.e., that the encrypting transposition step has been repeated at least once. Such a plaintext-plaintext compromise suggests a parallel to Kerkhoffs' method of superimposition.

21.3.1 Example. The method is based on the simple fact that equal encryption steps perform the same permutation of the plaintext. The cryptotexts are therefore written one below the other and the columns thus formed are kept together. Assume we have (in phase) the cryptotext fragments (Kahn) GHINT and OWLCN

This means that the five pairs $\genfrac{}{}{0pt}{}{\mathrm{G}}{\mathrm{O}}$ $\genfrac{}{}{0pt}{}{\mathrm{H}}{\mathrm{W}}$ $\genfrac{}{}{0pt}{}{\mathrm{I}}{\mathrm{L}}$ $\genfrac{}{}{0pt}{}{\mathrm{N}}{\mathrm{C}}$ $\genfrac{}{}{0pt}{}{\mathrm{T}}{\mathrm{N}}$ are to be anagrammed. Among the $5 \times 4 = 20$ combinations only the following twelve (in descending contact order)

TH	NG	GT	IN	TI	NI	GH	HI	IG	TN	HT	GI
NW	CO	ON	LC	NL	CL	OW	WL	LO	NC	WN	OL

have sufficiently large contact at both levels. Using only the first four combinations, there is just the meaningless solution $\genfrac{}{}{0pt}{}{\mathrm{INGTH}}{\mathrm{LCONW}}$ and even with the first eight combinations only cyclic shifts of this solution are obtained, as the following graph with five nodes and eight branches shows:

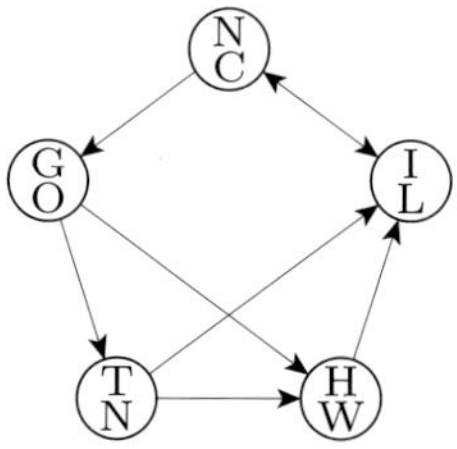

Using also the ninth combination, there is a further solution $\genfrac{}{}{0pt}{}{\mathrm{NIGTH}}{\mathrm{CLONW}}$ which is senseless. Only with the first eleven combinations is the meaningful solution $\genfrac{}{}{0pt}{}{\mathrm{NIGHT}}{\mathrm{CLOWN}}$ (and its cyclic shifts) obtained, which can be seen from the following graph with five nodes and eleven branches:

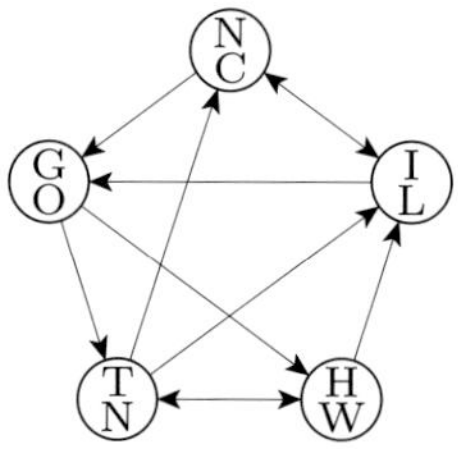

Thus, NIGHT and CLOWN are the solutions obtained by multiple anagramming. "There will be one order—and only one—in which the two messages will simultaneously make sense" was in 1879 the empirical finding of Edward S. Holden.

21.3.2 Practical use. The example shows that multiple anagramming of two or more cryptotexts can be treated as a graph-theoretic problem, where the number of nodes equals the length of the texts. Mechanized calculation of the feasible combinations is possible. The combinatorial complexity of the search problem of finding a path through all nodes without visiting a node twice limits the length of texts that can be treated this way. If about half a dozen texts of length say 25, 36, 49, 64, 81, or 100 are given, as was frequently the case with field ciphers, multiple anagramming with some computer support can be done quickly. Multiple anagramming is particularly important for transpositions made by means of grilles and route transcriptions, since these devices are prefabricated and as a rule destined for multiple use. Superencryption of a polyalphabetic substitution by transposition (*chiffre à triple* *clef* of Kerkhoffs) withstands the contact method of multiple anagramming. This is not to say that there are no other ways to attack it.

21.3.3 Hassard, Grosvenor, Holden. Transposition can also be done with words instead of with letters; then multiple anagramming of words is the cure. Multiple anagramming was invented or at least for the first time made public in 1878—five years before Kerckhoffs—by John R. G. Hassard and William M. Grosvenor, two editors of the *New York Tribune* (which co-operated), and independently by E. S. Holden, already mentioned, a mathematician at the U.S. Naval Observatory in Washington. The reason for such a massive effort was a scandal in the U.S. Senate, based on some hundreds of encrypted telegrams. An amateurish system was used: plaintext with disguised proper names and revealing words was written into a grille, and four such grilles were used with 15, 20, 25, and 30 words. The telegrams were decrypted independently and coinciding solutions were obtained which ensured their correctness and authenticity. Revealing the scandal had deep political consequences, moreover the American public became intensely informed about secret codes and how to break them. Possibly the preference in the U.S.A. for cryptograms as a pastime stems from this source.

In 1914, the French under Colonel François Cartier had learned their lesson when they were confronted with the *Heer* of the German *Kaiser*, which used a double columnar transposition as trench code. This was not new for the French, since the Germans stupidly had used the method already in peacetime to a great extent for drill messages. To make it more obvious, all signals were marked by the codegroup ÜBCHI (*Übungschiffre*) in the preamble; the French therefore called the system *ubchi*. With multiple anagramming they could pretty soon decrypt the messages at least in large fragments (a typical situation). This allowed them to reconstruct the password. On October 1, 1914 Cartier and his aides Adolphe Olivary, Henri Schwab, and Gustave Freyss gave the decryption rule to various French headquarters, enabling them to read the German wireless traffic as quickly as the Germans themselves. This situation lasted until mid-November 1914.

The generals of the *Kaiserliche Heer* then made a terrible blunder: they changed from the obstinate, time-consuming double columnar transposition to a simple columnar transposition, superencrypted by a VIGENÈRE addition with key *ABC*, which could be done in the head. This *complication illusoire*—stripping off the addition only needed a look at the frequency profile—allowed the French to use contacts in solving a single columnar transposition, which was a simple matter. The situation lasted until May 1915 and saved the French a lot of work.

Although *Le Matin* had published the story of the French success in October 1914, the German army returned to transposition at the end of 1916, this time using a turning grille. This lasted four months and caused no problem for the French, of course.

22 Concluding Remarks

> Insufficient cooperation in the development of one's own procedures, faulty production and distribution of key documents, incomplete keying procedures, overlooked possibilities for compromise during the introduction of keying procedures, and many other causes can provide the unauthorized decryptor with opportunities.
>
> *Erich Hüttenhain*[1] 1978

The history of cryptology shows that the unauthorized decryptor feasts on the mistakes of the adversary (Sect. 11.2.5). Encryption errors are made by crypto clerks. Tactical and strategic cryptographic faults occur in intelligence and communication organizations at all levels up to generals and directors. This even includes political questions of organization. The split among the services in Germany before and during the Second World War, ultimately but not merely a consequence of the rivalry between Ribbentrop, Göring, and Himmler, the division into *Sonderdienst Dahlem* in the *Abteilung Pers Z* of the *Auswärtiges Amt*, *Chiffrierabteilung* (*Chi*) in the *Oberkommando der Wehrmacht*, *B-Dienst* of the *Kriegsmarine*, *Forschungsamt* of the *Reichsluftfahrtministerium*, and *Amt VI* of the *Reichssicherheitshauptamt* was extremely counterproductive; the British concentrated their services from the very beginning of the tensions under the Foreign Office in the Government Code and Cypher School and even the military services did not feel badly served, not to mention the secret services, M.I.6 (under Stewart Menzies) and the American O.S.S. (under David Bruce); the partners all sat together in Winston Churchill's secret 'London Controlling Section' (L.C.S.) .

But both the Germans and the British (with their 'need to know' doctrine) maintained internal barriers for reasons of intelligence security; their effect was that no one division could learn enough from the others to be useful, which also allowed them sometimes to hush up flops and failures.

It is more a matter for historians than for cryptologists to judge to what extent results from cryptology have influenced war and peace. A voluminous journalistic record includes everything from serious discussion to sensationalist revelations.

---

[1] Dr. Erich Hüttenhain (26. 1. 1905 – 1. 12. 1990) studied mathematics (Heinrich Behnke) and astronomy in Münster. In 1936 he entered the Cipher Board (*Chi*) of the *Oberkommando der Wehrmacht* (OKW); he was finally head of group IV (analytic cryptanalysis) in the *Hauptgruppe Kryptanalyse* of *Ministerialrat* Wilhelm Fenner who worked there from 1922. After the war Hüttenhain directed from 1956 until 1973 an office of the Federal Gouvernment in Bad Godesberg, the *Zentralstelle für das Chiffrierwesen* ('German Cipher Board'). His successor (1973–1993) was Dr. Otto Leiberich.

22.1 Success in Breaking

Cryptography itself has its enemies. Generals and ambassadors sometimes consider the trouble not worthwhile. They may feel that depending on a crypto clerk is time-wasting and humiliating, and may doubt his or her honesty. The great philosopher Voltaire even went so far as to call codebreakers charlatans: *ceux qui se vantent de déchiffrer une lettre sans être instruit des affaires qu'on y traite ... sont de plus grands charlatans que ceux qui se vanteraient d'entendre une langue qu'ils n'ont point apprise* [Those who boast of being able to decrypt a letter without being informed on the affairs it deals with ... are greater charlatans than those who would boast of understanding a language they had not learned]. And the Earl of Clarendon, a hundred years earlier, wrote in a letter to the doctor John Barwick "I have heard of many of the pretenders of that skill, and have spoken with some of them, but have found them all to be mountebanks." In 1723, the British House of Commons spoke of the 'mystery of decyphering'. Public opinion on cryptographers has improved slightly since. Still, cryptanalysts cannot do miracles.

Some names of successful unauthorized decryptors have become known: in the First World War the British William R. Hall, Nigel de Grey, Malcolm Hay of Seaton, Oswald Thomas Hitchings, G. L. Brooke-Hunt, the French Georges Painvin, François Cartier, Marcel Givierge, E.-A. Soudart, the U.S. American Parker Hitt, J. Rives Childs, Frank Moorman, Joseph O. Mauborgne, Herbert Osborne Yardley, Charles J. Mendelsohn, the Italian Luigi Sacco, the Austrian Maximilian Ronge, Andreas Figl, Hermann Pokorny, and the Prussian Ludwig Deubner. Many more were forgotten and never publicized. In the Second World War, there was an even greater number of persons involved in codebreaking, many during wartime only. In a recent book edited by Francis Harry Hinsley and Alan Stripp, memoirs of some 30 Bletchleyites, as they were proudly called, are collected. It is partly accidental whether a cryptanalyst becomes known—Fedor Novopaschenny at the *Chi-Stelle*, Georg Schröder at the *Forschungsamt*, and Fritz Neeb at Army Group *Mitte* are examples. Cort Rave (see Sect. 19.3.2) remained totally unnoticed, and so did until recently Sergei Tolstoy, great codebreaker of the Soviet Union.

22.1.1 B-Dienst, Chi-Stelle, Sonderdienst Dahlem. A person who lived in the background until David Kahn made his name public is Wilhelm Tranow. A former radioman in the *Kriegsmarine*, he cracked Royal Navy signals in the First World War and was successful again in 1935. During the Second World War he was Head Cryptanalyst of the German Navy's *B-Dienst*[2]. Perhaps as a result of the way the Second World War ended, little has been known about German codebreaking (apart from Rohrbach's success), but it would be wrong to conclude that there were none. As an unbiased historian David Kahn describes the situation in mid-1943 of the

---
[2] Short for *Beobachtungsdienst*, originated from the *Beobachtungs- und Entzifferungsdienst* of the *Kaiserliche Marine*, therefore sometimes also called χ*B-Dienst* or *xB-Dienst*.

B-Dienst under the regime of the experienced and energetic Tranow as follows: "... the B-Dienst was at the height of its powers, solving 5 to 10% of its intercepts in time for Dönitz to use them in tactical decisions. Early information sometimes enabled him to move his U-boats so that a convoy would encounter the middle of the pack." 1000 men worked for Tranow at Berlin headquarters, and 4000, many imtercept operators, in the field (Kahn).

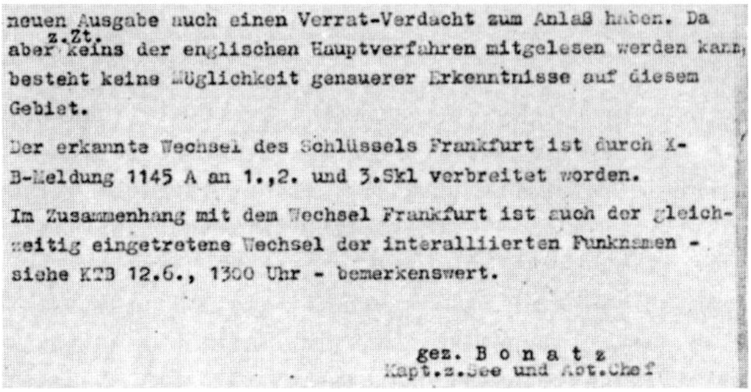
neuen Ausgabe auch einen Verrat-Verdacht zum Anlaß haben. Da aber z.Zt. keins der englischen Hauptverfahren mitgelesen werden kann, besteht keine Möglichkeit genauerer Erkenntnisse auf diesem Gebiet.

Der erkannte Wechsel des Schlüssels Frankfurt ist durch X-B-Meldung 1145 A an 1.,2. und 3.Skl verbreitet worden.

Im Zusammenhang mit dem Wechsel Frankfurt ist auch der gleichzeitig eingetretene Wechsel der interalliierten Funknamen - siehe KTB 12.6., 1300 Uhr - bemerkenswert.

gez. B o n a t z
Kapt.z.See und Abt.Chef

Fig. 161. Entry in the war diary (June 15, 1943) of the Chief, German Navy radio reconnaissance: *"Since at present none of the main English cyphers can be read ... The identified change of the 'Frankfurt' key was communicated by X-B-report 1145 A to ... "*

Indeed, from April 1940 on, the *B-Dienst* broke a third to a half of the current Naval Cypher, including the *Merchant Navy Code*. When the British introduced Naval Cypher No. 2 (German codename 'Köln') on August 20, 1940, it was partly broken towards the end of 1940 and fully in February 1941, and remained so more than two years. Thus during the climax of the U-boat war, Naval Cypher No. 3 (German codename 'Frankfurt'), introduced in June 1941 for the Allied Atlantic convoys, was compromised. The task included stripping off the superencryption of what actually was a code, and for this purpose six Hollerith tabulating machines were used to find parallels. By the end of 1942, 80% of the signals were deciphered, but only 10% in time to be operationally useful. Dönitz conceded that more than half of his total information came from this source. It ran dry only when Commander (later Vice Admiral Sir) Norman Denning's suspicions were roused by ENIGMA sources decrypted in Bletchley Park, whereupon the two Navies discontinued Naval Cypher No. 3 on June 10, 1943 (Fig. 161) and started to use Naval Cypher No. 5. Nevertheless, the Germans still obtained some decryptions (Fig. 162). The British command, like the German, did not like to believe that their codes might be insecure. Patrick Beesley blames the defeat on Bletchley Park, which was offensively minded and did little to defend the security of their own encryption methods. Colin Burke reports that there was rivalry and a choked flow of information between the United Kingdom and the United States of America in 1942. This stopped at the naval side in October 1942, continued on the Army side until about September 1943; finally, however, an unprecedented cooperation and relationship of trust developed.

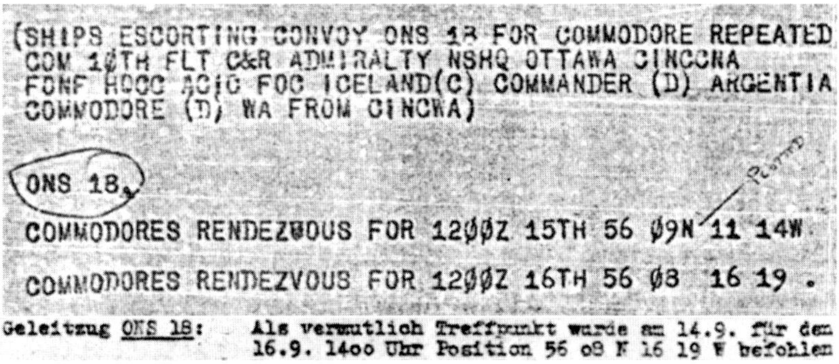
(SHIPS ESCORTING CONVOY ONS 18 FOR COMMODORE REPEATED
COM 10TH FLT C&R ADMIRALTY NSHQ OTTAWA CINCCNA
FONF HOCC AC|C FOC ICELAND(C) COMMANDER (D) ARGENTIA
COMMODORE (D) WA FROM CINCWA)

ONS 18.

COMMODORES RENDEZVOUS FOR 1200Z 15TH 56 09N 11 14W.

COMMODORES RENDEZVOUS FOR 1200Z 16TH 56 08 16 19 .

Geleitzug ONS 18: Als vermutlich Treffpunkt wurde am 14.9. für den 16.9. 1400 Uhr Position 56 08 N 16 19 W befohlen

Fig. 162. Comparison of a radio signal of CINCWA from September 14, 1943 to convoy ONS 18 with a decryption by the *B-Dienst* of the *Kriegsmarine*: "*... vermutlich Treffpunkt ... 16.9. 1400 Uhr* [German summer time] *Position 56 08 N 16 19 W*"

The question has been asked: Did cryptanalysis decide the Battle of the Atlantic? Jürgen Rohwer and Harry Hinsley have pointed out that the situation was quite balanced as long as the British were forced to wage a defensive warfare. Only after mid-1943, when the Allies were strong enough to turn to an offensive anti-U-boat war, was the German U-boat command crippled.

The success of the *B-Dienst* had tradition: When the war broke out, it could already read Naval Cypher No. 1, a 4-digit superencrypted code; this became possible following a compromise in 1935 in the Abyssinian War with a widely used 5-digit naval code that was already broken. During the attack on Norway in April–May 1940, the *Kriegsmarine* always had a precise picture of the situation in the British Admiralty. This explains somewhat the surprisingly fortunate course of events for the Germans. Tranow's great success went back to First World War experiences. Kahn cites an anonymous source: "If one man in German intelligence ever held the keys to victory in World War II, it was Wilhelm Tranow." With Hitler there was no key to this victory.

The High Command of the Armed Forces (OKW), according to Hüttenhain, broke the traffic between the French War Ministry and the French army departments early in the 1930s. However, their encryption was miserable: a numeral code which remained fixed for many years was superencrypted by a periodic VIGENÈRE *modulo* 10 with a period varying between 7 and 31. All signals could be read. Only between Paris and Savoy was a different method used, involving a transposition for superencryption. In 1938, OKW's *Chi* succeeded here, too. When war started on September 3, 1939, the French War Ministry ordered this method to be used throughout. Thus, the Germans could read the French wireless signals from the first day without delay, which explains the advantages they had in the battle of France in June 1940. France had made the mistake (Sect. 11.1.3) of adopting as her main method an encryption method that had already been in restricted use for some time.

Rohrbach's lasting success between 1942 and September 1944 against the strip cipher method of U.S. diplomacy was already discussed in Sect. 14.3.6. The so-called CQ radio signals of the State Department in Washington to all its diplomatic representations played an important role in creating cryptotext-cryptotext and even plaintext-cryptotext compromises.

Less important was the success mentioned in Sect. 19.4.1 against the Rumanian military attaché. But military attachés as a rule are quite promising goals, with a mixture of military and diplomatic habits leading to interferences. While Field Marshal Rommel in North Africa was fighting against the British 8th Army under Field Marshal Montgomery in the fall of 1941, the Germans of the *Forschungsamt* in Berlin succeeded in penetrating the traffic of the American Military Attaché in Cairo, Colonel Frank Bonner Fellers (later Brigadier General and military secretary to General Douglas MacArthur)—partly because Fellers had the persistent habit of beginning his signals with stereotypes, partly because Italy, then still at peace with the U.S.A., had its *penetrazione squadra* 'borrow' the codebook from the American Embassy in Rome in order to copy it. Fellers reported day by day in the brand new BLACK code among other things the plans of the 8th Army for the next day, which could always be forwarded to Rommel within a few hours. The U.S.A. was cryptologically an unsafe ally of Great Britain. Italy was more prudent, or at least she thought so: The Chief of the Italian Intelligence Service, General Cesare Amè, did not provide the Germans with the codebook, but gave them only decryptions. Since the Germans also recorded the cryptotext signals, they had a perfect cleartext-cryptotext compromise and could reconstruct the codebook, and could even check the trustworthiness of their ally. The break had catastrophic consequences in June 1942 for an Allied convoy bound for Malta. Bletchley Park once more cast suspicion on their friends and Fellers had to resign. Nevertheless, he was decorated with the Distinguished Service Medal. He was almost as bad as Murphy.

In using their Hagelin M-209 machines, the cipher clerks of the U.S. Army proved no more disciplined than their German colleagues: they chose for message keys preferably six initial letters from their girlfriends' names, and thus usually the same keys a whole day long. Such in-phase encryptions allowed Erich Hüttenhain to penetrate daily the none-too-secure Beaufort encryption. Field Marshal Erwin Rommel profited from this (Otto Leiberich).

22.1.2 Angō Kenkyū Han. Japan tried in the 1930s to break not only Chinese codes, but primarily American ones. This was not too difficult, since despite Yardley's warning (Sect. 8.5.6) U.S. diplomatic cryptology was still irresponsible. Under Roosevelt a new code, BROWN, was introduced, but it came into the hands of a gang of safe-crackers in Zagreb, and thus was probably compromised, yet it was not taken out of use since 'only' criminals were involved. And Stanley K. Hornbeck wrote to his boss, the Secretary of State Stimson: "Mr. Secretary: I have the feeling that it is altogether probable that the Japanese are 'breaking' every confidential telegram that goes to and from us." The unreliability of American diplomatic codes was accepted as inevitable. In this situation it is not astonishing that the decryption service of the Japanese Foreign Ministry, *Angō Kenkyū Han*, was sometimes successful in decrypting the simpler codes, e.g., GRAY. But it had no joy with BROWN and *Toku mu Han*, the decryption service of the admiralty, was also unlucky. Nothing but a raid could help; under the command of captain Hideya Morikawa, towards the end of 1937 the BROWN code and the strip cipher device M-138, the appearance of which was not known to the Japanese, were photographed in the American consulate in Kobe. Nevertheless, they did not manage to read the M-138 traffic. The *tokumu han* sailors then concentrated on the related strip cipher device CSP 642 of the U.S. Navy (Sect. 14.1). They got results only slowly, because their methods were behind the times. But they had received the BAMS Code (Sect. 4.4.5) from the Germans, whose raider ship *Atlantis* had captured it on July 10, 1940. The Japanese only had to strip the superencryption, which they managed, of course.

In the opposite direction there was more success. The U.S.A., carrying the main load of the Allied war effort in the Pacific theater, scored also the most solutions of Japanese wireless signals. Japan may have felt that she was protected by her language being strange and inpenetrable to Westerners; but this was not the case. The Americans broke Japanese codes and ciphers in the 1920s (Yardley), 1930s (Holtwick), and 1940s (Rosen). There are reliable reports (see Sect. 19.3.2) that also the German side broke continually the PURPLE traffic.

Little is known about U.S. American success in breaking Soviet encryption after the Venona breaks. In 1972, during the Strategic Arms Limitation Talks, N.S.A. made a hit, "but the solution was a fluke, made possible by a Soviet enciphering error" (Kahn). Such a thing may happen now and then. And if it happened more often, there were good reasons not to brag about it.

22.1.3 Glavnoye Razvedyvatelnoye Upravlenie (Razvedupr, GRU). The 'Chief Intelligence Directorate' of the Soviet Union—notwithstanding its reputation in eavesdropping, spying, theft, and blackmailing—also had cryptanalytical successes, e.g., against Swiss diplomacy working with Hagelin machines as well as against Italy, not to speak of smaller nations.

Toward the end of the Second World War, the number of ENIGMA cipher documents seized by the Red Army grew to such an extent that the percentage of successes against Wehrmacht ENIGMA traffic was considerable. However, it seems that there were no codebreaking machines comparable to the Polish, British, and U.S. American bombs. In the Cold War era, according to Louis Tordella, the Soviet Union was even successful against the rotor machine KW-7 used by NATO. In 1992, David Kahn found a Russian living at the time in England, Victor Makarov, who had worked as an interpreter in the 16th Directorate of the K.G.B. (Director: General Andrei Nicolayevich Andreyev) and was familiar with its work. From him and by later contact with Andreyev, Kahn learned some details, among which was the contention that, from the end of 1941, Soviet cryptanalysts under Sergei Tolstoy had success against the Japanese PURPLE machine. However, a technically complete picture of Soviet cryptanalysis is still lacking.

22.2 Mode of Operation of the Unauthorized Decryptor

> It is naturally only the work of the unauthorized decryptor which is of mathematical interest, where here an experienced decryptor is understood. Experience in decryption must be won through many years of practice. Thus the decryptor, depending on his situation and inclination, will develop either a more linguistic or a more mathematical orientation. Solutions of sufficiently complicated methods are the collective works of several decryptors of both orientations, either of which in turn has yet further specialists. The mathematicians in particular need specialists in machine methods.
>
> *Hans Rohrbach* 1949

> Deciphering is an affair of time, ingenuity, and patience.
>
> *Charles Babbage* 1864

> The cryptographer's main requisites are probably patience, accuracy, stamina, a reasonable clear hand, some experience, and an ability to work with others.
>
> *Christopher Morris* 1992

Since I am not a professional decryptor, I feel it difficult and easy at the same time to speak about the work of the unauthorized decryptor. Difficult, because I have collected my experiences without the pressure of the professional environment and without sweat and tears. Easy, because I am not in danger of being infatuated by success or embittered by failures. However, my mathematical approach has helped me to systematize cryptological attack and defense.

Anyhow, the literature and my personal contacts have shown me that professional decryptors do not have an easy life. Rejewski, for example, went back after the war to communist Poland and had to choose a job as a business director rather than to make a university career. Alastair Denniston advised his son Robin: "Do what you like to do, but don't do what I do." Robin Denniston became a publisher. Sometimes it may have been difficult to keep absolutely silent, and this also for twenty, thirty or forty years; especially in situations like the one I. J. Good experienced: he was stationed in a hotel near Bletchley Park and was treated by a retired banking clerk to a lively description of a commercial ENIGMA which his bank had used in his earlier days.

22.2.1 Glamour and misery. The work of the professional cryptologist is thankless; he is not allowed to celebrate his success in public or with his friends, and not even his family will be allowed to know what he is doing. He is permanently in danger of being abducted or blackmailed. Such restrictions usually persist even after active duty.

On the other hand, Ralph V. Anderson, who in 1940 entered the code room at the U.S. Navy Department in Washington, D.C. and in 1946 joined the Department of State in cryptography, where he served for almost twenty years, confessed "If I had been given the choice of any position I wanted, I would have chosen the one I had."

22.2.2 Personality, It seems to be most difficult to give general rules and advice for the attitude to be applied in unauthorized decryption. Bazeries, who was a very successful cryptanalyst, recommended *changer son fusil d'épaule* , trying a new line of attack, which will only help people with enough imagination. Not to be blinkered, not to follow the beaten track; this is more easily said than done. Fresh ideas will help. The example of Alan Turing and Gordon Welchman shows this: in their inexperience lay their strength. Therefore they were better than Dillwyn Knox, who was much more experienced, but also less daring. As a team, Turing and Knox were unbeatable, and even Turing and Welchman together achieved more than the mere sum of their working power.

One thing will not happen to the successful unauthorized decryptor: he will not be discouraged by the alleged complexity of the task. The Poles were so successful because they automated any analysis that was too time consuming for hand work after they had found the idea. The expectations of *Chi* how long it would take to break the ENIGMA were cut by parallelization by a factor of six and by mechanization by a factor of at least twenty. The Welchman Bombe even makes trillions of cross-plugging possibilities irrelevant, as Welchman remarked proudly, because the avalanche propagation of the voltage in a relatively simple feedback circuit is done "in less than a thousandth of a second."

22.2.3 Strategies. In principle there are infinitely many ways of cryptanalytic attacks. In the following, only a rough survey of the strategies of cryptanalysis is given.

22.2.3.1 The purest form of unauthorized decryption makes no assumption whatsoever. This pure cryptanalysis does not use and does not need the linguist, for it is mathematical in nature. As *David Kahn* said, it functions in principle even for a language that the unauthorized decryptor does not know, e.g., the last intermediate text of a composition of two or more encryption x, say superencrypted code where the codebook is not known. Pure cryptanalysis is directly suited for execution by a machine and it can be written in the form of a computer program. Pure cryptanalysis as a rule needs longer texts than any of the attacks below. In some cases of a plaintext-plaintext compromise, e.g., determination of the period of a polyalphabetic encryption (Chapter 18) or in-phase adjustment and superimposition of several polyalphabetic encryptions with different initial key settings, as well as in the case of a cryptotext-cryptotext compromise (Chapter 19), pure cryptanalysis performs a reduction to an intermediate language which is a monoalphabetic, possibly polygraphic encryption of the plaintext language.

22.2.3.2 Pure cryptanalysis is a special case of the *cryptotext-only attack* ('known cryptotext attack'), which allows only reflections and assumptions on the kind of language the plaintext is taken from. Typically, the distribution of the frequencies of the single characters in the cryptotext is investigated. If it is reasonably close to one of several natural languages that could come under consideration, all encryption methods can be excluded which level frequencies, in particular proper polygraphic ones (provided they do not feign frequencies, as discussed in Sect. 4.1.2) and proper polyalphabetic encryptions; among the remaining monoalphabetic ones are functional simple substitutions, transpositions and their compositions. If even the individual letter frequencies are close to those of some natural language, proper simple substitutions can be excluded; among the remaining ones are transpositions, as well as polygraphic encryptions feigning a transposition.

Thus a frequency examination (Chapter 15) may break a monoalphabetic encryption; it may also be used to strip a simple substitution from a transposition.

However, if the distribution of the frequencies of the single characters in the cryptotext is leveled, then (provided the use of polyphones can be excluded) suspicion about a polyalphabetic and/or polygraphic encryption is justified. Both possibilities are to be taken into consideration. In the first case, pure cryptanalysis may help to find a reduction to a monoalphabetic simple or proper polygraphic substitution, which may be treated with a frequency examination (Chapter 15) of single characters or of polygrams. These examinations are already linguistic in nature.

22.2.3.3 Much more linguistic are the methods based on a partial or complete plaintext-cryptotext compromise. They use probable words or phrases as starting points for pattern finding (Chapters 13, 14). There is the *known plaintext attack* and the *chosen plaintext attack*, which differ only in the way the compromise is achieved, passively or actively. The known plaintext attack needs sly, clever guesses of plaintext fragments. Sympathetic understanding of the adversary's feelings, of his ways of thinking, of his idioms and phraseology is required, and this is helped by knowing not only the adversary's language, but also his milieu. The British in Bletchley Park had champions in preparing the confrontations of plaintext fragments and cryptotext, the cribs (Sect. 19.7.1); only a few of them can be mentioned here. As well as the linguist Hilary Hinsley née Brett-Smith and the linguistically versed mathematician Shaun Wylie, there were also people with a kind of abstract ability for pattern finding in Bletchley Park: the chess champion Hugh Alexander and the formally gifted Germanic philologist Mavis Batey née Lever. Her abilities can be illustrated by the fact that she noticed one day the absence of the letter L in a long fragment of ENIGMA ciphertext. To notice this was already unheard of. But she also concluded that there was a long filling with plaintext /l/. This led to the determination of the setting, to a lasting break, and finally to the victory of the British fleet over the Italian on March 28, 1941 near Cape Matapán on the Greek coast.

Success in the known plaintext attack requires, moreover, that the unauthorized decryptor is in possession of all the results of intelligence, by combat reconnaissance, by interrogation of prisoners, by questioning civilians, by eavesdropping, by spies, and particularly by decryptions achieved by other decryptors. This requirement is very much in conflict with security measures ('need to know' doctrine) and also politically unrealistic—otherwise it would have been best if in the Second World War Churchill himself had prepared the cribs.

The chosen plaintext attack, on the contrary, needs cunning in producing a compromise. Cunning is inexhaustible. Events that have been reported vary from inducing a certain combat action, like artillery fire in the First World War and the '*erloschen ist leuchttonne*' trick in the Second World War (Sect. 11.1.3), to foisting a message on the adversary, like the Japanese cuckoo's egg and Figl's newspaper forage (Sect. 11.1.2).

A third case, *derived plaintext attack*, comes from a cryptotext-cryptotext compromise if one of the systems is already broken and the plaintext can thus be obtained. This 'continuation of a break' was a frequent stratagem in Bletchley Park, where the emergency situations leading to a cryptotext-cryptotext compromise were deliberately induced ('gardening', Sect. 19.4.1).

22.2.3.4 A particular sort of attack, the *chosen ciphertext attack*, may be used in the case of asymmetric methods with a public key for encryption, when the functioning of a tamper-proof 'black box' for decryption is wanted, i.e., the private key is to be revealed.

22.2.4 Hidden dangers. A cryptotext-cryptotext compromise is particularly insidious because it is so easily overlooked. It may be caused by the installation of many key nets if there is a lack of cipher discipline (see Sect. 19.4.1) or it may be a consequence of cryptological thoughtlessness (indicator-doubling with the ENIGMA until May 1940, Sect. 19.6.1). Specific methods of attack are also discussed in Sects. 19.4 and 19.5 . Cryptotext-cryptotext compromises allow pure cryptanalysis which can be done with supercomputers. Since for public keys, cryptotext-cryptotext compromise is inherent in the system, the danger hopefully prompts increased wariness.

22.2.5 Deciphering in layers. For a composition of encryption methods, one normally aims at stripping off one encryption after another. This is easier, if a superencryption is made over an encryption method that has been used for some time and has been broken in the meantime: the intermediate text is then a known language. It is particularly simple if the superencryption method is already broken, for then the composition is no more resistant than the newly introduced method (S.D. superencryption, Sect. 19.6.3.1). Quite generally it can be stated that the German Armed Forces could not have educated their adversaries better regarding the ENIGMA: they introduced refinements in small steps, each time late enough that the Poles and the British had mastered the last step. And this happened on other occasions too: When in April 1944 at Mykonos documents on a *Reserve-Handschlüssel-Verfahren* fell into the wrong hands, the method was not changed totally, but only slightly and stepwise, thus educating the British.

22.2.6 Violence. Cryptanalysis in the proper sense does not include procurement of the adversary's encryption documents and devices (up to whole machines) by illegal purchase, spying out at customs offices, theft and burglary, or combat missions and raids (Sect. 11.1.10). The experiences of the Second World War have fully confirmed Kerckhoffs' admonition and Shannon's maxim "The enemy knows the system being used." The SIGABA (ECM Mark II) of the U.S. Army was one of the few devices of the Second World War that did not fall into the hands of the enemy, and this perhaps only because after the D-Day landings the war in Europe was over in less than a year.

Moreover, the destruction of the wire-bound communication channels of the adversary, which the Allies executed before and during the landings in Normandy, assisted cryptanalysis: its effect was "to force a proportion of useful intelligence on to the air" (Ralph Bennett).

22.2.7 Prevention. What can be done to prevent cryptanalysis, to protect communication channels? The most important defensive weapon seems to be imagination. It is necessary to enter completely into the cryptanalytic thinking of the hypothetical unauthorized decryptor, and to be able psychologically to do so. Inhibitions are as out of place as arrogance is. The defender should not only have imagination, he or she must have enough imagination to sense the imagination of the attacker.

There are three cases of grave thoughtlessness from the rich story of the ENIGMA decryption (for 1. and 2., examples are to be found in Fig. 51b):

1. It was absolutely unnecessary to rigorously abstain from using the same rotor in the same position on two consecutive days, as the *Luftwaffe* did ('non-crashing wheel order'), or from using the same wheel order twice in the same month. This mock randomness saved the British a lot of work in finding the wheel order, once a continuous flow of encryptions was established.

2. It was bad to avoid the use of two consecutive letters, like /a/ and /b/, for steckering, since this reduced the number of plugboard connections to be tested in the bombes and even allowed the British to build a special catch circuit which they wittily called CSKO, 'consecutive stecker knock-out'.

3. It was stupid to make the entrance substitution (performed by the plugboard) involutory, thus allowing the diagonal board. Neither the British TYPEX imitation nor the Japanese PURPLE had this 'simplification'. In fact, the Germans occasionally used the *Uhr* box[3] (Plate M), an artificial and awkward attachment that made the plugboard substitution (but not the full ENIGMA encryption) non-involutory, putting the diagonal board 'out of business' (Welchman). The Uhr box was to be changed frequently, probably every full hour—a telling sign that some German authorities now had serious doubts about the security of the ENIGMA, but could not help it any more.

[3] The *Uhr* box, introduced in 1944, amounts to using 10 stecker pairs and has 40 positions, 10 of which (0, 4, 8, ... 36) preserve involution. The scrambler inside performs a permutation with the cycle representation (1 31 5 39 9 23 17 27 33 19 21 3 29 35 13 11) (0 6 16 26) (2 4 18 24) (12 38 32 22) (14 36 34 20) (7 25) (8 30) (10 28) (15 37).

The faults around the ENIGMA were called by Welchman "a comedy of errors." He wrote: "The German errors ... stemmed from not exploring the theory of the Enigma cipher machine in sufficient depth, from weakness in machine operating procedures, message-handling procedures, and radio net procedures; and above all from failure to monitor all procedures." Then he went on to mention the indicator doubling, the 'Cillis' and Herivel tips, 'Parkerism' (a habit of the German producer of operating instructions that flourished in 1942, of repeating entire monthly sequences of discriminants, ring settings, wheel orders, or steckers; e.g. SCORPION settings were copies of PRIMROSE settings for the previous month); and not least 'inadvertent assistance' of German staff members in providing cribs. All these faults may be blamed solely on people: he wrote "the [ENIGMA] machine as it was would have been impregnable if it had been used properly." Nevertheless, German cipher security improved throughout 1944 and the first half of 1945 (Philip Marks). But the worst mistake, repeated from 1933 until 1945, was "introducing cryptographic improvements in a piecemeal fashion".

22.3 Illusory Security

Welchman could have added ironically, that just this can never be expected—in line with Rohrbach's maxim (Sect. 11.2.5) that no machine and no crypto system will ever be used properly all the time. The wartime cryptanalyst and peacetime mathematician Hans Rohrbach knew it. Adolf Paschke, *Vortragender Legationsrat im AA* and nominal head of the linguistic group in Pers Z, knew it as well. He strongly disadvocated to use the ENIGMA in the diplomatic channels even for topics of lesser importance, like visa regulations. Cipher machines were frowned upon. The *Geheimschreiber* T 52a was considered unsafe; in fact it was discovered in the AA how it could be cryptanalyzed without much labor. This explains the parallel success Beurling had. And T 52e messages transmitted by Military Attachées on Foreign Office channels were decrypted by Pers Z people themselves. Only for non-secret traffic within Germany on wire lines was T52 considered acceptable. One exception was made in 1944 on the wireless line between Madrid and Berlin, where a SZ 42 *Schlüsselzusatz* was used for messages up to *Geheim*, but not for the top classification *Geheime Reichssache*. This reflects the caution Pers Z took. And Erich Fellgiebel, Chief of OKW Signal Communications, is said to have it expressed by exaggerating: '*Funken ist Landesverrat*'.

But otherwise and elsewhere, the spirit of illusory security blossomed. Wheresoever there was a chance for a quicker and less secure cryptographic method, it had good prospects. Wishful thinking prevailed. Apart from the rare cases when individual keys were used at all, let alone made properly and run with care, very few cryptological systems remained unbroken between 1900 and 1950. In the ENIGMA case, the Navy key nets ‚Neptun', ‚Thetis', ‚Aegir' and ‚Sleipnir' were impregnable, but some of them had very little traffic or carried messages considered not to be important enough by the Allies.

Supervision of the own side's traffic was occasionally done, for example when Rowlett found a weakness as Friedman's Converter M-228 SIGCUM went into operation in January 1943 (see Sect. 8.8.6). It would have paid back, if for example a OKW Special Group had supervised the ENIGMA traffic of Göring's undisciplined *Luftwaffe* key net RED; they would have found the leakages the British profited from early enough to stop further disaster.

But even supervision does not help, if it is done insufficiently. Paschke knew, of course, that the AA one-time pads were fabricated mechanically by an array of 48 five-digit counters; after every printing step, most of them were moved forward at an irregular interval ('complementary propulsion'). As a special precaution, consecutively printed sheets were never assembled into the same block. This seemed to be completely sufficient, but it was not, as the FLORADORA story (see Sect. 8.8.7) shows.

22.4 Importance of Cryptology

The reader who has read this book chapter by chapter may at first have found it difficult to suppress a smile from time to time. The history of cryptology is full of exciting, funny, personal stories. That makes it attractive even for the layman.

Little by little, however, somber shadows are cast over the scene. The battle of Tannenberg gives a first example. The entry of the U.S.A. into the First World War was triggered by a telegram on January 16, 1917 from the German Foreign Minister Arthur Zimmermann to the ambassador in Mexico, Heinrich von Eckardt, that was decrypted in London's Room 40 of the Admiralty by Nigel de Grey (1886–1951). Its content—a proposal to stir up Mexico against its northern neighbor—was brought to the notice of President Wilson, who concluded that "right is more precious than peace." And the events of the Second World War were played out in front of a hideous backdrop. The decades of the Cold War displayed a cruelty which the romantics of spy novels cannot wipe away.

Talking to a former cryptanalyst in his official capacity always needs tact and discretion. Sometimes one confronts the arrogance of the professional who shows that he knows something but does not show what he knows. However, to be prudent is good advice for the professional, as the example of Welchman shows, who faced persecution after publication of his book *The Hut Six Story*.

22.4.1 Scruples. Cryptanalysis was felt by many of the people involved as a heavy burden, not so much because of the nervous stress, but because of conflicts of conscience. Cryptology shares this hardship not only with other branches of mathematics and computer science which are in danger of misuse but to a large extent with other sciences like physics, chemistry, and biology—it may suffice to mention the keywords nuclear energy, poison gas, and genetic manipulation. The price our century has paid for the enormous progress of science—which nobody wishes to forfeit—must also be paid by the scientists themselves. They must measure up to high requirements of humanity. The decline of some communist systems of injustice and the increasing bewilderment of people faced with unlimited possibilities raises the hope that scientists show insight and discretion. Thus, cryptology no more deserves condemnation than do the natural sciences. With a positive accent the back-cover text of the book by Meyer and Matyas says: "Cryptography is the only known practical means for protecting information transmitted through large communication networks such as telephone lines, microwave, or satellite." Elsewhere we read: "Cryptology has metamorphosized from an arcane art to a respectable subdiscipline of Computer Science." Common saying puts it this way: "Today, code-making and code-breaking are games anybody can play."

In fact, original scientific papers on cryptological themes are found today not only in the few specialist journals and symposia, but here and there also in computer science, particularly in theoretical computer science. Contact and mutual fertilization occur mainly with the emerging theory of complexity and the theory of formal languages; moreover, from mathematics, number theory and combinatorics are confederates.

22.4.2 New ideas. Cryptology itself has developed new ways of thinking in connection with the information theory of Shannon and Rényi, and in connection with public keys has pushed new concepts such as asymmetric crypto systems and authentication to the fore. Authentication even widens the aspect of secrecy to more general perspectives of communication. The central concept is the *protocol*, an agreed-upon method and procedure of communication; a cryptographic protocol between two partners includes not only measures based on mistrust against a third party, but also on mutual partial mistrust.

The problem may be how two partners can share certain secrets without thus sacrificing other secrets. Another problem may be how two partners can build up confidence step by step without the risk of revealing some secrets. Applications in daily private, public, political, and economic life are obvious; they concern the behavior of spouses, powers, parties, and firms. Everyday examples are the certification procedure of the holder of a check card that he or she is its legitimate possessor and thus its legal owner, or a licensing negotiation, where the inventor has to convince the presumptive licensee about the usefulness and efficiency of his method, without compromising this before the contract is signed.

This is the idea of a *covert proof*, 'zero-knowledge' proof: the partner will not be told anything that he could not find out himself.

The problem is quite old: In the times of Tartaglia and Cardano, mathematicians tried to keep their methods secret. They were willing to apply a method, say for the solution of algebraic equations by radicals, secretly to examples they were given by an opponent, and then after a short while to present a solution like a rabbit from a hat, which everybody could easily check for correctness. Step by step, the spectators' confidence in the efficiency and correctness of the hidden method increased, until it was established without reasonable doubt, and still the method had not been given away. As we know, poor Niccolò Tartaglia was not successful at this game, and Cardano managed to trick him out of his method for cubic equations. He deserves our sympathy; today he would have found a better defense.

22.4.3 Deciphering the secrets of nature. Cryptanalysis in the widest sense even surpasses the frame of communication engineering. The scientific exploration of nature is frequently a cryptanalysis of her secrets.

To give an example: X-ray crystallography of proteins is a cryptanalytic task. To be determined is the phase function that belongs to a given amplitude function (in a three-dimensional space) measured in an X-ray refraction image. If and only if the phase function (the keytext) and the amplitude function (the cryptotext) fit such that they describe a physical reality (the plaintext) with positive electron density and the correct number of atoms, is the decryption (usually unique) successful. Assumptions on the structure of the molecules—e.g., the double helix structure of DNA as successfully guessed by Watson and Crick—play the role of probable words. This aspect was already discussed by Alan Turing and David Sayre in the 1950s.

Finding a needle in a haystack is an example of a coincidence examination. Smoothing of signals, discussed by Norbert Wiener in the 1950s, particularly stripping off noise from a signal, has a parallel in stripping an encryption. Even more serious is the cognitive task of detecting patterns of any hitherto unknown sort in a mass of data, which corresponds to advanced methods of cryptanalysis, like Friedman examination and Kullback examination.

It remains to be seen to what extent concepts surpassing Shannon's information theory, e.g., use of Kullback entropy, are relevant here. First steps in this area are outlined in the appendix *Axiomatic Information Theory*.

Finally, there is the main occupation of the thinking man: recognizing situations, forming concepts, elaborating abstractions. This, too, is in the widest sense a cryptanalytical task: It means finding something secret, something already existing in secrecy. Reading between the lines is the task, and intelligence is needed. Pure cryptanalysis tries to do this without further knowledge, without the help of intuition, but its results are limited, as is the reach of Artificial Intelligence. Where it works, it has the advantage of running automatically. The wide orchestra of cryptanalysis, however, uses intuition too, uses slyness and cunning. To show this interplay has been the main aim of this book. Cryptanalysis as a prototype for the methods in science: this has been my guiding principle in writing this book. Charles Babbage said (*Passages from the Life of a Philosopher*): "Deciphering is, in my opinion, one of the most fascinating of arts, and I fear I have wasted upon it more time than it deserves." I have spared no pains, but I hope I have not wasted my time.

Appendix: Axiomatic Information Theory

The logic of secrecy was the mirror-image
of the logic of information

Colin Burke 1994

Perfect security was promised at all times by the inventors of cryptosystems, particularly of crypto machines (Bazeries: *je suis indéchiffrable*). In 1945, Claude E. Shannon(1916–2001) gave in the framework of his information theory a clean definition of what could be meant by perfect security. We show in the following that it is possible to introduce the cryptologically relevant part of information theory axiomatically.

Shannon was in contact with cryptanalysis, since he worked 1936–1938 in the team of Vannevar Bush, who developed the COMPARATOR for determination of character coincidences. His studies in the Bell Laboratories, going back to the year 1940, led to a confidential report (*A Mathematical Theory of Cryptography*) dated Sept. 1, 1945, containing apart from the definition of Shannon entropy (Sect. 16.5) the basic relations to be discussed in this appendix. The report was published four years later: *Communication Theory of Secrecy Systems*, Bell System Technical Journal 28, 656-715 (1949).

A.1 Axioms of an Axiomatic Information Theory

It is expedient to begin with events, i.e., sets $\mathcal{X}, \mathcal{Y}, \mathcal{Z}$ of elementary events, and with the uncertainty ('equivocation') on events—the uncertainties expressed by non-negative real numbers. More precisely,

$H_{\mathcal{Y}}(\mathcal{X})$ denotes the uncertainty on $\mathcal{X}$, provided $\mathcal{Y}$ is known.

$H(\mathcal{X}) = H_{\emptyset}(\mathcal{X})$ denotes the uncertainty on $\mathcal{X}$, provided nothing is known.

A.1.1 Intuitively patent axioms for the real-valued binary set function H:

(0) $0 \leq H_{\mathcal{Y}}(\mathcal{X})$ ("Uncertainty is nonnegative.")

For $0 = H_{\mathcal{Y}}(\mathcal{X})$ we say "$\mathcal{Y}$ uniquely determines $\mathcal{X}$."

(1) $H_{\mathcal{Y}\cup\mathcal{Z}}(\mathcal{X}) \leq H_{\mathcal{Z}}(\mathcal{X})$ ("Uncertainty decreases, if more is known.")

For $H_{\mathcal{Y}\cup\mathcal{Z}}(\mathcal{X}) = H_{\mathcal{Z}}(\mathcal{X})$ we say "$\mathcal{Y}$ says nothing about $\mathcal{X}$."

The critical axiom on additivity is

(2) $H_{\mathcal{Z}}(\mathcal{X} \cup \mathcal{Y}) = H_{\mathcal{Y}\cup\mathcal{Z}}(\mathcal{X}) + H_{\mathcal{Z}}(\mathcal{Y})$.

This says that uncertainty can be built up additively over events.

The classical stochastic model for this axiomatic information theory is based on $p_X(a) = \Pr[X = a]$*, the probability that the random variable* X *assumes the value* a*, and defines*

$$H_\emptyset(\{X\}) = -\sum_{s\,:\,p_X(s)>0} p_X(s) \cdot \mathrm{ld}\; p_X(s)$$

$$H_\emptyset(\{X\} \cup \{Y\}) = -\sum_{s,t\,:\,p_{X,Y}(s,t)>0} p_{X,Y}(s,t) \cdot \mathrm{ld}\; p_{X,Y}(s,t)$$

$$H_{\{Y\}}(\{X\}) = -\sum_{s,t:\; p_{X/Y}(s/t)>0} p_{X,Y}(s,t) \cdot \mathrm{ld}\; p_{X/Y}(s/t)$$

where $p_{X,Y}(a,b) =_{\mathrm{def}} \Pr[(X = a) \wedge (Y = b)]$ *and* $p_{X/Y}(a/b)$ *obeys Bayes' rule for conditional probabilities:*

$$p_{X,Y}(s,t) = p_Y(t) \cdot p_{X/Y}(s/t) \quad , \textit{thus}$$

$$-\mathrm{ld}\; p_{X,Y}(s,t) = -\mathrm{ld}\; p_Y(t) - \mathrm{ld}\; p_{X/Y}(s/t) \ .$$

A.1.2 From the axioms (0), (1), and (2), all the other properties usually derived for the classical model can be obtained.

For $\mathcal{Y} = \emptyset$, (2) yields

(2a) $H_\mathcal{Z}(\emptyset) = 0$ ("There is no uncertainty on the empty event set.")

(1) and (2) imply

(3a) $H_\mathcal{Z}(\mathcal{X} \cup \mathcal{Y}) \le H_\mathcal{Z}(\mathcal{X}) + H_\mathcal{Z}(\mathcal{Y})$ ("Uncertainty is subadditive.")

(0) and (2) imply

(3b) $H_\mathcal{Z}(\mathcal{Y}) \le H_\mathcal{Z}(\mathcal{X} \cup \mathcal{Y})$ ("Uncertainty increases with larger event set.")

From (2) and the commutativity of $. \cup .$ follows

(4) $H_\mathcal{Z}(\mathcal{X}) - H_{\mathcal{Y}\cup\mathcal{Z}}(\mathcal{X}) = H_\mathcal{Z}(\mathcal{Y}) - H_{\mathcal{X}\cup\mathcal{Z}}(\mathcal{Y})$

(4) suggests the following definition:

The mutual information of $\mathcal{X}$ and $\mathcal{Y}$ under knowledge of $\mathcal{Z}$ is defined as

$$I_\mathcal{Z}(\mathcal{X}, \mathcal{Y}) =_{\mathrm{def}} H_\mathcal{Z}(\mathcal{X}) - H_{\mathcal{Y}\cup\mathcal{Z}}(\mathcal{X}) \ .$$

Thus, the mutual information $I_\mathcal{Z}(\mathcal{X}, \mathcal{Y})$ is a symmetric (and because of (1) nonnegative) function of the events $\mathcal{X}$ and $\mathcal{Y}$. From (2),

$$I_\mathcal{Z}(\mathcal{X}, \mathcal{Y}) = H_\mathcal{Z}(\mathcal{X}) + H_\mathcal{Z}(\mathcal{Y}) - H_\mathcal{Z}(\mathcal{X} \cup \mathcal{Y}) \ .$$

Because of (4), "$\mathcal{Y}$ says nothing about $\mathcal{X}$" and "$\mathcal{X}$ says nothing about $\mathcal{Y}$" are equivalent and are expressed by $I_\mathcal{Z}(\mathcal{X}, \mathcal{Y}) = 0$. Another way of saying this is that under knowledge of $\mathcal{Z}$, the events $\mathcal{X}$ and $\mathcal{Y}$ are mutually independent.

In the classical stochastic model, this situation is given if and only if X, Y *are independent random variables:* $p_{X,Y}(s,t) = p_X(s) \cdot_Y (t)$.

$I_\mathcal{Z}(\mathcal{X}, \mathcal{Y}) = 0$ is equivalent with the additivity of H under knowledge of $\mathcal{Z}$:

(5) $I_\mathcal{Z}(\mathcal{X}, \mathcal{Y}) = 0$ if and only if $H_\mathcal{Z}(\mathcal{X}) + H_\mathcal{Z}(\mathcal{Y}) = H_\mathcal{Z}(\mathcal{X} \cup \mathcal{Y})$.

A.2 Axiomatic Information Theory of Cryptosystems

For a cryptosystem $\mathbf{X}$, events in the sense of abstract information theory are sets of finite texts over Z_m as an alphabet. Let P be a plaintext(-event), C a cryptotext(-event), K a keytext(-event).[1] The uncertainties $H(K),\ H_C(K),\ H_P(K),\ H(C),\ H_P(C),\ H_K(C),\ H(P),\ H_K(P),\ H_C(P)$ are now called equivocations.

A.2.1 First of all, from (1) one obtains

$$\begin{aligned} H(K) &\geq H_P(K)\ , \quad & H(C) &\geq H_P(C)\ ,\\ H(C) &\geq H_K(C)\ , \quad & H(P) &\geq H_K(P)\ ,\\ H(P) &\geq H_C(P)\ , \quad & H(K) &\geq H_C(K)\ . \end{aligned}$$

A.2.1.1 If $\mathbf{X}$ is functional, then C is uniquely determined by P and K, thus

(CRYPT) $H_{P,K}(C) = 0$, i.e.,

$$I_K(P,C) = H_K(C)\ , \quad I_P(K,C) = H_P(C)$$

("plaintext and keytext together allow no uncertainty on the cryptotext.")

A.2.1.2 If $\mathbf{X}$ is injective, then P is uniquely determined by C and K, thus

(DECRYPT) $H_{C,K}(P) = 0$, i.e.,

$$I_C(K,P) = H_C(P)\ , \quad I_K(C,P) = H_K(P)$$

("cryptotext and keytext together allow no uncertainty on the plaintext.")

A.2.1.3 If $\mathbf{X}$ is Shannon, then K is uniquely determined by C and P, thus

(SHANN) $H_{C,P}(K) = 0$, i.e.,

$$I_P(C,K) = H_P(K)\ , \quad I_C(P,K) = H_C(K)$$

("cryptotext and plaintext together allow no uncertainty on the keytext.")

A.2.2 From (4) follows immediately

$$\begin{aligned} H_K(C) + H_{K,C}(P) &= H_K(P)\ , \quad & H_P(C) + H_{P,C}(K) &= H_P(K)\ ,\\ H_C(P) + H_{C,P}(K) &= H_C(K)\ , \quad & H_K(P) + H_{K,P}(C) &= H_K(C)\ ,\\ H_P(K) + H_{P,K}(C) &= H_P(C)\ , \quad & H_C(K) + H_{C,K}(P) &= H_C(P)\ . \end{aligned}$$

With (1) this gives

Theorem 1:

(CRYPT) implies $H_K(C) \leq H_K(P)\ , \quad H_P(C) \leq H_P(K)$,
(DECRYPT) implies $H_C(P) \leq H_C(K)\ , \quad H_K(P) \leq H_K(C)$,
(SHANN) implies $H_P(K) \leq H_P(C)\ , \quad H_C(K) \leq H_C(P)$.

A.2.3 In a cryptosystem, $\mathbf{X}$ is normally injective, i.e., (DECRYPT) holds. In Fig. 163, the resulting numerical relations are shown graphically. In the classical professional cryposystems, there are usually no homophones and the

[1] Following a widespread notational misusage, in the sequel we replace $\{X\}$ by X and $\{X\} \cup \{Y\}$ by X, Y ; we also omit $\emptyset$ as subscript.

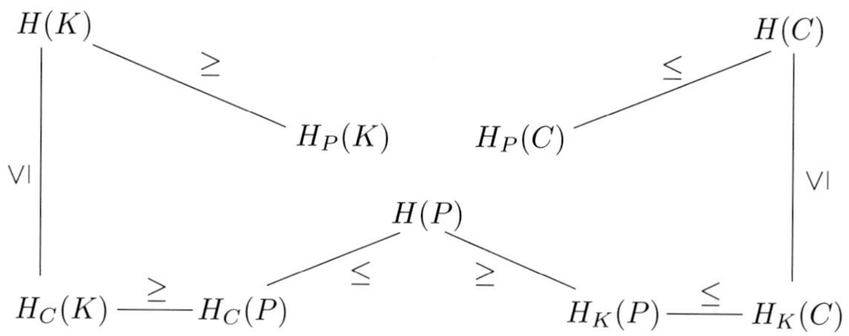

Fig. 163. Numerical equivocation relations for injective cryptosystems

Shannon condition (2.6.4) holds. Monoalphabetic simple substitution and transposition are trivial, and VIGENÈRE, BEAUFORT, and in particular VERNAM are serious examples of such classical cryptosystems.

The conjunction of any two of the three conditions (CRYPT), (DECRYPT), (SHANN) has far-reaching consequences in view of the antisymmetry of the numerical relations:

Theorem 2:
(CRYPT) $\wedge$ (DECRYPT) implies $H_K(C) = H_K(P)$

("Uncertainty on the cryptotext under knowledge of the keytext equals uncertainty on the plaintext under knowledge of the keytext,")

(DECRYPT) $\wedge$ (SHANN) implies $H_C(P) = H_C(K)$

("Uncertainty on the plaintext under knowledge of the cryptotext equals uncertainty on the keytext under knowledge of the cryptotext,")

(CRYPT) $\wedge$ (SHANN) implies $H_P(K) = H_P(C)$.

("Uncertainty on the keytext under knowledge of the plaintext equals uncertainty on the cryptotext under knowledge of the plaintext.")

In Fig. 164, the resulting numerical relations for classical cryptosystems with (CRYPT), (DECRYPT), and (SHANN) are shown graphically.

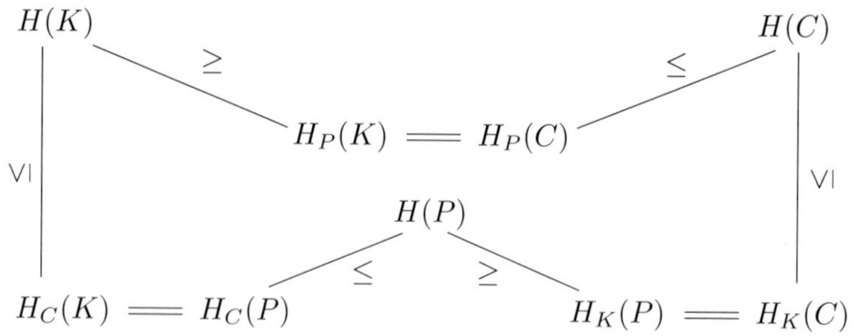

Fig. 164. Numerical equivocation relations for classical cryptosystems

A.3 Perfect and Independent Key Cryptosystems

A.3.1 A cryptosystem is called a perfect cryptosystem, if plaintext and cryptotext are mutually independent:

$$I(P, C) = 0 \ .$$

This is equivalent to $H(P) = H_C(P)$ and to $H(C) = H_P(C)$
("Without knowing the keytext: knowledge of the cryptotext does not change the uncertainty on the plaintext, and knowledge of the plaintext does not change the uncertainty on the cryptotext")

and is, according to (5), equivalent to $H(P, C) = H(P) + H(C)$.

A.3.2 A cryptosystem is called an independent key cryptosystem, if plaintext and keytext are mutually independent:

$$I(P, K) = 0 \ .$$

This is equivalent to $H(P) = H_K(P)$ and to $H(K) = H_P(K)$
("Without knowing the cryptotext: knowledge of the keytext does not change the uncertainty on the plaintext, and knowledge of the plaintext does not change the uncertainty on the keytext")

and, according to (5), is equivalent to $H(K, P) = H(K) + H(P)$.

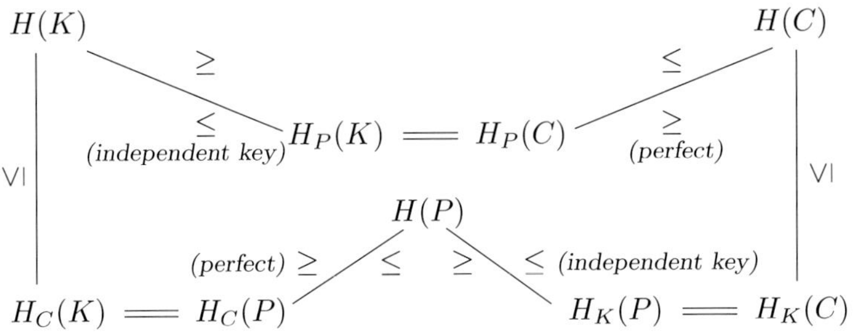

Fig. 165. Numerical equivocation relations for classical cryptosystems, with additional properties *perfect* and/or *independent key*

A.3.3 Shannon also proved a pessimistic inequality.

Theorem 3^K: In a perfect classical cryptosystem (Fig. 165),

$$H(P) \le H(K) \quad \text{and} \quad H(C) \le H(K) \ .$$

Proof: $H(P) \le H_C(P)$ *(perfect)*
$H_C(P) \le H_C(K)$ (DECRYPT), Theorem 1
$H_C(K) \le H(K)$ (1) .

Analogously with (CRYPT) for $H(C)$. ⋈

Thus, in a perfect classical cryptosystem, the uncertainty about the key is not smaller than the uncertainty about the plaintext, and not smaller than the uncertainty about the cryptotext.

From (SHANN) $\wedge$ (DECRYPT) with Theorem 1 we find $H_C(P) = H_C(K)$; after adding $H(C)$ on both sides, according to (2) we get $H(P,C) = H(K,C)$. In a perfect cryptosystem, $H(P,C) = H(P) + H(C)$.
Further, according to (2), $H(K,C) = H(K) + H_K(C)$. Thus

$$H_K(C) = H(P) - (H(K) - H(C)) = H(C) - (H(K) - H(P)) .$$

In Fig. 166, this result is displayed graphically.

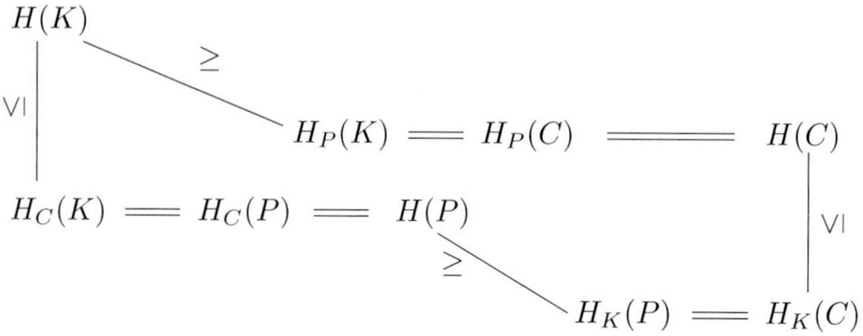

Fig. 166. Numerical equivocation relations for perfect classical cryptosystems

A.3.4 By a cyclic shift of K, C, P:

Theorem 3^C: In a classical cryptosystem with independent key,

$$H(K) \le H(C) \quad \text{and} \quad H(P) \le H(C) \quad \text{as well as}$$

$$H_C(P) = H(K) - (H(C) - H(P)) = H(P) - (H(C) - H(K)) .$$

A.4 Shannon's Main Theorem

A.4.1 For a classical cryptosystem which is both perfect and independent key, Theorems 3^K and 3^C imply immediately that $H(K) = H(C)$.

A.4.2 A cryptosystem with coinciding $H(K)$ and $H(C)$ shall be called a cryptosystem of Vernam type. Examples are given by encryptions with VIGENÈRE, BEAUFORT, and particularly VERNAM encryption steps, but also by linear polygraphic block encryptions.

In the stochastic model this condition is particularly fulfilled, if both C and K are texts of k characters with maximal $H(K)$ and maximal $H(C)$: $H(K) = H(C) = k \cdot \mathrm{ld}\, N$.

Main Theorem (Claude E. Shannon 1949):

In a classical cryptosystem, any two of the three properties

perfect ,
independent key ,
of Vernam type

imply the third one.

The proof is obvious from Fig. 165.

A.4.3 A sufficient condition for a classical cryptosystem to be perfect is that it is independent key and of Vernam type; these conditions can be guaranteed from outside. Then $H(P) \leq H(C) = H(K)$.

In the stochastic model, perfect security requires with $H(P) \leq H(K)$ that the key possesses at least as many characters as the plaintext, which means that every description of the key is at least as long as the key itself (Chaitin's requirement, Sect. 8.8.4).

Thus, perfect security requires safe distribution of an independent key which provides for every plaintext character a key character—an extreme requirement, which frequently cannot be fulfilled in practice. Non-perfect practical security is guaranteed only by the time required for breaking the encryption.

A.4.4 Shannon discussed a further property of a cryptosystem. We call a cryptosystem ideal (Shannon: strongly ideal), if cryptotext and keytext are mutually independent:

$$I(K, C) = 0 \ .$$

This is equivalent to $H(K) = H_C(K)$ and to $H(C) = H_K(C)$.

According to Shannon, ideal cryptosystems have practical disadvantages: for a perfect cryptosystem, $H(K) = H(P)$ must hold. Perfect ideal cryptosystems are necessarily adapted to the plaintext language, which usually is a natural language. In this case, rather complicated encryption algorithms are necessary. Also, transmission errors inevitably cause an avalanche effect. In fact, we have here a practically unattainable ideal.

A.5 Unicity Distance

The condition $H_C(P) > 0$ expresses that for known cryptotext there remains some uncertainty on the plaintext. For a classical cryptosystem with independent key (not necessarily perfect) this means, by Theorem 3^C,

$$H(K) > H(C) - H(P) \ .$$

We now use the stochastic model, with plaintext words V^* and cryptotext words W^* over a character set $V = W$ of N characters. We restrict our attention to words of length k.

Following Hellman (1975), we assume that N_P and N_C are numbers such that among the N^k words of length k the number of meaningful, i.e., possibly occurring, ones is just $(N_P)^k$ and the number of occurring cryptotexts is just $(N_C)^k$. Then $N_P \leq N$ and $N_C \leq N$. If all these texts occur with equal probability, then in the stochastic model

$$H(P) = k \cdot \mathrm{ld}\, N_P \ , \quad H(C) = k \cdot \mathrm{ld}\, N_C \ .$$

Furthermore, we assume that Z is the cardinality of the class of methods, i.e., the number of key words. Assume that all these key words occur with equal probability. Then

$$H(K) = \operatorname{ld} Z \ .$$

The inequality above, meaning the existence of an uncertainty, turns into

$$\operatorname{ld} Z > k \cdot (\operatorname{ld} N_C - \operatorname{ld} N_P)$$

or, provided $\operatorname{ld} N_C > \operatorname{ld} N_P$,

$$k < U, \quad \text{where } U = \frac{1}{\operatorname{ld} N_C - \operatorname{ld} N_P} \cdot \operatorname{ld} Z \ .$$

Thus, if $k \geq U$, there is no uncertainty. U is a unicity distance (Sect. 12.6).

If N_C is maximal, $N_C = N$, i.e., if all possible cryptotexts occur with equal probability, and if $N_P < N$, i.e., plaintexts are in a natural language, then the condition $\operatorname{ld} N_C > \operatorname{ld} N_P$ is certainly fulfilled, and the unicity distance is

$$U = \frac{1}{\operatorname{ld} N - \operatorname{ld} N_P} \cdot \operatorname{ld} Z \ ;$$

it is determined solely by the Shannon entropy $\operatorname{ld} N_P$ of the plaintext words. This depends in turn on the cryptanalytic procedure. If the analysis is limited to single-letter frequencies, then the Shannon entropy $\operatorname{ld} N_P^{(1)}$ is to be considered, the values of which are not very different in English, French, or German, and amount in the Meyer-Matyas count to $\operatorname{ld} N_P^{(1)} \approx 4.17$ [bit], where $N = 26$ and $\operatorname{ld} N = \operatorname{ld} 26 \approx 4.70$ [bit]. Furthermore, with $\operatorname{ld} N_P^{(2)} \approx 3.5$ [bit] for bigram frequencies and $\operatorname{ld} N_P^{(3)} \approx 3.2$ [bit] for trigram frequencies, we find

(1) $U \approx \frac{1}{0.53} \operatorname{ld} Z$ for decryption with single-letter frequencies,

(2) $U \approx \frac{1}{1.2} \operatorname{ld} Z$ for decryption with bigram frequencies,

(3) $U \approx \frac{1}{1.5} \operatorname{ld} Z$ for decryption with trigram frequencies.

For plaintext words, the average length is about 4.5 and the corresponding Shannon entropy about $\operatorname{ld} N_P^{(\mathrm{w})} \approx 2.6$ [bit], thus

(w) $U \approx \frac{1}{2.1} \operatorname{ld} Z$ for decryption with word frequencies.

The Shannon entropy of the English language under consideration of all, even grammatical and semantic, side conditions is considerably smaller; a value of about $\operatorname{ld} N_P^{(*)} \approx 1.2$ [bit] seems about right. This gives the unicity distance

(*) $U \approx \frac{1}{3.5} \operatorname{ld} Z$ for decryption in free style,

which is also given in Sect. 12.6 .

For simple (monographic) substitution with $Z = 26!$, we have $\operatorname{ld} Z = 88.38$ (Sect. 12.1.1.1); this leads to the values 167, 74, 59, 42, and 25 for the unicity distance, which are confirmed by practical experience. The situation is rather similar for the German, French, Italian, Russian, and related Indo-European languages.

A.6 Code Compression

Although Shannon was led to his information theory by his occupation with cryptological questions during the Second World War, information theory, in the form relevant and interesting for communication engineering, has no secrecy aspects. Its practical importance lies more in showing how to increase the transmission rate by suitable coding, up to a limit which corresponds to a message without any redundancy—say a message P of k characters with the maximal uncertainty $H(P) = k \cdot \mathrm{ld}\, N$.

The cryptological results above apply immediately to communication channels. Theoretically, a transmission requiring $\mathrm{ld}\, 26 = 4.70$ [bit/char] can be compressed by coding to one requiring only about 1.20 [bit/char]. A good approximation of this rate needs tremendous circuitry. The simplest case of a Huffman coding works on single characters only and reduces the transmission rate only to about 4.17 [bit/char], while Huffman coding for bigrams and trigrams, which needs a larger memory, does not bring a dramatic reduction. In future, however, economic and practical redundancy elimination by Huffman coding for tetragrams should be within reach using special chips.

The situation is different for the transmission of pictures. The compression obtainable by relatively simple methods is remarkable and finds increasingly practical use. For these applications, the truism of post-Shannon cryptology, that code compression of the plaintext is a useful step in improving the practical security of a cryptosystem, is particularly appropriate.

A.7 Impossibility of Complete Disorder

When in the 1920s the use of independent ("individual") keys was recommended, their fabrication did not seem to be a problem. That an individual key should be a random sequence of key characters was intuitively clear. After the work of Shannon and particularly of Chaitin in 1974, all attempts to produce a random sequence algorithmically had to be dropped. If keys were to be generated by algorithms, genuine random keytexts were not attainable. Thus, some order had to remain—the question was which one.

Consequently, 'pseudo random sequences' with a long period were increasingly suspected of having hidden regularities that would help cryptanalysis, although concrete examples are so far lacking in the open literature. The professionals responsible for the security of their own systems were faced with more and more headaches, while aspiring codebreakers could always hold out the hope of unexpected solutions.

Strangely, at about this time a similar development took place in mathematics. In 1973, H. Burkill and L. Mirsky wrote:

> There are numerous theorems in mathematics which assert, crudely speaking, that every system of a certain class possesses a large subsystem with a higher degree of organization than the original system.

We give a number of examples:

(1) Every graph of n nodes contains either a large subgraph of k nodes which is connected, or a large subgraph of k nodes which is unconnected. (k is the Ramsey number, e.g., $k = 6$ for $n = 102$, F. P. Ramsey 1930)

(2) Every bounded infinite sequence of complex numbers contains a convergent infinite subsequence. (K. Weierstrass 1865)

(3) If the natural numbers are partitioned into two classes, at least one of these classes contains an arithmetic series of arbitrarily large length. (Issai Schur about 1925, B. L. van der Waerden 1927)

(4) Every partial order of $n^2 + 1$ elements contains either a chain of length $n + 1$ or a set of $n + 1$ incomparable elements. (R. P. Dilworth 1950)

(5) Every sequence of $n^2 + 1$ natural numbers contains either a monotonically increasing or a monotonically decreasing subsequence of length $n + 1$. (P. Erdős, G. Szekeres 1950)

Between these and some other examples there seemed to be no connection, before P. Erdős, in 1950, tried a synopsis and found a general theorem which gave many single results by specialization. Under the name *Ramsey Theory*, this has led since 1970 to many subtle mathematical works on disorderly systems with orderly subsystems, for example in 1975 'on sets of integers containing no k in arithmetic progression' by E. Szemerédi. The fundamental impossibility of complete disorder should be interpreted as a warning to cryptologists, to be careful with the use of machine-produced keys—at the moment only a theoretical danger, but nevertheless a serious one.

Marian Rejewski, Polish hero of decryption, expressed the warning in 1978 in the following form:

Whenever there is arbitrariness, there is also a certain regularity.

Bibliography

Good introductions to classical cryptology for amateurs:

Gaines, Helen Fouché, *Cryptanalysis*. Dover, New York 1956 (new ed.)

Smith, Laurence Dwight, *Cryptography*. Dover, New York 1955 (new ed.)

Millikin, Donald D., *Elementary Cryptography and Cryptanalysis*. New York 1943 (3rd ed.)

An introduction which also appeals mathematically oriented readers:

Sinkov, Abraham, *Elementary Cryptanalysis*. Mathematical Association of America, Washington 1966

This book, written by a professional cryptologist, certainly does not reproduce the full knowledge of the author.

A classic of cryptanalysis:

Friedman, William Frederick, *Military Cryptanalysis*. Part I, II, III, IV. Washington, 1938, 1938, 1938, 1942 *(obtainable as reprint)*

A comprehensive historical study of cryptology according to the state of the open literature in 1967:

Kahn, David, *The Codebreakers*. Macmillan, New York 1967

This book, written with journalistic verve by a professional historian, also gives references to special, not easily accessible historical literature, particularly before the 19th century.

"Of primary importance for a knowledge of modern cryptology" (Kahn) is the article

Rohrbach, Hans, *Mathematische und Maschinelle Methoden beim Chiffrieren und Dechiffrieren*. FIAT Review of German Science, 1939–1946: Applied Mathematics, Vol. 3 Part I pp. 233–257, Wiesbaden: Office of Military Government for Germany, Field Information Agencies 1948

An English translation by Bradford Hardie is in Cryptologia II, 20–37, 101–121 (1978).

Results of the British codebreakers, including mention of the BOMBE and COLOSSUS machines, may be found in:

Bertrand, Gustave, *Enigma ou la plus grande énigme de la guerre 1939–1945*. Librairie Plon, Paris 1973

Winterbotham, Frederick W., *The Ultra Secret*. Weidenfeld and Nicolson, London 1974

Beesly, Patrick, *Very Special Intelligence.* Hamish Hamilton, London 1977

Lewin, Ronald, *Ultra Goes to War.* Hutchinson, London 1978

Johnson, Brian,*The Secret War.* Methuen, London 1978

Rohwer, Jürgen; Jäckel, Eberhard, *Die Funkaufklärung und ihre Rolle im Zweiten Weltkrieg.* Motorbuch-Verlag, Stuttgart 1979

Randell, Brian, *The COLOSSUS.* In: N. Metropolis et al., *A History of Computing in the Twentieth Century.* Academic Press, New York 1980

Later detailed and reliable reports on the ENIGMA break:

Welchman, Gordon., *The Hut Six Story: Breaking the Enigma Codes.* McGraw-Hill, New York 1982

Garliński, Józef, *Intercept. The Enigma War.* Scribner, New York 1980

Kozaczuk, Władysław, *Enigma.* Arms and Armour Press, London 1984
Polish original edition: W kregu Enigmy, 1979

Hinsley, Francis H. et al., *British Intelligence in the Second World War.* Volumes I – IV, Cambridge University Press 1979–1988

Bloch, Gilbert, *Enigma avant Ultra.* 'Texte definitive', September 1988
An English translation by Cipher A. Deavours is in Cryptologia XI, 142–155, 227–234 (1987), XII, 178–184 (1988)

Kahn, David, *Seizing the Enigma.* Houghton-Mifflin, Boston 1991

Hinsley, Francis H., Stripp, Alan (eds.), *Codebreakers. The inside story of Bletchley Park.* Oxford University Press 1993

A biography of the life and work of Alan Turing:

Hodges, Andrew, *Alan Turing: The Enigma.* Simon and Schuster New York, 1983

First-hand information on statistical methods is found in:

Kullback, Solomon, *Statistical Methods in Cryptanalysis.* Aegean Park Press, Laguna Hills, CA 1976

World War II cryptanalysis in the U.S.A. is treated in:

Rowlett, Frank B., *The Story of Magic.* Aegean Park Press, Laguna Hills, CA 1998

Cryptological devices and machines are investigated in:

Türkel, Siegfried, *Chiffrieren mit Geräten und Maschinen.* Graz 1927

Deavours, Cipher A. and Kruh, Louis, *Machine Cryptography and Modern Cryptanalysis.* Artech House, Dedham, MA 1985

Specialist works on modern cryptology:

Konheim, Alan G., *Cryptography*. Wiley, New York 1981

Meyer, C. H., Matyas, St. M., *Cryptography*. Wiley, New York 1982

Brassard, G., *Modern Cryptology*. Lecture Notes in Computer Science, Vol. 325 , Springer, Berlin 1988

Beker, H. and Piper, F., *Cipher Systems*. Northwood Books, London 1982

Salomaa, Arto, *Public-Key Cryptography*. Springer, Berlin 1990

The last two books also cover cryptanalysis of the Hagelin machines in some detail.

Schneier, Bruce, *Applied Cryptography*. Wiley, New York 1993, 1995 (2nd ed.)

This book contains protocols, algorithms, and source code in C.

Goldreich, Oded, *Modern Cryptography, Probabilistic Proofs and Pseudorandomness*. Springer, Berlin 1999

A book typical for the modern part of scientific cryptology.

Questions of cryptology and civil rights are discussed in:

Hoffman, Lance J. (ed.), *Building in Big Brother*. Springer, New York 1995

Denning, Dorothy E. R., *Cryptography and Data Security*. Addison-Wesley, Reading, MA 1983

More recent elementary books of general interest:

Bamford, James, *The Puzzle Palace*. Penguine Books, New York 1983

Kippenhahn, Rudolf, *Verschlüsselte Botschaften*. Rowohlt, Reinbek 1997

Sebag-Montefiori, Hugh, *ENIGMA. The Battle for the Code* Weidenfeld & Nicolson, London 2000

Budiansky, Stephen, *Battle of Wits*. Simon & Schuster, New York 2000

Among the specialist journals are:

Cryptologia. A Quarterly Journal Devoted to Cryptology. Editors: David Kahn, Louis Kruh, Cipher A. Deavours, Brian J. Winkel, Greg Mellen. ISSN 0161-1194. Terre Haute, Indiana

Journal of Cryptology. The Journal of the International Association for Cryptologic Research. Editor-in-Chief: Gilles Brassard. ISSN 0933-2790. Springer, New York

Under the heading *Advances in Cryptology* appear proceedings of the annual *International Cryptology Conference* (CRYPTO), *International Conference on the Theory and Application of Cryptographic Techniques* (EUROCRYPT) and *International Conference on the Theory and Application of Cryptology and Information Security* (ASIACRYPT), sponsored by the International Association for Cryptologic Research (IACR), in the Lecture Notes in Computer Science series, Springer, Berlin.

Of interest mainly to historians are the works of

Breithaupt 1737; Hindenburg 1795, 1796; Andres 1799; Klüber 1809; Lindenfels 1819; Vesin de Romanini 1838, 1844; Kasiski 1863; Myer 1866; Koehl 1876; Fleißner von Wostrowitz 1881; Kerckhoffs 1883; Josse 1885; de Viaris 1888, 1893; Valério 1892; Carmona 1894; Gioppi di Türkheim 1897; Bazeries 1901; Myszkowski 1902; Delastelle 1902; Meister 1902, 1906; Schneickert 1900, 1905, 1913; Hitt 1916; Langie 1918; Friedman 1918, 1922, 1924, 1925; Givierge 1925; Lange-Soudart 1925; Sacco 1925, 1947; Figl 1926; Gyldén 1931; Yardley 1931; Ohaver 1933; Baudouin 1939; d'Agapayeff 1939; Pratt 1939; Eyraud 1953; Weiss 1956; Callimahos 1962, Muller 1971; Konheim 1981; Meyer-Matyas 1982.

A rather complete bibliography may be found in:

Shulman, David, *An Annotated Bibliography of Cryptography.* Garland, New York 1976

The works of Hitt, Langie, Friedman, Givierge, Lange-Soudart, Sacco, Gyldén, Ohaver, Callimahos, Kullback, are obtainable as reprints from:

Aegean Park Press, P.O. Box 2837, Laguna Hills, CA 92654-0837, USA.

William Friedman (seated) with (from left) Solomon Kullback, Frank Rowlett and Abraham Sinkov (Arlington Hall, 1944)

Nigel de Grey

Dillwyn Knox

Hans Rohrbach

Fig. 167. Famous 20th century cryptanalysts

Index

A-1 (code), 75
A-21, 119, 120
A B C, 26, 127, 423
ABC 6th edition (code), 74
Abel, Rudolf, 9, 55
ABNER, 412
Abwehr (OKW), 16, 135, 246, 396, 418
acknowledgment, 16
Acme (code), 74
acrophony, 67
acrostics, 18, 19
Adair, Gilbert, 240
ADAM and EVE, 411
addition, 78, 115, 166, 222, 223
– *modulo* 2, 129–130, 364
– *modulo* 2^n, 129
– *modulo* 10, 157–158
– *modulo* 10^n, 158
–, polygraphic, 129, 158, 222, 223
–, symbolic, 157, 166
additive *(adj)*, 79
additive *(n)*, 158, 351, 378
ADFGVX system, 39, 51, 159
Adleman, Leonard M., 189, 190
A.E.F., *see* American Expeditionary Force
‚Aegir', FOREIGN (key net), 435
Æneas, 11
affine, 78
Afrika-Korps, German, 200, 298
AGAT, 150
'Agnes', 401, 409
'Agnus Dei', 409
agony column, 26, 321
Airenti (code), 73
Aktiebolaget Cryptograph, 133
alarm, 16, 195
Albam, 45
Albert, A. Adrian, 2, 84, 413
Alberti, Leon Battista, 38, 39, 126–129, 271
Alberti disc, 40, 50, 126–128, 227, 321, 360
ALBERTI encryption step, 104, 114, 128, 141, 156, 224, 225, 227, 263, 331, 336, 340, 343, 350, 351, 355, 358, 360, 362
alemania, 14
Alexander, Conel Hugh O'Donel, 90, 270, 432
algebraic alphabet, 78, 84
algorithmic definition, 35, 150
alignment, 331, 334–343, 353
allegorical code, 15
alphabet, 34, 38
–, accompanying, 48, 50, 101, 331
–, complementary, 45, 88
–, decimated, 88, 115, 221
–, inverse, 45
–, involutory *see* self-reciprocal
–, mixed, 44, 47
–, powers of a mixed, 48, 50
–, reversed, 45, 88, 117, 159, 235
–, rotated, 103, 104, 105
–, shifted, 48, 50, 102–105, 117
–, self-reciprocal, 45, 46, 117
–, vertically continued, 102, 103
Alpha-AXP (211 64) 178
alphabet ring, 112, 134, 135
alphabets, unrelated, 117–123, 339, 357
al-Qalqashandi, 43
amalgamation, 159, 161, 164–166, 170, 175, 211
Amè, Cesare, 428
American Expeditionary Force (A.E.F.), 30, 66, 75, 200
Amt VI of the *R.S.H.A.*, 58–59, 424
'Amt für Militärkunde', 31
anagram, 98–100, 420
anagramming, 418–423
–, multiple, 97, 421, 422, 423
ananym, 91
Anderson, Ralph V., 430
Andree, Richard V., 247, 250
Andres, Johann Baptist, 119, 123, 452
Andrew, Christopher, 3, 90
Andreyev, Andrei Nicolayevich, 429
Angō Kenkyū Han, 31, 428–429
angō kikai taipu, 132, 141
/anx/, 394
'appen', 373

Archer, Philip E., 204
Argenti, Giovanni Battista, 35, 45, 48, 52, 54, 68, 117, 147, 197, 201, 237, 320
–, Matteo, 35, 53, 54, 62, 117, 131, 147, 201, 237
argot, 14, 19
aristocrats, 239, 240, 251, 291
arithmetic *modulo* 2, 84, 156, 363–364
arithmetical operations, 166, 169
Arlington Hall, 315
Armed Forces Security Agency, 30
Army Security Agency, 30
Arnold, Benedict, 158
ars occulte scribendi, 8
Arthold, J., 277
Asché, 390, 394
ASCII (code), 53, 291
astragal, 11
asymmetric methods, 5, 181–183
Athbash, 45
'Atlanta' 411
Atlantis (ship), 61, 429
ATLAS, ATLAS II, 320, 412
Augustus, Roman emperor, 47
Auriol, L. J. d', 84
Auswärtiges Amt, 30, 149–151, 418, 424
authentication, 25, 29, 174, 177, 180, 181, 183, 187, 194, 195, 212, 437
autokey, 146
AUTOSCRITCHER, 269, 411
autostereogram, 11, 12
AVA factory, 394, 396
avalanche effect, 165, 171, 174
Ave Maria code, 14, 16
Axiomatic Information Theory, 230, 438, 439–446
AZ (code), 74

B Talk, 20
B-1 (code), 75
B-21, B-211, 165
BC-543, 133
Babbage, Charles, 4, 26, 36, 115, 116, 147, 156, 217, 261, 263, 298, 311, 321, 430, 438
Babbage, Dennis, 90, 269, 403, 412
BACH, 59, 204
Bach, Johann Sebastian, 19
back slang, 20, 91
BACKWARDS BEAUFORT, 117
BACKWARDS VIGENÈRE, 117
Bacon, Sir Francis, 8, 9, 29, 39, 53
Baker, Stewart A., 207
Balzac, Honoré de, 29
Bammel, S. E., 124
Bamford, James E., 451
BAMS (code), 61, 74, 429
[ban], 220, 401
Banburism, 401, 432, 438
Banbury sheets, 316, 401
bar code, 74
bar drum, 133
barrier, 164–165
Baravelli (code), 73, 198
basic wheel setting, *see* ground setting
basis analysis, 143, 413–415
bâtons, 50
bâtons, méthode de, 264–269
Baudot, Jean Maurice Émile, 39
Baudouin, Roger, 452
Bauer, Bernhard, 12
Bauer, Friedrich Ludwig, 162, 277, 288, 313
Bayes' rule (Bayes, Thomas), 440
Bazeries, Étienne, 4, 8, 28, 29, 36–38, 47, 49, 69, 91, 117, 121–124, 155, 165, 207, 217, 227, 239, 255, 256, 258, 272, 277, 298, 325, 418, 431, 439, 452
Bazeries (code), 73
Bazeries' cylinder, 36, 47, 49, 120, 121, 209, 234, 260
BC 543, 133
B-Dienst, 61, 204, 205–206, 424, 425, 426
Beaufort, Sir Francis, 115
BEAUFORT encryption step, 114–116, 129, 133, 143–144, 148, 159, 165, 226, 251, 263, 303, 305, 309, 321, 325, 360, 362, 428, 442, 444
Befehlstafel, 76
Beesly, Patrick, 218, 237, 426, 450
Behnke, Heinrich, 424
Beiler, Albert H., 186
Beker, H., 451
Belaso, Giovanni Battista, 127, 128, 129, 132, 145–146
Bell numbers, 237
Bennet, Ralph, 434
Bentley's (code), 74
Bernstein, David S., 212–213, 214
Bernstein, Paul, 112, 133
Berry, Duchesse de, 35
Berthold, Hugo A., 200, 201
Bertrand, Gustave, 388, 390, 398, 402, 449
Beth, Thomas, 215, 417
Beurling, Arne, 2, 364, 372–373, 435
Bevan, John Henry, 30
Bi language, 20

Bibo, Major, 59
bifide, 34
bigram, 34
– coincidences, 313, 316
– frequencies, 285, 286, 287, 288, 337, 418
– repetitions, 325, 328, 346
Bigrammbewertungsgerät, 418
Biham, Eli, 175
biliteral, 9, 39
binary, 39, 84
– addition, 130
– alphabet, 39, 46, 129, 144
– cipher, 39
– circuit, 84, 130
– code, – encoding, 53, 129, 227
– digits (Z_2, $\mathbb{Z}_2$), 23, 39, 89, 129–130
– linear substitution, 84
– numbers, 23, 89, 129–130
bipartite, 34, 51, 57, 65
Bischoff, Bernhard, 100
bit, binary digit (Z_2, $\mathbb{Z}_2$), 39, 129–130
[bit], 220, 401
bitwise binary encryption, 129–130
Biuro Szyfrów, 108, 252, 388, 393, 394
Black Chamber, 69, 71, 138–139, 212, 215
BLACK (code), 40, 75, 428
Blair, William, 131
Blakely, Bob and G. R., 169
Bletchley Park, 3, 30, 64, 90, 107, 135, 142, 151, 158, 159, 200, 204, 211, 252, 269, 270, 316, 319, 364–372, 377–378, 369–412, 425–433
Bloch, Gilbert, 450
block encryption, 34, 166, 170
block transposition, 95–97, 156, 159, 421
blockdiagonal, 85
BLUE (code), 75, 199
Böll, Heinrich, 95, 228, 229, 274, 275, 278, 282, 285, 420
Boetzel, A., 10, 165
'boldest' line, 338
'Bolek' (Gustave Bertrand), 390
Bolton (code), 72
bomba, 396, 397, 398–399, 400, 401, 402
BOMBE, 402, 403–411
Bonatz, Heinz, 205, 426
book cipher, 8, 44
bookseller's price cipher, 27, 43
Boolean algebra, 84, 130, 310, 370
B.P. *see* Bletchley Park
Brachet (code), 73
Branstad, D. K., 170
Braquenié, Henri, 398
Brassard, G., 451
Brett-Smith, Hilary, 432
Britzelmayr, Wilhelm, VI
Broadhurst, S. W., 369
Broadway buildings, 398
Brooke-Hunt, G. L., 425
BROWN (code), 75, 428–429
Brown, Cave, 90
Browne, Thomas, 8
Broy, Manfred, VII, 82
Bruce, David, 424
Brunswick (code), 73
BRUSA pact, 411
brute force attack, 175–177, 191, 207, 234
Brynielsson, L. B., 143
B.S.-4, 388, 395
BSI, 31, 213
Buck, F. J., 84
Budiansky, F. J., 451
Buell (code), 73
Bulldog (ship), 203
Bundesstelle für Fernmeldestatistik, 31
Bundesamt für Sicherheit in der Informationstechnik (BSI), 31, 213
Bundesnachrichtendienst (BND), 30, 31
Burgess, Guy, 151
Burke, Colin, 319, 370, 411, 418, 426, 439
Burkill, H., 447
Bush, Vannevar, 2, 319, 325, 371, 439
Byron, Lord George Gordon Noel, 22, 29
byte (Z_{256}), 39, 129, 166

C-35/C-36, 77, 133, 134, Plate G
C-38 *see* M-209
C-38m, 208
C-41, 133
Cabinet noir, 69
'Cadix', 402
Cadogan, 21
cadran, 50
CAESAR addition, simple, 47, 85, 116, 158, 222, 229, 230–233, 235, 246, 249, 272, 274, 275, 286, 291, 331
–, polygraphic, 79, 148, 158, 223, 224–226
CAESAR encryption step, 47, 116, 148, 159, 230–233
calão, 14
Callimahos, Lambros D., 24, 85, 452
Campaigne, Howard H., 412
Campbell, Lucille, 150
Canaris, Wilhelm, 16, 135, 246
Candela, Rosario, 264
cant, 14, 20
Cantor, Georg, 29

canvasses, 398
caption code, 71, 76
Caramuel y Lobkowitz, Giovanni, 39
Cardano, Geronimo, 22, 94, 129, 145, 437
Carlet, Jean Robert du, 1
Carmichael's ψ function, 189
Carmichael's theorem, 189, 192
Carmona, J. G., 452
carry device, 89, 130, 226, 363
Carter, James Earl, 212
Cartier, François, 97, 261, 423, 425
Cartouche, 20
Casanova, Giacomo Girolamo Chevalier de Seingalt, *27*
Casement, Hugh, VII, 100, 236
category of methods, 35
CBC, 174, 190, 214, 219
CCITT 2, 129, 363
CCM, 138
CD-55, CD-57, 77
censor, 14, 18, 24
Central Intelligence Agency (C.I.A.), 29
CESA, 216
CHAFFINCH (key net) 403
Chaitin, Gregory J., 150, 151, 445, 447
Chandler, W. W., 369
Chanel, Coco, 31
character, 31
– coincidences, 301
characteristic, 394
Charlemagne, 43
Charles I, 68
Charles II, 69
Chase, Pliny Earle, 65, 149, 169
Château de Vignolles, 402
Château des Fouzes, 402
Chess, Abraham P., 4
Chi, 304, 305, 307, 335–336, 337, 342–343
– test, 262, 308, 320, 325, 339, 352, 357
Chi, OKW Abt., 31, 252, 316, 325, 353, 376, 377, 406, 418, 424, 427, 431
Chi-Stelle, 390, 425, 427
Chi-wheel, 368, 371
Chiang Kai-shek, 140
chiffre carré, 101
chiffre à damier, 64
Childs, J. Rives, 378, 425
Chinese remainder theorem, 188, 194
chip, 164, 169, 176, 177, 178, 214
choice operator, nondeterministic, 32
Chorrin, 131
Chorukor, 241
chosen ciphertext attack 433
chosen plaintext attack, 432, 433
chronogram, 19
chronostichon, chronodistichon, 19
Church, Alonzo (Church thesis), 150
Churchill, Winston, 3, 9, 30, 55, 98, 122, 209, 380, 424, 433
C.I.A., 29
Cicero (code), 74
Ciężki, Maksymilian, 388, 390, 398
cifrario tascabile, 38
Cillis, 403, 435
cipher, 34
– disk, 42, 50, 126–127, Plate B
– security, 196, 197, 435
– slide, rods, 50
– teletype machine, 53, 148, 156, 157, 319, 353, 364, 366, 368, 372
Cipher Block Chaining, *see* CBC
ciphering device, 118–119, 120–123
ciphering machine, 5, 28, 105
ciphertext-only attack, 248, 431
cipher wheel, 156–157
City of Bagdad (ship), 61
Clarendon, Edward Hyde Earl of, 425
class of families, 262
classification of cryptography, 23, 24
Clausen, Max, 55
Clausen-Thue, William, 72
cleartext, 31
clef, 40, 95, 128
clef principale, 128
CLIPPER, 177, 214
clique, 278, 281, 282, 291, 295
cliques on the rods, 264
clock method, 401
closing, 54, 55, 157
closure, 404
COBRA, 411
code, 17, 34, 66, 354
– book, 5, 35, 40, 66
– compression, 166, 447
–, decryption of, 354
– groups, 39, 66, 242
–, literal, 72
–, numeral, 71
–, one-part, 70, 354
–, two-part, 71
Code Compilation Section, U.S. Signal Corps, 30
Collange, Gabriel de, 38
Collon, Auguste L. A., 64
COLOSSUS, 2, 218, 319, 218, 320, 353, 369, 370, 372, 402, 409, 449
columnar transposition, 26, 95, 96, 97

–, double, 26, 97, 421
–, U.S. Army double, 97
columns, 326–329, 332–343, 352–353, 374–376
combinatorial complexity, 40, 62, 127–128, 196, 207, 220, 227
Combined Cipher Machine (CCM) 138
commuting encryptions, 152, 156, 360
companion matrix, 143
COMPARATOR, 319, 326, 371
complementary alphabet, 88
complete-unit transposition, 95
complexity theory, 183, 184, 185, 196
–, subexponential, 184, 187
complication illusoire, 26, 61, 65, 95, 135, 251, 257, 374, 394, 418, 421, 423
composition of classes of methods, 155
compression, 166, 447
compromise, 151, 154, 197, 209, 270, 356, 377, 378, 381, 388, 403
computable irrational numbers, 150
concealment cipher, 8, 18, 23
conceptual word, 287
confrontation, 278, 280, 282
confusion, 160
congruence root, primitive, 186
conjugated encryption step, 65
contact, 293, 418–423
Coombs, A. W. M., 369
COPPERHEAD, 353
Coppersmith, Donald, 175, 176, 187
CORAL, 143
Cot, Pierre, 151
coupled pattern finding method, 248
Coventry raid, 3
covert proof, 437
CQ signals, 270, 427
crab, 91, 100
– (Knox), 136
crash, 251
Crawford, David J., 269
CRAY-1, CRAY X-MP, CRAY C90, 144, 184, 320, 412, Plate Q
Cray, Seymour, 412, Plate Q
crib, 242, 378, 403, 404, 406–409, 432, 435
criminology, 4, 51, 239
Croissant, Klaus, 8
Croix Grecque transposition, 93
Cromwell, Oliver, 18
cross-plugging, 46, 110, 112, 390, 394, 397–400, 404, 406, 407, 412, 431
cross-ruff, 378
(CRYPT), 441
crypt width, 33
cryptanalysis, V, VI, VII, 3, 4, 24, 26, 31, 65, 197, 217ff., 368, 438
Crypto AG, VII, 119, 150, 178
crypto board, 178, Plate P
cryptozilla, 178
crypto clerk, 196, 201, 209, 210, 291, 425
cryptographic equation, 41, 110, 115, 116, 117
cryptographic fault, 197, 209, 210, 424
cryptography, VII, 1, 2ff., 7, 8, 24, 29–31
cryptology, V, VII, 2, 3, 4, 5, 7, 24, 29–31
Cryptoquip, 5, 239, 240, 471
cryptosystem, 33, 101
–, classical, 441, 442, 443, 444
–, fixed, 40, 127, 175
–, ideal, 445
–, independent key, 443–445
–, of Vernam type, 444
–, perfect, 443, 444
–, pure, 156, 358, 359
–, Shannon, 41, 156, 251, 359, 441, 442
–, transitive, 260, 359
cryptotext, 31, 34, 41
– vocabulary, 31
cryptotext-cryptotext-compromise, 151, 197, 200, 209, 270, 356, 376–380, 388, 433
cryptotext-only attack (known cryptotext attack), 291, 431
CSKO, 434
CSP 642, 123, 252, 263, 339, 429
CSP 845 *see* M-138-A
CSP 889 *see* M-134-C
CSP 1500 *see* M-209
CSP 1700 138
cue, 16, 17, 23
Culpeper, Edmund, 239
CULPER, 69, 239
CVCCV, 76, 421
CVCVC, 70, 421
CX-52, 134
cycle, 404–406
– decomposition, 105, 394
– notation, 44, 105
– numbers, cyclotomic numbers, 78
cyclic group, 360, 362, 363
cyclometer, 394
cylinder and strip devices, 28, 36, 49, 50, 120–123, 132, 209, 215, 252, 256–260, 339
Cynthia, 99
cypher, 30
Cyrillic alphabet, 38
Czech alphabet, 38

D'Agapayeff, Alexander, 452
damier, 62, 64
Damm, Arvid Gerhard, 106, *107*, 119, 120, 133, 140, 148
Darhan (code), 74
Dato, Leonardo, 126
Deavours, Cipher A., 112, 113, 205, 265, 267, 269, 394, 401, 404, 450, 451
[deciban], 220, 401
decimal digits (Z_{10}), 39, 316
decimation, 88, 115, 221
(DECRYPT), 441
decryption key, 153, 179, 181
decryption of ancient scripts, 25, Plate A
decryption step, 41, 47, 180
Defense Calculator (IBM 701), 412
Defense Intelligence Agency (DIA), 30
definal, 32
de Grey, Nigel, 151, 425, 436, *452*
Delastelle, Félix Marie, 65, 165, 322, 452
DEMON, 320
denary, 39, 52
– cipher, 39
Denning, Dorothy E. R., 215, 451
Denning, Norman, 426
Denniston, Alastair G., 158, 398, 430
–, Robin, 356, 430
depth, 256, 326, 331, 338, 356
Dershavin, Gavrila Romanovich, 241
DES, 35, 170–176, 177, 179, 180, 183, 211, 214, 215, 219
–, modes of operation for, 174, 219
Desch, Joseph, 409–411
Desch BOMBE, 409
Deubner, Ludwig, 47, 425
–, Ottfried, 261
Deutsches Museum, Munich, VII
de Viaris (code), 73
de Viaris, Gaëtan Henri Léon , 28, 72, 116, 123, 207, 255–263, 276, 322, 452
de Vries, Mauritius, 2, 85
D.I.A., 29
diagonal board, 159, 407–408, 409, 412
diagonally continued alphabets, 104
Diccionario Cryptographico (code), 73
Dickinson, Velvalee, 16
Dickson, L. E., 81
dieder group, 363
difference method, 350, 351–353, 374–376
difference table, 352, 353, 374, 379
differential cryptanalysis, 175, 197, 373
Differenzenrechengerät, 353
Diffie, Whitfield, 5, 153, 181, 195, 212
diffusion, 160
Digital Signature Algorithm (DSA), 195–196
Digital Signature Standard (DSS), 195
digraphic substitution, 56–61, 76, 222, 231
Dilworth, R. P., 448
direction-finding, VII, 206
directory, 181–183
discrete logarithm function, 186–188
discriminant, 159, 374
disinformation, 3, 206
disk, 42–43, 50, 114, 126
division algorithm, 89
DNA, 438
Dodgson, Charles Lutwidge [Lewis Carroll], 117, 165, 381
DOLPHIN, *see* ‚Hydra' (key net),
Donelly, Ignatius, 29
Dönitz, Karl, 206, 377, 426
Doppelkassettenverfahren, 64
Doppelwürfelverfahren, 97
Doppler, 316, 328 *see also* repetitions
double casket, double PLAYFAIR, 64
double cipher, 128, 345
double columnar transposition, 26, 97, 421
double cross, 14, 17
double encipherment of text setting, 388, 389, 396, 398
double-ended scrambler, 399, 400, 403, 404, 408, 409
'double key', 128, 345
doublet, 201, 395
doubling, 201, 237, 388, 396, 433
doubly safe primes, 193, 196
Douglas, Chevalier, 15
'dragging', 369
DRAGON, 319
Dreher, 287
Dreyfus, Alfred, 198, 199
Driscoll, Agnes Meyer, 141, 319, 409
du Carlet, Jean Robert, 1
dual encryption steps, 360
Ducros, Oliver, 121
Dudeney, Henry Ernest, 92
DUENNA, 269, 412
Dulles, Allen Welsh, 40, 261, 376
dummy, 20, 33, 43, 200, 201, 209
– text, 32
Dunning, Mary Jo, 142
duplication, 50, 105
Dyer, Thomas H., 314

Eachus, Joseph, 409–412
Eastman 5202, 325, 371
eavesdropping, 25, 194, 263, 429, 433
EBCDIC (code), 53
Eckardt, Heinrich von, 436
ECM Mark II, III, 112, 138, 434
Edward Prince of Wales, 27
efficiency boundary, 184
efficient (method), 182, 183
Ehler, Herbert, VII
electric contact realization, 105
Electronic Code Book (ECB), 35, 174–175, 190, 214
ElGamal, Tahir, 195
elliptic curve method (ECM), 187
Ellis, James H., 153
"emergency clear" device, 154
encicode, 351
enciphering step, encoding step, 34
encryption, fixed monoalphabetic, 40, 175
–, non-periodic, 34, 131, 144
–, periodic, 34, 127, 311, 356
–, polyalphabetic, 131, 144, 225–226, 251, 303
–, progressive, 127, 132, 257
–, quasi-nonperiodic, 131, 144, 414
encryption and signature methods, 181
encryption block, 34
– error, 25, 197, 210, 424
– key, open, 179
– philosophy, 208
– security, 31, 123, 151, 196, 197–216
– step, 33–35, 40–41, 42, 56, 78, 101
– table, 34, 125, 349, 364, 387
– width, 33–34, 101
endomorphic encryption, 33, 35, 39, 44, 48
endomorphic linear substitution, 79, 80
endomorphic substitution, 102
Engineering Research Associates, Inc. (E.R.A.), 412
Engstrom, Howard Theodore, 319, 412
ENIGMA, 3, 28, 31, 46, 107–114, 133–136, 140, 200–206, 210, 252, 264, 377, 378, 388–411, 430, 434
ENIGMA A, B (1923), 107, 133, 152
ENIGMA C (1926), 107–108, 152
ENIGMA D (1927), 108, 113, 135, 140, 152
ENIGMA G (1928), 110, 152
ENIGMA I (1930), 110, 134, 152
ENIGMA II, 110
ENIGMA K (INDIGO), 108
ENIGMA M4, 112, 200, 204, 388, Plate I
ENIGMA (Abwehr), 135, 396
–, 3-rotor-, 110–114, 152, 200–206
–, 4-rotor-, 112, 200, 204, 205, Plate I
–, commercial, 90, 107, 108, 264, 267
– equation, 110
–, ground setting, 59, 390, 394, 395, 398, 402
–, *Grundstellung* *see* ground setting
–, message setting *see* text setting
–, number of, 112
– replicas, 394, 398
–, ring setting, 112, 204, 389, 394–396, 397–398, 406
–, *Ringstellung* *see* ring setting
–, rotor order, 111, 112, 389, 394–396, 400, 404, 434
–, rotors, 107–114, 132–138, Plate K
– –, numbering of, 110, 113
–, *Spruchschlüssel* *see* text setting
–, *Tagesschlüssel* *see* ground setting
–, text setting, 388–396, 404
entropy, 310, 439
equifrequency cipher, 272
equivocation, 439, 441
ERA 1101, ERA 1101 A, ERA 1103, 412
Erdős, Pál, 125, 448
Eriksson, Bertil E. G., 54
'*Erloschen ist Leuchttonne*', 199, 251, 252
error-detecting and -correcting codes, 25
Erskine, Ralph, VII, 4, 113, 412
Escrowed Encryption Standard (EES), 7, 213
Euler, Leonhard, 10, 82
Euler's 36-officer problem, 57
Euler's totient function φ, 88, 189, 221
European Union, 214
Euwe, Max, 145
event, 209, 439
–, elementary, 439
'*évitez les courants d'air*', 49, 120
Ewing, Sir Alfred, 377
exclusion of encryption methods, 271
exclusive or, 130
exhaustion of probable word position,
–, non-coincidence, 251–253, 260
–, binary non-coincidence, 254–255, 260
–, zig-zag, 263–264
exhaustive search, 207, 219, 220–234
exponentiation in $\mathbb{F}(p)$, 185, 186
'Expositur 3000', 402
Eyraud, Charles, 38, 45, 61, 88, 95, 96, 102, 115, 117, 123, 126, 276, 277, 287, 288, 303, 421, 452
EYRAUD encryption step, 115
Eytan [Ettinghausen], Walter, 159

Fabyan, George, 29
Fabian, Rudolph J. 353
factorization, 184, 191
'Fall Gelb', 402
falsification, 25
family code, 18
family of accompanying alphabets, 50
family of message blocks, 262
Fano, Robert M., Fano condition, 35
Fano condition, 35, 54
feedback cycle system, 403–407
Feinstein-Grotjan, Genevieve, 142, 150
Feistel, Horst, 170, 171
Fellers, Frank Bonner, 428
Fellgiebel, Erich, 31, 110, 435
female, 46, 209, 391, 398
Fenner, Wilhelm, 424
Fermat prime, 169, 186
Fermat's theorem, 168
Ferner, Robert, 142
Fersen, Axel Graf von, 131
Fetterlein, Ernst C. ('Felix'), 3, 149, 158
FIALKA, 137
fiber, 32
Fibonacci numbers, 162
FIDNET, 216
Figl, Andreas, 58, 199, 242, 277, 425, 452
Filby, P. William 151, 158, 366
finitely generated, 33
First Amendment, 6
first character, 78
fist, 203
five-digit code, 73, 74, 75
five-letter code, 39, 72
fixpoint, 163, 164, 169, 192, 396, 397, 405
FLB's resurrection, 161
Fleissner grille, 94
Fleissner von Wostrowitz, Eduard, 94, 276, 452
Fliegerführer Trondheim, 402
FLORADORA 151, 158, 378, 436
Flowers, T. H., 369
Floyd, Robert W., 144
FLUSS, 59, 60
FOREIGN, *see* ‚Aegir‘ (key net)
formal cipher, 36, 291
Forschungsamt des RLM, 263, 424, 425, 428
Forschungsstelle der Reichspost, 9
Forster, Otto, 186
Försvarets Radioanstalt (FRA), 31, 372
fourbesque, 14
four-digit code, 73
Fox, Philip E., 269
$\mathbb{F}(p)$, 79, 167, 185
fractionating method, 64
'Frankfurt', 206, 426
Franksen, Ole Immanuel, 116, 322
FREAK, 418
free-style methods, 298, 344, 446
Freemasons' cipher, 43
frequency count, 274, 277, 279, 280, 283
frequency distribution, 57, 271, 276
frequency ordering, 276, 277
frequency profile, 273–275, 333–334, 341
Freyss, Gustave, 423
Friderici, Joannes Balthasar, 39
'Fried Reports', 270
Friedman, Elizebeth Smith, 4, 29, 239
Friedman, William Frederick, 2, 4, 24, 29, 30, 66, 72, 76, 77, 84, 112, 120, 121, 123, 127, 137, 139, 142, 150, 215, 218, 230, 239, 255, 260, 263, 264, 298, 301, 312, 314, 317, 320, 322, 324, 325, 329, 330, 354, 355, 381, 386, 401, 435, 438, 449, *452*
Friedman examination, 312, 322, 324, 325, 329, 330, 438
Friedmann (code), 73
Friedrich, J., 25
Friedrichs, Asta, 67, 261, 376
Fuchs, Klaus 151
function inversion, 182–183
functional, 32
'Funf', 401
Funkspiel, 195, 203
'Für GOD', 120

G.2 A.6, 30, 67, 201
GADFLY (key net) 403
Gagliardi, Francesco, 99
Gaines, Helen Fouché, 58, 61, 95, 235, 238, 240, 247, 271, 276, 277, 287, 291, 293, 449
Galilei, Galileo, 99
Galland (code), 73
Galois field, 79, 167
garbling and corruption, 25
gardening, 378, 433
Gardner, Martin, 335
Garliński, Józef, 450
Gaujac, Paul, 199
Gauss, Carl Friedrich, 188
Gaussin, Joseph, 116
GC&CS, 3, 30, 90, 398, 399, 424
Geheimklappe, 61, 67, 76
Geheimschreiber, 156, 157, 368, 372–373
Geiger counter, 150
gematria, 12, 36

generating relation, 33
generation of a quasi-nonperiodic key, 145
generatrix, 121, 256, 257–260, 262
Gerold, Anton, VII, 196
Gherardi, Loris, 40
GIANT, 269, 412
Giant-Step-Baby-Step algorithm, 186
Gioppi di Türkheim, Luigi Count, 66, 452
Gisevius, Hans Bernd, 376
Givierge, Marcel, 96, 128, 202, 208, 218, 256, 261, 263, 276, 277, 421, 425, 452
Gleason, Andrew M., 2
Gold, Harry, 151
Goldbach, Christian, 69
Goldberg, Emanuel, 9, 320, 326
GOLDBERG, 320
Gold-Bug, 43, 298–299
Goldbach, Christian, 69
Goldreich, Oded, 451
Good, Irving John [Isidor Jacob Gudak], VI, 243, 366, 368, 430
Gordon, D. M., 187
Göring, Hermann, 263, 376, 403, 424, 436
Government Code and Cypher School, (G.C.&C.S.) 3, 30, 90, 398, 399, 411, 424
Government Communications Headquarters (G.C.H.Q.), 30
graph, 11, 12, 448
GRAY (code), 68, 75, 199, 429
'Greek rotors' α β γ, 112, 134, 200, 204
Greek-Latin square, 57
GREEN (code), 70, 75, 199, 421
GREEN (key net), 402
GREEN (machine), 140
Greenglass, David, 151
Gretz-Armainvilliers, 402
Grew, Joseph C., 75, 198, 199
Griechenwalze, 112, 134, 200, 204
grill (grid) method ('metoda rusztu'), 394
grille, 18, 22, 23, 93, 94
Gripenstierna, Fredrik, 49, 52, 119, 120
GRONSFELD encryption step, 117
Grosvenor, William M., 423
ground setting, 59, 112, 154, 389, 395, 397, 398, 403
group property, 155
'Group Z', 402
Groves, Leslie R., 52
GRU, 429
Grundstellung, *see* ground setting
Grunsky, Helmut, VI, 261
Güntsch, Fritz-Rudolf, VII
Gyldén, Yves, 4, 24, 133, 165, 452
Hagelin, Boris Caesar Wilhelm, 77, 107, 133, 140, 165, 208, *215*, 252, 311, 356, 360, 394, 428–429, 451
half-adder, 130
half-rotor, 107, 114, 141, 165
Hall, Marshall, 2
Hall, William Reginald, 425
Hallock, Richard, 150
Hamming, Richard W., 25, 74
Hand-Duenna 270
Handschlüssel, 64, 97, 433
Harmon, John M., 5, 6
Harriot, Thomas, 39
Harris, Martha, 214
Hartfield, John Charles, 72
HARVEST, 412
Harvey (code), 72
Hašek, Jaroslav, 40, 55, 94
Hasenjäger, Gisbert, VI, 406, 408
Hassard, John R. G., 423
Havel, Václav, 241
Hawkins, Charles A., 212
Hay of Seaton, Malcolm V., 425
Hayhanen, Reino, 55
HCM (Hebern), 137
H.E. [Hans-Thilo Schmidt], 390
HEATH ROBINSON, 318, 319, 320, 353, 369
Hebern, Edward Hugh, 106, *107*, 112, 132, 137, 140
HECATE, 320
Heidenberg, Johannes (Trithemius), 114
‚Heimische Gewässer', DOLPHIN (key net), 403
Heimsoeth & Rinke, 90, 107
Heise, Werner, 125
Hellman, M. E., V, 5, 153, 181, 195, 445
Helmich, Joseph G. 137
Henkels, M., 66
Henri IV of France, 68
Herivel, John, 403
Herivel tip, 403, 435
Hermann, Arthur J., 122
heterogeneous encryption, 43
Hettler, Eberhard, 366
hiatus, 35, 54, 256, 267, 298
hieroglyphs, 67
Hildegard von Bingen 43
Hill, Lester S., 2, 84, 85, 130
HILL encryption step, 80, 84, 222, 223
Hilton, Peter J., 92
Himmler, Heinrich, 9, 31, 376, 424
Hindenburg, C. F., 93, 452
Hindenburg, Paul von, 202

Hinsley, Francis Harry, 90, 371, 425, 427, 450
Histiæus, 9
Hitchings, Oswald Thomas, 425
Hitler, Adolf, 31, 159, 201, 246, 427
Hitt, Parker, *122*, 126, 131, 148, 202, 230, 295, 297, 298, 425, 452
Hitts admonition 126, 131, 154
Hodges, Andrew, 366, 401, 450
Hoffman, Lance J., 451
Holden, Carl F., 411
Holden, Edward S., 421, 423
Holmes, Sherlock, 11
Holtwick, Jack S., 141, 429
holocryptic, 149–150, 151, 231
Homan, W. B., 272
homogenous linear substitution, 79, 80–81, 84
homophones, 32–36, 43, 44, 52, 53, 65, 68, 71, 76, 126, 183, 201, 209, 221, 255, 385
Hooper, Stanford Caldwell, 319
Hoover, Herbert, 139, 212
Hopf, Eberhard, 160
Hopkins, Johns, 386, *387*
Horak, Otto, VII, 1, 176, 207
horizontally shifted alphabets, 102–105
Hornbeck, Stanley K., 428
HORNET (key net), 403
Hotel-Telegraphenschlüssel, 74
Höttl, Wilhelm, 58, 59
Houdin, Robert, 13
Howard, John E., 412
Huffman coding, 447
Hünke, Anneliese, 261
Hüttenhain [Hammerschmidt], Erich, VI, 150, 200, 210, 252, 270, 277, 316, 317, 377, 424, 427 , 428
Huyghens, Christiaan, 98
‚Hydra', DOLPHIN (key net), 403
'Hypothetical Machine' (HYPO), 319
index of coincidence, 4, 137, 301
indicator, 59, 120, 126, 154, 204, 267, 368, 373–374, 388–389, 392, 393, 395
indicator doubling, 388, 401, 433, 435
INDIGO, 108, 267
individual key, 148–151, 203, 211, 337, 378
induced, 33
inflation of cryptotext, 211
influence letter, 148
informal cipher, 290–291
Informatik (*Deutsches Museum*), VII
information, mutual, 440
information theory, V, 3, 230, 310, 447
–, axiomatic, 439–446
inhomogenous linear substitution, 81
injective, 32, 33, 41, 78, 169, 227, 441
Inman, Bobby Ray, 30
in-phase adjustment, 356, 357, 373, 431
Institute for Defense Analyses (I.D.A.), 30
International Traffic in Arms Regulations (ITAR) 6, 178, 213
intermediary cryptotext, 340, 343, 344, 351, 375, 380, 384
invariance theorems, 234, 235, 246, 271, 272, 274, 286, 303, 305, 309
inverse alphabet, 45
involution, 45
involutory *see* self-reciprocal
irrational numbers, 35
irregular wheel movement, 132–138, 157
ISBN (code), 74
isolog, 209
isomorphism, 41, 108
isomorphs, method of, 264–269
isopsephon, 36
italic capitals for key characters, 40
iteration, 160–164, 191–193
iteration exponent, 191, 193
i-Wurm, 148

International Business Machines (IBM), 39, 170, 314–315, 326, 412
IBM 701, 412
ICKY, 325
I.D.A., 30
IDEA, 177
ideal, 445
identity, 79, 84, 105, 129, 130, 208, 359
ideogram, 67
idiomorph, 235, 236
independent key, 443–445
Index Calculus method, 186–187
Jäckel, Eberhard, 450,
JADE, 143
Jäger, Lieutenant, 67, 198, 243
jargon code, 14, 16, 23
Javanais, 20
je suis indéchiffrable, 28, 439
Jefferson, Thomas, 47, 69, *121*, 207, 227
Jeffreys, John R. F., 398
Jensen, Willi, VI, 277, 316, 325, 376, 418
Jipp, August, 156, 366
JN-25A, JN-25B (code), 74
JN-157 (code) 411

Johnson, Brian, 5, 112, 366, 370, 450
Johnson, Esther, 21
Josse, Henri, 452
Joyce, James, 36, 241
Julius Caesar, 47
'Jumbo', 401

Kaeding, F. W., 276, 277, 285
Kahn, David, V, VII, 5, 9, 24, 25, 65–67, 69, 78, 92, 106, 111, 116, 118, 127–129, 132, 135, 142, 149, 206–208, 240, 270, 276, 291, 292, 300, 301, 311, 316, 321, 323, 331, 348, 374, 376, 421, 425–427, 429, 431, 449–451
Kāma-sūtra, 45
Kaplanski, N., 125
Kappa, 218, 301–308, 311–320
– test, 315–320, 325, 327
Kappa-Chi Theorem, 306, 307
Kappa-Phi Theorem, 307, 308, 327
Kappa-Phi$^{(u)}$ Theorem, 327
Kasiski, Friedrich W., 230, 276, 320, 322, 452
Kasiski examination, 320–327, 330, 340, 346, 369
Katscher (code), 72
Keen, Harold ('Doc'), 400
Kenngruppe, 159, 374
Kepler, Johannes, 99
Kerckhoffs, Auguste, 96, 197, 207, 237, 276, 322, 346, 349–350, 356–358, 421, 434, 452
Kerckhoffs' maxim, VII, 207, 227, 270, 394
Kesselring, Albert, 371
key distribution, 40, 211
– escrow system, 7, 213
– exchange algorithm, 154
– generator, 145–148, 152, 154, 368
– group, 156, 169, 358–360, 363
– management, 151–154
– negotiation, risk of, 151–154
– net, 377, 389, 397, 402, 403
– vocabulary, 40
– words, 26, 131, 132, 148, 154, 174, 197, 202, 325, 335, 343, 345, 348, 351, 387
'key phrase' cipher, 43
key-symmetric, 41, 173, 177
keytext, 41, 127, 131, 132, 154, 174, 263, 264, 322, 356, 378, 415, 441–445
Kinsey, Alfred C., 8
Kippenhahn, Rudolf, 451
Kircher, Athanasius, 8, 74
Kirchhofer, Kirk H., VII
kiss, 377
Klar, Christian, 8
Klartextfunktion, 148, 157, 372
KL-7 rotor machine, 136, 137
Klüber, J. L., 452
knapsack problem, 196
knight's tour transcription, 92, 93
knock-cipher, 51
known cryptotext attack, 431
known plaintext attack, 432
Knox, Alfred Dillwyn ('Dilly'), 90, 136, 264, 390, 398, 399, 431, *452*
Koblitz, Neil, 187
Koch, Hugo Alexander, 106–107
Koch, Ignaz Baron de, 69
Koehl, Alexis, 65, 165, 452
Kolmogorov, Andrei Nikolaevich, 150
Komet (ship), 204
Konheim, Alan G., V, 253, 276, 451
Korn, Willi, 107, 108, 135, 394
Kovalevskaya, Sofia, 203
Kowalewski, Jan 138
Kozaczuk, Władysław, 388, 390, 394, 450
Kraitchik, M., 184
Krause, Reinhard, IV
Kratzer, Uwe, 243
Krebs (ship), 203
Krivitsky, Walter [Samuel Ginsberg], 206
Krohn (code), 72
/krkr/, 201
Krug, Hansgeorg, 261, 262, 376
Kruh, Louis, 112, 113, 205, 265, 401, 404, 408, 450, 451
KRU, KRUS, KRUSA, KRUSÄ, 26, 76
Kryha, Alexander von, 133, 140, Plate F
Kulissenverfahren, 65, 166
Kullback, Solomon, 85, 137, 141, 158, 234, 282, 288, 303, 304, 308, 310, 313, 450, *452*
Kullback entropy, 310, 438
Kullback examination, 314, 320, 321, 326–329, 330, 331, 340, 401, 438
Kulp, G. W., 312, 313, 326–329, 330, 332–335, 336, 350
Kunze, Werner, 85, 141, 143, 261, 376
Küpfmüller, Karl, 200, 277
Kursk, Battle of 371
Kurzsignalheft, 71, 204
KW-7, 136, 429
KWIC Index, 247, 248

Lange, André, 40, 277, 281, 452
Langer, Gwido ('Luc'), 388, 398, 402
Langie, André, 8, 297, 452
Langlotz, Erich, 149, 151, 376

largondu, –jem, –ji, 21
LARRABEE, 114–115, 120, 148
last character, 78
Latin square, 123–125, 260, 264, 359, 364
Lauenburg (ship), 204
'Law Enforcement Access Field' (LEAF), 214
L.C.S., 30, 424
left-univalent, 32
Legendre, Adrien-Marie, 188
Leiberich, Otto, VII, 142, 376, 424, 428
Leibniz, Gottfried Wilhelm von, 39, 69, 98
Lemoine, Rodolphe [Stallmann, Rudolf] ('Rex') 390
Lenin, Vladimir Ilych, 31
Lenstra, Hendrik W., 187
Léotard, François, 199
Leutbecher, Armin, 186
leveling of frequencies, 52, 57, 432
Lever, Mavis, 432
Levine, Isaac Don, 206
Levine, Jack, 66, 247
Lewin, Ronald, 237, 450
Lewinski, Richard (pseud.), 90
Lewis Carroll [Charles Lutwidge Dodgson], 116, 165, 380
lexicalization, 21
Lieber (code), 72
Lindenfels, J. B., 452
Lindenmayer, Aristide, 145
linear substitution, 78, 79, 80, 84, 85, 88
linear shift register, 143, 415
–, reconstruction of a, 415–417
LINOTYPE, 277
lipogram, 240, 275
Lisicki, Tadeusz, 397
literary English, 282
lobster (Knox), 136
logic switching panel, 370
logograms, 32
Lombard (code), 74
London Controlling Section (L.C.S.), 30, 424
longueur de sériation, 165
Lonsdale, Gordon [Konon Molody], 9
Lorenz, Mr. 151
LORENZ *Schlüsselzusatz* SZ40/42, 28, 53, 134, 148, 150, 157, 353, 366, 372, 378
Los Alamos, 52, 53
Louis (code), 73
Louis XIV, 69
Louis XV, 15
Louis XVI, 131
lower-case letters (plain characters), 40
Loyd, Sam, 92
'Luc' (Gwido Langer), 388
Lucan, Henno, 205
LUCIFER, 170, 171, 175
Ludendorff, Erich, 51, 159, 202
Ludwig II, King of Bavaria 99
Ludwig the Severe, Duke of Bavaria 19
lug cage, 133

M-94 ≙ CSP 488, 76, 122, 125, 207, 260, 339, Plate D
M-134-A (SIGMYC), 112, 150
M-134-C (SIGABA) ≙ CSP 889 (ECM Mark II), 112, 138, 205, 434
M-138, 123
M-138-A ≙ CSP 845, 123, 205, 207, 260, 270, 429
M-138-T4, 123, Plate E
M-209 ≙ CSP 1500 ≙ C 38, 77, 122, 133, 202–203, 208, 360, 395, 428, Plate H
M-228 (SIGCUM), 150, 435
M-325 (SIGFOY), 137
Macbeth, James C. H., 74
MacLean, Donald Duart, 151
machine key period, 133, 135, 157, 357
Mackensen telegram, 378
MacPhail, Malcolm, 90
MADAME X (X-68003), 409
Madison, James 69
Maertens, Eberhard, 204, 206
Magdeburg (cruiser) 70
MAGIC, 4, 142
Makarov, Victor, 429
Malik, Rex, 372
Mamert-Gallian (code), 72
MAMMOTH 400
Mandelbrot, Benoît, 283
Mann, Paul August, VI
Mann, Thomas, 13, 200
Mantua, Duke of, 43
map grid, 39
Marconi (code), 74, 75
Marie Antoinette, 27, 131
Marks, Leo, 98
masking, 12–16
Massey, J. L., 177
Matapán, 432
matching, 44, 203, 278, 291, 295, 297, 331
–, optimal, 284
Matton, Pierre-Ernest, 198
Matyas, S. M., 219, 276, 280, 282, 437, 451
Mauborgne, Joseph O., 85, *122*, 125, 148, 425

Maul, Michael, 326
maximal length of message, 126, 134, 135, 147, 202
Mayer, Stefan, 398
maze, 11, 12
McCurley, K. S., 187
Meader, Ralph, 412
Meaker, O. Phelps, 276, 285
Medical Greek, 92
Meister, Aloys, 126, 452
Mellen, Greg, 451
'Memex', 326
Mencken, Henry Louis 303
Mendelsohn, Charles J., 118, 425
menu, 403, 404, 406
Menzies, Sir Stewart Graham, 3, 424
'Mephisto Polka', 145
Merchant Navy Code, 426
Mergenthaler, Ottmar, 276, 277
Mersenne prime, 144
message length, maximal, 126, 134, 135, 147, 202, 202
message setting 389
metaphor, 14
méthode des bâtons 264
'metoda rusztu', 394
Meurling, Per, 54
Meyer, C. H., 219, 276, 280, 282, 437, 451
Meyer, Helmuth, 17
MI-8, 30, 75, 138
Mi-544 (Lorenz), 150
M.I.1 (b), 30, 149
M.I.6, 3, 29, 30, 424
M.I.8, 30
Micali, Silvio, 215
microdot, 9
Michie, Donald, 318, 319
microprocessor, 89, 167, 169, 178
MIKE, 418
Military Intelligence Code, 75
Miller, V. S., 187
Millikin, Donald D., 449
Milner-Barry, Sir Stuart, 90, 210
'Ming', 401, 409
Minocyclin, 99
Mirabeau, Honoré Gabriel Riqueti Count of, 51, 165, 169
Mirsky, L., 447
mishit, 243, 252, 254, 258, 397, 406, 415
Mitchel, William J., 74
mixed alphabet, 44, 47
mixed rows block transposition, 96, 421
– – columnar transposition, 96, 421
mnemonic key *see* password
mobile telephone, 177
modular transformation, 160, 162–164
Monnier, Sophie Marquise de, 51, 165
monoalphabetic, 34, 42, 95, 127, 151, 174
monocyclic permutation, 46
monographic, 34, 42, 68, 151
Montgomery, P., 184, 427
Moorman, Frank, 67, 201, 425
Morehouse, Lyman F., 130
Moreo, Juan de, 68
Morikawa, Hideya, 429
Morris, Christopher, 430
Morse, Marston, 145
Morse code, 9, 11, 39, 71, 159, 166, 277
mot convenue, 16
mot vide, 287
mozilla, 178
Mullard EF 36, 370
Muller, André, 93, 452
Müller, Hans-Kurt, 261, 376
MULTIPLEX encryption step, 123, 225, 260
multiplex systems, 121, 123, 224, 227, 252
multiplication, symbolic, 88, 115, 166
multiplication of primes, 185
München (ship), 203
Murphy, Robert D., 67, 200, 209, 243, 428
Murray, Donald, 363
Murray, Joan, 399
Myer, Albert J., 323, 331, 452
MYK-78 (Mykotronx), 177
Myszkowski, Emile V. T., 452

Nabokov, Vladimir, 241
Napier, John Laird of Merchiston, 39
National Bureau of Standards (U.S.), 35, 170, 174
National Defense Research Committee, 319
National Institute of Standards and Technology (N.I.S.T.), 174, 176, 195, 196
national security, 6, 175
National Security Agency (N.S.A.), 30, 170, 175, 207, 211, 215, 412, 429
NATO, 136
Naval Cipher Nr. 1, 427
Naval Cipher Nr. 2, 376, 426
Naval Cipher Nr. 3, 206, 376, 426
Naval Cipher Nr. 5, 206, 376
Navy Code Box (NCB), 122
NCR, 409, 411
Nebel, Fritz, 51, 159
Neeb, Fritz, 425
'need to know' doctrine, 424, 433

NEMA, 113
‚Neptun' (key net), 435
Nero, Roman emperor, 36
Netscape Communicator, 178
Newman, Maxwell Herman Alexander, 369
newspaper German, 279
Newton, Isaac, 98
Niethe (code), 73
Nihilist cipher, 10, 51, 165
Nihilist transposition, 96, 421
Nilac (code), 73
NKVD, 54, 55
'no letter may represent itself', 108, 208, 240, 251, 252, 262
nomenclator, 68–70
non-carry binary addition, 130
non-computable real numbers, 150
non-content words, 287
non-pattern words, 249
non-periodic, 34
Norris, William C., 412
Noskwith, Rolf, 377
notch, 134–135, 267, 402
Notschlüssel, 97
Novopaschenny, Fedor, 425
null, 18, 21, 33, 43, 53, 94, 201, 209, 272
– cipher, 18, 23
– text, 32
number field sieve (NFS), 184, 187
Nuovo Cifrario Mengarini (code), 73

O-2, 123, 261–263
Oberkommando der Wehrmacht (OKW), 110, 135, 246, 418
octopartite simple substitution, 53
Office of Strategic Services (OSS), 261, 424
offizier, 159, 320
Ohaver, M. E., 166, 324, 452
O'Keenan, Charles, 10, 165
OKW Cipher Branch (*Chi*), 31, 252, 316, 325, 353, 376, 389, 406, 418, 424, 427, 467
Olivary, Adolphe, 423
Olivetti, 157
OMALLEY, 320
OMI, 113, 136
1-cycle, 44, 252, 391, 396
one-part code, 70, 354
one-time key, 148–151, 203, 211
one-time pad, –tape, 149–151, Plate O
one-to-one function, 32, 44
one-way function, 182, 183
OP-20-G, 30, 319, 325, 411, 412
OP-20-G, 30, 319
open code, 9, 12–22, 23
open encryption key system, 179, 196
open-letter cipher, *see* null cipher
Operational Intelligence Centre, 30
ORANGE (machine), 85, 141, 376
ORANGE (key net), 403
order of a matrix, 143
order-preserving, 69, 78
OSS, 261, 424
ostensibly secret messages, 29
OTP (one-time pad), 148–151
Ottico Meccanica Italiana, 112–113, 136
overlay sheets, 316, 317, 318

P = NP 182
Painvin, Georges-Jean, 85, 159, 261, 425
PA-K2 system, 98
palindrome, 91, 92, 241
pangram, 249–250
Panizzardi, Alessandro, 19
Pannwitz, Erika, 261
parallels *see* repetitions
Parkerism, 435
partition, 47, 105, 272, 273, 278–285
Paschke, Adolf, 376, 435, 436
'passport control officers', 3
password, 43, 45, 48, 49, 51, 52, 54, 55, 57, 62, 92, 95, 97, 98, 103, 118, 121, 128, 155, 165, 178, 183, 197, 219, 239, 259, 262, 295, 296, 345, 414, 417
–, reconstruction of, 354–355, 385–387
Pastoure, 8
pastry dough mixing, 160
Patrick, J. N. H., 158
Patronen-Geheimschrift, 94
pattern, 59, 235–237
– finding method, 218, 235–242, 243, 247–249, 394, 421, 432
–, normal form of –s, 235
pawl, 134
P. C. Bruno, 402
Pearl Harbor, 17, 199
Pendergrass, James T., 412
penetrazione squadra, 199, 428
Pepys, Samuel, 8
Perec, Georges, 240
perfect security, 439, 443
period length, 35, 131, 133, 144, 321, 329
Perioden- und Phasensuchgerät, 316, 325
periodic, 34, 126, 132, 311, 320
permutation, 20, 44, 57, 129
–, self-reciprocal, 45

PERMUTE encryption step, 118, 123, 225
Pers Z, 141, 376, 418, 421, 424, 435
Petard (destroyer), 219
PETER ROBINSON, 318
Peter the Great, 43, 69
Peterson's (code), 74
Phaistos, disc of, 25, Plate A
Phi, 308, 309, 326
Phi-Test, 326–330, 401
Philby, Harold, 151
Philip II of Spain, 68
Phillips, Cecil James, 150
photoelectric sensing, 316, 320, 326, 353
PHYTON, 411
picture encryption, 162, 164
picture transmission, 447
pig Latin, 21
'pigpen' cipher, 43
Piper, F., 451
placode, plain code, 351, 352, 354, 375
plaintext, 31, 32, 34, 38, 41, 44
– vocabulary, 31
plaintext attack, chosen, 432, 433
plaintext attack, derived, 433
plaintext attack, known, 432
'Plaintext Bit 5 Two steps back', 372
plaintext-cryptotext compromise, 209, 270, 356, 394, 403–412, 432
plaintext-plaintext compromise, 197, 209, 356, 366, 378, 421, 431
planned obsolescence, 26, 74, 76, 219
'playback', 203
Playfair, Lyon Baron of St. Andrews, 26, 62
PLAYFAIR encryption step, 62–65, 200, 209, 222, 231, 295, 297, 298, 313
plugboard, 46, 90, 110, 112, 135, 136, 158, 165, 265, 399, 406, 434
'pluggable' reflecting rotor, 110, 112, 269
Poe, Edgar Allan, 5, 43, 298, 312, 321, 335
Pokorny, Hermann, 47, 242, 425
Polares, VP 26 (ship), 203
Polheim, Christopher, 52
Pollard, John, 184, 187
POLLUX, 166
polyalphabetic encryption, 14, 34, 85, 101, 118, 126, 131, 132, 144, 202, 226, 251, 311
Polybios square, 39, 51, 54, 65, 159, 165, 363
polygraphic, 34, 56, 78, 95, 189
– substitution, 56, 128, 164, 223, 224
polyphones, polyphony, 35–37, 48, 71, 121, 234, 238, 414
Pomerance, C., 184
Poore, Ralph Spencer, 176
Porta, Giovanni Battista, 38, 42, 43, 45, 56, 117, 118, 127–129, 131, 197, 200, 201, 202, 207, 243, 263, 320, 321, 363
PORTA encryption step, 117, 146, 251, 254
Poste de Commandement Bruno, 402
powers of a mixed alphabet, 48, 50, 102, 143, 355
ppmpqs, 184
Pratt, Fletcher, 287, 452
preamble, 389, 395, 423
Pretty Good Privacy (PGP), 177–178, 213
primary alphabet, 101, 129, 154, 331, 335, 336, 338, 339, 344–346, 354
primitive root, 186
priming key, 146, 147, 148, 174
private key, 180–182, 433
probable word, 199, 199, 208, 209, 228, 242, 248, 251, 254, 256, 260, 263, 413, 415, 432
progressive encryption, 127, 132, 133, 257
properly self-reciprocal, 45, 108, 221, 390
protocol, 195, 215, 437
pseudo random keytext, 154, 174, 447
Psi, 305–306, 308, 309–310
Ptydepe, 241
public cryptography, 6, 7, 31, 196, 219
– key system, V, 179, 181–194, 196
punched card machines, 262, 314–316, 353
punched sheets, 397, 398–400
punctuation, 18, 19, 36, 208, 290, 291
pure (cryptosystem, cipher), 156, 358, 359
pure cryptanalysis, 203, 249, 295, 319, 368, 431, 433, 438
PURPLE, 4, 141–142, 376, 429, 434
Pyry, 398, 400
PYTHON, 320, 411

Qalqashandi, 43
quadratic sieve, 184, 185, 187
Quadratic Reciprocity, 188
quasi-nonperiodic key, 131, 154, 202
quaternary cipher, 39, 127
quatsch, 201
'Quex', 3
quinary cipher, 39
Quine, Willard Van Orman, 2
quinpartite simple substitution, 53
QWERTZU... 390

Rabin, M. O., 188
radio call signal, 58, 85

rail fence transposition, 93
raising to a power *modulo q*, 168, 188
Ramsey, Frank Plumpton, 448
Randell, Brian, VI, 450
random key sequence, 148, 150, 447
– text, 200, 237
– variable, 440
'Rapid Analytical Machine' (RAM), 320
RAPID SELECTOR, 326
Rasterschlüssel 44, 98
RATTLER, 320, 411
Rave, Cort, 376, 425
Raven, Frank, 143
Razvedupr, 429
RC2, RC4, 176, 177
rebus, 67
reciphering, 157
reciprocal pairs, 86–88, 168, 169
recovery exponent, 191–193
RED (code), 75, 199
RED (key net), 399, 402, 403, 436
RED (machine), 85, 140–142, 376
redundancy, 195, 230, 264, 362, 447
reflecting rotor, reflector 107, 110
–, pluggable, 110, 269, 412
reflection, 45, 129
–, genuine, 46
reglette, 50
regular matrix, 73, 81, 83, 84, 88, 414
Reichling, Walter, 17
Reichsbahn, Deutsche, 110, 356
Reichsluftfahrtministerium, 31
Reichspost, Deutsche, 9, 110
Rejewski, Marian, *2*, 90, 152, 388, 390–393, 394–396, 398, 399, 402, 430, 448
Remmert, Reinhold, 82
Rényi, Alfred, 310, 437
Rényi-α-entropy, 310
Reparto crittografico, 242
repetitions, 316, 320–326, 332, 340, 346
–, accidental, 322–325, 346
repetition pattern, 235–242
residual classes, 78, 81, 83, 352
reversed alphabet, 45, 88, 116, 159, 235
reversal, 272, 275
reverses, 74, 287
'Rex' (Rodolphe Lemoine), 390
Ribbentrop, Joachim von, 31, 376, 424
Richelieu, Armand Jean du Plessis Duc de, 22, 95
Riesel, Hans, 189
right-univalent, 32
Ringelnatz, Joachim, 20
ring setting, 112, 135, 154, 204, 389, 394, 396–397, 400, 406
Rittler, Franz, 241
Rivest, Ronald L., 189, 190
Robinson, Ralph M., 144
ROBINSON AND CLEAVER, 318
ROCKEX, 150
'rodding', 264
Rogers, Henry, 72
Rohrbach, Hans, VI, 2, 65, 123, 202, 205, 219, 261–262, 336, 418, 425–429, 435, *452*
Rohrbach's maxim, 210, 239, 249, 335, 346, 388, 414
Rohwer, Jürgen, 112, 427, 450
Romanini, Ch. Vesin de, 276
Rommel, Erwin, 208, 371, 403, 428
Ronge, Maximilian, 242, 425
Room 40 , 377, 399, 436
Room 47 Foreign Office, 30, 398
Roosevelt, Franklin Delano, 9, 75, 123, 139, 199, 206, 209, 380, 411, 428
Rosen, Leo, 3, 142, 269, 429
Rosenberg, Julius and Ethel 151
Rossberg, Ehrhard, 156, 366
Rosser, Barkley, 2
Rossignol, Antoine, 69, 70, 127
rotated alphabets, 102–105, 124, 264
Rothstein, J., 124
rotor, 106
–, 'fast' (rightmost), 134, 265, 267, 394, 401
–, 'medium' (middle), 134, 267, 268, 408
–, 'slow' (leftmost), 134
ROTOR encryption step, 104–108, 156, 264, 338
rotor movement, irregular, 132–135, 137
–, regular, 132–134, 136
rotor order *see* wheel order
Rotscheidt, Wilhelm, 376
Rotterdam-Gerät, 206
rotwelsch, 14
route transcription, 92
row-column transcription, 92
Rowlett, Frank, 112, 137, 141, 150, 435, *452*
Różicki, Jerzy, 152, 388, 394, 398, 401, 402
RSA method, 189–194, 196
R.S.H.A. Amt VI, 58, 59, 424
R.S.H.A. radio observation post, 58
Rudolf Mosse (code), 74
Rundstedt, Gerd von 371
running key, 34, 35, 127, 145, 165, 196
Russian copulation, 55, 138, 200, 243, 404
Russians, 3, 9, 10, 27, 41, 47, 51, 69, 137, 139, 149, 202, 212, 263, 429

SA Cipher, 35, 72
Sacco, Luigi, VI, 117, 126, 201, 208, 219, 242, 276, 277, 425, 452
safe primes, 169, 193, 196
Safford, Laurance, 3, 137
Saint-Cyr slide, 40, 50, 116
Salomaa, Arto, 133, 156, 182, 185, 189, 190, 193, 451
Sandherr, Jean, 198
Sandwith, Humphrey, 398
'saw-buck' principle, 316, 319, 369, 370
Satzbuch, 70, 71, 76, 158
Sayre, David 438
S-box, 170, 175
Schauffler, Rudolf, 149, 376
Schellenberg, Walter, 9, 31, 58
Scherbius, Arthur, 106, *107*, 132–134, 399
shift register, 143, 200, 415, 417
–, linear, 143, 415
Schilling von Cannstatt, Paul, 27
Schimmel, Annemarie, 261
Schleyer, Johann Martin, 349
Schlüsselheft, 67, 158
Schlüsselzusatz, 157, 319, 353, 366–372, Plate N
Schmidt, Arno, 29
–, Hans-Thilo (Asché), 390, 394
Schneickert, Hans, 452
Schneier, Bruce, 451
Schnorr, Claus-Peter, 150, 195
Schoeneberg, Bruno, 189
Scholz, Arnold, 189
Schott, Caspar, 8, 117
Schröder, Georg, 425
Schur, Issai, 448
Schwab, Henri, 423
scissors-and-paste method, 234, 418
SCORPION (key net) 403
scrambler, 399, 403–408, 411, 434
'scrambling' (audio), 9, 409
scritch, scritchmus, 265, 269, 411–412
SD *see Sicherheitsdienst*
Sebag-Montefiori, Hugh, 451
secrecy, 2, 8, 24, 25, 26, 29, 68, 180
Secret Intelligence Service (S.I.S.), 30
secret marks, 13, 27, 42, 67
secret writing, covert, 2, 8, 23
–, masked, 12–16, 18, 23
–, overt, 8, 23, 25
–, veiled, 18, 22, 23
secure socket layer (SSL), 178
security check, 195, 203
Selchow, Kurt, 376
self-reciprocal linear substitution, 80
self-reciprocal matrix, 83, 84
self-reciprocal permutation, 44, 45, 46
Selmer, Ernst S., 2, 143
semagram, 9, 10, 11, 23, 51
senary cipher, 39
senseless, 130, 148
Service de Renseignements (S.R., 2bis), 31
Servizio degli Informazione Militare (S.I.M.), 31, 40
Sestri, Giovanni, 116
session key, 174
Shakespeare, William, 29
Shamir, Adi, 175, 189, 190
Shanks, Daniel, 186
(SHANN), 441
Shannon, Claude Elwood, VI, 2, *24*, 25, 26, 100, 126, 146, 147, 156, 159, 170, 210, 230, 235, 250, 310, 401, 409, 437
Shannon cryptosystem, 41, 251, 261, 263, 270, 359, 360, 441
Shannon entropy, 310, 401, 446
Shannon's Information Theory, 439–447
Shannon's Main Theorem, 444
Shannon's maxim, 28, 29, 137, 145, 196, 197, 207, 210, 220, 227, 339, 434
Shannon's Mixing, 160, 164
SHARK, *see* ‚Triton' (key net)
Shaw, Harold R., 11
shift number, 114
shifted alphabets, 50, 102, 125
shorthand symbols, 32
shrdlu, 277
shredder, 149, 203
Shulman, David, 452
Sicherheitsdienst (SD), 31, 64, 395, 433
Siemens Geheimschreiber T 52, 28, 53, 148, 156, 157, 368, 370, 372, 435
SIGABA ≙ M-134-C, 112, 136, 137, 138, 205, 434
SIGCUM ≙ M-228, 150, 435
SIGFOY ≙ M-325, 137
SIGINT, VII
SIGMYC ≙ M-134-A, 112, 150
Signal Corps U.S. Army, 30, 77, 122, 137, 139, 323
Signal Intelligence Service *see* SIS
Signal Security Agency U.S. Army, 205, 269
signature, 181, 182, 189, 195
SIGPIK, SIGSYG, 76
SIGTOT, 123, 130, 150, 263
Silverman, R. D., 184
Simeone de Crema, 43, 53
similarity transformation, 97, 105

'simultaneous scanning', 399
simple linear substitution, 85
simple substitution, 42, 43
Sinclair, Sir Hugh P. F., 3
Sinkov, Abraham, 3, 85, 137, 288, 330, 339, 342, 381, 449, *452*
S.I.S. (Secret Intelligence Service), 30
SIS (U.S. Army Signal Intelligence Service), 314, 409, 411, 412, 413, 418
Sittler, F. J., 73, 158
Sittler (code), 73
SKIPJACK, 177, 214
skytale, 100
Slater (code), 74
‚Sleipnir' (key net), 435
slide, 40, 50, 104, 116
slide method, 338–339
slip-ring, 106
Small, Albert, 142
small capitals (crypto characters), 40
smart card, 194
Smid, M. E., 170
Smith, Francis O. J., 71
Smith, Laurence Dwight, 93, 197, 276, 277, 281, 449
Smith, W. W., 297
Smith-Corona Co., 133
SNEGOPAD 55
'snowfall', 55
Snyder, Samuel S., 142, 412
S.O.E., 98
Solzhenitsyn, Alexander, 51
Sonderdienst Dahlem, 261, 263, 353, 424
Sora, Iacomo Boncampagni Duke of, 320
Sorge, Richard, 55
Soudart, E.-A., 40, 277, 281, 425, 452
Sperr-Ring, 112
Spezialvergleicher, 418
Spets otdel, 31
'spider', 401
spreading of encryption errors, 147, 211
spoonerism, 64, 92
Spruchschlüssel *see* text setting
–, *chiffrierter*, 389
spurt, 9
spy cipher, 54
squadra penetrazione, 199, 428
squaring *modulo q*, 169, 186, 188
standard alphabet, 44, 47, 78, 102, 114
Station X, 30, 398
stator, 108, 113
Steckerbrett, 46, 110
Steckerverbindung, 111, 112, 389, 400, 434
steganography, linguistic, 9, 11, 23, 24
–, technical, 8, 9, 23
Stein, Karl, VI, VII, 406
Steinbrüggen, Ralf, 36
Steiner & Stern (code), 73
Steiner's theorem, 305
Stimson, Henry L., 139, 212, 428
Stirling's formula, 220
stochastic source, 149, 273, 303, 306
Strachey, Christopher, 399
Strachey, Oliver, 140, 399
straddling, 33, 35, 53–55
strategy of alignment, 342
stream encryption, -cipher, 34, 148
STRETCH, 412
strip devices, 50, 122, 123, 132, 208, 252, 256, 261, 427, 429
strip method, 233, 234, 265–267
Stripp, Alan, 425, 450
stripping off superencryption, 61, 315, 319, 351, 376, 426, 433, 438
Stuart, Mary, 68
Stummel, Ludwig, 204, 206, 377
'sturgeon', 156, 372
subexponential complexity, 184, 187
substitution, 17, 24, 42
–, binary linear, 84
–, bipartite digraphic, 57, 59, 65
–, bipartite simple, 51, 65
–, decomposed linear, 85
–, digraphic, 56, 231
–, linear polygraphic, 78–88, 160, 223, 231, 413–417
–, monocyclic simple, 46, 221
–, multipartite simple, 51–53, 160
–, polygraphic, 56–66, 91, 155, 222–223, 235
–, properly self-reciprocal, 45, 108, 221, 390, 394, 406
–, self-reciprocal, 45
–, self-reciprocal linear, 80
–, simple, 42, 43, 85
–, tripartite digraphic, 61
–, tripartite simple, 52
–, unipartite simple, 42–43, 221
substitution à double clef, 128, 343
– *à triple clef*, 129
substitution notation, 44
Suetonius, 47
suffixing, parasitic, 19
superencryption, -enciphering, 157–159, 208, 351–354, 373–376, 378, 388, 425–429
superimposition, 356–376, 378, 398, 421, 431

–, in-phase, 373–376, 401
SUPER ROBINSON, 318
SUPERSCRITCHER, 269, 412
surjective, 32, 41
swap, 46
swapping of roles, 360, 378, 414
Swift, Jonathan, 21, 99, 420
Swiss ENIGMA (ENIGMA K), 108, 373
switchboard, 105
SYKO, 120
symbolic addition, 157, 166
symmetric functions, 309
symmetric methods, 180–182
symmetry of position, 346–351, 352, 358, 378–381
synthetic language, 200, 201
SZ 40, SZ 42, SZ 42a (Lorenz), 28, 148, 157 319, 353, 366–372, 435, Plate N
SZ3-92, 280, 286, 300
Szekeres, G., 448
Szemerédi, E. 448

T 43 (Siemens SFM 43), 157
T 52a, T 52b, T 52c, T 52d, T 52e (Siemens), 28, 148, 156–157, 368, 372–373
T-52, T-55 (Hagelin), 150
tabula recta, 103, 114, 124, 127, 128, 132, 348, 349, 385–387
– with permuted headline, 387
Tagesschlüssel, 154, 389, 390
Takagi, Shiro, 316
Tallmadge, Benjamin, 69
tamperproof key carriers, 154, 178, 433
Tannenberg, battle of, 202, 436
Tartaglia, Niccolò, 437–438
Taunt, Derek, 269, 412
Technical Operations Division, 16
telegraphic English, 282
Telescand (code), 73
teletype code ($\mathbb{Z}_2^5$), 363
ten-letter code, 39
ternary substitution, 52
TESSIE, 354
test register, 400, 404, 408
test texts for teletype lines, 250
TETRA, 325
tetragram, 34
tetragraphic substitution, 66, 222, 231
text setting, 388, 389, 390, 392, 394–398, 401, 404
theta series, 82
‚Thetis‘ (key net), 435
thieves' Latin, 14
Thomas, E. E., 206
Thompson, Eric, 219
'thrasher', 157
three-digit code, 76
three-letter code, 72, 76
Thue, Axel, 145
Tibbals, Cyrus (code), 74
Tiltman, John H., 64, 356, 366, 367
Tokumu Han, 31, 429
Tolstoy, Sergei, 425, 429
Tomash, Erwin, 412
tomographic methods, 62–65, 165–166
Tompkins, Charles B., 412
Tordella, Louis, 429
totient function φ, 88, 189, 221
Townsend, Robert, 69, 239
traffic analysis, VII
traffic padding, 200, 201
Tranow, Wilhelm, 426, 427
transitive cryptosystem, 260, 359
translation, 79, 223
transposition, 23, 91, 95, 160, 222, 223, 228, 229, 231, 232, 234, 271, 303, 309
–, block–, 95, 155, 418
–, double columnar, 26, 97, 227, 421
–, feigning a, 57
–, mixed-rows block–, 96, 97, 421
–, mixed-rows columnar, 96, 97, 164, 421
–, Nihilist, 96
–, simple columnar, 95
transposition double, 96, 97
trapdoor, 175, 183–188
– one-way functions, 183
'trasher', 157
'treble key', 129, 387
trellis cipher, 93
trench codes, 76, 418
Trevanion, Sir John, 8, 18
trifide, 34
trigram, 34
– coincidences, 313, 317
– frequencies, 286, 289
– repetitions, 325, 346
trigraphic substitution, 66, 222, 231
tripartite, 34
triple-DES, 176
Trithemius, 8, 14–15, 16, 38, 39, 52, 114, 127, 129, 132, 146
‚Triton‘, SHARK (key net), 112, 204, 411
Trotzki, Lev Davidovich, 31
'tunny', 157, 367, 368
Turing, Alan Mathison, *2*, 89, 90, 185, 220, 372, 398–401, 409, 431, 438, 450
Turing BOMBE, 159, 402, 403–409, 411

turning grille, 76, 93, 94
turnover, 135–136
Türkel, Siegfried, 450
Tut Latin, 20
Tutte, William Thomas, 319, 368
'Twenty Committee', 14
Twinn, Peter, 90, 136, 390, 396
two-character differential, 74
2-cycle, 46, 391–393
two-part code, – nomenclator, 70
TYPEX, 112, 136, 138, 434, Plate L

ubchi, 97, 423
U-boat *U-13*, *U-33*, 203
U-boat *U-110*, 203
U-boat *U-559*, 204, 378
U-boat *U-570*, 204
Uhr box, 110, 434, Plate M
Ulbricht, Heinz, 113
ULTRA, 4, 142, 211
Umkehrwalze, Umkerwaltz, 107
unambiguous, 2, 8, 25, 32, 41
unauthorized decryption, 3, 31, 101, 123, 197, 217ff., 430
uncertainty, 439
'Uncle Walter', 107
unicity distance, 100, 126, 229–231, 240, 249, 299–300, 445–446
unipartite, 34, 42–43, 85
Universal Trade Code (Yardley), 138
UNIX, 183
Urfé, Madame d', 27
U.S. Intelligence Board (U.S.I.B.), 29
Uzès, 402

vacuum tube noise, 150
Valério, P., 276, 288, 354, 452
van der Waerden, Bartel Leendert, 448
van Wijngaarden, Adrian, 10
variants, 21, 32, 33, 36, 227–234
variante à l'allemande, 116
variante de Richelieu, 95
Vātsāyana, 45
Vaz Subtil (code), 73
Venona breaks, 150–151
Verlaine, Paul, 16
Verlan, 21
Vernam, Gilbert S., 28, 53, *122*, 123, 130, 148, 149, 311, 364
VERNAM encryption step, 129, 130, 144, 145, 148, 150, 152, 156–157, 417, 442, 444
Vernam type, 444, 445
Wylie, Shaun, 432
Wynn-Williams, C. E., 319
Verne, Jules, 93, 117,
vertically continued alphabets, 102–105
Vesin de Romanini, Charles François, 276, 452
Vetterlein, Kurt, 9
Viaris, Gaëtan Henri Léon Marquis de, 28, 72, 116, 123, 207, 255–263, 276, 322, 452
'Victory', 401, 409
Viète, François, 2, 68
Vigenère, Blaise de, 9, 27, 114, 117, 128, 129, 146, 148, 202
VIGENÈRE encryption step, 114–115, 128–130, 143, 144, 146–148, 152, 157, 159, 160, 165, 179, 208, 222, 224, 226, 227, 231, 257, 263, 272, 303, 305, 309, 321, 331, 332, 335, 336, 340, 350, 351, 360, 362, 378, 415, 417, 423, 427, 442, 444
Vinay, Émile, 116
VIPER, 320, 411
Volapük, 349–350
Voltaire [François Marie Arouet], 425
von dem Bach-Zelewski, Erich, 64
vowel distances, 290
'vowel-solution method', 293
VULTURE I (Schlüsselnetz) 403

Waggoner, Thomas A. Jr., 158
Wake and Kiska islands, 252
Wallis, John, 2, 8, 69
WALRUS 411
Walter (code), 72
Walze, 106
Walzenlage, 111, 208, 389
Wanderer-Werke, 134
WARLOCK, 320
War Station, 30
Washington, George, 69
WASP (key net) 403
Wassenaar Arrangement, 214
Wäsström, Sven, 119
Watt, Donald Cameron, 388
Weaver, Warren, 2, 310
Wehrmacht ENIGMA, 46, 109–111, 113, 134, 135, 152, 154, 267

Wehrmachtnachrichtenverbindungen Chiffrierwesen, 31, 201, 252, 316, 325, 353, 406, 424, 427
Weierstrass, Karl, 203, 448
Weierud, Frode, VII, 113
Weigel, Erhard, 39
'weight of evidence', 401, 438
Weisband, William, 151
Weiss, Georg, 452
Welchman, Gordon, 5, 90, 108, 135, 205, 209, 388, 399, 400, 402, 403, *406*, 407, 408, 409, 411, 431, 434, 435, 436, 450
Wenger, Joseph N., 319–320, 409
Werftschlüssel, 377
West, Nigel [Rupert Allason], 3
Western Union Code, 74
Wetterkurzschlüssel, 71, 204
Wheatstone, Charles, 4, 26, 48, 62, 63, 68, 132, 209, Plate C
wheel movement, irregular, 133, 137, 138
–, regular, 132, 133, 134, 136
wheel setting, basic, *see* ground setting
wheel order, 110–112, 154, 389, 394–397, 400
White, Harry C., 151
Whitelaw's Telegraph Cyphers 39
Widman, Kjell-Ove, VII
Wiener, M. J., 177, 194
Wiener, Norbert, 438
Wilkins, John, 8, 24, 39
Williams, H. C., 188
Williams, Sam B., 409
Wills, John, 72
Willson, Russell, 122
Wilson, Woodrow, 114, 436
Winkel, Brian J., 335, 451
Winterbotham, Frederick W., 90, 211, 449
Witt, Ernst, VI, 336, 353
Witzleben, Erwin von, 159
Wolseley, Lord Garnet J., 346
Women's Royal Naval Service, 159
Woodhull, Sam, 69, 239
word frequency, 287
word length frequency, 290
word spacing, 36–37, 39, 190, 201, 209, 238, 239, 247, 250, 290–291
Wright, Ernest Vincent, 240, 241
Wüsteney, Herbert, 157
WWW Software Communicator, 178

xB-Dienst see *B-Dienst*
X-ray crystallography, 438
X-68003 (MADAME X), 409

Yardley, Herbert O., 30, 138–140, 399, 425, 429, 452
YELLOW (key net), 402

Zeichenvergleichslabyrinth, 316
Zellweger A.G., 113
Zemanek, Heinz, 277
Zener diode, 150
Zentralstelle für das Chiffrierwesen (ZfCh), 31, 424
zero-knowledge proof, 437
Ziegenrücker, Joachim, 261
zig-zag method, 263, 359–362
Zimmermann, Arthur, 436
Zimmermann, Philip R., 177–178, 213, 219
Zimmermann telegram, 436
Zipf, George K., 283
ZMUG, 368
Zuse, Konrad, 28, 370
Zygalski, Henryk, 388, 394, 397, 402
Zygalski sheets, 159, 397, 398–400, 402

Photo Credits

Kahn, David, *The Codebreakers*. Macmillan, New York 1967:
Figs. 1, 4, 5, 12, 23, 30, 31, 33, 34, 35, 36, 37, 38, 40, 57

Smith, Laurence Dwight, *Cryptography*. Dover, New York 1955:
Figs. 3, 16, 24, 53

Bayerische Staatsbibliothek München
Figs. 10, 52

Lange, André and E.-A. Soudart, *Traité de cryptographie*. Paris 1925:
Fig. 26

Crypto AG, Zug, Switzerland:
Figs. 48, 54, 55, 60a, 60b, 62, 64, 65, Plates L, O, P

Deavours, Cipher A. and Kruh, Louis, *Machine Cryptography and Modern Cryptanalysis*. Artech House, Dedham, MA 1985:
Figs. 63*ℓ*, 66, 67, 68

Bundesamt für Sicherheit in der Informationstechnik, Bonn:
Fig. 63*r*

Public Record Office, London:
Figs. 143, 160a

FRA, Bromma, Sweden:
Fig. 145

National Museum of American History, Washington:
Fig. 160b

Deutsches Museum (Reinhard Krause), Munich, Germany:
Plates A, B, C, D, F, G, I, K, N, Q

Russell, Francis, *The Secret War*. Time-Life Books, Chicago, IL 1981:
Plates E, H, M

Solution for the second Cryptoquip of Fig. 87:

I=m. Entry with search for patterns results in *1211234* (KRKKRLH) and *53675* (ULZIU). Among the one or two dozen possibilities only very few are not too weird. Among those, /peppery/ , followed by (for *5*r*6*m*5*) /aroma/ are suitable and lead to further success:

KRKKRLH	PLRUI	OZGK	AYMMGORA	U	LYPQ,	QRUAH	ULZIU
peppery	re m	p	e		r ,	e y	r m
peppery	ream	o p	e	a	r ,	ea y	aroma
peppery	cream	soup	use	a	r c ,	ea y	aroma
peppery	cream	soup	i use	a	rich,	hea y	aroma
peppery	cream	soup	di used	a	rich,	heady	aroma
peppery	cream	soup	diffused	a	rich,	heady	aroma

Printing: Mercedes-Druck, Berlin
Binding: Stürtz AG, Würzburg